MW00973397

BNi Building News

BNi Building News

PUBLIC WORKS 2000 COSTBOOK

SEVENTH EDITION

BNi Building News

Los Angeles • Anaheim

Boston • Washington, D.C.

BNi® Building News

EDITOR-IN-CHIEF
William D. Mahoney, P.E.

TECHNICAL SERVICES
Edward B. Wetherill, CSI
Rod Yabut
Gayle Enriques

GRAPHIC DESIGN
Robert O. Wright

BNI Publications, Inc.

LOS ANGELES
10801 NATIONAL BLVD. SUITE 100
LOS ANGELES, CA 90064

ANAHEIM
1612 S. CLEMENTINE STREET
ANAHEIM, CA 92802

BOSTON
629 HIGHLAND AVENUE
NEEDHAM HEIGHTS, MA 02494

WASHINGTON, D.C.
502 MAPLE AVENUE WEST
VIENNA, VA 22180

1-888-BNI-BOOK *(1-888-264-2665)*

ISBN 1-55701-2946

PREFACE

For the past 50 years, Building News has been dedicated to providing construction professionals with timely and reliable information. Based on this experience, our staff has researched and compiled thousands of up-to-the-minute costs for the **Building News 2000 Costbooks**. This book is an essential reference for contractors, engineers, architects, facilities managers — any construction professional who must provide an estimate or any type of building project.

Whether working up a preliminary estimate or submitting a formal bid, the costs listed here can quickly and easily be tailored to your needs. All costs are based on national averages, while a table of modifiers is provided for regional adjustments. Overhead and profit are included in all costs.

Complete man-hour tables follow the unit costs to provide data on typical durations of specific tasks. This information can be used to schedule projects as well as to determine specific labor costs based on local labor rates.

All data is categorized according to the MASTERFORMAT of the Construction Specifications Institute (CSI). This industry standard provides an all-inclusive checklist to ensure that no element of a project is overlooked. In addition, to make specific items even easier to locate, there is a complete alphabetical index.

This costbook contains an appendix with reference charts and tables taken from an array of sources. Text explains the costs in certain categories and provides helpful pointers that should be taken into account with every estimate.

A section on square foot costs provides an overview of project costs for different building types — commercial, residential, etc. — with summaries of the actual projects. Square foot costs are invaluable for making budget estimates and checking prices when time is a factor.

The "Features in this Book" section presents a clear overview of the many features of this book. Included is an explanation of the data, sample page layout and discussion of how to best use the information in the book.

Of course, all buildings and construction projects are unique. The costs provided in this book are based on averages from well-managed projects with good labor productivity under normal working conditions (eight hours a day). Other circumstances affecting costs such as overtime, unusual working conditions, savings from buying bulk quantities for large projects, and unusual or hidden costs must be factored in as they arise.

The data provided in this book is for estimating purposes only. Check all applicable federal, state and local codes and regulations for specific requirements.

BNi Building News

TABLE OF CONTENTS

CSI MASTERFORMAT

All data in the Costbook pages and Man-Hour tables is organized according to the CSI MASTERFORMAT — the industry standard numbering/classification system. The data is divided into the 16 divisions as shown below. The five digit numbers within each division correspond to the MASTERFORMAT Broadscope designations. Each section is further broken down into MASTERFORMAT Mediumscope designations. These numbers, in most cases, are the same as those used in architectural and engineering specifications.

The construction estimating information in this book is divided into two main sections: Costbook Pages and Man-Hour Tables. Each is organized to the 16 divisions of the CSI MASTERFORMAT as shown below. In addition, there are extensive Construction Reference Tables, Geographic Cost Modifiers, Square Foot Tables and a detailed Index. Sample pages with graphic explanations are included before the Costbook pages, Man-Hour Tables and Square Foot Tables. These explanations, along with the discussions below, will provide a good understanding of what is included in this book and how it can best be used for construction estimating.

GENERAL REQUIREMENTS — *Division 1*

- Summary of Work 01010
- Allowances 01020
- Measurement and Payment 01025
- Alternates/Alternatives 01030
- Modification Procedures 01035
- Coordination 01040
- Field Engineering 01050
- Regulatory Requirements 01060
- Identification Systems 01070
- References 01090
- Special Project Procedures 01100
- Project Meetings 01200
- Submittals 01300
- Quality Control 01400
- Construction Facilities & Temporary Controls 01500
- Material and Equipment 01600
- Facility Startup/Commissioning 01650
- Contract Closeout 01700
- Maintenance 01800

SITEWORK — *Division 2*

- Subsurface Investigation 02010
- Demolition 02050
- Site Preparation 02100
- Dewatering 02140
- Shoring and Underpinning 02150
- Excavation Support Systems 02160
- Cofferdams 02170
- Earthwork 02200
- Tunneling 02300
- Piles and Caissons 02350
- Railroad Work 02450
- Marine Work 02480
- Paving and Surfacing 02500
- Utility Piping Materials 02600
- Water Distribution 02660
- Fuel and Steam Distribution 02680
- Sewerage and Drainage 02700
- Restoration and Underground Pipe 02760
- Ponds and Reservoirs 02770
- Power and Communications 02780
- Site Improvements 02800
- Landscaping 02900

CONCRETE — *Division 3*

- Concrete Formwork 03100
- Concrete Reinforcement 03200
- Concrete Accessories 03250
- Cast-In-Place Concrete 03300
- Concrete Curing 03370
- Precast Concrete 03400
- Cementitious Decks and Toppings 03500
- Grout 03600
- Concrete Restoration and Cleaning 03700
- Mass Concrete 03800

MASONRY — *Division 4*

- Mortar and Masonry Grout 04100
- Masonry Accessories 04150
- Unit Masonry 04200
- Stone 04400
- Masonry Restoration and Cleaning 04500
- Refractories 04550
- Corrosion Resistant Masonry 04600
- Simulated Masonry 04700

METALS — *Division 5*

- Metal Materials 05010
- Metal Coatings 05030
- Metal Fastening 05050
- Structural Metal Framing 05100
- Metal Joists 05200
- Metal Decking 05300
- Cold Formed Metal Framing 05400

FEATURES IN THIS BOOK

The construction estimating information in this book is divided into two main sections: Costbook Pages and Man-Hour Tables. Each section is organized according to the 16 divisions of the MASTERFORMAT as shown on the previous pages. In addition, there are extensive Supporting Construction Reference tables, Geographic Costs Modifiers, Square Foot tables and a detailed Index.

Sample pages with graphic explanations are included before the Costbook pages and Man-Hour tables. These explanations, along with the discussions below, will provide a good understanding of what is included in this book and how it can best be used in construction estimating.

Material Costs

The material costs used in this book represent national averages for prices that a contractor would expect to pay plus an allowance for freight (if applicable), handling and storage. These costs reflect neither the lowest or highest prices, but rather a typical average cost over time. Periodic fluctuations in availability and in certain commodities (e.g. copper, lumber) can significantly affect local material pricing. In the final estimating and bidding stages of a project when the highest degree of accuracy is required, it is best to check local, current prices.

Labor Costs

Labor costs include the basic wage, plus commonly applicable taxes, insurance and markups for overhead and profit. The labor rates used here to develop the costs are typical average prevailing wage rates. Rates for different trades are used where appropriate for each type of work.

Taxes and insurance which are most often applied to labor rates include employer-paid Social Security/Medicare taxes (FICA), Worker's Compensation insurance, state and federal unemployment taxes, and business insurance. Fixed government rates as well as average allowances are included in the labor costs. However, most of these items vary significantly from state to state and within states. For more specific data, local agencies and sources should be consulted.

Equipment Costs

Costs for various types and pieces of equipment are included in Division 1 - General Requirements and can be included in an estimate when required either as a total "Equipment" category or with specific appropriate trades. Costs for equipment are included when appropriate in the installation costs in the Costbook pages.

Overhead And Profit

Included in the labor costs are allowances for overhead and profit for the contractor/employer whose workers are performing the specific tasks. No cost allowances or fees are included for management of subcontractors by the general contractor or construction manager. These costs, where appropriate, must be added to the costs as listed in the book.

The allowance for overhead is included to account for office overhead, the contractors' typical costs of doing business. These costs normally include in-house office staff salaries and benefits, office rent and operating expenses, professional fees, vehicle costs and other operating costs which are not directly applicable to specific jobs. It should be noted for this book that office overhead as included should be distinguished from project overhead, the General Requirements (CSI Division 1) which are specific to particular projects. Project overhead should be included on an item by item basis for each job.

Depending on the trade, an allowance of 10-15 percent is incorporated into the labor/installation costs to account for typical profit of the installing contractor. See Division 1, General Requirements, for a more detailed review of typical profit allowances.

Adjustments to Costs

The costs as presented in this book attempt to represent national averages. Costs, however, vary among regions, states and even between adjacent localities.

In order to more closely approximate the probable costs for specific locations throughout the U.S., a table of Geographic Cost Modifiers is provided. These adjustment factors are used to modify costs obtained from this book to help account for regional variations of construction costs. Whenever local current costs are known, whether material or equipment prices or labor rates, they should be used if more accuracy is required.

Man-Hour Tables

The man-hour data used to develop the labor costs are listed in the second main section of this book, the "Man-Hour Tables". These productivties represent typical installation labor for thousands of construction items. The data takes into account all activities involved in normal construction under commonly experienced working conditions such as site movement, material handling, start-up, etc. As with the Costbook pages, these items are listed according to the CSI MASTERFORMAT.

Square Foot Tables

Included as an additional reference are Square Foot Tables which list hundreds of actual projects for dozens of building types, each with associated building size and total square foot building cost. This data provides an overview of construction costs by building type. These costs are for actual projects. The variations within similar building types may be due, among other factors, to size, location, quality and specified components, material and processes. Depending upon all such factors, specific building costs can vary significantly and may not necessarily fall within the range of costs as presented.

Editor's Note: The **Building News 2000 Costbooks** are intended to provide accurate, reliable, average costs and typical productivities for thousands of common construction components. The data is developed and compiled from various industry sources, including government, manufacturers, suppliers and working professionals. The intent of the information is to provide assistance and guidelines to construction professionals in estimating. The user should be aware that local conditions, material and labor availability and cost variations, economic considerations, weather, local codes and regulations, etc., all affect the actual cost of construction. These and other such factors must be considered and incorporated into any and all construction estimates.

Sample Costbook Page

In order to best use the information in this book, please review this sample page and read the "Features In This Book" section.

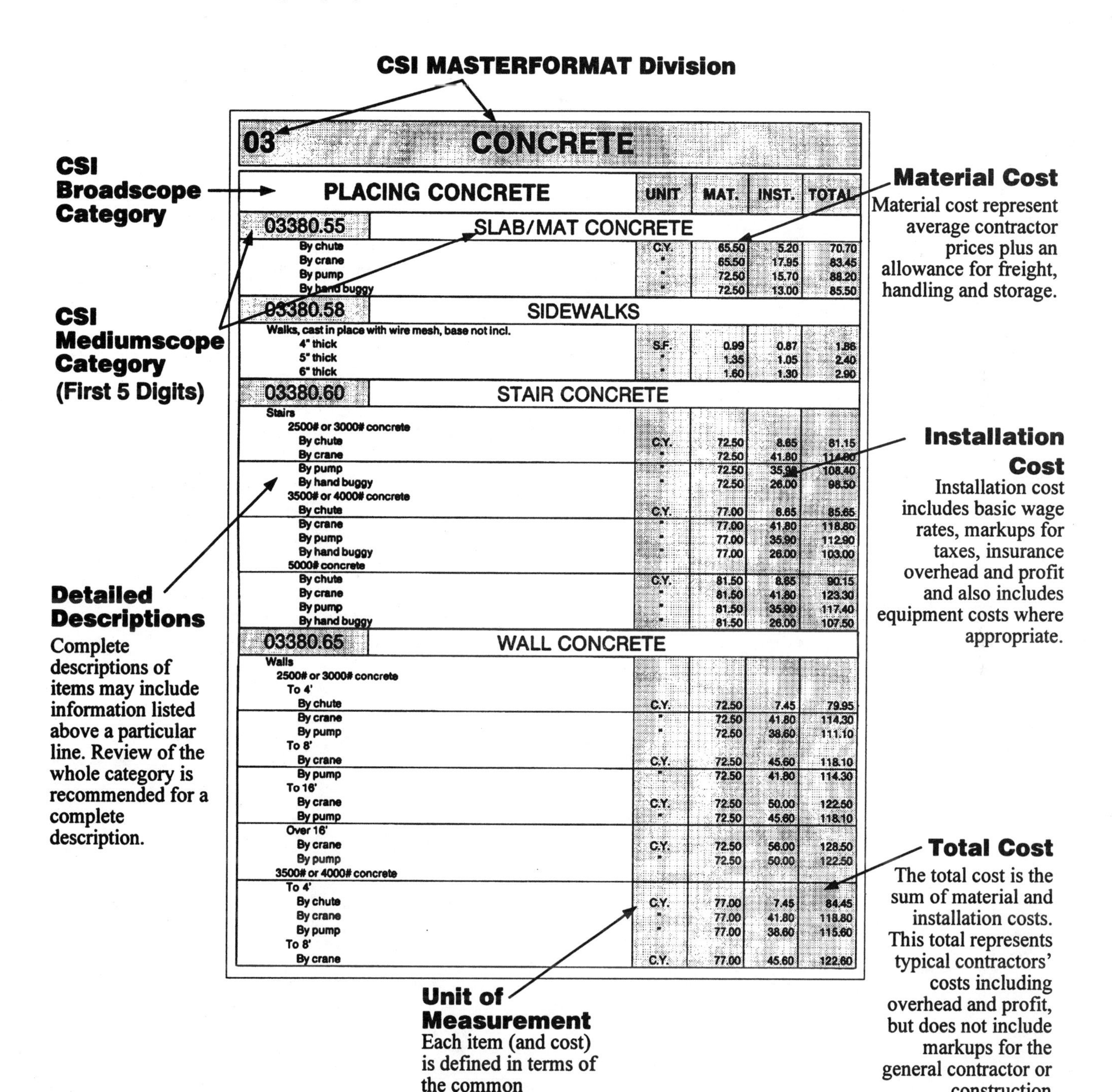

03 CONCRETE

PLACING CONCRETE	UNIT	MAT.	INST.	TOTAL
03380.55 SLAB/MAT CONCRETE				
By chute	C.Y.	65.50	5.20	70.70
By crane	"	65.50	17.95	83.45
By pump	"	72.50	15.70	88.20
By hand buggy	"	72.50	13.00	85.50
03380.58 SIDEWALKS				
Walks, cast in place with wire mesh, base not incl.				
4" thick	S.F.	0.99	0.87	1.86
5" thick	"	1.35	1.05	2.40
6" thick	"	1.60	1.30	2.90
03380.60 STAIR CONCRETE				
Stairs				
2500# or 3000# concrete				
By chute	C.Y.	72.50	8.65	81.15
By crane	"	72.50	41.80	114.80
By pump	"	72.50	35.90	108.40
By hand buggy	"	72.50	26.00	98.50
3500# or 4000# concrete				
By chute	C.Y.	77.00	8.65	85.65
By crane	"	77.00	41.80	118.80
By pump	"	77.00	35.90	112.90
By hand buggy	"	77.00	26.00	103.00
5000# concrete				
By chute	C.Y.	81.50	8.65	90.15
By crane	"	81.50	41.80	123.30
By pump	"	81.50	35.90	117.40
By hand buggy	"	81.50	26.00	107.50
03380.65 WALL CONCRETE				
Walls				
2500# or 3000# concrete				
To 4'				
By chute	C.Y.	72.50	7.45	79.95
By crane	"	72.50	41.80	114.30
By pump	"	72.50	38.60	111.10
To 8'				
By crane	C.Y.	72.50	45.60	118.10
By pump	"	72.50	41.80	114.30
To 16'				
By crane	C.Y.	72.50	50.00	122.50
By pump	"	72.50	45.60	118.10
Over 16'				
By crane	C.Y.	72.50	56.00	128.50
By pump	"	72.50	50.00	122.50
3500# or 4000# concrete				
To 4'				
By chute	C.Y.	77.00	7.45	84.45
By crane	"	77.00	41.80	118.80
By pump	"	77.00	38.60	115.60
To 8'				
By crane	C.Y.	77.00	45.60	122.60

BNi® Building News

01 GENERAL

REQUIREMENTS

REQUIREMENTS	UNIT	MAT.	INST.	TOTAL
01020.10 ALLOWANCES				
Overhead				
$20,000 project				
Minimum	PCT.			15.00
Average	"			20.00
Maximum	"			40.00
$100,000 project				
Minimum	PCT.			12.00
Average	"			15.00
Maximum	"			25.00
$500,000 project				
Minimum	PCT.			10.00
Average	"			12.00
Maximum	"			20.00
$1,000,000 project				
Minimum	PCT.			6.00
Average	"			10.00
Maximum	"			12.00
$10,000,000 project				
Minimum	PCT.			1.50
Average	"			5.00
Maximum	"			8.00
Profit				
$20,000 project				
Minimum	PCT.			10.00
Average	"			15.00
Maximum	"			25.00
$100,000 project				
Minimum	PCT.			10.00
Average	"			12.00
Maximum	"			20.00
$500,000 project				
Minimum	PCT.			5.00
Average	"			10.00
Maximum	"			15.00
$1,000,000 project				
Minimum	PCT.			3.00
Average	"			8.00
Maximum	"			15.00
Professional fees				
Architectural				
$100,000 project				
Minimum	PCT.			5.00
Average	"			10.00
Maximum	"			20.00
$500,000 project				
Minimum	PCT.			5.00
Average	"			8.00
Maximum	"			12.00
$1,000,000 project				
Minimum	PCT.			3.50
Average	"			7.00
Maximum	"			10.00

01 GENERAL

REQUIREMENTS	UNIT	MAT.	INST.	TOTAL
01020.10 ALLOWANCES				
Structural engineering				
Minimum	PCT.			2.00
Average	"			3.00
Maximum	"			5.00
Mechanical engineering				
Minimum	PCT.			4.00
Average	"			5.00
Maximum	"			15.00
Electrical engineering				
Minimum	PCT.			3.00
Average	"			5.00
Maximum	"			12.00
Taxes				
Sales tax				
Minimum	PCT.			4.00
Average	"			5.00
Maximum	"			8.25
Unemployment				
Minimum	PCT.			2.00
Average	"			5.00
Maximum	"			7.00
Social security (FICA)	"			7.85
01050.10 FIELD STAFF				
Superintendent				
Minimum	YEAR			50,985
Average	"			73,645
Maximum	"			107,635
Field engineer				
Minimum	YEAR			39,655
Average	"			58,916
Maximum	"			84,975
Foreman				
Minimum	YEAR			28,325
Average	"			45,320
Maximum	"			67,980
Bookkeeper/timekeeper				
Minimum	YEAR			16,995
Average	"			21,527
Maximum	"			33,990
Watchman				
Minimum	YEAR			11,330
Average	"			14,163
Maximum	"			22,660
01310.10 SCHEDULING				
Scheduling for				
$100,000 project				
Minimum	PCT.			1.00
Average	"			2.00
Maximum	"			4.00
$500,000 project				

01 GENERAL

REQUIREMENTS	UNIT	MAT.	INST.	TOTAL
01310.10 SCHEDULING				
Minimum	PCT.			0.50
Average	"			1.00
Maximum	"			2.00
$1,000,000 project				
Minimum	PCT.			0.30
Average	"			0.80
Maximum	"			1.50
Scheduling software, not including computer				
Minimum	EA.			500.00
Average	"			2,500
Maximum	"			40,000
01330.10 SURVEYING				
Surveying				
Small crew	DAY	0.00	570.00	570.00
Average crew	"	0.00	870.00	870.00
Large crew	"	0.00	1,140	1,140
Lot lines and boundaries				
Minimum	ACRE	0.00	410.00	410.00
Average	"	0.00	870.00	870.00
Maximum	"	0.00	1,430	1,430
01380.10 JOB REQUIREMENTS				
Job photographs, small jobs				
Minimum	EA.			57.75
Average	"			115.50
Maximum	"			254.10
Large projects				
Minimum	EA.			404.25
Average	"			577.50
Maximum	"			1,386
01410.10 TESTING				
Testing concrete, per test				
Minimum	EA.			12.71
Average	"			23.10
Maximum	"			46.20
Soil, per test				
Minimum	EA.			23.10
Average	"			69.30
Maximum	"			184.80
Welding, per test				
Minimum	EA.			11.55
Average	"			23.10
Maximum	"			86.63
01500.10 TEMPORARY FACILITIES				
Barricades, temporary				
Highway				
Concrete	L.F.	8.10	2.85	10.95

01 GENERAL

REQUIREMENTS	UNIT	MAT.	INST.	TOTAL
01500.10 TEMPORARY FACILITIES				
Wood	L.F.	2.30	1.15	3.45
Steel	"	2.55	0.95	3.50
Pedestrian barricades				
Plywood	S.F.	1.60	0.95	2.55
Chain link fence	"	1.95	0.95	2.90
Trailers, general office type, per month				
Minimum	EA.			144.38
Average	"			231.00
Maximum	"			462.00
Crew change trailers, per month				
Minimum	EA.			80.85
Average	"			92.40
Maximum	"			138.60
01505.10 MOBILIZATION				
Equipment mobilization				
Bulldozer				
Minimum	EA.			113.30
Average	"			237.93
Maximum	"			396.55
Backhoe/front-end loader				
Minimum	EA.			67.98
Average	"			118.97
Maximum	"			260.59
Crane, crawler type				
Minimum	EA.			1,246
Average	"			3,059
Maximum	"			6,571
Truck crane				
Minimum	EA.			294.58
Average	"			441.87
Maximum	"			759.11
Pile driving rig				
Minimum	EA.			5,665
Average	"			11,330
Maximum	"			20,394
01525.10 CONSTRUCTION AIDS				
Scaffolding/staging, rent per month				
Measured by lineal feet of base				
10' high	L.F.			8.16
20' high	"			14.73
30' high	"			20.63
40' high	"			23.79
50' high	"			28.32
Measured by square foot of surface				
Minimum	S.F.			0.36
Average	"			0.62
Maximum	"			1.11
Safety nets, heavy duty, per job				

01 GENERAL

REQUIREMENTS	UNIT	MAT.	INST.	TOTAL
01525.10 CONSTRUCTION AIDS				
Minimum	S.F.			0.23
Average	"			0.29
Maximum	"			0.63
Tarpaulins, fabric, per job				
Minimum	S.F.			0.17
Average	"			0.29
Maximum	"			0.74
01570.10 SIGNS				
Construction signs, temporary				
Signs, 2' x 4'				
Minimum	EA.			23.10
Average	"			55.44
Maximum	"			196.35
Signs, 4' x 8'				
Minimum	EA.			48.51
Average	"			127.05
Maximum	"			542.85
Signs, 8' x 8'				
Minimum	EA.			62.37
Average	"			196.35
Maximum	"			2,021
01600.10 EQUIPMENT				
Air compressor				
60 cfm				
By day	EA.			58.92
By week	"			176.75
By month	"			532.51
300 cfm				
By day	EA.			124.63
By week	"			373.89
By month	"			1,122
600 cfm				
By day	EA.			339.90
By week	"			1,020
By month	"			3,059
Air tools, per compressor, per day				
Minimum	EA.			22.66
Average	"			28.32
Maximum	"			39.66
Generators, 5 kw				
By day	EA.			56.65
By week	"			169.95
By month	"			521.18
Heaters, salamander type, per week				
Minimum	EA.			67.98
Average	"			96.31
Maximum	"			203.94
Pumps, submersible				

REQUIREMENTS	UNIT	MAT.	INST.	TOTAL
01600.10 EQUIPMENT				
50 gpm				
By day	EA.			45.32
By week	"			135.96
By month	"			407.88
100 gpm				
By day	EA.			56.65
By week	"			169.95
By month	"			509.85
500 gpm				
By day	EA.			90.64
By week	"			271.92
By month	"			815.76
Diaphragm pump, by week				
Minimum	EA.			79.31
Average	"			135.96
Maximum	"			283.25
Pickup truck				
By day	EA.			84.97
By week	"			249.26
By month	"			770.44
Dump truck				
6 cy truck				
By day	EA.			226.60
By week	"			679.80
By month	"			2,039
10 cy truck				
By day	EA.			283.25
By week	"			849.75
By month	"			2,549
16 cy truck				
By day	EA.			453.20
By week	"			1,360
By month	"			4,079
Backhoe, track mounted				
1/2 cy capacity				
By day	EA.			464.53
By week	"			1,416
By month	"			4,192
1 cy capacity				
By day	EA.			736.45
By week	"			2,209
By month	"			6,628
2 cy capacity				
By day	EA.			1,246
By week	"			3,739
By month	"			11,217
3 cy capacity				
By day	EA.			2,379
By week	"			7,138
By month	"			21,414
Backhoe/loader, rubber tired				
1/2 cy capacity				

REQUIREMENTS

01600.10 EQUIPMENT

REQUIREMENTS	UNIT	MAT.	INST.	TOTAL
By day	EA.			283.25
By week	"			849.75
By month	"			2,549
3/4 cy capacity				
By day	EA.			339.90
By week	"			1,020
By month	"			3,059
Bulldozer				
75 hp				
By day	EA.			396.55
By week	"			1,190
By month	"			3,569
200 hp				
By day	EA.			1,133
By week	"			3,399
By month	"			10,197
400 hp				
By day	EA.			1,700
By week	"			5,099
By month	"			15,296
Cranes, crawler type				
15 ton capacity				
By day	EA.			509.85
By week	"			1,530
By month	"			4,589
25 ton capacity				
By day	EA.			623.15
By week	"			1,869
By month	"			5,608
50 ton capacity				
By day	EA.			1,133
By week	"			3,399
By month	"			10,197
100 ton capacity				
By day	EA.			1,700
By week	"			5,099
By month	"			15,296
Truck mounted, hydraulic				
15 ton capacity				
By day	EA.			481.52
By week	"			1,445
By month	"			4,334
Loader, rubber tired				
1 cy capacity				
By day	EA.			339.90
By week	"			1,020
By month	"			3,059
2 cy capacity				
By day	EA.			509.85
By week	"			1,983
By month	"			5,948
3 cy capacity				

01 GENERAL

REQUIREMENTS	UNIT	MAT.	INST.	TOTAL
01600.10 EQUIPMENT				
By day	EA.			906.40
By week	"			2,719
By month	"			8,158
01740.10 BONDS				
Performance bonds				
Minimum	PCT.			0.60
Average	"			1.90
Maximum	"			3.00

02 SITEWORK

SOIL TESTS

SOIL TESTS	UNIT	MAT.	INST.	TOTAL
02010.10 SOIL BORING				
Borings, uncased, stable earth				
2-1/2" dia.				
Minimum	L.F.	0.00	12.45	12.45
Average	"	0.00	18.70	18.70
Maximum	"	0.00	29.90	29.90
4" dia.				
Minimum	L.F.	0.00	13.60	13.60
Average	"	0.00	21.35	21.35
Maximum	"	0.00	37.40	37.40
Cased, including samples				
2-1/2" dia.				
Minimum	L.F.	0.00	14.95	14.95
Average	"	0.00	24.90	24.90
Maximum	"	0.00	49.80	49.80
4" dia.				
Minumum	L.F.	0.00	29.90	29.90
Average	"	0.00	42.70	42.70
Maximum	"	0.00	60.00	60.00
Drilling in rock				
No sampling				
Minimum	L.F.	0.00	27.20	27.20
Average	"	0.00	39.30	39.30
Maximum	"	0.00	53.50	53.50
With casing and sampling				
Minimum	L.F.	0.00	37.40	37.40
Average	"	0.00	49.80	49.80
Maximum	"	0.00	74.50	74.50
Test pits				
Light soil				
Minimum	EA.	0.00	190.00	190.00
Average	"	0.00	250.00	250.00
Maximum	"	0.00	500.00	500.00
Heavy soil				
Minimum	EA.	0.00	300.00	300.00
Average	"	0.00	370.00	370.00
Maximum	"	0.00	750.00	750.00

DEMOLITION

DEMOLITION	UNIT	MAT.	INST.	TOTAL
02060.10 BUILDING DEMOLITION				
Building, complete with disposal				
Wood frame	C.F.	0.00	0.21	0.21
Concrete	"	0.00	0.31	0.31
Steel frame	"	0.00	0.42	0.42

DEMOLITION

02060.10 BUILDING DEMOLITION

DEMOLITION	UNIT	MAT.	INST.	TOTAL
Partition removal				
Concrete block partitions				
4" thick	S.F.	0.00	1.45	1.45
8" thick	"	0.00	1.90	1.90
12" thick	"	0.00	2.60	2.60
Brick masonry partitions				
4" thick	S.F.	0.00	1.45	1.45
8" thick	"	0.00	1.80	1.80
12" thick	"	0.00	2.40	2.40
16" thick	"	0.00	3.60	3.60
Cast in place concrete partitions				
Unreinforced				
6" thick	S.F.	0.00	9.95	9.95
8" thick	"	0.00	10.70	10.70
10" thick	"	0.00	12.45	12.45
12" thick	"	0.00	14.95	14.95
Reinforced				
6" thick	S.F.	0.00	11.50	11.50
8" thick	"	0.00	14.95	14.95
10" thick	"	0.00	16.60	16.60
12" thick	"	0.00	19.95	19.95
Terra cotta				
To 6" thick	S.F.	0.00	1.45	1.45
Stud partitions				
Metal or wood, with drywall both sides	S.F.	0.00	1.45	1.45
Metal studs, both sides, lath and plaster	"	0.00	1.90	1.90
Door and frame removal				
Hollow metal in masonry wall				
Single				
2'6"x6'8"	EA.	0.00	35.80	35.80
3'x7'	"	0.00	47.70	47.70
Double				
3'x7'	EA.	0.00	57.00	57.00
4'x8'	"	0.00	57.00	57.00
Wood in framed wall				
Single				
2'6"x6'8"	EA.	0.00	20.45	20.45
3'x6'8"	"	0.00	23.85	23.85
Double				
2'6"x6'8"	EA.	0.00	28.60	28.60
3'x6'8"	"	0.00	31.80	31.80
Remove for re-use				
Hollow metal	EA.	0.00	71.50	71.50
Wood	"	0.00	47.70	47.70
Floor removal				
Brick flooring	S.F.	0.00	1.15	1.15
Ceramic or quarry tile	"	0.00	0.64	0.64
Terrazzo	"	0.00	1.25	1.25
Heavy wood	"	0.00	0.76	0.76
Residential wood	"	0.00	0.82	0.82
Resilient tile or linoleum	"	0.00	0.29	0.29
Ceiling removal				

02 SITEWORK

DEMOLITION

02060.10 BUILDING DEMOLITION	UNIT	MAT.	INST.	TOTAL
Acoustical tile ceiling				
Adhesive fastened	S.F.	0.00	0.29	0.29
Furred and glued	"	0.00	0.24	0.24
Suspended grid	"	0.00	0.18	0.18
Drywall ceiling				
Furred and nailed	S.F.	0.00	0.32	0.32
Nailed to framing	"	0.00	0.29	0.29
Plastered ceiling				
Furred on framing	S.F.	0.00	0.71	0.71
Suspended system	"	0.00	0.95	0.95
Roofing removal				
Steel frame				
Corrugated metal roofing	S.F.	0.00	0.57	0.57
Built-up roof on metal deck	"	0.00	0.95	0.95
Wood frame				
Built up roof on wood deck	S.F.	0.00	0.88	0.88
Roof shingles	"	0.00	0.48	0.48
Roof tiles	"	0.00	0.95	0.95
Concrete frame	C.F.	0.00	1.90	1.90
Concrete plank	S.F.	0.00	1.45	1.45
Built-up roof on concrete	"	0.00	0.82	0.82
Cut-outs				
Concrete, elevated slabs, mesh reinforcing				
Under 5 cf	C.F.	0.00	28.60	28.60
Over 5 cf	"	0.00	23.85	23.85
Bar reinforcing				
Under 5 cf	C.F.	0.00	47.70	47.70
Over 5 cf	"	0.00	35.80	35.80
Window removal				
Metal windows, trim included				
2'x3'	EA.	0.00	28.60	28.60
2'x4'	"	0.00	31.80	31.80
2'x6'	"	0.00	35.80	35.80
3'x4'	"	0.00	35.80	35.80
3'x6'	"	0.00	40.90	40.90
3'x8'	"	0.00	47.70	47.70
4'x4'	"	0.00	47.70	47.70
4'x6'	"	0.00	57.00	57.00
4'x8'	"	0.00	71.50	71.50
Wood windows, trim included				
2'x3'	EA.	0.00	15.90	15.90
2'x4'	"	0.00	16.80	16.80
2'x6'	"	0.00	17.85	17.85
3'x4'	"	0.00	19.05	19.05
3'x6'	"	0.00	20.45	20.45
3'x8'	"	0.00	22.00	22.00
6'x4'	"	0.00	23.85	23.85
6'x6'	"	0.00	26.00	26.00
6'x8'	"	0.00	28.60	28.60
Walls, concrete, bar reinforcing				
Small jobs	C.F.	0.00	19.05	19.05
Large jobs	"	0.00	15.90	15.90

02 SITEWORK

DEMOLITION	UNIT	MAT.	INST.	TOTAL
02060.10 BUILDING DEMOLITION				
Brick walls, not including toothing				
4" thick	S.F.	0.00	1.45	1.45
8" thick	"	0.00	1.80	1.80
12" thick	"	0.00	2.40	2.40
16" thick	"	0.00	3.60	3.60
Concrete block walls, not including toothing				
4" thick	S.F.	0.00	1.60	1.60
6" thick	"	0.00	1.70	1.70
8" thick	"	0.00	1.80	1.80
10" thick	"	0.00	2.05	2.05
12" thick	"	0.00	2.40	2.40
Rubbish handling				
Load in dumpster or truck				
Minimum	C.F.	0.00	0.64	0.64
Maximum	"	0.00	0.95	0.95
For use of elevators, add				
Minimum	C.F.	0.00	0.14	0.14
Maximum	"	0.00	0.29	0.29
Rubbish hauling				
Hand loaded on trucks, 2 mile trip	C.Y.	0.00	22.25	22.25
Machine loaded on trucks, 2 mile trip	"	0.00	14.95	14.95

HIGHWAY DEMOLITION	UNIT	MAT.	INST.	TOTAL
02065.10 PAVEMENT DEMOLITION				
Bituminous pavement, up to 3" thick				
On streets				
Minimum	S.Y.	0.00	4.25	4.25
Average	"	0.00	6.00	6.00
Maximum	"	0.00	9.95	9.95
On pipe trench				
Minimum	S.Y.	0.00	6.00	6.00
Average	"	0.00	7.45	7.45
Maximum	"	0.00	14.95	14.95
Concrete pavement, 6" thick				
No reinforcement				
Minimum	S.Y.	0.00	7.45	7.45
Average	"	0.00	9.95	9.95
Maximum	"	0.00	14.95	14.95
With wire mesh				
Minimum	S.Y.	0.00	11.50	11.50
Average	"	0.00	14.95	14.95
Maximum	"	0.00	18.70	18.70

HIGHWAY DEMOLITION

HIGHWAY DEMOLITION	UNIT	MAT.	INST.	TOTAL
02065.10 PAVEMENT DEMOLITION				
With rebars				
Minimum	S.Y.	0.00	14.95	14.95
Average	"	0.00	18.70	18.70
Maximum	"	0.00	24.90	24.90
9" thick				
No reinforcement				
Minimum	S.Y.	0.00	9.95	9.95
Average	"	0.00	12.45	12.45
Maximum	"	0.00	14.95	14.95
With wire mesh				
Minimum	S.Y.	0.00	15.75	15.75
Average	"	0.00	18.70	18.70
Maximum	"	0.00	23.00	23.00
With rebars				
Minimum	S.Y.	0.00	19.95	19.95
Average	"	0.00	24.90	24.90
Maximum	"	0.00	33.20	33.20
12" thick				
No reinforcement				
Minimum	S.Y.	0.00	12.45	12.45
Average	"	0.00	14.95	14.95
Maximum	"	0.00	18.70	18.70
With wire mesh				
Minimum	S.Y.	0.00	17.60	17.60
Average	"	0.00	21.35	21.35
Maximum	"	0.00	27.20	27.20
With rebars				
Minimum	S.Y.	0.00	24.90	24.90
Average	"	0.00	29.90	29.90
Maximum	"	0.00	37.40	37.40
Sidewalk, 4" thick, with disposal				
Minimum	S.Y.	0.00	3.55	3.55
Average	"	0.00	5.00	5.00
Maximum	"	0.00	7.10	7.10
Removal of pavement markings by waterblasting				
Minimum	S.F.	0.00	0.12	0.12
Average	"	0.00	0.14	0.14
Maximum	"	0.00	0.29	0.29
02065.15 SAW CUTTING PAVEMENT				
Pavement, bituminous				
2" thick	L.F.	0.00	1.10	1.10
3" thick	"	0.00	1.40	1.40
4" thick	"	0.00	1.70	1.70
5" thick	"	0.00	1.85	1.85
6" thick	"	0.00	2.00	2.00
Concrete pavement, with wire mesh				
4" thick	L.F.	0.00	2.15	2.15
5" thick	"	0.00	2.30	2.30
6" thick	"	0.00	2.55	2.55
8" thick	"	0.00	2.80	2.80
10" thick	"	0.00	3.10	3.10

02 SITEWORK

HIGHWAY DEMOLITION	UNIT	MAT.	INST.	TOTAL
02065.15 SAW CUTTING PAVEMENT				
Plain concrete, unreinforced				
4" thick	L.F.	0.00	1.85	1.85
5" thick	"	0.00	2.15	2.15
6" thick	"	0.00	2.30	2.30
8" thick	"	0.00	2.55	2.55
10" thick	"	0.00	2.80	2.80
02065.80 CURB & GUTTER				
Curb removal				
Concrete, unreinforced				
Minimum	L.F.	0.00	3.00	3.00
Average	"	0.00	3.75	3.75
Maximum	"	0.00	4.65	4.65
Reinforced				
Minimum	L.F.	0.00	4.80	4.80
Average	"	0.00	5.35	5.35
Maximum	"	0.00	6.00	6.00
Combination curb and 2'gutter				
Unreinforced				
Minimum	L.F.	0.00	3.95	3.95
Average	"	0.00	5.15	5.15
Maximum	"	0.00	7.45	7.45
Reinforced				
Minimum	L.F.	0.00	6.25	6.25
Average	"	0.00	8.30	8.30
Maximum	"	0.00	14.95	14.95
Granite curb				
Minimum	L.F.	0.00	4.25	4.25
Average	"	0.00	5.00	5.00
Maximum	"	0.00	5.75	5.75
Asphalt curb				
Minimum	L.F.	0.00	2.50	2.50
Average	"	0.00	3.00	3.00
Maximum	"	0.00	3.55	3.55
02065.85 GUARDRAILS				
Remove standard guardrail				
Steel				
Minimum	L.F.	0.00	3.75	3.75
Average	"	0.00	5.00	5.00
Maximum	"	0.00	7.45	7.45
Wood				
Minimum	L.F.	0.00	3.25	3.25
Average	"	0.00	3.85	3.85
Maximum	"	0.00	6.25	6.25
02075.80 CORE DRILLING				
Concrete				
6" thick				
3" dia.	EA.	0.00	26.00	26.00
4" dia.	"	0.00	30.30	30.30

HIGHWAY DEMOLITION

02075.80 CORE DRILLING	UNIT	MAT.	INST.	TOTAL
6" dia.	EA.	0.00	36.40	36.40
8" dia.	"	0.00	60.50	60.50
8" thick				
3" dia.	EA.	0.00	36.40	36.40
4" dia.	"	0.00	45.50	45.50
6" dia.	"	0.00	52.00	52.00
8" dia.	"	0.00	73.00	73.00
10" thick				
3" dia.	EA.	0.00	45.50	45.50
4" dia.	"	0.00	52.00	52.00
6" dia.	"	0.00	60.50	60.50
8" dia.	"	0.00	91.00	91.00
12" thick				
3" dia.	EA.	0.00	60.50	60.50
4" dia.	"	0.00	73.00	73.00
6" dia.	"	0.00	91.00	91.00
8" dia.	"	0.00	120.00	120.00

HAZARDOUS WASTE

02080.10 ASBESTOS REMOVAL	UNIT	MAT.	INST.	TOTAL
Enclosure using wood studs & poly, install & remove	S.F.	0.77	0.71	1.49
Trailer (change room)	DAY			71.75
Disposal suits (4 suits per man day)	"			29.69
Type C respirator mask, includes hose & filters, per man	"			14.84
Respirator mask & filter, light contamination	"			5.81
Air monitoring test, 12 tests per day				
Off job testing	DAY			766.92
On the job testing	"			1,014
Asbestos vacuum with attachments	EA.			445.37
Hydraspray piston pump	"			569.07
Negative air pressure system	"			519.53
Grade D breathing air equipment	"			1,299
Glove bag, 44" x 60" x 6 mil plastic	"			3.95
40 CY asbestos dumpster				
Weekly rental	EA.			494.34
Pick up/delivery	"			235.05
Asbestos dump fee	"			123.58

02080.12 DUCT INSULATION REMOVAL	UNIT	MAT.	INST.	TOTAL
Remove duct insulation, duct size				
6" x 12"	L.F.	0.00	1.60	1.60

HAZARDOUS WASTE

HAZARDOUS WASTE	UNIT	MAT.	INST.	TOTAL
02080.12 DUCT INSULATION REMOVAL				
x 18"	L.F.	0.00	2.20	2.20
x 24"	"	0.00	3.20	3.20
8" x 12"	"	0.00	2.40	2.40
x 18"	"	0.00	2.60	2.60
x 24"	"	0.00	3.60	3.60
12" x 12"	"	0.00	2.40	2.40
x 18"	"	0.00	3.20	3.20
x 24"	"	0.00	4.10	4.10
02080.15 PIPE INSULATION REMOVAL				
Removal, asbestos insulation				
2" thick, pipe				
1" to 3" dia.	L.F.	0.00	2.40	2.40
4" to 6" dia.	"	0.00	2.70	2.70
3" thick				
7" to 8" dia.	L.F.	0.00	2.85	2.85
9" to 10" dia.	"	0.00	3.00	3.00
11" to 12" dia.	"	0.00	3.20	3.20
13" to 14" dia.	"	0.00	3.35	3.35
15" to 18" dia.	"	0.00	3.60	3.60

SITE DEMOLITION

SITE DEMOLITION	UNIT	MAT.	INST.	TOTAL
02105.10 CATCH BASINS/MANHOLES				
Abandon catch basin or manhole (fill with sand)				
Minimum	EA.	0.00	190.00	190.00
Average	"	0.00	300.00	300.00
Maximum	"	0.00	500.00	500.00
Remove and reset frame and cover				
Minimum	EA.	0.00	100.00	100.00
Average	"	0.00	150.00	150.00
Maximum	"	0.00	250.00	250.00
Remove catch basin, to 10' deep				
Masonry				
Minumum	EA.	0.00	300.00	300.00
Average	"	0.00	370.00	370.00
Maximum	"	0.00	500.00	500.00
Concrete				
Minimum	EA.	0.00	370.00	370.00
Average	"	0.00	500.00	500.00
Maximum	"	0.00	600.00	600.00

02 SITEWORK

SITE DEMOLITION

	UNIT	MAT.	INST.	TOTAL
02105.20 FENCES				
Remove fencing				
Chain link, 8' high				
For disposal	L.F.	0.00	1.45	1.45
For reuse	"	0.00	3.60	3.60
Wood				
4' high	S.F.	0.00	0.95	0.95
6' high	"	0.00	1.15	1.15
8' high	"	0.00	1.45	1.45
Masonry				
8" thick				
4' high	S.F.	0.00	2.85	2.85
6' high	"	0.00	3.60	3.60
8' high	"	0.00	4.10	4.10
12" thick				
4' high	S.F.	0.00	4.75	4.75
6' high	"	0.00	5.70	5.70
8' high	"	0.00	7.15	7.15
12' high	"	0.00	9.55	9.55
02105.30 HYDRANTS				
Remove fire hydrant				
Minimum	EA.	0.00	190.00	190.00
Average	"	0.00	250.00	250.00
Maximum	"	0.00	370.00	370.00
Remove and reset fire hydrant				
Minimum	EA.	0.00	500.00	500.00
Average	"	0.00	750.00	750.00
Maximum	"	0.00	1,490	1,490
02105.42 DRAINAGE PIPING				
Remove drainage pipe, not including excavation				
12" dia.				
Minimum	L.F.	0.00	5.00	5.00
Average	"	0.00	6.25	6.25
Maximum	"	0.00	7.85	7.85
18" dia.				
Minimum	L.F.	0.00	6.80	6.80
Average	"	0.00	7.85	7.85
Maximum	"	0.00	9.95	9.95
24" dia.				
Minimum	L.F.	0.00	8.30	8.30
Average	"	0.00	9.95	9.95
Maximum	"	0.00	12.45	12.45
36" dia.				
Minimum	L.F.	0.00	9.95	9.95
Average	"	0.00	12.45	12.45
Maximum	"	0.00	15.75	15.75

02 SITEWORK

SITE DEMOLITION

02105.43 GAS PIPING

Description	UNIT	MAT.	INST.	TOTAL
Remove welded steel pipe, not including excavation				
4" dia.				
Minimum	L.F.	0.00	7.45	7.45
Average	"	0.00	9.35	9.35
Maximum	"	0.00	12.45	12.45
5" dia.				
Minimum	L.F.	0.00	12.45	12.45
Average	"	0.00	14.95	14.95
Maximum	"	0.00	18.70	18.70
6" dia.				
Minimum	L.F.	0.00	15.75	15.75
Average	"	0.00	18.70	18.70
Maximum	"	0.00	24.90	24.90
8" dia.				
Minimum	L.F.	0.00	23.00	23.00
Average	"	0.00	29.90	29.90
Maximum	"	0.00	39.30	39.30
10" dia.				
Minimum	L.F.	0.00	29.90	29.90
Average	"	0.00	37.40	37.40
Maximum	"	0.00	49.80	49.80

02105.45 SANITARY PIPING

Description	UNIT	MAT.	INST.	TOTAL
Remove sewer pipe, not including excavation				
4" dia.				
Minimum	L.F.	0.00	4.15	4.15
Average	"	0.00	6.00	6.00
Maximum	"	0.00	9.95	9.95
6" dia.				
Minimum	L.F.	0.00	4.65	4.65
Average	"	0.00	6.80	6.80
Maximum	"	0.00	12.45	12.45
8" dia.				
Minimum	L.F.	0.00	5.00	5.00
Average	"	0.00	7.45	7.45
Maximum	"	0.00	14.95	14.95
10" dia.				
Minimum	L.F.	0.00	5.35	5.35
Average	"	0.00	7.85	7.85
Maximum	"	0.00	16.60	16.60
12" dia.				
Minimum	L.F.	0.00	5.75	5.75
Average	"	0.00	8.30	8.30
Maximum	"	0.00	18.70	18.70
15" dia.				
Minimum	L.F.	0.00	6.25	6.25
Average	"	0.00	8.80	8.80
Maximum	"	0.00	21.35	21.35
18" dia.				
Minimum	L.F.	0.00	6.80	6.80
Average	"	0.00	9.95	9.95

SITE DEMOLITION	UNIT	MAT.	INST.	TOTAL
02105.45 SANITARY PIPING				
Maximum	L.F.	0.00	24.90	24.90
24" dia.				
Minimum	L.F.	0.00	7.45	7.45
Average	"	0.00	12.45	12.45
Maximum	"	0.00	29.90	29.90
30" dia.				
Minimum	L.F.	0.00	8.30	8.30
Average	"	0.00	14.95	14.95
Maximum	"	0.00	37.40	37.40
36" dia.				
Minimum	L.F.	0.00	9.95	9.95
Average	"	0.00	18.70	18.70
Maximum	"	0.00	49.80	49.80
02105.48 WATER PIPING				
Remove water pipe, not including excavation				
4" dia.				
Minimum	L.F.	0.00	6.00	6.00
Average	"	0.00	6.80	6.80
Maximum	"	0.00	7.85	7.85
6" dia.				
Minimum	L.F.	0.00	6.25	6.25
Average	"	0.00	7.10	7.10
Maximum	"	0.00	8.30	8.30
8" dia.				
Minimum	L.F.	0.00	6.80	6.80
Average	"	0.00	7.85	7.85
Maximum	"	0.00	9.35	9.35
10" dia.				
Minimum	L.F.	0.00	7.10	7.10
Average	"	0.00	8.30	8.30
Maximum	"	0.00	9.95	9.95
12" dia.				
Minimum	L.F.	0.00	7.45	7.45
Average	"	0.00	8.80	8.80
Maximum	"	0.00	10.70	10.70
14" dia.				
Minimum	L.F.	0.00	7.85	7.85
Average	"	0.00	9.35	9.35
Maximum	"	0.00	11.50	11.50
16" dia.				
Minimum	L.F.	0.00	8.30	8.30
Average	"	0.00	9.95	9.95
Maximum	"	0.00	12.45	12.45
18" dia.				
Minimum	L.F.	0.00	8.80	8.80
Average	"	0.00	10.70	10.70
Maximum	"	0.00	13.60	13.60
20" dia.				
Minimum	L.F.	0.00	9.35	9.35
Average	"	0.00	11.50	11.50

02 SITEWORK

SITE DEMOLITION	UNIT	MAT.	INST.	TOTAL
02105.48 WATER PIPING				
Maximum	L.F.	0.00	14.95	14.95
Remove valves				
6"	EA.	0.00	74.50	74.50
10"	"	0.00	83.00	83.00
14"	"	0.00	93.50	93.50
18"	"	0.00	120.00	120.00
02105.60 UNDERGROUND TANKS				
Remove underground storage tank, and backfill				
50 to 250 gals	EA.	0.00	500.00	500.00
600 gals	"	0.00	500.00	500.00
1000 gals	"	0.00	750.00	750.00
4000 gals	"	0.00	1,200	1,200
5000 gals	"	0.00	1,200	1,200
10,000 gals	"	0.00	1,990	1,990
12,000 gals	"	0.00	2,490	2,490
15,000 gals	"	0.00	2,990	2,990
20,000 gals	"	0.00	3,740	3,740
02105.66 SEPTIC TANKS				
Remove septic tank				
1000 gals	EA.	0.00	120.00	120.00
2000 gals	"	0.00	150.00	150.00
5000 gals	"	0.00	190.00	190.00
15,000 gals	"	0.00	1,490	1,490
25,000 gals	"	0.00	1,990	1,990
40,000 gals	"	0.00	2,990	2,990
02105.80 WALLS, EXTERIOR				
Concrete wall				
Light reinforcing				
6" thick	S.F.	0.00	7.45	7.45
8" thick	"	0.00	7.85	7.85
10" thick	"	0.00	8.30	8.30
12" thick	"	0.00	9.35	9.35
Medium reinforcing				
6" thick	S.F.	0.00	7.85	7.85
8" thick	"	0.00	8.30	8.30
10" thick	"	0.00	9.35	9.35
12" thick	"	0.00	10.70	10.70
Heavy reinforcing				
6" thick	S.F.	0.00	8.80	8.80
8" thick	"	0.00	9.35	9.35
10" thick	"	0.00	10.70	10.70
12" thick	"	0.00	12.45	12.45
Masonry				
No reinforcing				
8" thick	S.F.	0.00	3.30	3.30
12" thick	"	0.00	3.75	3.75
16" thick	"	0.00	4.25	4.25
Horizontal reinforcing				

02 SITEWORK

SITE DEMOLITION

SITE DEMOLITION	UNIT	MAT.	INST.	TOTAL
02105.80 WALLS, EXTERIOR				
8" thick	S.F.	0.00	3.75	3.75
12" thick	"	0.00	4.05	4.05
16" thick	"	0.00	4.80	4.80
Vertical reinforcing				
8" thick	S.F.	0.00	4.80	4.80
12" thick	"	0.00	5.55	5.55
16" thick	"	0.00	6.80	6.80
Remove concrete headwall				
15" pipe	EA.	0.00	110.00	110.00
18" pipe	"	0.00	120.00	120.00
24" pipe	"	0.00	140.00	140.00
30" pipe	"	0.00	150.00	150.00
36" pipe	"	0.00	170.00	170.00
48" pipe	"	0.00	210.00	210.00
60" pipe	"	0.00	300.00	300.00
02110.10 CLEARING AND GRUBBING				
Clear wooded area				
Light density	ACRE	0.00	3,740	3,740
Medium density	"	0.00	4,980	4,980
Heavy density	"	0.00	5,980	5,980
02110.50 TREE CUTTING & CLEARING				
Cut trees and clear out stumps				
9" to 12" dia.	EA.	0.00	300.00	300.00
To 24" dia.	"	0.00	370.00	370.00
24" dia. and up	"	0.00	500.00	500.00
Loading and trucking				
For machine load, per load, round trip				
1 mile	EA.	0.00	60.00	60.00
3 mile	"	0.00	68.00	68.00
5 mile	"	0.00	74.50	74.50
10 mile	"	0.00	100.00	100.00
20 mile	"	0.00	150.00	150.00
Hand loaded, round trip				
1 mile	EA.	0.00	140.00	140.00
3 mile	"	0.00	160.00	160.00
5 mile	"	0.00	190.00	190.00
10 mile	"	0.00	220.00	220.00
20 mile	"	0.00	280.00	280.00
Tree trimming for pole line construction				
Light cutting	L.F.	0.00	0.75	0.75
Medium cutting	"	0.00	1.00	1.00
Heavy cutting	"	0.00	1.50	1.50

02 SITEWORK

DEWATERING

02144.10 WELLPOINT SYSTEMS	UNIT	MAT.	INST.	TOTAL
Pumping, gas driven, 50' hose				
3" header pipe	DAY	0.00	560.00	560.00
6" header pipe	"	0.00	700.00	700.00
Wellpoint system per job				
6" header pipe	L.F.	1.15	2.25	3.40
8" header pipe	"	1.45	2.80	4.25
10" header pipe	"	1.95	3.70	5.65
Jetting wellpoint system				
14' long	EA.	38.10	37.10	75.20
18' long	"	45.00	46.40	91.40
Sand filter for wellpoints	L.F.	1.90	0.93	2.83
Replacement of wellpoint components	EA.	0.00	11.15	11.15

SHORING AND UNDERPINNING

02162.10 TRENCH SHEETING	UNIT	MAT.	INST.	TOTAL
Closed timber, including pull and salvage, excavation				
8' deep	S.F.	2.35	5.00	7.35
10' deep	"	2.35	5.25	7.60
12' deep	"	2.35	5.55	7.90
14' deep	"	2.35	5.85	8.20
16' deep	"	2.35	6.25	8.60
18' deep	"	2.35	7.15	9.50
20' deep	"	2.35	7.65	10.00

02170.10 COFFERDAMS	UNIT	MAT.	INST.	TOTAL
Cofferdam, steel, driven from shore				
15' deep	S.F.	10.10	12.55	22.65
20' deep	"	10.10	11.70	21.80
25' deep	"	10.10	10.95	21.05
30' deep	"	10.10	10.35	20.45
40' deep	"	10.10	9.75	19.85
Driven from barge				
20' deep	S.F.	10.10	13.50	23.60
30' deep	"	10.10	12.55	22.65
40' deep	"	10.10	11.70	21.80
50' deep	"	10.10	10.95	21.05

02 SITEWORK

EARTHWORK	UNIT	MAT.	INST.	TOTAL
02210.10 HAULING MATERIAL				
Haul material by 10 cy dump truck, round trip distance				
1 mile	C.Y.	0.00	3.10	3.10
2 mile	"	0.00	3.70	3.70
5 mile	"	0.00	5.05	5.05
10 mile	"	0.00	5.55	5.55
20 mile	"	0.00	6.20	6.20
30 mile	"	0.00	7.40	7.40
Site grading, cut & fill, sandy clay, 200' haul, 75 hp dozer	"	0.00	2.25	2.25
Spread topsoil by equipment on site	"	0.00	2.45	2.45
Site grading (cut and fill to 6") less than 1 acre				
75 hp dozer	C.Y.	0.00	3.70	3.70
1.5 cy backhoe/loader	"	0.00	5.55	5.55
02210.30 BULK EXCAVATION				
Excavation, by small dozer				
Large areas	C.Y.	0.00	1.10	1.10
Small areas	"	0.00	1.85	1.85
Trim banks	"	0.00	2.80	2.80
Drag line				
1-1/2 cy bucket				
Sand or gravel	C.Y.	0.00	2.50	2.50
Light clay	"	0.00	3.30	3.30
Heavy clay	"	0.00	3.75	3.75
Unclassified	"	0.00	4.00	4.00
2 cy bucket				
Sand or gravel	C.Y.	0.00	2.30	2.30
Light clay	"	0.00	3.00	3.00
Heavy clay	"	0.00	3.30	3.30
Unclassified	"	0.00	3.50	3.50
2-1/2 cy bucket				
Sand or gravel	C.Y.	0.00	2.15	2.15
Light clay	"	0.00	2.70	2.70
Heavy clay	"	0.00	3.00	3.00
Unclassified	"	0.00	3.15	3.15
3 cy bucket				
Sand or gravel	C.Y.	0.00	1.85	1.85
Light clay	"	0.00	2.50	2.50
Heavy clay	"	0.00	2.70	2.70
Unclassified	"	0.00	2.85	2.85
Hydraulic excavator				
1 cy capacity				
Light material	C.Y.	0.00	2.50	2.50
Medium material	"	0.00	3.00	3.00
Wet material	"	0.00	3.75	3.75
Blasted rock	"	0.00	4.25	4.25
1-1/2 cy capacity				
Light material	C.Y.	0.00	0.97	0.97
Medium material	"	0.00	1.30	1.30
Wet material	"	0.00	1.55	1.55
Blasted rock	"	0.00	1.95	1.95
2 cy capacity				
Light material	C.Y.	0.00	0.86	0.86

EARTHWORK	UNIT	MAT.	INST.	TOTAL
02210.30 BULK EXCAVATION				
Medium material	C.Y.	0.00	1.10	1.10
Wet material	"	0.00	1.30	1.30
Blasted rock	"	0.00	1.55	1.55
Wheel mounted front-end loader				
7/8 cy capacity				
Light material	C.Y.	0.00	1.95	1.95
Medium material	"	0.00	2.20	2.20
Wet material	"	0.00	2.60	2.60
Blasted rock	"	0.00	3.10	3.10
1-1/2 cy capacity				
Light material	C.Y.	0.00	1.10	1.10
Medium material	"	0.00	1.20	1.20
Wet material	"	0.00	1.30	1.30
Blasted rock	"	0.00	1.40	1.40
2-1/2 cy capacity				
Light material	C.Y.	0.00	0.92	0.92
Medium material	"	0.00	0.97	0.97
Wet material	"	0.00	1.05	1.05
Blasted rock	"	0.00	1.10	1.10
3-1/2 cy capacity				
Light material	C.Y.	0.00	0.86	0.86
Medium material	"	0.00	0.92	0.92
Wet material	"	0.00	0.97	0.97
Blasted rock	"	0.00	1.05	1.05
6 cy capacity				
Light material	C.Y.	0.00	0.52	0.52
Medium material	"	0.00	0.56	0.56
Wet material	"	0.00	0.60	0.60
Blasted rock	"	0.00	0.65	0.65
Track mounted front-end loader				
1-1/2 cy capacity				
Light material	C.Y.	0.00	1.30	1.30
Medium material	"	0.00	1.40	1.40
Wet material	"	0.00	1.55	1.55
Blasted rock	"	0.00	1.75	1.75
2-3/4 cy capacity				
Light material	C.Y.	0.00	0.78	0.78
Medium material	"	0.00	0.86	0.86
Wet material	"	0.00	0.97	0.97
Blasted rock	"	0.00	1.10	1.10
02220.10 BORROW				
Borrow fill, F.O.B. at pit				
Sand, haul to site, round trip				
10 mile	C.Y.	9.10	7.80	16.90
20 mile	"	9.10	12.95	22.05
30 mile	"	9.10	19.45	28.55
Place borrow fill and compact				
Less than 1 in 4 slope	C.Y.	9.10	3.90	13.00
Greater than 1 in 4 slope	"	9.10	5.20	14.30

02 SITEWORK

EARTHWORK	UNIT	MAT.	INST.	TOTAL
02220.20 GRAVEL AND STONE				
F.O.B. PLANT				
No. 21 crusher run stone	C.Y.			21.10
No. 26 crusher run stone	"			20.53
No. 57 stone	"			21.69
No. 67 gravel	"			10.55
No. 68 stone	"			21.10
No. 78 stone	"			21.10
No. 78 gravel, (pea gravel)	"			12.31
No. 357 or B-3 stone	"			21.69
Structural & foundation backfill				
No. 21 crusher run stone	TON			18.75
No. 26 crusher run stone	"			19.35
No. 57 stone	"			21.10
No. 67 gravel	"			9.38
No. 68 stone	"			18.75
No. 78 stone	"			18.75
No. 78 gravel, (pea gravel)	"			15.25
No. 357 or B-3 stone	"			21.69
02220.40 BUILDING EXCAVATION				
Structural excavation, unclassified earth				
3/8 cy backhoe	C.Y.	0.00	10.40	10.40
3/4 cy backhoe	"	0.00	7.80	7.80
1 cy backhoe	"	0.00	6.50	6.50
Foundation backfill and compaction by machine	"	0.00	15.55	15.55
02220.50 UTILITY EXCAVATION				
Trencher, sandy clay, 8" wide trench				
18" deep	L.F.	0.00	1.25	1.25
24" deep	"	0.00	1.40	1.40
36" deep	"	0.00	1.60	1.60
Trench backfill, 95% compaction				
Tamp by hand	C.Y.	0.00	17.85	17.85
Vibratory compaction	"	0.00	14.30	14.30
Trench backfilling, with borrow sand, place & compact	"	9.10	14.30	23.40
02220.60 TRENCHING				
Trenching and continuous footing excavation				
By gradall				
1 cy capacity				
Light soil	C.Y.	0.00	2.20	2.20
Medium soil	"	0.00	2.40	2.40
Heavy/wet soil	"	0.00	2.60	2.60
Loose rock	"	0.00	2.85	2.85
Blasted rock	"	0.00	3.00	3.00
By hydraulic excavator				
1/2 cy capacity				
Light soil	C.Y.	0.00	2.60	2.60
Medium soil	"	0.00	2.85	2.85
Heavy/wet soil	"	0.00	3.10	3.10
Loose rock	"	0.00	3.45	3.45

EARTHWORK	UNIT	MAT.	INST.	TOTAL
02220.60 TRENCHING				
Blasted rock	C.Y.	0.00	3.90	3.90
1 cy capacity				
Light soil	C.Y.	0.00	1.85	1.85
Medium soil	"	0.00	1.95	1.95
Heavy/wet soil	"	0.00	2.10	2.10
Loose rock	"	0.00	2.20	2.20
Blasted rock	"	0.00	2.40	2.40
1-1/2 cy capacity				
Light soil	C.Y.	0.00	1.65	1.65
Medium soil	"	0.00	1.75	1.75
Heavy/wet soil	"	0.00	1.85	1.85
Loose rock	"	0.00	1.95	1.95
Blasted rock	"	0.00	2.10	2.10
2 cy capacity				
Light soil	C.Y.	0.00	1.55	1.55
Medium soil	"	0.00	1.65	1.65
Heavy/wet soil	"	0.00	1.75	1.75
Loose rock	"	0.00	1.85	1.85
Blasted rock	"	0.00	1.95	1.95
2-1/2 cy capacity				
Light soil	C.Y.	0.00	1.40	1.40
Medium soil	"	0.00	1.50	1.50
Heavy/wet soil	"	0.00	1.55	1.55
Loose rock	"	0.00	1.65	1.65
Blasted rock	"	0.00	1.75	1.75
Trencher, chain, 1' wide to 4' deep				
Light soil	C.Y.	0.00	1.40	1.40
Medium soil	"	0.00	1.60	1.60
Heavy soil	"	0.00	1.85	1.85
Hand excavation				
Bulk, wheeled 100'				
Normal soil	C.Y.	0.00	31.80	31.80
Sand or gravel	"	0.00	28.60	28.60
Medium clay	"	0.00	40.90	40.90
Heavy clay	"	0.00	57.00	57.00
Loose rock	"	0.00	71.50	71.50
Trenches, up to 2' deep				
Normal soil	C.Y.	0.00	35.80	35.80
Sand or gravel	"	0.00	31.80	31.80
Medium clay	"	0.00	47.70	47.70
Heavy clay	"	0.00	71.50	71.50
Loose rock	"	0.00	95.50	95.50
Trenches, to 6' deep				
Normal soil	C.Y.	0.00	40.90	40.90
Sand or gravel	"	0.00	35.80	35.80
Medium clay	"	0.00	57.00	57.00
Heavy clay	"	0.00	95.50	95.50
Loose rock	"	0.00	140.00	140.00
Backfill trenches				
With compaction				
By hand	C.Y.	0.00	23.85	23.85
By 60 hp tracked dozer	"	0.00	1.40	1.40

EARTHWORK	UNIT	MAT.	INST.	TOTAL
02220.60 TRENCHING				
By 200 hp tracked dozer	C.Y.	0.00	0.86	0.86
By small front-end loader	"	0.00	1.60	1.60
Spread dumped fill or gravel, no compaction				
6" layers	S.Y.	0.00	0.93	0.93
12" layers	"	0.00	1.10	1.10
Compaction in 6" layers				
By hand with air tamper	S.Y.	0.00	0.73	0.73
Backfill trenches, sand bedding, no compaction				
By hand	C.Y.	9.10	23.85	32.95
By small front-end loader	"	9.10	2.20	11.30
02220.70 ROADWAY EXCAVATION				
Roadway excavation				
1/4 mile haul	C.Y.	0.00	1.55	1.55
2 mile haul	"	0.00	2.60	2.60
5 mile haul	"	0.00	3.90	3.90
Excavation of open ditches	"	0.00	1.10	1.10
Trim banks, swales or ditches	S.Y.	0.00	1.30	1.30
Bulk swale excavation by dragline				
Small jobs	C.Y.	0.00	3.75	3.75
Large jobs	"	0.00	2.15	2.15
Spread base course	"	0.00	1.95	1.95
Roll and compact	"	0.00	2.60	2.60
02220.71 BASE COURSE				
Base course, crushed stone				
3" thick	S.Y.	1.95	0.39	2.34
4" thick	"	2.55	0.42	2.97
6" thick	"	3.85	0.46	4.31
8" thick	"	4.90	0.52	5.42
10" thick	"	6.05	0.56	6.61
12" thick	"	6.60	0.65	7.25
Base course, bank run gravel				
4" deep	S.Y.	1.40	0.41	1.81
6" deep	"	2.05	0.45	2.50
8" deep	"	2.65	0.49	3.14
10" deep	"	3.40	0.52	3.92
12" deep	"	4.05	0.60	4.65
Prepare and roll sub base				
Minimum	S.Y.	0.00	0.39	0.39
Average	"	0.00	0.49	0.49
Maximum	"	0.00	0.65	0.65
02220.90 HAND EXCAVATION				
Excavation				
To 2' deep				
Normal soil	C.Y.	0.00	31.80	31.80
Sand and gravel	"	0.00	28.60	28.60
Medium clay	"	0.00	35.80	35.80
Heavy clay	"	0.00	40.90	40.90

02 SITEWORK

EARTHWORK

EARTHWORK	UNIT	MAT.	INST.	TOTAL
02220.90 HAND EXCAVATION				
Loose rock	C.Y.	0.00	47.70	47.70
To 6' deep				
Normal soil	C.Y.	0.00	40.90	40.90
Sand and gravel	"	0.00	35.80	35.80
Medium clay	"	0.00	47.70	47.70
Heavy clay	"	0.00	57.00	57.00
Loose rock	"	0.00	71.50	71.50
Backfilling foundation without compaction, 6" lifts	"	0.00	17.85	17.85
Compaction of backfill around structures or in trench				
By hand with air tamper	C.Y.	0.00	20.45	20.45
By hand with vibrating plate tamper	"	0.00	19.05	19.05
1 ton roller	"	0.00	27.80	27.80
Miscellaneous hand labor				
Trim slopes, sides of excavation	S.F.	0.00	0.05	0.05
Trim bottom of excavation	"	0.00	0.06	0.06
Excavation around obstructions and services	C.Y.	0.00	95.50	95.50
02240.05 SOIL STABILIZATION				
Straw bale secured with rebar	L.F.	1.25	0.95	2.20
Filter barrier, 18" high filter fabric	"	1.25	2.85	4.10
Sediment fence, 36" fabric with 6" mesh	"	2.90	3.60	6.50
Soil stabilization with tar paper, burlap, straw and stakes	S.F.	0.24	0.04	0.28
02240.30 GEOTEXTILE				
Filter cloth, light reinforcement				
Woven				
12'-6" wide x 50' long	S.F.	0.22	0.04	0.26
Various lengths	"	0.34	0.04	0.38
Non-woven				
14'-8" wide x 430' long	S.F.	0.12	0.04	0.16
Various lengths	"	0.18	0.04	0.22
02270.10 SLOPE PROTECTION				
Gabions, stone filled				
6" deep	S.Y.	17.55	13.90	31.45
9" deep	"	21.55	15.90	37.45
12" deep	"	28.40	18.55	46.95
18" deep	"	36.40	22.25	58.65
36" deep	"	64.50	37.10	101.60
02270.40 RIPRAP				
Riprap				
Crushed stone blanket, max size 2-1/2"	TON	19.75	41.60	61.35
Stone, quarry run, 300 lb. stones	"	24.70	38.40	63.10
400 lb. stones	"	24.70	35.60	60.30
500 lb. stones	"	24.70	33.30	58.00
750 lb. stones	"	24.70	31.20	55.90
Dry concrete riprap in bags 3" thick, 80 lb. per bag	BAG	3.95	2.10	6.05

02 SITEWORK

EARTHWORK

02280.20 SOIL TREATMENT	UNIT	MAT.	INST.	TOTAL
Soil treatment, termite control pretreatment				
Under slabs	S.F.	0.09	0.16	0.25
By walls	"	0.09	0.19	0.28

02290.30 WEED CONTROL	UNIT	MAT.	INST.	TOTAL
Weed control, bromicil, 15 lb./acre, wettable powder	ACRE	190.00	140.00	330.00
Vegetation control, by application of plant killer	S.Y.	0.02	0.11	0.13
Weed killer, lawns and fields	"	0.14	0.06	0.20

TUNNELING

02300.10 PIPE JACKING	UNIT	MAT.	INST.	TOTAL
Pipe casing, horizontal jacking				
18" dia.	L.F.	52.00	55.50	107.50
21" dia.	"	61.00	59.50	120.50
24" dia.	"	64.50	62.50	127.00
27" dia.	"	71.50	62.50	134.00
30" dia.	"	78.00	65.50	143.50
36" dia.	"	89.50	71.50	161.00
42" dia.	"	130.00	78.00	208.00
48" dia.	"	160.00	83.00	243.00

PILES AND CAISSONS

02360.50 PRESTRESSED PILING	UNIT	MAT.	INST.	TOTAL
Prestressed concrete piling, less than 60' long				
10" sq.	L.F.	7.95	3.65	11.60
12" sq.	"	11.10	3.80	14.90
14" sq.	"	11.55	3.90	15.45
16" sq.	"	14.20	4.00	18.20
18" sq.	"	19.65	4.30	23.95
20" sq.	"	26.60	4.40	31.00
24" sq.	"	34.00	4.50	38.50

PILES AND CAISSONS

02360.50 PRESTRESSED PILING

PILES AND CAISSONS	UNIT	MAT.	INST.	TOTAL
More than 60' long				
12" sq.	L.F.	11.35	3.15	14.50
14" sq.	"	12.45	3.20	15.65
16" sq.	"	15.00	3.25	18.25
18" sq.	"	19.70	3.30	23.00
20" sq.	"	26.60	3.40	30.00
24" sq.	"	31.80	3.45	35.25
Straight cylinder, less than 60' long				
12" dia.	L.F.	10.30	4.00	14.30
14" dia.	"	13.95	4.10	18.05
16" dia.	"	17.00	4.20	21.20
18" dia.	"	21.55	4.30	25.85
20" dia.	"	25.50	4.40	29.90
24" dia.	"	31.60	4.50	36.10
More than 60' long				
12" dia.	L.F.	10.30	3.20	13.50
14" dia.	"	13.95	3.25	17.20
16" dia.	"	17.00	3.30	20.30
18" dia.	"	21.55	3.40	24.95
20" dia.	"	25.50	3.45	28.95
24" dia.	"	32.20	3.50	35.70
Concrete sheet piling				
12" thick x 20' long	S.F.	11.80	8.80	20.60
25' long	"	11.80	8.00	19.80
30' long	"	11.80	7.30	19.10
35' long	"	11.80	6.75	18.55
40' long	"	11.80	6.25	18.05
16" thick x 40' long	"	16.15	4.90	21.05
45' long	"	16.15	4.60	20.75
50' long	"	16.15	4.40	20.55
55' long	"	16.15	4.20	20.35
60' long	"	16.15	4.00	20.15

02360.60 STEEL PILES

STEEL PILES	UNIT	MAT.	INST.	TOTAL
H-section piles				
8x8				
36 lb/ft				
30' long	L.F.	12.10	7.30	19.40
40' long	"	12.10	5.85	17.95
50' long	"	12.10	4.90	17.00
10x10				
42 lb/ft				
30' long	L.F.	13.75	7.30	21.05
40' long	"	13.75	5.85	19.60
50' long	"	13.75	4.90	18.65
57 lb/ft				
30' long	L.F.	17.60	7.30	24.90
40' long	"	17.60	5.85	23.45
50' long	"	17.60	4.90	22.50
12x12				
53 lb/ft				
30' long	L.F.	16.50	8.00	24.50

PILES AND CAISSONS

02360.60 STEEL PILES

	UNIT	MAT.	INST.	TOTAL
40' long	L.F.	16.50	6.25	22.75
50' long	"	16.50	4.90	21.40
74 lb/ft				
30' long	L.F.	22.00	8.00	30.00
40' long	"	22.00	6.25	28.25
50' long	"	22.00	4.90	26.90
14x14				
73 lb/ft				
40' long	L.F.	22.00	8.00	30.00
50' long	"	22.00	6.25	28.25
60' long	"	22.00	4.90	26.90
89 lb/ft				
40' long	L.F.	27.50	8.00	35.50
50' long	"	27.50	6.25	33.75
60' long	"	27.50	4.90	32.40
102 lb/ft				
40' long	L.F.	31.90	8.00	39.90
50' long	"	31.90	6.25	38.15
60' long	"	31.90	4.90	36.80
117 lb/ft				
40' long	L.F.	36.30	8.35	44.65
50' long	"	36.30	6.50	42.80
60' long	"	36.30	5.00	41.30
Splice				
8"	EA.	60.50	47.70	108.20
10"	"	66.00	57.00	123.00
12"	"	90.00	57.00	147.00
14"	"	110.00	71.50	181.50
Driving cap				
8"	EA.	34.10	28.60	62.70
10"	"	34.10	35.80	69.90
12"	"	34.10	35.80	69.90
14"	"	33.30	40.90	74.20
Standard point				
8"	EA.	44.00	28.60	72.60
10"	"	55.00	35.80	90.80
12"	"	77.50	40.90	118.40
14"	"	100.00	47.70	147.70
Heavy duty point				
8"	EA.	50.50	31.80	82.30
10"	"	67.50	40.90	108.40
12"	"	90.00	47.70	137.70
14"	"	110.00	57.00	167.00
Tapered friction piles, with fluted steel casing, up to 50'				
With 4000 psi concrete no reinforcing				
12" dia.	L.F.	14.35	4.40	18.75
14" dia.	"	16.55	4.50	21.05
16" dia.	"	19.85	4.60	24.45
18" dia.	"	22.05	5.15	27.20

PILES AND CAISSONS

PILES AND CAISSONS	UNIT	MAT.	INST.	TOTAL
02360.65 STEEL PIPE PILES				
Concrete filled, 3000# concrete, up to 40'				
8" dia.	L.F.	12.70	6.25	18.95
10" dia.	"	15.60	6.50	22.10
12" dia.	"	19.05	6.75	25.80
14" dia.	"	21.00	7.00	28.00
16" dia.	"	23.80	7.30	31.10
18" dia.	"	28.90	7.65	36.55
Pipe piles, non-filled				
8" dia.	L.F.	10.50	4.90	15.40
10" dia.	"	12.40	5.00	17.40
12" dia.	"	15.25	5.15	20.40
14" dia.	"	16.25	5.50	21.75
16" dia.	"	19.10	5.65	24.75
18" dia.	"	21.00	5.85	26.85
Splice				
8" dia.	EA.	35.00	57.00	92.00
10" dia.	"	37.10	57.00	94.10
12" dia.	"	41.70	71.50	113.20
14" dia.	"	46.30	71.50	117.80
16" dia.	"	62.00	95.50	157.50
18" dia.	"	82.50	95.50	178.00
Standard point				
8" dia.	EA.	44.90	57.00	101.90
10" dia.	"	50.50	57.00	107.50
12" dia.	"	79.50	71.50	151.00
14" dia.	"	110.00	71.50	181.50
16" dia.	"	120.00	95.50	215.50
18" dia.	"	160.00	95.50	255.50
Heavy duty point				
8" dia.	EA.	56.00	71.50	127.50
10" dia.	"	67.50	71.50	139.00
12" dia.	"	90.00	95.50	185.50
14" dia.	"	120.00	95.50	215.50
16" dia.	"	160.00	110.00	270.00
18" dia.	"	200.00	110.00	310.00
02360.70 STEEL SHEET PILING				
Steel sheet piling,12" wide				
20' long	S.F.	9.35	8.80	18.15
35' long	"	9.35	6.25	15.60
50' long	"	9.35	4.40	13.75
Over 50' long	"	9.35	4.00	13.35
02360.80 WOOD AND TIMBER PILES				
Treated wood piles, 12" butt, 8" tip				
25' long	L.F.	7.00	8.80	15.80
30' long	"	7.50	7.30	14.80
35' long	"	7.50	6.25	13.75
40' long	"	7.50	5.50	13.00
12" butt, 7" tip				

PILES AND CAISSONS

PILES AND CAISSONS	UNIT	MAT.	INST.	TOTAL
02360.80 WOOD AND TIMBER PILES				
40' long	L.F.	8.45	5.50	13.95
45' long	"	8.45	4.90	13.35
50' long	"	10.30	4.40	14.70
55' long	"	10.30	4.00	14.30
60' long	"	10.30	3.65	13.95
02360.90 PILE TESTING				
Pile test				
50 ton to 100 ton	EA.			14,362
To 200 ton	"			21,318
To 300 ton	"			23,562
To 400 ton	"			35,904
To 600 ton	"			41,514
02380.10 CAISSONS				
Caisson, including 3000# concrete, in stable ground				
18" dia.	L.F.	11.00	17.55	28.55
24" dia.	"	18.35	18.30	36.65
30" dia.	"	28.90	21.95	50.85
36" dia.	"	40.40	25.10	65.50
48" dia.	"	69.50	29.30	98.80
60" dia.	"	120.00	39.90	159.90
72" dia.	"	170.00	48.80	218.80
84" dia.	"	210.00	62.50	272.50
Wet ground, casing required but pulled				
18" dia.	L.F.	11.00	21.95	32.95
24" dia.	"	18.35	24.40	42.75
30" dia.	"	28.90	27.40	56.30
36" dia.	"	40.40	29.30	69.70
48" dia.	"	69.50	36.60	106.10
60" dia.	"	120.00	48.80	168.80
72" dia.	"	170.00	73.00	243.00
84" dia.	"	210.00	110.00	320.00
Soft rock				
18" dia.	L.F.	11.00	62.50	73.50
24" dia.	"	18.35	110.00	128.35
30" dia.	"	28.90	150.00	178.90
36" dia.	"	40.40	220.00	260.40
48" dia.	"	69.50	290.00	359.50
60" dia.	"	120.00	440.00	560.00
72" dia.	"	170.00	490.00	660.00
84" dia.	"	210.00	550.00	760.00

RAILROAD WORK

02450.10 RAILROAD WORK

RAILROAD WORK	UNIT	MAT.	INST.	TOTAL
Rail				
90 lb	L.F.	16.85	0.60	17.45
100 lb	"	19.25	0.60	19.85
115 lb	"	21.65	0.60	22.25
132 lb	"	24.05	0.60	24.65
Rail relay				
90 lb	L.F.	6.70	0.60	7.30
100 lb	"	7.45	0.60	8.05
115 lb	"	9.00	0.60	9.60
132 lb	"	10.90	0.60	11.50
New angle bars, per pair				
90 lb	EA.	62.50	0.75	63.25
100 lb	"	67.50	0.75	68.25
115 lb	"	86.50	0.75	87.25
132 lb	"	100.00	0.75	100.75
Angle bar relay				
90 lb	EA.	25.90	0.75	26.65
100 lb	"	26.30	0.75	27.05
115 lb	"	27.60	0.75	28.35
132 lb	"	29.50	0.75	30.25
New tie plates				
90 lb	EA.	8.40	0.53	8.93
100 lb	"	8.80	0.53	9.33
115 lb	"	9.60	0.53	10.13
132 lb	"	10.20	0.53	10.73
Tie plate relay				
90 lb	EA.	2.65	0.53	3.18
100 lb	"	3.75	0.53	4.28
115 lb	"	3.75	0.53	4.28
132 lb	"	4.60	0.53	5.13
Track accessories				
Wooden cross ties, 8'	EA.	33.70	3.75	37.45
Concrete cross ties, 8'	"	200.00	7.45	207.45
Tie plugs, 5"	"	13.25	0.37	13.62
Track bolts and nuts, 1"	"	3.25	0.37	3.62
Lockwashers, 1"	"	0.83	0.25	1.08
Track spikes, 6"	"	0.77	1.50	2.27
Wooden switch ties	B.F.	1.20	0.37	1.57
Rail anchors	EA.	3.30	1.35	4.65
Ballast	TON	9.50	7.45	16.95
Gauge rods	EA.	25.90	6.00	31.90
Compromise splice bars	"	320.00	9.95	329.95
Turnout				
90 lb	EA.	9,020	1,490	10,510
100 lb	"	9,500	1,490	10,990
110 lb	"	10,220	1,490	11,710
115 lb	"	10,460	1,490	11,950
132 lb	"	11,420	1,490	12,910
Turnout relay				
90 lb	EA.	5,770	1,490	7,260
100 lb	"	6,370	1,490	7,860
110 lb	"	6,610	1,490	8,100

RAILROAD WORK	UNIT	MAT.	INST.	TOTAL
02450.10 RAILROAD WORK				
115 lb	EA.	6,970	1,490	8,460
132 lb	"	7,580	1,490	9,070
Railroad track in place, complete				
New rail				
90 lb	L.F.	110.00	14.95	124.95
100 lb	"	110.00	14.95	124.95
110 lb	"	110.00	14.95	124.95
115 lb	"	120.00	14.95	134.95
132 lb	"	120.00	14.95	134.95
Rail relay				
90 lb	L.F.	63.00	14.95	77.95
100 lb	"	69.50	14.95	84.45
110 lb	"	69.50	14.95	84.45
115 lb	"	75.50	14.95	90.45
132 lb	"	78.00	14.95	92.95
No. 8 turnout				
90 lb	EA.	22,600	1,990	24,590
100 lb	"	25,130	1,990	27,120
110 lb	"	27,530	1,990	29,520
115 lb	"	28,130	1,990	30,120
132 lb	"	28,740	1,990	30,730
No. 8 turnout relay				
90 lb	EA.	18,760	1,990	20,750
100 lb	"	18,760	1,990	20,750
110 lb	"	18,760	1,990	20,750
115 lb	"	21,280	1,990	23,270
132 lb	"	21,280	1,990	23,270
Railroad crossings, asphalt, based on 8" thick x 20'				
Including track and approach				
12' roadway	EA.	540.00	370.00	910.00
15' roadway	"	640.00	430.00	1,070
18' roadway	"	750.00	500.00	1,250
21' roadway	"	830.00	600.00	1,430
24' roadway	"	940.00	750.00	1,690
Precast concrete inserts				
12' roadway	EA.	780.00	150.00	930.00
15' roadway	"	940.00	190.00	1,130
18' roadway	"	1,120	250.00	1,370
21' roadway	"	1,440	300.00	1,740
24' roadway	"	1,740	330.00	2,070
Molded rubber, with headers				
12' roadway	EA.	4,710	150.00	4,860
15' roadway	"	5,890	190.00	6,080
18' roadway	"	6,970	250.00	7,220
21' roadway	"	7,700	300.00	8,000
24' roadway	"	9,380	330.00	9,710

02 SITEWORK

PAVING AND SURFACING

PAVING AND SURFACING	UNIT	MAT.	INST.	TOTAL
02510.20 ASPHALT SURFACES				
Asphalt wearing surface, for flexible pavement				
1" thick	S.Y.	1.75	1.45	3.20
1-1/2" thick	"	2.70	1.75	4.45
2" thick	"	3.60	2.20	5.80
3" thick	"	5.30	2.95	8.25
Binder course				
1-1/2" thick	S.Y.	2.30	1.65	3.95
2" thick	"	3.10	2.00	5.10
3" thick	"	4.60	2.65	7.25
4" thick	"	6.15	2.95	9.10
5" thick	"	7.90	3.25	11.15
6" thick	"	9.45	3.65	13.10
Bituminous sidewalk, no base				
2" thick	S.Y.	3.85	1.75	5.60
3" thick	"	5.70	1.85	7.55
02520.10 CONCRETE PAVING				
Concrete paving, reinforced, 5000 psi concrete				
6" thick	S.Y.	17.40	13.70	31.10
7" thick	"	19.55	14.65	34.20
8" thick	"	22.00	15.65	37.65
9" thick	"	22.70	16.90	39.60
10" thick	"	26.00	18.30	44.30
11" thick	"	28.90	19.95	48.85
12" thick	"	30.60	21.95	52.55
15" thick	"	38.60	27.40	66.00
Concrete paving, for pipe trench, reinforced				
7" thick	S.Y.	36.40	14.95	51.35
8" thick	"	41.50	16.60	58.10
9" thick	"	43.90	18.70	62.60
10" thick	"	49.40	21.35	70.75
Fibrous concrete				
5" thick	S.Y.	17.35	16.90	34.25
8" thick	"	22.05	18.30	40.35
Roller compacted concrete, (RCC), place and compact				
8" thick	S.Y.	21.40	21.95	43.35
12" thick	"	32.50	27.40	59.90
Steel edge forms up to				
12" deep	L.F.	0.56	0.95	1.51
15" deep	"	0.69	1.15	1.84
Paving finishes				
Belt dragged	S.Y.	0.00	1.45	1.45
Curing	"	0.26	0.29	0.55
02545.10 ASPHALT REPAIR				
Coal tar emulsion seal coat, rubber additive, fuel resistant	S.Y.	0.84	0.41	1.25
Bituminous surface treatment, single	"	0.77	0.29	1.06
Double	"	1.00	0.03	1.03
Bituminous prime coat	"	0.52	0.04	0.56
Tack coat	"	0.25	0.03	0.28
Crack sealing, concrete paving	L.F.	0.43	0.19	0.62

02 SITEWORK

PAVING AND SURFACING

02545.10 ASPHALT REPAIR	UNIT	MAT.	INST.	TOTAL
Bituminous paving for pipe trench, 4" thick	S.Y.	4.90	9.95	14.85
Polypropylene, nonwoven paving fabric	"	0.65	0.14	0.79
Rubberized asphalt	"	1.00	2.60	3.60
Asphalt slurry seal	"	2.50	1.70	4.20

02580.10 PAVEMENT MARKINGS	UNIT	MAT.	INST.	TOTAL
Pavement line marking, paint				
4" wide	L.F.	0.06	0.07	0.13
6" wide	"	0.08	0.16	0.24
8" wide	"	0.09	0.24	0.33
Reflective paint, 4" wide	"	0.23	0.24	0.47
Airfield markings, retro-reflective				
White	L.F.	0.37	0.24	0.61
Yellow	"	0.40	0.24	0.64
Preformed tape, 4" wide				
Inlaid reflective	L.F.	1.05	0.04	1.09
Reflective paint	"	0.73	0.07	0.80
Thermoplastic				
White	L.F.	0.41	0.14	0.55
Yellow	"	0.41	0.14	0.55
12" wide, thermoplastic, white	"	0.87	0.41	1.28
Directional arrows, reflective preformed tape	EA.	76.00	28.60	104.60
Messages, reflective preformed tape (per letter)	"	38.10	14.30	52.40
Handicap symbol, preformed tape	"	63.50	28.60	92.10
Parking stall painting	"	1.00	5.70	6.70

UTILITIES

02605.30 MANHOLES	UNIT	MAT.	INST.	TOTAL
Precast sections, 48" dia.				
Base section	EA.	140.00	120.00	260.00
1'0" riser	"	38.10	100.00	138.10
1'4" riser	"	46.30	110.00	156.30
2'8" riser	"	68.50	110.00	178.50
4'0" riser	"	130.00	120.00	250.00
2'8" cone top	"	83.00	150.00	233.00
Precast manholes, 48" dia.				
4' deep	EA.	310.00	300.00	610.00
6' deep	"	480.00	370.00	850.00
7' deep	"	540.00	430.00	970.00
8' deep	"	610.00	500.00	1,110
10' deep	"	690.00	600.00	1,290
Cast-in-place, 48" dia., with frame and cover				

UTILITIES	UNIT	MAT.	INST.	TOTAL
02605.30 MANHOLES				
5' deep	EA.	410.00	750.00	1,160
6' deep	"	540.00	850.00	1,390
8' deep	"	790.00	1,000	1,790
10' deep	"	920.00	1,200	2,120
Brick manholes, 48" dia. with cover, 8" thick				
4' deep	EA.	510.00	350.00	860.00
6' deep	"	640.00	390.00	1,030
8' deep	"	820.00	440.00	1,260
10' deep	"	1,020	510.00	1,530
12' deep	"	1,280	590.00	1,870
14' deep	"	1,560	710.00	2,270
Inverts for manholes				
Single channel	EA.	82.50	140.00	222.50
Triple channel	"	94.00	180.00	274.00
Frames and covers, 24" diameter				
300 lb	EA.	250.00	28.60	278.60
400 lb	"	270.00	31.80	301.80
500 lb	"	310.00	40.90	350.90
Watertight, 350 lb	"	320.00	95.50	415.50
For heavy equipment, 1200 lb	"	700.00	140.00	840.00
Steps for manholes				
7" x 9"	EA.	8.85	5.70	14.55
8" x 9"	"	11.20	6.35	17.55
Curb inlet, 4' throat, cast in place				
12"-30" pipe	EA.	260.00	750.00	1,010
36"-48" pipe	"	280.00	850.00	1,130
Raise exist frame and cover, when repaving	"	0.00	300.00	300.00
02610.10 CAST IRON FLANGED PIPE				
Cast iron flanged sections				
4" pipe, with one bolt set				
3' section	EA.	110.00	13.60	123.60
4' section	"	120.00	14.95	134.95
5' section	"	130.00	16.60	146.60
6' section	"	140.00	18.70	158.70
8' section	"	170.00	21.35	191.35
10' section	"	180.00	29.90	209.90
12' section	"	200.00	49.80	249.80
15' section	"	230.00	74.50	304.50
18' section	"	260.00	100.00	360.00
6" pipe, with one bolt set				
3' section	EA.	140.00	14.95	154.95
4' section	"	150.00	17.60	167.60
5' section	"	160.00	19.95	179.95
6' section	"	180.00	23.00	203.00
8' section	"	200.00	33.20	233.20
10' section	"	230.00	37.40	267.40
12' section	"	250.00	49.80	299.80
15' section	"	290.00	74.50	364.50
18' section	"	380.00	110.00	490.00
8" pipe, with one bolt set				

UTILITIES

02610.10 CAST IRON FLANGED PIPE

	UNIT	MAT.	INST.	TOTAL
3' section	EA.	190.00	18.70	208.70
4' section	"	210.00	21.35	231.35
5' section	"	230.00	24.90	254.90
6' section	"	240.00	29.90	269.90
8' section	"	270.00	42.70	312.70
10' section	"	300.00	49.80	349.80
12' section	"	330.00	74.50	404.50
15' section	"	390.00	100.00	490.00
18' section	"	470.00	120.00	590.00
10" pipe, with one bolt set				
3' section	EA.	240.00	19.15	259.15
4' section	"	280.00	22.00	302.00
5' section	"	300.00	25.80	325.80
6' section	"	330.00	31.10	361.10
8' section	"	360.00	45.30	405.30
10' section	"	390.00	53.50	443.50
12' section	"	450.00	83.00	533.00
15' section	"	490.00	110.00	600.00
18' section	"	600.00	150.00	750.00
12" pipe, with one bolt set				
3' section	EA.	330.00	20.75	350.75
4' section	"	350.00	24.10	374.10
5' section	"	380.00	28.70	408.70
6' section	"	400.00	34.00	434.00
8' section	"	460.00	49.80	509.80
10' section	"	510.00	57.50	567.50
12' section	"	570.00	93.50	663.50
15' section	"	640.00	120.00	760.00
18' section	"	680.00	170.00	850.00

02610.11 CAST IRON FITTINGS

	UNIT	MAT.	INST.	TOTAL
Mechanical joint, with 2 bolt kits				
90 deg bend				
4"	EA.	80.00	19.05	99.05
6"	"	100.00	22.00	122.00
8"	"	130.00	28.60	158.60
10"	"	190.00	40.90	230.90
12"	"	290.00	57.00	347.00
14"	"	400.00	71.50	471.50
16"	"	460.00	95.50	555.50
45 deg bend				
4"	EA.	68.50	19.05	87.55
6"	"	91.50	22.00	113.50
8"	"	130.00	28.60	158.60
10"	"	160.00	40.90	200.90
12"	"	220.00	57.00	277.00
14"	"	260.00	71.50	331.50
16"	"	340.00	95.50	435.50
Tee, with 3 bolt kits				
4" x 4"	EA.	120.00	28.60	148.60
6" x 6"	"	170.00	35.80	205.80

02 SITEWORK

UTILITIES

	UNIT	MAT.	INST.	TOTAL
02610.11 CAST IRON FITTINGS				
8" x 8"	EA.	230.00	47.70	277.70
10" x 10"	"	340.00	71.50	411.50
12" x 12"	"	460.00	95.50	555.50
Wye, with 3 bolt kits				
6" x 6"	EA.	230.00	35.80	265.80
8" x 8"	"	340.00	47.70	387.70
10" x 10"	"	460.00	71.50	531.50
12" x 12"	"	630.00	95.50	725.50
Reducer, with 2 bolt kits				
6" x 4"	EA.	86.00	35.80	121.80
8" x 6"	"	110.00	47.70	157.70
10" x 8"	"	210.00	71.50	281.50
12" x 10"	"	250.00	95.50	345.50
Flanged, 90 deg bend, 125 lb.				
4"	EA.	51.50	23.85	75.35
6"	"	71.00	28.60	99.60
8"	"	100.00	35.80	135.80
10"	"	180.00	47.70	227.70
12"	"	250.00	71.50	321.50
14"	"	490.00	95.50	585.50
16"	"	730.00	95.50	825.50
Tee				
4"	EA.	89.50	35.80	125.30
6"	"	130.00	40.90	170.90
8"	"	190.00	47.70	237.70
10"	"	350.00	57.00	407.00
12"	"	480.00	71.50	551.50
14"	"	1,060	95.50	1,156
16"	"	1,600	140.00	1,740
02610.13 GATE VALVES				
Gate valve, (AWWA) mechanical joint, with adjustable box				
4" valve	EA.	470.00	49.80	519.80
6" valve	"	530.00	60.00	590.00
8" valve	"	710.00	74.50	784.50
10" valve	"	1,060	88.00	1,148
12" valve	"	1,410	110.00	1,520
14" valve	"	3,540	120.00	3,660
16" valve	"	4,720	140.00	4,860
18" valve	"	5,890	150.00	6,040
Flanged, with box, post indicator (AWWA)				
4" valve	EA.	420.00	60.00	480.00
6" valve	"	500.00	68.00	568.00
8" valve	"	710.00	83.00	793.00
10" valve	"	1,060	100.00	1,160
12" valve	"	1,530	120.00	1,650
14" valve	"	3,530	150.00	3,680
16" valve	"	4,710	190.00	4,900

UTILITIES

	UNIT	MAT.	INST.	TOTAL
02610.15 WATER METERS				
Water meter, displacement type				
1"	EA.	170.00	40.20	210.20
1-1/2"	"	590.00	44.70	634.70
2"	"	890.00	50.50	940.50
02610.17 CORPORATION STOPS				
Stop for flared copper service pipe				
3/4"	EA.	19.25	20.10	39.35
1"	"	26.10	22.35	48.45
1-1/4"	"	70.00	26.80	96.80
1-1/2"	"	84.00	33.50	117.50
2"	"	120.00	40.20	160.20
02610.40 DUCTILE IRON PIPE				
Ductile iron pipe, cement lined, slip-on joints				
4"	L.F.	6.75	4.15	10.90
6"	"	8.15	4.40	12.55
8"	"	10.65	4.65	15.30
10"	"	14.65	5.00	19.65
12"	"	18.10	6.00	24.10
14"	"	22.85	7.45	30.30
16"	"	28.00	8.30	36.30
18"	"	31.50	9.35	40.85
20"	"	35.90	10.70	46.60
Mechanical joint pipe				
4"	L.F.	8.00	5.75	13.75
6"	"	9.55	6.25	15.80
8"	"	12.60	6.80	19.40
10"	"	16.50	7.45	23.95
12"	"	20.95	9.95	30.90
14"	"	26.30	11.50	37.80
16"	"	28.80	13.60	42.40
18"	"	32.50	14.95	47.45
20"	"	37.40	16.60	54.00
Fittings, mechanical joint				
90 degree elbow				
4"	EA.	100.00	19.05	119.05
6"	"	130.00	22.00	152.00
8"	"	190.00	28.60	218.60
10"	"	270.00	40.90	310.90
12"	"	360.00	57.00	417.00
14"	"	560.00	71.50	631.50
16"	"	700.00	95.50	795.50
18"	"	1,050	110.00	1,160
20"	"	1,170	140.00	1,310
45 degree elbow				
4"	EA.	100.00	19.05	119.05
6"	"	120.00	22.00	142.00
8"	"	160.00	28.60	188.60
10"	"	230.00	40.90	270.90
12"	"	340.00	57.00	397.00
14"	"	470.00	71.50	541.50

UTILITIES

02610.40 DUCTILE IRON PIPE	UNIT	MAT.	INST.	TOTAL
16"	EA.	560.00	95.50	655.50
18"	"	820.00	140.00	960.00
20"	"	970.00	140.00	1,110
Tee				
4"x3"	EA.	140.00	35.80	175.80
4"x4"	"	150.00	35.80	185.80
6"x3"	"	180.00	40.90	220.90
6"x4"	"	180.00	40.90	220.90
6"x6"	"	200.00	40.90	240.90
8"x4"	"	250.00	47.70	297.70
8"x6"	"	280.00	47.70	327.70
8"x8"	"	260.00	47.70	307.70
10"x4"	"	330.00	57.00	387.00
10"x6"	"	360.00	57.00	417.00
10"x8"	"	370.00	57.00	427.00
10"x10"	"	420.00	57.00	477.00
12"x4"	"	510.00	71.50	581.50
12"x6"	"	400.00	71.50	471.50
12"x8"	"	430.00	71.50	501.50
12"x10"	"	490.00	71.50	561.50
12"x12"	"	530.00	76.50	606.50
14"x4"	"	640.00	81.50	721.50
14"x6"	"	680.00	81.50	761.50
14"x8"	"	690.00	81.50	771.50
14"x10"	"	720.00	81.50	801.50
14"x12"	"	740.00	88.00	828.00
14"x14"	"	730.00	88.00	818.00
16"x4"	"	830.00	95.50	925.50
16"x6"	"	850.00	95.50	945.50
16"x8"	"	750.00	95.50	845.50
16"x10"	"	760.00	95.50	855.50
16"x12"	"	750.00	95.50	845.50
16"x14"	"	790.00	95.50	885.50
16"x16"	"	810.00	95.50	905.50
18"x6"	"	960.00	100.00	1,060
18"x8"	"	980.00	100.00	1,080
18"x10"	"	1,000	100.00	1,100
18"x12"	"	1,020	100.00	1,120
18"x14"	"	1,130	100.00	1,230
18"x16"	"	1,120	100.00	1,220
18"x18"	"	1,270	100.00	1,370
20"x6"	"	1,230	110.00	1,340
20"x8"	"	1,240	110.00	1,350
20"x10"	"	1,260	110.00	1,370
20"x12"	"	1,280	110.00	1,390
20"x14"	"	1,320	110.00	1,430
20"x16"	"	1,540	110.00	1,650
20"x18"	"	1,610	110.00	1,720
20"x20"	"	1,650	110.00	1,760
Cross				
4"x3"	EA.	180.00	47.70	227.70
4"x4"	"	190.00	47.70	237.70

UTILITIES	UNIT	MAT.	INST.	TOTAL
02610.40 DUCTILE IRON PIPE				
6"x3"	EA.	200.00	57.00	257.00
6"x4"	"	210.00	57.00	267.00
6"x6"	"	230.00	57.00	287.00
8"x4"	"	270.00	63.50	333.50
8"x6"	"	290.00	63.50	353.50
8"x8"	"	320.00	63.50	383.50
10"x4"	"	370.00	71.50	441.50
10"x6"	"	400.00	71.50	471.50
10"x8"	"	430.00	71.50	501.50
10"x10"	"	510.00	71.50	581.50
12"x4"	"	480.00	81.50	561.50
12"x6"	"	540.00	81.50	621.50
12"x8"	"	530.00	81.50	611.50
12"x10"	"	610.00	88.00	698.00
12"x12"	"	650.00	88.00	738.00
14"x4"	"	620.00	95.50	715.50
14"x6"	"	700.00	95.50	795.50
14"x8"	"	730.00	95.50	825.50
14"x10"	"	780.00	95.50	875.50
14"x12"	"	840.00	100.00	940.00
14"x14"	"	920.00	100.00	1,020
16"x4"	"	810.00	110.00	920.00
16"x6"	"	830.00	110.00	940.00
16"x8"	"	880.00	110.00	990.00
16"x10"	"	930.00	110.00	1,040
16"x12"	"	970.00	110.00	1,080
16"x14"	"	1,060	110.00	1,170
16"x16"	"	1,120	110.00	1,230
18"x6"	"	1,040	130.00	1,170
18"x8"	"	1,070	130.00	1,200
18"x10"	"	1,120	130.00	1,250
18"x12"	"	1,180	130.00	1,310
18"x14"	"	1,400	130.00	1,530
18"x16"	"	1,490	130.00	1,620
18"x18"	"	1,580	130.00	1,710
20"x6"	"	1,250	140.00	1,390
20"x8"	"	1,280	140.00	1,420
20"x10"	"	1,340	140.00	1,480
20"x12"	"	1,400	140.00	1,540
20"x14"	"	1,470	140.00	1,610
20"x16"	"	1,700	140.00	1,840
20"x18"	"	1,820	140.00	1,960
20"x20"	"	1,930	140.00	2,070
02610.60 PLASTIC PIPE				
PVC, class 150 pipe				
4" dia.	L.F.	3.05	3.75	6.80
6" dia.	"	5.70	4.05	9.75
8" dia.	"	9.10	4.25	13.35
10" dia.	"	12.90	4.65	17.55
12" dia.	"	19.05	5.00	24.05

UTILITIES

02610.60 PLASTIC PIPE

UTILITIES	UNIT	MAT.	INST.	TOTAL
Schedule 40 pipe				
1-1/2" dia.	L.F.	1.70	1.70	3.40
2" dia.	"	1.95	1.80	3.75
2-1/2" dia.	"	2.20	1.90	4.10
3" dia.	"	3.15	2.05	5.20
4" dia.	"	4.50	2.40	6.90
6" dia.	"	7.60	2.85	10.45
90 degree elbows				
1"	EA.	1.10	4.75	5.85
1-1/2"	"	1.90	4.75	6.65
2"	"	2.15	5.20	7.35
2-1/2"	"	5.40	5.70	11.10
3"	"	6.50	6.35	12.85
4"	"	17.95	7.15	25.10
6"	"	40.40	9.55	49.95
45 degree elbows				
1"	EA.	1.25	4.75	6.00
1-1/2"	"	2.00	4.75	6.75
2"	"	2.70	5.20	7.90
2-1/2"	"	5.50	5.70	11.20
3"	"	6.70	6.35	13.05
4"	"	19.65	7.15	26.80
6"	"	42.60	9.55	52.15
Tees				
1"	EA.	1.45	5.70	7.15
1-1/2"	"	2.15	5.70	7.85
2"	"	2.90	6.35	9.25
2-1/2"	"	5.25	7.15	12.40
3"	"	7.20	8.15	15.35
4"	"	20.20	9.55	29.75
6"	"	40.40	11.45	51.85
Couplings				
1"	EA.	0.67	4.75	5.42
1-1/2"	"	0.78	4.75	5.53
2"	"	1.35	5.20	6.55
2-1/2"	"	2.70	5.70	8.40
3"	"	5.05	6.35	11.40
4"	"	11.20	7.15	18.35
6"	"	14.60	9.55	24.15
Drainage pipe				
PVC schedule 80				
1" dia.	L.F.	1.05	1.70	2.75
1-1/2" dia.	"	1.25	1.70	2.95
ABS, 2" dia.	"	1.60	1.80	3.40
2-1/2" dia.	"	2.30	1.90	4.20
3" dia.	"	2.70	2.05	4.75
4" dia.	"	3.65	2.40	6.05
6" dia.	"	6.05	2.85	8.90
8" dia.	"	8.10	3.95	12.05
10" dia.	"	10.85	4.65	15.50
12" dia.	"	17.75	5.00	22.75
90 degree elbows				

UTILITIES	UNIT	MAT.	INST.	TOTAL
02610.60 PLASTIC PIPE				
1"	EA.	1.85	4.75	6.60
1-1/2"	"	2.30	4.75	7.05
2"	"	2.75	5.20	7.95
2-1/2"	"	6.60	5.70	12.30
3"	"	6.80	6.35	13.15
4"	"	12.00	7.15	19.15
6"	"	26.30	9.55	35.85
45 degree elbows				
1"	EA.	2.95	4.75	7.70
1-1/2"	"	3.80	4.75	8.55
2"	"	4.70	5.20	9.90
2-1/2"	"	8.80	5.70	14.50
3"	"	9.40	6.35	15.75
4"	"	17.75	7.15	24.90
6"	"	41.20	9.55	50.75
Tees				
1"	EA.	1.95	5.70	7.65
1-1/2"	"	6.20	5.70	11.90
2"	"	7.55	6.35	13.90
2-1/2"	"	8.80	7.15	15.95
3"	"	9.60	8.15	17.75
4"	"	18.30	9.55	27.85
6"	"	36.00	11.45	47.45
Couplings				
1"	EA.	1.60	4.75	6.35
1-1/2"	"	2.75	4.75	7.50
2"	"	4.05	5.20	9.25
2-1/2"	"	8.60	5.70	14.30
3"	"	8.80	6.35	15.15
4"	"	9.15	7.15	16.30
6"	"	15.45	9.55	25.00
Pressure pipe				
PVC, class 200 pipe				
3/4"	L.F.	1.00	1.45	2.45
1"	"	1.15	1.50	2.65
1-1/4"	"	1.25	1.60	2.85
1-1/2"	"	1.50	1.70	3.20
2"	"	1.65	1.80	3.45
2-1/2"	"	2.40	1.90	4.30
3"	"	3.05	2.05	5.10
4"	"	5.25	2.40	7.65
6"	"	9.95	2.85	12.80
8"	"	15.10	4.25	19.35
90 degree elbows				
3/4"	EA.	0.51	4.75	5.26
1"	"	0.63	4.75	5.38
1-1/4"	"	1.00	4.75	5.75
1-1/2"	"	1.20	4.75	5.95
2"	"	2.40	5.20	7.60
2-1/2"	"	3.45	5.70	9.15
3"	"	7.85	6.35	14.20
4"	"	20.15	7.15	27.30

UTILITIES	UNIT	MAT.	INST.	TOTAL
02610.60 PLASTIC PIPE				
6"	EA.	38.30	9.55	47.85
8"	"	75.50	14.30	89.80
45 degree elbows				
3/4"	EA.	0.75	4.75	5.50
1"	"	0.95	4.75	5.70
1-1/4"	"	1.35	4.75	6.10
1-1/2"	"	1.65	4.75	6.40
2"	"	2.35	5.20	7.55
2-1/2"	"	3.85	5.70	9.55
3"	"	8.80	6.35	15.15
4"	"	16.50	7.15	23.65
6"	"	40.60	9.55	50.15
8"	"	82.50	14.30	96.80
Tees				
3/4"	EA.	0.63	5.70	6.33
1"	"	0.81	5.70	6.51
1-1/4"	"	1.20	5.70	6.90
1-1/2"	"	1.65	5.70	7.35
2"	"	2.40	6.35	8.75
2-1/2"	"	3.80	7.15	10.95
3"	"	11.60	8.15	19.75
4"	"	16.60	9.55	26.15
6"	"	57.50	11.45	68.95
8"	"	120.00	15.90	135.90
Couplings				
3/4"	EA.	0.39	4.75	5.14
1"	"	0.57	4.75	5.32
1-1/4"	"	0.75	4.75	5.50
1-1/2"	"	0.81	4.75	5.56
2"	"	1.15	5.20	6.35
2-1/2"	"	2.40	5.70	8.10
3"	"	3.80	6.35	10.15
4"	"	5.30	6.35	11.65
6"	"	16.60	7.15	23.75
8"	"	29.80	9.55	39.35
02610.90 VITRIFIED CLAY PIPE				
Vitrified clay pipe, extra strength				
6" dia.	L.F.	2.95	6.80	9.75
8" dia.	"	3.55	7.10	10.65
10" dia.	"	5.40	7.45	12.85
12" dia.	"	7.80	9.95	17.75
15" dia.	"	14.15	14.95	29.10
18" dia.	"	21.20	16.60	37.80
24" dia.	"	38.90	21.35	60.25
30" dia.	"	66.00	29.90	95.90
36" dia.	"	94.50	42.70	137.20
02630.10 TAPPING SADDLES & SLEEVES				
Tapping saddle, tap size to 2"				
4" saddle	EA.	40.80	14.30	55.10

02 SITEWORK

UTILITIES

UTILITIES	UNIT	MAT.	INST.	TOTAL
02630.10 TAPPING SADDLES & SLEEVES				
6" saddle	EA.	47.60	17.85	65.45
8" saddle	"	55.50	23.85	79.35
10" saddle	"	64.50	28.60	93.10
12" saddle	"	77.00	40.90	117.90
14" saddle	"	86.00	57.00	143.00
Tapping sleeve				
4x4	EA.	430.00	19.05	449.05
6x4	"	560.00	22.00	582.00
6x6	"	560.00	22.00	582.00
8x4	"	580.00	28.60	608.60
8x6	"	590.00	28.60	618.60
10x4	"	940.00	60.00	1,000
10x6	"	1,360	60.00	1,420
10x8	"	1,420	60.00	1,480
10x10	"	1,450	62.50	1,513
12x4	"	1,460	62.50	1,523
12x6	"	1,470	68.00	1,538
12x8	"	1,510	74.50	1,585
12x10	"	1,590	83.00	1,673
12x12	"	1,640	93.50	1,734
Tapping valve, mechanical joint				
4" valve	EA.	410.00	190.00	600.00
6" valve	"	490.00	250.00	740.00
8" valve	"	730.00	370.00	1,100
10" valve	"	1,160	500.00	1,660
12" valve	"	2,040	750.00	2,790
Tap hole in pipe				
4" hole	EA.	0.00	35.80	35.80
6" hole	"	0.00	57.00	57.00
8" hole	"	0.00	95.50	95.50
10" hole	"	0.00	110.00	110.00
12" hole	"	0.00	140.00	140.00
02640.15 VALVE BOXES				
Valve box, adjustable, for valves up to 20"				
3' deep	EA.	100.00	9.55	109.55
4' deep	"	100.00	11.45	111.45
5' deep	"	100.00	14.30	114.30
02640.19 THRUST BLOCKS				
Thrust block, 3000# concrete				
1/4 c.y.	EA.	76.00	61.00	137.00
1/2 c.y.	"	97.00	73.00	170.00
3/4 c.y.	"	110.00	120.00	230.00
1 c.y.	"	140.00	240.00	380.00
02645.10 FIRE HYDRANTS				
Standard, 3 way post, 6" mechanical joint				
2' deep	EA.	1,000	500.00	1,500
4' deep	"	1,070	600.00	1,670
6' deep	"	1,190	750.00	1,940

UTILITIES	UNIT	MAT.	INST.	TOTAL
02645.10 FIRE HYDRANTS				
8' deep	EA.	1,340	850.00	2,190
02665.10 CHILLED WATER SYSTEMS				
Chilled water pipe, 2" thick insulation, w/casing				
Align and tack weld on sleepers				
1-1/2" dia.	L.F.	11.05	1.35	12.40
3" dia.	"	17.75	2.15	19.90
4" dia.	"	20.40	3.00	23.40
6" dia.	"	23.30	3.75	27.05
8" dia.	"	32.80	4.25	37.05
10" dia.	"	41.90	5.00	46.90
12" dia.	"	49.80	6.00	55.80
14" dia.	"	67.50	6.50	74.00
16" dia.	"	86.00	7.45	93.45
Align and tack weld on trench bottom				
18" dia.	L.F.	88.50	8.30	96.80
20" dia.	"	120.00	9.35	129.35
Preinsulated fittings				
Align and tack weld on sleepers				
Elbows				
1-1/2"	EA.	230.00	25.20	255.20
3"	"	290.00	40.20	330.20
4"	"	380.00	50.50	430.50
6"	"	520.00	67.00	587.00
8"	"	740.00	80.50	820.50
Tees				
1-1/2"	EA.	350.00	26.80	376.80
3"	"	500.00	44.70	544.70
4"	"	380.00	57.50	437.50
6"	"	810.00	80.50	890.50
8"	"	1,080	100.00	1,180
Reducers				
3"	EA.	380.00	33.50	413.50
4"	"	590.00	40.20	630.20
6"	"	800.00	50.50	850.50
8"	"	840.00	67.00	907.00
Anchors, not including concrete				
4"	EA.	160.00	50.50	210.50
6"	"	240.00	50.50	290.50
Align and tack weld on trench bottom				
Elbows				
10"	EA.	970.00	93.50	1,064
12"	"	1,170	110.00	1,280
14"	"	1,460	110.00	1,570
16"	"	1,580	120.00	1,700
18"	"	1,760	140.00	1,900
20"	"	2,120	150.00	2,270
Tees				
10"	EA.	1,580	93.50	1,674
12"	"	2,000	110.00	2,110

UTILITIES	UNIT	MAT.	INST.	TOTAL
02665.10 CHILLED WATER SYSTEMS				
14"	EA.	2,120	110.00	2,230
16"	"	2,250	120.00	2,370
18"	"	2,550	140.00	2,690
20"	"	2,910	150.00	3,060
Reducers				
10"	EA.	1,200	62.50	1,263
12"	"	1,460	68.00	1,528
14"	"	1,580	74.50	1,655
16"	"	1,940	83.00	2,023
18"	"	2,120	93.50	2,214
20"	"	2,250	110.00	2,360
Anchors, not including concrete				
10"	EA.	340.00	62.50	402.50
12"	"	370.00	68.00	438.00
14"	"	430.00	74.50	504.50
16"	"	600.00	83.00	683.00
18"	"	840.00	93.50	933.50
20"	"	1,130	110.00	1,240
02670.10 WELLS				
Domestic water, drilled and cased				
4" dia.	L.F.	9.90	43.90	53.80
6" dia.	"	10.05	48.80	58.85
8" dia.	"	12.85	55.00	67.85
02685.10 GAS DISTRIBUTION				
Gas distribution lines				
Polyethylene, 60 psi coils				
1-1/4" dia.	L.F.	0.85	2.70	3.55
1-1/2" dia.	"	1.15	2.85	4.00
2" dia.	"	1.45	3.35	4.80
3" dia.	"	3.10	4.00	7.10
30' pipe lengths				
3" dia.	L.F.	3.05	4.45	7.50
4" dia.	"	4.75	5.05	9.80
6" dia.	"	7.50	6.70	14.20
8" dia.	"	13.85	8.05	21.90
Steel, schedule 40, plain end				
1" dia.	L.F.	3.05	3.35	6.40
2" dia.	"	4.25	3.65	7.90
3" dia.	"	7.30	4.00	11.30
4" dia.	"	9.00	9.95	18.95
5" dia.	"	16.85	10.70	27.55
6" dia.	"	22.45	12.45	34.90
8" dia.	"	28.10	13.60	41.70
Natural gas meters, direct digital reading, threaded				
250 cfh @ 5 lbs	EA.	110.00	80.50	190.50
425 cfh @ 10 lbs	"	280.00	80.50	360.50
800 cfh @ 20 lbs	"	390.00	100.00	490.00
1000 cfh @ 25 lbs	"	1,150	100.00	1,250
1,400 cfh @ 100 lbs	"	2,660	130.00	2,790
2,300 cfh @ 100 lbs	"	3,700	200.00	3,900

UTILITIES

UTILITIES	UNIT	MAT.	INST.	TOTAL
02685.10 GAS DISTRIBUTION				
5,000 cfh @ 100 lbs	EA.	5,430	400.00	5,830
Gas pressure regulators				
Threaded				
3/4"	EA.	48.00	50.50	98.50
1"	"	50.50	67.00	117.50
1-1/4"	"	52.50	67.00	119.50
1-1/2"	"	340.00	67.00	407.00
2"	"	350.00	80.50	430.50
Flanged				
3"	EA.	1,000	100.00	1,100
4"	"	1,490	130.00	1,620
02690.10 STORAGE TANKS				
Oil storage tank, underground				
Steel				
500 gals	EA.	520.00	190.00	710.00
1,000 gals	"	900.00	250.00	1,150
4,000 gals	"	2,470	500.00	2,970
5,000 gals	"	3,370	750.00	4,120
10,000 gals	"	4,820	1,490	6,310
Fiberglass, double wall				
550 gals	EA.	1,870	250.00	2,120
1,000 gals	"	2,420	250.00	2,670
2,000 gals	"	2,860	370.00	3,230
4,000 gals	"	3,850	750.00	4,600
6,000 gals	"	4,400	1,000	5,400
8,000 gals	"	5,940	1,490	7,430
10,000 gals	"	7,480	1,870	9,350
12,000 gals	"	8,910	2,490	11,400
15,000 gals	"	10,560	3,320	13,880
20,000 gals	"	13,420	3,740	17,160
Above ground				
Steel				
275 gals	EA.	210.00	150.00	360.00
500 gals	"	670.00	250.00	920.00
1,000 gals	"	1,010	300.00	1,310
1,500 gals	"	1,680	370.00	2,050
2,000 gals	"	2,120	500.00	2,620
5,000 gals	"	5,500	750.00	6,250
Fill cap	"	27.50	40.20	67.70
Vent cap	"	27.50	40.20	67.70
Level indicator	"	63.00	40.20	103.20
02695.40 STEEL PIPE				
Steel pipe, extra heavy, A 53, grade B, seamless				
1/2" dia.	L.F.	1.35	4.00	5.35
3/4" dia.	"	2.15	4.25	6.40
1" dia.	"	2.75	4.45	7.20
1-1/4" dia.	"	3.35	5.05	8.40
1-1/2" dia.	"	3.80	5.75	9.55

02 SITEWORK

UTILITIES

02695.40 STEEL PIPE	UNIT	MAT.	INST.	TOTAL
2" dia.	L.F.	4.40	6.70	11.10
3" dia.	"	8.40	7.45	15.85
4" dia.	"	14.60	8.30	22.90
6" dia.	"	34.80	9.35	44.15
8" dia.	"	44.90	10.70	55.60
10" dia.	"	67.50	12.45	79.95
12" dia.	"	83.00	14.95	97.95

02695.80 STEAM METERS	UNIT	MAT.	INST.	TOTAL
In-line turbine, direct reading, 300 lb, flanged				
2"	EA.	3,170	50.50	3,221
3"	"	3,400	67.00	3,467
4"	"	3,740	80.50	3,821
Threaded, 2"				
5" line	EA.	6,000	400.00	6,400
6" line	"	6,120	400.00	6,520
8" line	"	6,230	400.00	6,630
10" line	"	6,570	400.00	6,970
12" line	"	6,680	400.00	7,080
14" line	"	6,910	400.00	7,310
16" line	"	7,360	400.00	7,760

SEWERAGE AND DRAINAGE

02720.10 CATCH BASINS	UNIT	MAT.	INST.	TOTAL
Standard concrete catch basin				
Cast in place, 3'8" x 3'8", 6" thick wall				
2' deep	EA.	210.00	370.00	580.00
3' deep	"	290.00	370.00	660.00
4' deep	"	370.00	500.00	870.00
5' deep	"	440.00	500.00	940.00
6' deep	"	490.00	600.00	1,090
4'x4', 8" thick wall, cast in place				
2' deep	EA.	230.00	370.00	600.00
3' deep	"	320.00	370.00	690.00
4' deep	"	420.00	500.00	920.00
5' deep	"	490.00	500.00	990.00
6' deep	"	540.00	600.00	1,140
Frames and covers, cast iron				
Round				
24" dia.	EA.	180.00	71.50	251.50
26" dia.	"	200.00	71.50	271.50
28" dia.	"	240.00	71.50	311.50

SEWERAGE AND DRAINAGE

02720.10 CATCH BASINS

SEWERAGE AND DRAINAGE	UNIT	MAT.	INST.	TOTAL
Rectangular				
23"x23"	EA.	160.00	71.50	231.50
27"x20"	"	200.00	71.50	271.50
24"x24"	"	190.00	71.50	261.50
26"x26"	"	210.00	71.50	281.50
Curb inlet frames and covers				
27"x27"	EA.	390.00	71.50	461.50
24"x36"	"	240.00	71.50	311.50
24"x25"	"	220.00	71.50	291.50
24"x22"	"	190.00	71.50	261.50
20"x22"	"	250.00	71.50	321.50
Airfield catch basin frame and grating, galvanized				
2'x4'	EA.	460.00	71.50	531.50
2'x2'	"	330.00	71.50	401.50

02720.40 STORM DRAINAGE

SEWERAGE AND DRAINAGE	UNIT	MAT.	INST.	TOTAL
Headwalls, cast in place, 30 deg wingwall				
12" pipe	EA.	270.00	91.00	361.00
15" pipe	"	330.00	91.00	421.00
18" pipe	"	400.00	100.00	500.00
24" pipe	"	600.00	100.00	700.00
30" pipe	"	720.00	120.00	840.00
36" pipe	"	780.00	180.00	960.00
42" pipe	"	980.00	180.00	1,160
48" pipe	"	1,040	240.00	1,280
54" pipe	"	1,180	300.00	1,480
60" pipe	"	1,410	360.00	1,770
4" cleanout for storm drain				
4" pipe	EA.	370.00	35.80	405.80
6" pipe	"	450.00	35.80	485.80
8" pipe	"	620.00	35.80	655.80
Connect new drain line				
To existing manhole	EA.	86.50	95.50	182.00
To new manhole	"	75.00	57.00	132.00

02720.45 STORM DRAINAGE, CON. PIPE

SEWERAGE AND DRAINAGE	UNIT	MAT.	INST.	TOTAL
Concrete pipe				
Unreinforced, plain, bell and spigot, Class II				
Pipe diameter 6"	L.F.	3.90	6.80	10.70
8"	"	4.30	7.45	11.75
10"	"	4.40	7.85	12.25
12"	"	5.20	8.30	13.50
15"	"	6.30	8.80	15.10
18"	"	8.55	9.35	17.90
21"	"	10.70	9.95	20.65
24"	"	13.35	10.70	24.05
Reinforced, bell and spigot, Class III				
Pipe diameter 12"	L.F.	9.25	8.30	17.55
15"	"	10.40	8.80	19.20
18"	"	11.55	9.35	20.90
21"	"	15.00	9.95	24.95
24"	"	19.65	10.70	30.35

SEWERAGE AND DRAINAGE

02720.45 STORM DRAINAGE, CON. PIPE

	UNIT	MAT.	INST.	TOTAL
27"	L.F.	23.10	11.50	34.60
30"	"	25.40	12.45	37.85
36"	"	38.10	13.60	51.70
42"	"	47.40	14.95	62.35
48"	"	66.00	16.60	82.60
54"	"	77.50	18.70	96.20
60"	"	83.00	21.35	104.35
66"	"	110.00	24.90	134.90
72"	"	130.00	29.90	159.90
78"	"	160.00	33.20	193.20
84"	"	180.00	37.40	217.40
90"	"	200.00	40.40	240.40
96"	"	230.00	42.70	272.70
Class IV				
Pipe diameter 12"	L.F.	8.45	8.30	16.75
15"	"	8.65	8.80	17.45
18"	"	11.55	9.35	20.90
21"	"	16.15	9.95	26.10
24"	"	20.20	10.70	30.90
27"	"	23.70	11.50	35.20
30"	"	25.40	12.45	37.85
36"	"	38.80	13.60	52.40
42"	"	49.70	14.95	64.65
48"	"	67.00	16.60	83.60
54"	"	87.50	18.70	106.20
60"	"	100.00	21.35	121.35
66"	"	120.00	24.90	144.90
72"	"	160.00	29.90	189.90
78"	"	180.00	33.20	213.20
84"	"	220.00	37.40	257.40
90"	"	230.00	40.40	270.40
96"	"	270.00	42.70	312.70
Class V				
Pipe diameter 12"	L.F.	10.40	8.30	18.70
15"	"	10.65	8.80	19.45
18"	"	14.45	9.35	23.80
21"	"	18.50	9.95	28.45
24"	"	23.10	10.70	33.80
27"	"	28.90	11.50	40.40
30"	"	33.50	12.45	45.95
36"	"	46.20	13.60	59.80
42"	"	60.00	14.95	74.95
48"	"	78.50	16.60	95.10
54"	"	100.00	18.70	118.70
60"	"	120.00	21.35	141.35
66"	"	140.00	24.90	164.90
72"	"	180.00	29.90	209.90
78"	"	220.00	33.20	253.20
84"	"	250.00	37.40	287.40
90"	"	250.00	40.40	290.40
96"	"	290.00	42.70	332.70
Eliptical pipe, reinforced				

SEWERAGE AND DRAINAGE

02720.45 STORM DRAINAGE, CON. PIPE

SEWERAGE AND DRAINAGE	UNIT	MAT.	INST.	TOTAL
Class III				
20" x 30"	L.F.	28.90	10.70	39.60
22" x 34"	"	32.30	11.50	43.80
24" x 38"	"	43.90	12.45	56.35
27" x 42"	"	55.50	13.60	69.10
29" x 45"	"	69.50	14.25	83.75
32" x 49"	"	75.00	14.95	89.95
34" x 54"	"	77.50	16.60	94.10
38" x 60"	"	86.50	18.70	105.20
43" x 68"	"	92.50	21.35	113.85
48" x 76"	"	110.00	23.00	133.00
53" x 83"	"	150.00	24.90	174.90
62" x 98"	"	210.00	29.90	239.90
82" x 128"	"	290.00	42.70	332.70
Flared end section pipe				
Pipe diameter 12"	L.F.	42.10	8.30	50.40
15"	"	49.50	8.80	58.30
18"	"	55.50	9.35	64.85
24"	"	68.50	10.70	79.20
30"	"	83.50	12.45	95.95
36"	"	120.00	13.60	133.60
42"	"	120.00	14.95	134.95
48"	"	140.00	16.60	156.60
54"	"	150.00	18.70	168.70
60"	"	160.00	21.35	181.35
Porous concrete pipe, standard strength				
Pipe diameter 4"	L.F.	2.65	5.75	8.40
6"	"	2.90	6.00	8.90
8"	"	3.30	6.25	9.55
10"	"	6.95	6.50	13.45
12"	"	9.25	6.50	15.75

02720.50 STORM DRAINAGE, STEEL PIPE

SEWERAGE AND DRAINAGE	UNIT	MAT.	INST.	TOTAL
Steel pipe				
Coated,corrugated metal pipe, paved invert				
16 gauge, pipe diameter 8"	L.F.	6.60	5.00	11.60
12"	"	9.90	5.35	15.25
15"	"	12.10	5.75	17.85
18"	"	14.30	6.25	20.55
21"	"	17.60	6.80	24.40
24"	"	20.90	7.45	28.35
30"	"	27.50	8.30	35.80
36"	"	37.40	9.35	46.75
42"	"	54.50	9.95	64.45
48"	"	74.00	10.70	84.70
14 gauge, pipe diameter 12"	"	12.20	5.35	17.55
15"	"	14.50	5.75	20.25
18"	"	16.95	6.25	23.20
21"	"	20.55	6.80	27.35
24"	"	26.40	7.45	33.85
30"	"	29.70	8.30	38.00

SEWERAGE AND DRAINAGE

02720.50 STORM DRAINAGE, STEEL PIPE	UNIT	MAT.	INST.	TOTAL
36"	L.F.	42.40	9.35	51.75
42"	"	57.00	9.95	66.95
48"	"	78.50	10.70	89.20
54"	"	91.00	11.50	102.50
60"	"	100.00	12.45	112.45
66"	"	120.00	13.60	133.60
12 gauge, pipe diameter 18"	"	23.00	6.25	29.25
21"	"	27.80	6.80	34.60
24"	"	32.70	7.45	40.15
30"	"	44.80	8.30	53.10
36"	"	51.50	9.35	60.85
42"	"	69.00	9.95	78.95
48"	"	67.00	10.70	77.70
54"	"	100.00	11.50	111.50
60"	"	110.00	12.45	122.45
66"	"	130.00	13.60	143.60
72"	"	140.00	14.95	154.95
78"	"	150.00	15.75	165.75
10 gauge, pipe diameter 24"	"	42.40	7.45	49.85
30"	"	54.50	8.30	62.80
36"	"	66.50	9.35	75.85
42"	"	78.50	9.95	88.45
48"	"	97.00	10.70	107.70
54"	"	110.00	11.50	121.50
60"	"	110.00	12.45	122.45
66"	"	130.00	13.60	143.60
72"	"	110.00	14.95	124.95
78"	"	170.00	15.75	185.75
84"	"	180.00	16.60	196.60
90"	"	190.00	17.60	207.60
8 gauge, pipe diameter 48"	"	100.00	10.70	110.70
54"	"	120.00	11.50	131.50
60"	"	130.00	12.45	142.45
66"	"	140.00	13.60	153.60
72"	"	150.00	14.95	164.95
78"	"	180.00	15.75	195.75
84"	"	190.00	16.60	206.60
90"	"	200.00	17.60	217.60
96"	"	230.00	18.70	248.70
Plain,corrugated metal pipe				
16 gauge, pipe diameter 8"	L.F.	5.50	5.00	10.50
12"	"	7.70	5.35	13.05
15"	"	8.80	5.75	14.55
18"	"	13.20	6.25	19.45
21"	"	19.80	6.80	26.60
24"	"	19.80	7.45	27.25
30"	"	26.40	8.30	34.70
36"	"	33.00	9.35	42.35
42"	"	43.60	9.95	53.55
48"	"	60.50	10.70	71.20
14 gauge, pipe diameter 12"	"	11.25	5.35	16.60
15"	"	14.15	5.75	19.90

02 SITEWORK

SEWERAGE AND DRAINAGE

02720.50 STORM DRAINAGE, STEEL PIPE	UNIT	MAT.	INST.	TOTAL
18"	L.F.	16.20	6.25	22.45
21"	"	18.75	6.80	25.55
24"	"	21.80	7.45	29.25
30"	"	26.60	8.30	34.90
36"	"	36.30	9.35	45.65
42"	"	50.00	9.95	59.95
48"	"	72.50	10.70	83.20
54"	"	82.50	11.50	94.00
60"	"	94.50	12.45	106.95
66"	"	110.00	13.60	123.60
12 gauge, pipe diameter 18"	"	18.15	6.25	24.40
21"	"	20.55	6.80	27.35
24"	"	24.20	7.45	31.65
30"	"	29.00	8.30	37.30
36"	"	42.40	9.35	51.75
42"	"	58.00	9.95	67.95
48"	"	60.50	10.70	71.20
54"	"	90.00	11.50	101.50
60"	"	100.00	12.45	112.45
66"	"	120.00	13.60	133.60
72"	"	130.00	14.95	144.95
78"	"	140.00	15.75	155.75
10 gauge, pipe diameter 24"	"	29.00	7.45	36.45
30"	"	32.70	8.30	41.00
36"	"	48.40	9.35	57.75
42"	"	63.00	9.95	72.95
48"	"	86.00	10.70	96.70
54"	"	95.00	11.50	106.50
60"	"	82.50	12.45	94.95
66"	"	130.00	13.60	143.60
72"	"	140.00	14.95	154.95
78"	"	150.00	15.75	165.75
84"	"	170.00	16.60	186.60
90"	"	180.00	17.60	197.60
8 gauge, pipe diameter 48"	"	91.00	10.70	101.70
54"	"	100.00	11.50	111.50
60"	"	110.00	12.45	122.45
66"	"	130.00	13.60	143.60
72"	"	150.00	14.95	164.95
78"	"	160.00	15.75	175.75
84"	"	180.00	16.60	196.60
90"	"	210.00	17.60	227.60
96"	"	220.00	18.70	238.70
Steel arch				
Coated, corrugated				
16 gauge, 17" x 13"	L.F.	15.55	6.80	22.35
21" x 15"	"	20.90	7.45	28.35
14 gauge, 29" x 18"	"	30.80	8.30	39.10
36" x 22"	"	39.60	10.70	50.30
12 gauge, 43" x 28"	"	53.00	12.45	65.45
50" x 30"	"	60.50	13.60	74.10
58" x 36"	"	69.50	14.95	84.45

SEWERAGE AND DRAINAGE

02720.50 STORM DRAINAGE, STEEL PIPE

	UNIT	MAT.	INST.	TOTAL
66" x 40"	L.F.	81.50	15.75	97.25
72" x 44"	"	100.00	16.60	116.60
Plain, corrugated				
16 gauge, 17" x 13"	L.F.	9.15	6.80	15.95
21" x 15"	"	13.20	7.45	20.65
14 gauge, 29" x 18"	"	25.30	8.30	33.60
36" x 22"	"	30.80	10.70	41.50
12 gauge, 43" x 28"	"	35.20	12.45	47.65
50" x 30"	"	41.80	13.60	55.40
58" x 36"	"	46.20	14.95	61.15
66" x 40"	"	53.00	15.75	68.75
72" x 44"	"	81.50	16.60	98.10
Nestable corrugated metal pipe				
16 gauge, pipe diameter 10"	L.F.	8.80	5.15	13.95
12"	"	11.00	5.35	16.35
15"	"	14.30	5.75	20.05
18"	"	16.50	6.25	22.75
24"	"	23.10	7.45	30.55
30"	"	28.60	8.30	36.90
14 gauge, pipe diameter 12"	"	11.00	5.35	16.35
15"	"	14.30	5.75	20.05
18"	"	20.90	6.25	27.15
24"	"	26.40	7.45	33.85
30"	"	28.60	8.30	36.90
36"	"	33.00	9.35	42.35

02720.70 UNDERDRAIN

	UNIT	MAT.	INST.	TOTAL
Drain tile, clay				
6" pipe	L.F.	2.85	3.30	6.15
8" pipe	"	4.50	3.50	8.00
12" pipe	"	9.05	3.75	12.80
Porous concrete, standard strength				
6" pipe	L.F.	2.55	3.30	5.85
8" pipe	"	3.70	3.50	7.20
12" pipe	"	6.95	3.75	10.70
15" pipe	"	8.15	4.15	12.30
18" pipe	"	10.60	5.00	15.60
Corrugated metal pipe, perforated type				
6" pipe	L.F.	4.85	3.75	8.60
8" pipe	"	5.70	3.95	9.65
10" pipe	"	6.95	4.15	11.10
12" pipe	"	9.80	4.40	14.20
18" pipe	"	12.10	4.65	16.75
Perforated clay pipe				
6" pipe	L.F.	3.40	4.25	7.65
8" pipe	"	4.55	4.40	8.95
12" pipe	"	7.95	4.55	12.50
Drain tile, concrete				
6" pipe	L.F.	2.10	3.30	5.40
8" pipe	"	3.30	3.50	6.80
12" pipe	"	6.60	3.75	10.35

SEWERAGE AND DRAINAGE

SEWERAGE AND DRAINAGE	UNIT	MAT.	INST.	TOTAL
02720.70 UNDERDRAIN				
Perforated rigid PVC underdrain pipe				
4" pipe	L.F.	0.94	2.50	3.44
6" pipe	"	1.70	3.00	4.70
8" pipe	"	2.40	3.30	5.70
10" pipe	"	3.75	3.75	7.50
12" pipe	"	6.25	4.25	10.50
Underslab drainage, crushed stone				
3" thick	S.F.	0.20	0.50	0.70
4" thick	"	0.24	0.57	0.81
6" thick	"	0.26	0.62	0.88
8" thick	"	0.32	0.65	0.97
Plastic filter fabric for drain lines	"	0.08	0.29	0.37
Gravel fill in trench, crushed or bank run, 1/2" to 3/4"	C.Y.	20.55	37.40	57.95
02730.10 SANITARY SEWERS				
Clay				
6" pipe	L.F.	3.40	5.00	8.40
8" pipe	"	4.55	5.35	9.90
10" pipe	"	5.65	5.75	11.40
12" pipe	"	9.05	6.25	15.30
PVC				
4" pipe	L.F.	1.10	3.75	4.85
6" pipe	"	2.20	3.95	6.15
8" pipe	"	3.30	4.15	7.45
10" pipe	"	4.40	4.40	8.80
12" pipe	"	6.60	4.65	11.25
Cleanout				
4" pipe	EA.	4.60	35.80	40.40
6" pipe	"	10.15	35.80	45.95
8" pipe	"	30.00	35.80	65.80
Connect new sewer line				
To existing manhole	EA.	60.00	95.50	155.50
To new manhole	"	41.60	57.00	98.60
02740.10 DRAINAGE FIELDS				
Perforated PVC pipe, for drain field				
4" pipe	L.F.	0.99	3.30	4.29
6" pipe	"	1.85	3.55	5.40
02740.50 SEPTIC TANKS				
Septic tank, precast concrete				
1000 gals	EA.	540.00	250.00	790.00
2000 gals	"	1,020	370.00	1,390
5000 gals	"	4,450	750.00	5,200
25,000 gals	"	18,020	2,990	21,010
40,000 gals	"	28,880	4,980	33,860
Leaching pit, precast concrete, 72" diameter				
3' deep	EA.	400.00	190.00	590.00
6' deep	"	500.00	210.00	710.00
8' deep	"	620.00	250.00	870.00

SEWERAGE AND DRAINAGE

02760.10 PIPELINE RESTORATION

	UNIT	MAT.	INST.	TOTAL
Relining existing water main				
6" dia.	L.F.	5.05	21.95	27.00
8" dia.	"	5.70	23.10	28.80
10" dia.	"	6.35	24.40	30.75
12" dia.	"	6.95	25.80	32.75
14" dia.	"	7.45	27.40	34.85
16" dia.	"	8.10	29.30	37.40
18" dia.	"	8.70	31.30	40.00
20" dia.	"	9.60	33.80	43.40
24" dia.	"	10.20	36.60	46.80
36" dia.	"	11.10	43.90	55.00
48" dia.	"	12.50	48.80	61.30
72" dia.	"	15.75	55.00	70.75
Replacing in line gate valves				
6" valve	EA.	500.00	290.00	790.00
8" valve	"	790.00	370.00	1,160
10" valve	"	1,190	440.00	1,630
12" valve	"	2,060	550.00	2,610
16" valve	"	4,670	630.00	5,300
18" valve	"	7,070	730.00	7,800
20" valve	"	9,720	880.00	10,600
24" valve	"	13,880	1,100	14,980
36" valve	"	37,850	1,460	39,310

POWER & COMMUNICATIONS

02780.20 HIGH VOLTAGE CABLE

	UNIT	MAT.	INST.	TOTAL
High voltage XLP copper cable, shielded, 5000v				
#6 awg	L.F.	0.94	0.61	1.55
#4 awg	"	1.05	0.75	1.80
#2 awg	"	1.40	0.89	2.29
#1 awg	"	1.45	0.98	2.43
#1/0 awg	"	1.75	1.10	2.85
#2/0 awg	"	2.95	1.35	4.30
#3/0 awg	"	3.00	1.60	4.60
#4/0 awg	"	3.15	1.70	4.85
#250 awg	"	3.65	2.00	5.65
#300 awg	"	4.20	2.25	6.45
#350 awg	"	4.70	2.50	7.20
#500 awg	"	6.70	3.40	10.10
#750 awg	"	9.90	3.75	13.65
Ungrounded, 15,000v				
#1 awg	L.F.	2.25	1.45	3.70

POWER & COMMUNICATIONS

02780.20 HIGH VOLTAGE CABLE	UNIT	MAT.	INST.	TOTAL
#1/0 awg	L.F.	2.65	1.60	4.25
#2/0 awg	"	3.05	1.70	4.75
#3/0 awg	"	3.50	1.85	5.35
#4/0 awg	"	3.85	2.15	6.00
#250 awg	"	4.35	2.25	6.60
#300 awg	"	4.90	2.50	7.40
#350 awg	"	5.45	2.90	8.35
#500 awg	"	7.20	3.75	10.95
#750 awg	"	10.65	4.55	15.20
#1000 awg	"	15.70	5.75	21.45
Aluminum cable, shielded, 5000v				
#6 awg	L.F.	1.20	0.51	1.71
#4 awg	"	1.35	0.61	1.96
#2 awg	"	1.45	0.70	2.15
#1 awg	"	1.60	0.80	2.40
#1/0 awg	"	1.75	0.89	2.64
#2/0 awg	"	1.95	0.94	2.88
#3/0 awg	"	2.25	0.98	3.23
#4/0 awg	"	2.50	1.10	3.60
#250 awg	"	2.65	1.20	3.85
#300 awg	"	2.90	1.45	4.35
#350 awg	"	3.10	1.60	4.70
#500 awg	"	3.75	1.70	5.45
#750 awg	"	4.85	2.10	6.95
#1000 awg	"	5.50	2.35	7.85
Ungrounded, 15,000v				
#1 awg	L.F.	2.20	0.98	3.18
#1/0 awg	"	2.35	1.15	3.50
#2/0 awg	"	2.50	1.25	3.75
#3/0 awg	"	2.75	1.30	4.05
#4/0 awg	"	3.00	1.35	4.35
#250 awg	"	3.20	1.45	4.65
#300 awg	"	3.45	1.50	4.95
#350 awg	"	3.75	1.70	5.45
#500 awg	"	4.35	2.00	6.35
#750 awg	"	5.75	2.40	8.15
#1000 awg	"	7.55	3.00	10.55
Indoor terminations, 5000v				
#6 - #4	EA.	37.80	7.35	45.15
#2 - #2/0	"	44.10	7.35	51.45
#3/0 - #250	"	56.50	7.35	63.85
#300 - #750	"	63.00	130.00	193.00
#1000	"	75.50	180.00	255.50
In-line splice, 5000v				
#6 - #4/0	EA.	82.00	180.00	262.00
#250 - #500	"	88.00	470.00	558.00
#750 - #1000	"	120.00	610.00	730.00
T-splice, 5000v				
#2 - #4/0	EA.	82.00	560.00	642.00
#250 - #500	"	88.00	940.00	1,028
#750 - #1000	"	120.00	1,170	1,290
Indoor terminations, 15,000v				

POWER & COMMUNICATIONS

	UNIT	MAT.	INST.	TOTAL
02780.20 HIGH VOLTAGE CABLE				
#2 - #2/0	EA.	50.50	160.00	210.50
#3/0 - #500	"	63.00	250.00	313.00
#750 - #1000	"	75.50	290.00	365.50
In-line splice, 15,000v				
#2 - #4/0	EA.	88.00	420.00	508.00
#250 - #500	"	120.00	560.00	680.00
#750 - #1000	"	150.00	840.00	990.00
T-splice, 15,000v				
#4	EA.	88.00	840.00	928.00
#250 - #500	"	120.00	1,400	1,520
#750 - #1000	"	150.00	2,100	2,250
Compression lugs, 15,000v				
#4	EA.	5.95	18.70	24.65
#2	"	6.85	24.95	31.80
#1	"	7.50	24.95	32.45
#1/0	"	11.45	31.20	42.65
#2/0	"	12.00	31.20	43.20
#3/0	"	13.50	39.80	53.30
#4/0	"	14.90	39.80	54.70
#250	"	17.40	44.50	61.90
#300	"	20.15	44.50	64.65
#350	"	21.35	54.00	75.35
#500	"	31.50	58.50	90.00
#750	"	50.00	70.50	120.50
#1000	"	73.50	89.00	162.50
Compression splices, 15,000v				
#4	EA.	6.70	31.20	37.90
#2	"	7.35	34.00	41.35
#1	"	8.35	42.00	50.35
#1/0	"	8.90	46.80	55.70
#2/0	"	9.55	54.00	63.55
#3/0	"	10.40	58.50	68.90
#4/0	"	11.35	65.50	76.85
#250	"	12.40	70.50	82.90
#350	"	14.10	81.50	95.60
#500	"	20.90	93.50	114.40
#750	"	34.30	120.00	154.30
02780.40 SUPPORTS & CONNECTORS				
Cable supports for conduit				
1-1/2"	EA.	43.80	16.25	60.05
2"	"	61.00	16.25	77.25
2-1/2"	"	68.00	18.70	86.70
3"	"	87.50	18.70	106.20
3-1/2"	"	110.00	23.35	133.35
4"	"	140.00	23.35	163.35
5"	"	250.00	31.20	281.20
6"	"	530.00	34.00	564.00
Split bolt connectors				
#10	EA.	1.35	9.35	10.70
#8	"	1.60	9.35	10.95

POWER & COMMUNICATIONS

02780.40 SUPPORTS & CONNECTORS	UNIT	MAT.	INST.	TOTAL
#6	EA.	1.75	9.35	11.10
#4	"	2.05	18.70	20.75
#3	"	2.90	18.70	21.60
#2	"	3.30	18.70	22.00
#1/0	"	4.30	31.20	35.50
#2/0	"	6.80	31.20	38.00
#3/0	"	10.40	31.20	41.60
#4/0	"	11.75	31.20	42.95
#250	"	12.15	46.80	58.95
#350	"	21.50	46.80	68.30
#500	"	28.20	46.80	75.00
#750	"	47.70	70.50	118.20
#1000	"	65.50	70.50	136.00
Single barrel lugs				
#6	EA.	0.47	11.70	12.17
#1/0	"	0.94	23.35	24.29
#250	"	2.25	31.20	33.45
#350	"	2.90	31.20	34.10
#500	"	5.65	31.20	36.85
#600	"	6.00	42.00	48.00
#800	"	6.80	42.00	48.80
#1000	"	8.15	42.00	50.15
Double barrel lugs				
#1/0	EA.	1.85	42.00	43.85
#250	"	5.35	60.50	65.85
#350	"	7.60	60.50	68.10
#600	"	11.60	89.00	100.60
#800	"	13.20	89.00	102.20
#1000	"	13.55	89.00	102.55
Three barrel lugs				
#2/0	EA.	14.75	60.50	75.25
#250	"	28.30	89.00	117.30
#350	"	46.30	89.00	135.30
#600	"	50.50	120.00	170.50
#800	"	82.00	120.00	202.00
#1000	"	120.00	120.00	240.00
Four barrel lugs				
#250	EA.	31.50	130.00	161.50
#350	"	52.50	130.00	182.50
#600	"	57.50	160.00	217.50
#800	"	90.00	160.00	250.00
Compression conductor adapters				
#6	EA.	4.05	13.85	17.90
#4	"	4.30	16.25	20.55
#2	"	4.50	20.80	25.30
#1	"	5.15	20.80	25.95
#1/0	"	5.35	24.95	30.30
#250	"	10.20	37.40	47.60
#350	"	12.15	39.80	51.95
#500	"	15.80	51.00	66.80
#750	"	21.60	53.50	75.10
Terminal blocks, 2 screw				

POWER & COMMUNICATIONS

02780.40 SUPPORTS & CONNECTORS	UNIT	MAT.	INST.	TOTAL
3 circuit	EA.	11.70	9.35	21.05
6 circuit	"	16.15	9.35	25.50
8 circuit	"	18.90	9.35	28.25
10 circuit	"	21.85	13.85	35.70
12 circuit	"	24.65	13.85	38.50
18 circuit	"	33.30	13.85	47.15
24 circuit	"	41.80	16.25	58.05
36 circuit	"	59.00	16.25	75.25
Compression splice				
#8 awg	EA.	2.00	17.80	19.80
#6 awg	"	2.35	12.90	15.25
#4 awg	"	2.50	12.90	15.40
#2 awg	"	3.85	24.95	28.80
#1 awg	"	5.45	24.95	30.40
#1/0 awg	"	6.65	24.95	31.60
#2/0 awg	"	7.10	39.80	46.90
#3/0 awg	"	8.30	39.80	48.10
#4/0 awg	"	8.80	39.80	48.60
#250 awg	"	9.30	63.50	72.80
#300 awg	"	10.15	63.50	73.65
#350 awg	"	10.35	65.50	75.85
#400 awg	"	14.05	65.50	79.55
#500 awg	"	16.40	70.50	86.90
#600 awg	"	24.90	70.50	95.40
#750 awg	"	26.40	81.50	107.90
#1000 awg	"	34.70	81.50	116.20

SITE IMPROVEMENTS

02810.40 LAWN IRRIGATION	UNIT	MAT.	INST.	TOTAL
Residential system, complete				
Minimum	ACRE			11,220
Maximum	"			22,440
Commercial system, complete				
Minimum	ACRE			16,830
Maximum	"			28,050
Components				
Pop-up head	EA.	9.50	17.85	27.35
Impact head	"	9.50	17.85	27.35
Rotary impact head	"	26.20	35.80	62.00
Shrub head	"	2.50	17.85	20.35
Hose bibb	"	4.65	35.80	40.45

SITE IMPROVEMENTS

02830.10 CHAIN LINK FENCE

SITE IMPROVEMENTS	UNIT	MAT.	INST.	TOTAL
Chain link fence, 9 ga., galvanized, with posts 10' o.c.				
4' high	L.F.	4.35	2.05	6.40
5' high	"	5.85	2.60	8.45
6' high	"	6.60	3.60	10.20
7' high	"	7.50	4.40	11.90
8' high	"	8.70	5.70	14.40
For barbed wire with hangers, add				
3 strand	L.F.	1.60	1.45	3.05
6 strand	"	2.70	2.40	5.10
Corner or gate post, 3" post				
4' high	EA.	51.00	9.55	60.55
5' high	"	56.50	10.60	67.10
6' high	"	62.50	12.45	74.95
7' high	"	75.50	14.30	89.80
8' high	"	79.00	15.90	94.90
4" post				
4' high	EA.	87.50	10.60	98.10
5' high	"	100.00	12.45	112.45
6' high	"	110.00	14.30	124.30
7' high	"	130.00	15.90	145.90
8' high	"	140.00	17.85	157.85
Gate with gate posts, galvanized, 3' wide				
4' high	EA.	56.50	71.50	128.00
5' high	"	72.50	95.50	168.00
6' high	"	87.00	95.50	182.50
7' high	"	100.00	140.00	240.00
8' high	"	110.00	140.00	250.00
Fabric, galvanized chain link, 2" mesh, 9 ga.				
4' high	L.F.	2.00	0.95	2.95
5' high	"	2.40	1.15	3.55
6' high	"	2.55	1.45	4.00
8' high	"	3.90	1.90	5.80
Line post, no rail fitting, galvanized, 2-1/2" dia.				
4' high	EA.	15.05	8.15	23.20
5' high	"	16.40	8.95	25.35
6' high	"	17.95	9.55	27.50
7' high	"	20.40	11.45	31.85
8' high	"	22.75	14.30	37.05
1-7/8" H beam				
4' high	EA.	20.65	8.15	28.80
5' high	"	23.05	8.95	32.00
6' high	"	27.60	9.55	37.15
7' high	"	31.30	11.45	42.75
8' high	"	33.90	14.30	48.20
2-1/4" H beam				
4' high	EA.	15.05	8.15	23.20
5' high	"	18.75	8.95	27.70
6' high	"	21.55	9.55	31.10
7' high	"	25.00	11.45	36.45
8' high	"	29.00	14.30	43.30
Vinyl coated, 9 ga., with posts 10' o.c.				
4' high	L.F.	4.75	2.05	6.80

02 SITEWORK

SITE IMPROVEMENTS

02830.10 CHAIN LINK FENCE	UNIT	MAT.	INST.	TOTAL
5' high	L.F.	5.65	2.60	8.25
6' high	"	6.75	3.60	10.35
7' high	"	7.35	4.40	11.75
8' high	"	8.40	5.70	14.10
For barbed wire w/hangers, add				
3 strand	L.F.	1.70	1.45	3.15
6 Strand	"	2.75	2.40	5.15
Corner, or gate post, 4' high				
3" dia.	EA.	62.50	9.55	72.05
4" dia.	"	96.00	9.55	105.55
6" dia.	"	110.00	11.45	121.45
Gate, with posts, 3' wide				
4' high	EA.	68.50	71.50	140.00
5' high	"	81.00	95.50	176.50
6' high	"	93.50	95.50	189.00
7' high	"	110.00	140.00	250.00
8' high	"	120.00	140.00	260.00
Line post, no rail fitting, 2-1/2" dia.				
4' high	EA.	26.30	8.15	34.45
5' high	"	35.80	8.95	44.75
6' high	"	43.10	9.55	52.65
7' high	"	50.50	11.45	61.95
8' high	"	55.50	14.30	69.80
Corner post, no top rail fitting, 4" dia.				
4' high	EA.	100.00	9.55	109.55
5' high	"	120.00	10.60	130.60
6' high	"	140.00	12.45	152.45
7' high	"	150.00	14.30	164.30
8' high	"	160.00	15.90	175.90
Fabric, vinyl, chain link, 2" mesh, 9 ga.				
4' high	L.F.	3.60	0.95	4.55
5' high	"	4.35	1.15	5.50
6' high	"	5.15	1.45	6.60
8' high	"	6.85	1.90	8.75
Swing gates, galvanized, 4' high				
Single gate				
3' wide	EA.	130.00	71.50	201.50
4' wide	"	140.00	71.50	211.50
Double gate				
10' wide	EA.	340.00	110.00	450.00
12' wide	"	360.00	110.00	470.00
14' wide	"	370.00	110.00	480.00
16' wide	"	420.00	110.00	530.00
18' wide	"	450.00	160.00	610.00
20' wide	"	470.00	160.00	630.00
22' wide	"	530.00	160.00	690.00
24' wide	"	540.00	190.00	730.00
26' wide	"	560.00	190.00	750.00
28' wide	"	600.00	230.00	830.00
30' wide	"	640.00	230.00	870.00
5' high				
Single gate				

SITE IMPROVEMENTS

02830.10 CHAIN LINK FENCE

SITE IMPROVEMENTS	UNIT	MAT.	INST.	TOTAL
3' wide	EA.	140.00	95.50	235.50
4' wide	"	160.00	95.50	255.50
Double gate				
10' wide	EA.	360.00	140.00	500.00
12' wide	"	390.00	140.00	530.00
14' wide	"	670.00	140.00	810.00
16' wide	"	440.00	140.00	580.00
18' wide	"	450.00	160.00	610.00
20' wide	"	510.00	160.00	670.00
22' wide	"	530.00	160.00	690.00
24' wide	"	560.00	190.00	750.00
26' wide	"	580.00	190.00	770.00
28' wide	"	650.00	230.00	880.00
30' wide	"	670.00	230.00	900.00
6' high				
Single gate				
3' wide	EA.	160.00	95.50	255.50
4' wide	"	170.00	95.50	265.50
Double gate				
10' wide	EA.	380.00	140.00	520.00
12' wide	"	420.00	140.00	560.00
14' wide	"	450.00	140.00	590.00
16' wide	"	490.00	140.00	630.00
18' wide	"	520.00	160.00	680.00
20' wide	"	540.00	160.00	700.00
22' wide	"	580.00	160.00	740.00
24' wide	"	620.00	190.00	810.00
26' wide	"	640.00	190.00	830.00
28' wide	"	690.00	230.00	920.00
30' wide	"	730.00	230.00	960.00
7' high				
Single gate				
3' wide	EA.	170.00	140.00	310.00
4' wide	"	190.00	140.00	330.00
Double gate				
10' wide	EA.	450.00	190.00	640.00
12' wide	"	490.00	190.00	680.00
14' wide	"	530.00	190.00	720.00
16' wide	"	560.00	190.00	750.00
18' wide	"	600.00	230.00	830.00
20' wide	"	640.00	230.00	870.00
22' wide	"	670.00	230.00	900.00
24' wide	"	720.00	290.00	1,010
26' wide	"	770.00	290.00	1,060
28' wide	"	810.00	360.00	1,170
30' wide	"	930.00	360.00	1,290
8' high				
Single gate				
3' wide	EA.	190.00	140.00	330.00
4' wide	"	200.00	140.00	340.00
Double gate				
10' wide	EA.	490.00	190.00	680.00

SITE IMPROVEMENTS	UNIT	MAT.	INST.	TOTAL
02830.10 CHAIN LINK FENCE				
12' wide	EA.	520.00	190.00	710.00
14' wide	"	560.00	190.00	750.00
16' wide	"	610.00	190.00	800.00
18' wide	"	640.00	230.00	870.00
20' wide	"	660.00	230.00	890.00
22' wide	"	700.00	230.00	930.00
24' wide	"	770.00	290.00	1,060
26' wide	"	800.00	290.00	1,090
28' wide	"	850.00	360.00	1,210
30' wide	"	840.00	360.00	1,200
Vinyl coated swing gates, 4' high				
Single gate				
3' wide	EA.	200.00	71.50	271.50
4' wide	"	220.00	71.50	291.50
Double gate				
10' wide	EA.	470.00	110.00	580.00
12' wide	"	600.00	110.00	710.00
14' wide	"	640.00	110.00	750.00
16' wide	"	710.00	110.00	820.00
18' wide	"	800.00	160.00	960.00
20' wide	"	950.00	160.00	1,110
22' wide	"	1,050	160.00	1,210
24' wide	"	1,130	190.00	1,320
26' wide	"	1,180	190.00	1,370
28' wide	"	1,380	230.00	1,610
30' wide	"	1,440	230.00	1,670
5' high				
Single gate				
3' wide	EA.	220.00	95.50	315.50
4' wide	"	250.00	95.50	345.50
Double gate				
10' wide	EA.	640.00	140.00	780.00
12' wide	"	690.00	140.00	830.00
14' wide	"	770.00	140.00	910.00
16' wide	"	800.00	140.00	940.00
18' wide	"	970.00	160.00	1,130
20' wide	"	1,050	160.00	1,210
22' wide	"	1,150	160.00	1,310
24' wide	"	1,310	190.00	1,500
26' wide	"	1,370	190.00	1,560
28' wide	"	1,530	230.00	1,760
30' wide	"	1,650	230.00	1,880
6' high				
Single gate				
3' wide	EA.	230.00	95.50	325.50
4' wide	"	240.00	95.50	335.50
Double gate				
10' wide	EA.	570.00	140.00	710.00
12' wide	"	660.00	140.00	800.00
14' wide	"	750.00	140.00	890.00
16' wide	"	880.00	140.00	1,020
18' wide	"	970.00	160.00	1,130

SITE IMPROVEMENTS

SITE IMPROVEMENTS	UNIT	MAT.	INST.	TOTAL
02830.10 CHAIN LINK FENCE				
20' wide	EA.	1,050	160.00	1,210
22' wide	"	1,140	160.00	1,300
24' wide	"	1,270	190.00	1,460
26' wide	"	1,350	190.00	1,540
28' wide	"	1,510	230.00	1,740
30' wide	"	1,650	230.00	1,880
7' high				
Single gate				
3' wide	EA.	260.00	140.00	400.00
4' wide	"	330.00	140.00	470.00
Double gate				
10' wide	EA.	640.00	190.00	830.00
12' wide	"	760.00	190.00	950.00
14' wide	"	860.00	190.00	1,050
16' wide	"	970.00	190.00	1,160
18' wide	"	1,090	230.00	1,320
20' wide	"	1,250	230.00	1,480
22' wide	"	1,380	230.00	1,610
24' wide	"	1,520	290.00	1,810
26' wide	"	1,650	290.00	1,940
28' wide	"	1,770	360.00	2,130
30' wide	"	1,910	360.00	2,270
8' high				
Single gate				
3' wide	EA.	270.00	140.00	410.00
4' wide	"	330.00	140.00	470.00
Double gate				
10' wide	EA.	660.00	190.00	850.00
12' wide	"	760.00	190.00	950.00
14' wide	"	870.00	190.00	1,060
16' wide	"	620.00	190.00	810.00
18' wide	"	1,100	230.00	1,330
20' wide	"	1,280	230.00	1,510
22' wide	"	1,470	230.00	1,700
24' wide	"	1,570	290.00	1,860
28' wide	"	1,820	290.00	2,110
30' wide	"	1,950	360.00	2,310
Motor operator for gates, no wiring	"			3,764
Drilling fence post holes				
In soil				
By hand	EA.	0.00	14.30	14.30
By machine auger	"	0.00	9.10	9.10
In rock				
By jackhammer	EA.	0.00	120.00	120.00
By rock drill	"	0.00	36.40	36.40
Aluminum privacy slats, installed vertically	S.F.	0.68	0.71	1.40
Post hole, dig by hand	EA.	0.00	19.05	19.05
Set fence post in concrete	"	6.70	14.30	21.00
02830.70 RECREATIONAL COURTS				
Walls, galvanized steel				
8' high	L.F.	8.75	5.70	14.45

SITE IMPROVEMENTS

02830.70 RECREATIONAL COURTS

SITE IMPROVEMENTS	UNIT	MAT.	INST.	TOTAL
10' high	L.F.	10.35	6.35	16.70
12' high	"	11.90	7.55	19.45
Vinyl coated				
8' high	L.F.	8.40	5.70	14.10
10' high	"	10.25	6.35	16.60
12' high	"	11.40	7.55	18.95
Gates, galvanized steel				
Single, 3' transom				
3'x7'	EA.	210.00	140.00	350.00
4'x7'	"	220.00	160.00	380.00
5'x7'	"	300.00	190.00	490.00
6'x7'	"	320.00	230.00	550.00
Double, 3' transom				
10'x7'	EA.	500.00	570.00	1,070
12'x7'	"	640.00	640.00	1,280
14'x7'	"	760.00	720.00	1,480
Double, no transom				
10'x10'	EA.	530.00	480.00	1,010
12'x10'	"	640.00	570.00	1,210
14'x10'	"	740.00	640.00	1,380
Vinyl coated				
Single, 3' transom				
3'x7'	EA.	400.00	140.00	540.00
4'x7'	"	440.00	160.00	600.00
5'x7'	"	440.00	190.00	630.00
6'x7'	"	450.00	230.00	680.00
Double, 3'				
10'x7'	EA.	1,180	570.00	1,750
12'x7'	"	1,210	640.00	1,850
14'x7'	"	1,310	720.00	2,030
Double, no transom				
10'x10'	EA.	1,180	480.00	1,660
12'x10'	"	1,200	570.00	1,770
14'x10'	"	1,310	640.00	1,950
Baseball backstop, regulation				
Galvanized	EA.			5,100
Vinyl coated	"			7,102
Softball backstop, regulation				
14' high				
Galvanized	EA.			4,736
Vinyl coated	"			7,042
18' high				
Galvanized	EA.			5,585
Vinyl coated	"			8,074
20' high				
Galvanized	EA.			6,617
Vinyl coated	"			9,531
22' high				
Galvanized	EA.			7,650
Vinyl coated	"			11,170
24' high				
Galvanized	EA.			9,300

SITE IMPROVEMENTS	UNIT	MAT.	INST.	TOTAL
02830.70 RECREATIONAL COURTS				
Vinyl coated	EA.			15,298
Wire and miscellaneous metal fences				
Chicken wire, post 4' o.c.				
2" mesh				
4' high	L.F.	0.99	1.45	2.44
6' high	"	1.10	1.90	3.00
Galvanized steel				
12 gauge, 2" by 4" mesh, posts 5' o.c.				
3' high	L.F.	1.60	1.45	3.05
5' high	"	2.25	1.80	4.05
14 gauge, 1" by 2" mesh, posts 5' o.c.				
3' high	L.F.	1.35	1.45	2.80
5' high	"	2.15	1.80	3.95
02840.30 GUARDRAILS				
Pipe bollard, steel pipe, concrete filled, painted				
6" dia.	EA.	110.00	23.85	133.85
8" dia.	"	170.00	35.80	205.80
12" dia.	"	270.00	95.50	365.50
Corrugated steel, guardrail, galvanized	L.F.	15.95	2.50	18.45
End section, wrap around or flared	EA.	44.60	28.60	73.20
Timber guardrail, 4" x 8"	L.F.	19.95	1.85	21.80
Guard rail, 3 cables, 3/4" dia.				
Steel posts	L.F.	8.50	7.45	15.95
Wood posts	"	9.70	6.00	15.70
Steel box beam				
6" x 6"	L.F.	35.10	8.30	43.40
6" x 8"	"	37.40	9.35	46.75
Concrete posts	EA.	22.65	14.30	36.95
Barrel type impact barrier	"	290.00	28.60	318.60
Light shield, 6' high	L.F.	19.10	5.70	24.80
02840.40 PARKING BARRIERS				
Timber, treated, 4' long				
4" x 4"	EA.	13.30	23.85	37.15
6" x 6"	"	24.20	28.60	52.80
Precast concrete, 6' long, with dowels				
12" x 6"	EA.	24.25	14.30	38.55
12" x 8"	"	26.60	15.90	42.50
02840.60 SIGNAGE				
Traffic signs				
Reflectorized signs per OSHA standards, including post				
Stop, 24"x24"	EA.	43.30	19.05	62.35
Yield, 30" triangle	"	25.20	19.05	44.25
Speed limit, 12"x18"	"	28.80	19.05	47.85
Directional, 12"x18"	"	39.70	19.05	58.75
Exit, 12"x18"	"	39.70	19.05	58.75
Entry, 12"x18"	"	39.70	19.05	58.75
Warning, 24"x24"	"	50.50	19.05	69.55
Informational, 12"x18"	"	18.05	19.05	37.10

SITE IMPROVEMENTS

SITE IMPROVEMENTS	UNIT	MAT.	INST.	TOTAL
02840.60 SIGNAGE				
Handicap parking, 12"x18"	EA.	19.25	19.05	38.30
02860.40 RECREATIONAL FACILITIES				
Bleachers, outdoor, portable, per seat				
10 tiers				
Minimum	EA.	22.90	9.35	32.25
Maximum	"	44.60	12.45	57.05
20 tiers				
Minimum	EA.	28.00	8.80	36.80
Maximum	"	53.50	11.50	65.00
Grandstands, fixed, wood seat, steel frame, per seat				
15 tiers				
Minimum	EA.	39.50	14.95	54.45
Maximum	"	67.50	24.90	92.40
30 tiers				
Minimum	EA.	40.70	13.60	54.30
Maximum	"	86.50	21.35	107.85
Seats				
Seat backs only				
Fiberglass	EA.	24.20	2.85	27.05
Steel and wood seat	"	33.10	2.85	35.95
Seat restoration, fiberglass on wood				
Seats	EA.	16.55	5.70	22.25
Plain bench, no backs	"	10.20	2.40	12.60
Benches				
Park, precast concrete with backs				
4' long	EA.	640.00	95.50	735.50
8' long	"	1,400	140.00	1,540
Fiberglass, with backs				
4' long	EA.	510.00	71.50	581.50
8' long	"	970.00	95.50	1,066
Wood, with backs and fiberglass supports				
4' long	EA.	280.00	71.50	351.50
8' long	"	290.00	95.50	385.50
Steel frame, 6' long				
All steel	EA.	240.00	71.50	311.50
Hardwood boards	"	170.00	71.50	241.50
Players bench (no back), steel frame, fir seat, 10' long	"	180.00	95.50	275.50
Backstops				
Handball or squash court, outdoor				
Wood	EA.			22,000
Masonry	"			22,550
Soccer goal posts	PAIR			1,980
Running track				
Gravel and cinders over stone base	S.Y.	5.95	3.75	9.70
Rubber-cork base resilient pavement	"	8.40	29.90	38.30
For colored surfaces, add	"	5.30	3.00	8.30
Colored rubberized asphalt	"	11.25	37.40	48.65
Artificial resilient mat over asphalt	"	26.50	74.50	101.00
Tennis courts				

02 SITEWORK

SITE IMPROVEMENTS	UNIT	MAT.	INST.	TOTAL
02860.40 RECREATIONAL FACILITIES				
Bituminous pavement, 2-1/2" thick	S.Y.	8.50	9.35	17.85
Colored sealer, acrylic emulsion				
3 coats	S.Y.	4.45	1.90	6.35
For 2 color seal coating, add	"	0.80	0.29	1.09
For preparing old courts, add	"	1.90	0.19	2.09
Net, nylon, 42' long	EA.	260.00	35.80	295.80
Paint markings on asphalt, 2 coats	"	67.00	290.00	357.00
Complete court with fence, etc., bituminous				
Minimum	EA.			10,441
Average	"			17,969
Maximum	"			25,497
Clay court				
Minimum	EA.			10,867
Average	"			14,812
Maximum	"			22,461
Playground equipment				
Basketball backboard				
Minimum	EA.	460.00	71.50	531.50
Maximum	"	870.00	81.50	951.50
Bike rack, 10' long	"	350.00	57.00	407.00
Golf shelter, fiberglass	"	1,760	71.50	1,832
Ground socket for movable posts				
Minimum	EA.	79.00	17.85	96.85
Maximum	"	160.00	17.85	177.85
Horizontal monkey ladder, 14' long	"	520.00	47.70	567.70
Posts, tether ball	"	250.00	14.30	264.30
Multiple purpose, 10' long	"	260.00	28.60	288.60
See-saw, steel				
Minimum	EA.	430.00	110.00	540.00
Average	"	800.00	140.00	940.00
Maximum	"	1,210	190.00	1,400
Slide				
Minimum	EA.	770.00	230.00	1,000
Maximum	"	1,340	260.00	1,600
Swings, plain seats				
8' high				
Minimum	EA.	570.00	190.00	760.00
Maximum	"	1,080	220.00	1,300
12' high				
Minimum	EA.	870.00	220.00	1,090
Maximum	"	1,580	320.00	1,900
02870.10 PREFABRICATED PLANTERS				
Concrete precast, circular				
24" dia., 18" high	EA.	210.00	28.60	238.60
42" dia., 30" high	"	280.00	35.80	315.80
Fiberglass, circular				
36" dia., 27" high	EA.	380.00	14.30	394.30
60" dia., 39" high	"	880.00	15.90	895.90
Tapered, circular				
24" dia., 36" high	EA.	300.00	13.00	313.00

SITE IMPROVEMENTS

02870.10 PREFABRICATED PLANTERS

	UNIT	MAT.	INST.	TOTAL
40" dia., 36" high	EA.	500.00	14.30	514.30
Square				
2' by 2', 17" high	EA.	250.00	13.00	263.00
4' by 4', 39" high	"	880.00	15.90	895.90
Rectangular				
4' by 1', 18" high	EA.	280.00	14.30	294.30

LANDSCAPING

02910.10 SHRUB & TREE MAINTENANCE

	UNIT	MAT.	INST.	TOTAL
Moving shrubs on site				
12" ball	EA.	0.00	35.80	35.80
24" ball	"	0.00	47.70	47.70
3' high	"	0.00	28.60	28.60
4' high	"	0.00	31.80	31.80
5' high	"	0.00	35.80	35.80
18" spread	"	0.00	40.90	40.90
30" spread	"	0.00	47.70	47.70
Moving trees on site				
24" ball	EA.	0.00	74.50	74.50
48" ball	"	0.00	100.00	100.00
Trees				
3' high	EA.	0.00	29.90	29.90
6' high	"	0.00	33.20	33.20
8' high	"	0.00	37.40	37.40
10' high	"	0.00	49.80	49.80
Palm trees				
7' high	EA.	0.00	37.40	37.40
10' high	"	0.00	49.80	49.80
20' high	"	0.00	150.00	150.00
40' high	"	0.00	300.00	300.00
Guying trees				
4" dia.	EA.	6.15	14.30	20.45
8" dia.	"	6.15	17.85	24.00

02920.10 TOPSOIL

	UNIT	MAT.	INST.	TOTAL
Spread topsoil, with equipment				
Minimum	C.Y.	0.00	7.80	7.80
Maximum	"	0.00	9.75	9.75
By hand				
Minimum	C.Y.	0.00	28.60	28.60
Maximum	"	0.00	35.80	35.80
Area preparation for seeding (grade, rake and clean)				
Square yard	S.Y.	0.00	0.23	0.23

LANDSCAPING	UNIT	MAT.	INST.	TOTAL
02920.10 TOPSOIL				
By acre	ACRE	0.00	1,140	1,140
Remove topsoil and stockpile on site				
4" deep	C.Y.	0.00	6.50	6.50
6" deep	"	0.00	6.00	6.00
Spreading topsoil from stock pile				
By loader	C.Y.	0.00	7.10	7.10
By hand	"	0.00	78.00	78.00
Top dress by hand	S.Y.	0.00	0.78	0.78
Place imported top soil				
By loader				
4" deep	S.Y.	0.00	0.78	0.78
6" deep	"	0.00	0.86	0.86
By hand				
4" deep	S.Y.	0.00	3.20	3.20
6" deep	"	0.00	3.60	3.60
Plant bed preparation, 18" deep				
With backhoe/loader	S.Y.	0.00	1.95	1.95
By hand	"	0.00	4.75	4.75
02930.30 SEEDING				
Mechanical seeding, 175 lb/acre				
By square yard	S.Y.	0.47	0.07	0.54
By acre	ACRE	1,910	360.00	2,270
450 lb/acre				
By square yard	S.Y.	0.76	0.09	0.85
By acre	ACRE	2,920	460.00	3,380
Seeding by hand, 10 lb per 100 s.y.				
By square yard	S.Y.	0.51	0.10	0.60
By acre	ACRE	2,020	480.00	2,500
Reseed disturbed areas	S.F.	0.45	0.14	0.59
02950.10 PLANTS				
Euonymus coloratus, 18" (purple wintercreeper)	EA.	1.45	4.75	6.20
Hedera Helix, 2-1/4" pot (English ivy)	"	0.58	4.75	5.33
Liriope muscari, 2" clumps	"	2.55	2.85	5.40
Santolina, 12"	"	2.90	2.85	5.75
Vinca major or minor, 3" pot	"	0.46	2.85	3.31
Cortaderia argentia, 2 gallon (pampas grass)	"	9.25	2.85	12.10
Ophiopogan japonicus, 1 quart (4" pot)	"	2.55	2.85	5.40
Ajuga reptans, 2-3/4" pot (carpet bugle)	"	0.46	2.85	3.31
Pachysandra terminalis, 2-3/4" pot (Japanese spurge)	"	0.64	2.85	3.49
02950.30 SHRUBS				
Juniperus conferia litoralis, 18"-24" (Shore Juniper)	EA.	21.40	11.45	32.85
Horizontalis plumosa, 18"-24" (Andorra Juniper)	"	22.65	11.45	34.10
Sabina tamar-iscfolia-tamarix juniper, 18"-24"	"	22.65	11.45	34.10
Chin San Hose, 18"-24" (San Hose Juniper)	"	22.65	11.45	34.10
Sargenti, 18"-24" (Sargent's Juniper)	"	21.40	11.45	32.85
Nandina domestica, 18"-24" (Heavenly Bamboo)	"	14.35	11.45	25.80
Raphiolepis Indica Springtime, 18"-24" Indian Hawthorn	"	15.45	11.45	26.90
Osmanthus Heterophyllus Gulftide, 18"-24" (Osmanthus)	"	16.50	11.45	27.95

02 SITEWORK

LANDSCAPING	UNIT	MAT.	INST.	TOTAL
02950.30 SHRUBS				
Ilex Cornuta Burfordi Nana, 18"-24" (Dwarf Burford Holly)	EA.	18.80	11.45	30.25
Glabra, 18"-24" (Inkberry Holly)	"	17.70	11.45	29.15
Azalea, Indica types, 18"-24"	"	19.95	11.45	31.40
Kurume types, 18"-24"	"	22.30	11.45	33.75
Berberis Julianae, 18"-24" (Wintergreen Barberry)	"	13.05	11.45	24.50
Pieris Japonica Japanese, 18"-24" (Japanese Pieris)	"	13.05	11.45	24.50
Ilex Cornuta Rotunda, 18"-24" (Dwarf Chinese Holly)	"	15.50	11.45	26.95
Juniperus Horizontalis Plumosa, 24"-30" (Andorra Juniper)	"	14.30	14.30	28.60
Rhodopendrow Hybrids, 24"-30"	"	38.00	14.30	52.30
Aucuba Japonica Varigata, 24"-30" (Gold Dust Aucuba)	"	13.00	14.30	27.30
Ilex Crenata Willow Leaf, 24"-30" (Japanese Holly)	"	14.30	14.30	28.60
Cleyera Japonica, 30"-36" (Japanese Cleyera)	"	16.70	17.85	34.55
Pittosporum Tobira, 30"-36"	"	19.05	17.85	36.90
Prumus Laurocerasus, 30"-36"	"	35.80	17.85	53.65
Ilex Cornuta Burfordi, 30"-36" (Burford Holly)	"	19.05	17.85	36.90
Abelia Grandiflora, 24"-36" (Yew Podocarpus)	"	13.05	14.30	27.35
Podocarpos Macrophylla, 24"-36" (Yew Podocarpus)	"	21.40	14.30	35.70
Pyracantha Coccinea Lalandi, 3'-4' (Firethorn)	"	12.20	17.85	30.05
Photinia Frazieri, 3'-4' (Red Photinia)	"	19.35	17.85	37.20
Forsythia Suspensa, 3'-4' (Weeping Forsythia)	"	12.20	17.85	30.05
Camellia Japonica, 3'-4' (Common Camellia)	"	21.50	17.85	39.35
Juniperus Chin Torulosa, 3'-4' (Hollywood Juniper)	"	22.85	17.85	40.70
Cupressocyparis Leylandi, 3'-4'	"	19.20	17.85	37.05
Ilex Opaca Fosteri, 5'-6' (Foster's Holly)	"	77.50	23.85	101.35
Opaca, 5'-6' (American Holly)	"	110.00	23.85	133.85
Nyrica Cerifera, 4'-5' (Southern Wax Myrtles)	"	24.25	20.45	44.70
Ligustrum Japonicum, 4'-5' (Japanese Privet)	"	19.05	20.45	39.50
02950.60 TREES				
Cornus Florida, 5'-6' (White flowering Dogwood)	EA.	53.50	23.85	77.35
Prunus Serrulata Kwanzan, 6'-8' (Kwanzan Cherry)	"	59.50	28.60	88.10
Caroliniana, 6'-8' (Carolina Cherry Laurel)	"	70.50	28.60	99.10
Cercis Canadensis, 6'-8' (Eastern Redbud)	"	48.50	28.60	77.10
Koelreuteria Paniculata, 8'-10' (Goldenrain tree)	"	83.00	35.80	118.80
Acer Platanoides, 1-3/4"-2" (11'-13') (Norway Maple)	"	110.00	47.70	157.70
Rubrum, 1-3/4"-2" (11'-13') (Red Maple)	"	83.00	47.70	130.70
Saccharum, 1-3/4"-2" (Sugar Maple)	"	150.00	47.70	197.70
Fraxinus Pennsylvanica, 1-3/4"-2" Laneolata-Green Ash	"	71.50	47.70	119.20
Celtis Occidentalis, 1-3/4"-2" (American Hackberry)	"	110.00	47.70	157.70
Glenditsia Triacantos Inermis, 2"	"	98.00	47.70	145.70
Prunus Cerasifera 'Thundercloud', 6'-8'	"	56.50	28.60	85.10
Yeodensis, 6'-8' (Yoshino Cherry)	"	60.00	28.60	88.60
Lagerstroemia Indica, 8'-10' (Crapemyrtle)	"	95.50	35.80	131.30
Crataegus Phaenopyrum, 8'-10' Washington Hawthorn	"	150.00	35.80	185.80
Quercus Borealis, 1-3/4"-2" (Northern Red Oak)	"	89.00	47.70	136.70
Quercus Acutissima, 1-3/4"-2" (8'-10') (Sawtooth Oak)	"	83.00	47.70	130.70
Saliz Babylonica, 1-3/4"-2" (Weeping Willow)	"	41.60	47.70	89.30
Tilia Cordata Greenspire, 1-3/4"-2" (10'-12')	"	180.00	47.70	227.70
Malus, 2"-2-1/2" (8'-10') (Flowering Crabapple)	"	89.00	47.70	136.70
Platanus Occidentalis, (12'-14')	"	140.00	57.00	197.00
Pyrus Calleryana Bradford, 2"-2-1/2" (Bradford Pear)	"	110.00	47.70	157.70

LANDSCAPING

LANDSCAPING	UNIT	MAT.	INST.	TOTAL
02950.60 TREES				
Quercus Palustris, 2"-2-1/2" (12'-14') (Pin Oak)	EA.	120.00	47.70	167.70
Phellos, 2-1/2"-3" (Willow Oak)	"	130.00	57.00	187.00
Nigra, 2"-2-1/2" (Water Oak)	"	110.00	47.70	157.70
Magnolia Soulangeana, 4'-5' (Saucer Magnolia)	"	64.50	23.85	88.35
Grandiflora, 6'-8' (Southern Magnolia)	"	89.00	28.60	117.60
Cedrus Deodara, 10'-12' (Deodare Cedar)	"	150.00	47.70	197.70
Ginkgo Biloba, 10'-12' (2"-2-1/2") (Maidenhair Tree)	"	140.00	47.70	187.70
Pinus Thunbergi, 5'-6' (Japanese Black Pine)	"	54.50	23.85	78.35
Strobus, 6'-8' (White Pine)	"	60.00	28.60	88.60
Taeda, 6'-8' (Loblolly Pine)	"	51.00	28.60	79.60
Quercus Virginiana, 2"-2-1/2" (live oak)	"	130.00	57.00	187.00
02970.10 FERTILIZING				
Fertilizing (23#/1000 sf)				
By square yard	S.Y.	0.02	0.09	0.11
By acre	ACRE	93.00	460.00	553.00
Liming (70#/1000 sf)				
By square yard	S.Y.	0.02	0.12	0.14
By acre	ACRE	93.00	610.00	703.00
02980.10 LANDSCAPE ACCESSORIES				
Steel edging, 3/16" x 4"	L.F.	0.39	0.36	0.75
Landscaping stepping stones, 15"x15", white	EA.	3.70	1.45	5.15
Wood chip mulch	C.Y.	31.20	19.05	50.25
2" thick	S.Y.	1.90	0.57	2.47
4" thick	"	3.60	0.82	4.42
6" thick	"	5.40	1.05	6.45
Gravel mulch, 3/4" stone	C.Y.	24.00	28.60	52.60
White marble chips, 1" deep	S.F.	0.45	0.29	0.74
Peat moss				
2" thick	S.Y.	2.25	0.64	2.89
4" thick	"	4.30	0.95	5.25
6" thick	"	6.60	1.20	7.80
Landscaping timbers, treated lumber				
4" x 4"	L.F.	0.97	0.95	1.92
6" x 6"	"	1.90	1.00	2.90
8" x 8"	"	3.10	1.20	4.30

FORMWORK

	UNIT	MAT.	INST.	TOTAL
03110.05 BEAM FORMWORK				
Beam forms, job built				
Beam bottoms				
1 use	S.F.	3.45	6.20	9.65
2 uses	"	2.00	5.80	7.80
3 uses	"	1.55	5.60	7.15
4 uses	"	1.30	5.35	6.65
5 uses	"	1.15	5.20	6.35
Beam sides				
1 use	S.F.	2.45	4.05	6.50
2 uses	"	1.45	3.85	5.30
3 uses	"	1.30	3.65	4.95
4 uses	"	1.15	3.45	4.60
5 uses	"	1.05	3.30	4.35
03110.10 BOX CULVERT FORMWORK				
Box culverts, job built				
6' x 6'				
1 use	S.F.	2.35	3.65	6.00
2 uses	"	1.30	3.45	4.75
3 uses	"	1.05	3.30	4.35
4 uses	"	0.89	3.15	4.04
5 uses	"	0.79	3.05	3.84
8' x 12'				
1 use	S.F.	2.35	3.05	5.40
2 uses	"	1.30	2.90	4.20
3 uses	"	1.05	2.80	3.85
4 uses	"	0.89	2.70	3.59
5 uses	"	0.79	2.60	3.39
03110.15 COLUMN FORMWORK				
Column, square forms, job built				
8" x 8" columns				
1 use	S.F.	2.70	7.30	10.00
2 uses	"	1.45	7.00	8.45
3 uses	"	1.25	6.75	8.00
4 uses	"	1.10	6.50	7.60
5 uses	"	0.96	6.30	7.26
12" x 12" columns				
1 use	S.F.	2.45	6.65	9.10
2 uses	"	1.35	6.40	7.75
3 uses	"	1.10	6.20	7.30
4 uses	"	0.96	6.00	6.96
5 uses	"	0.80	5.80	6.60
16" x 16" columns				
1 use	S.F.	2.35	6.10	8.45
2 uses	"	1.25	5.90	7.15
3 uses	"	0.99	5.70	6.69
4 uses	"	0.90	5.55	6.45
5 uses	"	0.74	5.35	6.09
24" x 24" columns				

03 CONCRETE

FORMWORK	UNIT	MAT.	INST.	TOTAL
03110.15 COLUMN FORMWORK				
1 use	S.F.	2.35	5.60	7.95
2 uses	"	1.10	5.45	6.55
3 uses	"	0.89	5.30	6.19
4 uses	"	0.74	5.15	5.89
5 uses	"	0.68	5.00	5.68
36" x 36" columns				
1 use	S.F.	2.35	5.20	7.55
2 uses	"	1.10	5.05	6.15
3 uses	"	0.92	4.95	5.87
4 uses	"	0.79	4.80	5.59
5 uses	"	0.74	4.70	5.44
Round fiber forms, 1 use				
10" dia.	L.F.	3.40	7.30	10.70
12" dia.	"	4.20	7.45	11.65
14" dia.	"	5.55	7.75	13.30
16" dia.	"	7.25	8.10	15.35
18" dia.	"	11.90	8.70	20.60
24" dia.	"	14.50	9.35	23.85
30" dia.	"	21.75	10.15	31.90
36" dia.	"	27.10	11.05	38.15
42" dia.	"	49.50	12.15	61.65
03110.18 CURB FORMWORK				
Curb forms				
Straight, 6" high				
1 use	L.F.	1.55	3.65	5.20
2 uses	"	0.95	3.45	4.40
3 uses	"	0.70	3.30	4.00
4 uses	"	0.64	3.15	3.79
5 uses	"	0.57	3.05	3.62
Curved, 6" high				
1 use	L.F.	1.70	4.55	6.25
2 uses	"	1.05	4.30	5.35
3 uses	"	0.81	4.05	4.86
4 uses	"	0.75	3.90	4.65
5 uses	"	0.70	3.70	4.40
03110.20 ELEVATED SLAB FORMWORK				
Elevated slab formwork				
Slab, with drop panels				
1 use	S.F.	2.65	2.90	5.55
2 uses	"	1.55	2.80	4.35
3 uses	"	1.20	2.70	3.90
4 uses	"	1.05	2.60	3.65
5 uses	"	0.95	2.50	3.45
Floor slab, hung from steel beams				
1 use	S.F.	2.15	2.80	4.95
2 uses	"	1.15	2.70	3.85
3 uses	"	1.05	2.60	3.65
4 uses	"	0.92	2.50	3.42

03 CONCRETE

FORMWORK	UNIT	MAT.	INST.	TOTAL
03110.20 ELEVATED SLAB FORMWORK				
5 uses	S.F.	0.79	2.45	3.24
Floor slab, with pans or domes				
1 use	S.F.	3.80	3.30	7.10
2 uses	"	2.55	3.15	5.70
3 uses	"	2.30	3.05	5.35
4 uses	"	2.15	2.90	5.05
5 uses	"	1.90	2.80	4.70
Equipment curbs, 12" high				
1 use	L.F.	2.00	3.65	5.65
2 uses	"	1.30	3.45	4.75
3 uses	"	1.10	3.30	4.40
4 uses	"	1.00	3.15	4.15
5 uses	"	0.85	3.05	3.90
03110.25 EQUIPMENT PAD FORMWORK				
Equipment pad, job built				
1 use	S.F.	2.60	4.55	7.15
2 uses	"	1.55	4.30	5.85
3 uses	"	1.25	4.05	5.30
4 uses	"	0.95	3.85	4.80
5 uses	"	0.79	3.65	4.44
03110.35 FOOTING FORMWORK				
Wall footings, job built, continuous				
1 use	S.F.	1.25	3.65	4.90
2 uses	"	0.86	3.45	4.31
3 uses	"	0.72	3.30	4.02
4 uses	"	0.63	3.15	3.78
5 uses	"	0.55	3.05	3.60
Column footings, spread				
1 use	S.F.	1.30	4.55	5.85
2 uses	"	0.95	4.30	5.25
3 uses	"	0.68	4.05	4.73
4 uses	"	0.57	3.85	4.42
5 uses	"	0.52	3.65	4.17
03110.50 GRADE BEAM FORMWORK				
Grade beams, job built				
1 use	S.F.	2.00	3.65	5.65
2 uses	"	1.15	3.45	4.60
3 uses	"	0.89	3.30	4.19
4 uses	"	0.74	3.15	3.89
5 uses	"	0.62	3.05	3.67
03110.53 PILE CAP FORMWORK				
Pile cap forms, job built				
Square				
1 use	S.F.	2.20	4.55	6.75
2 uses	"	1.30	4.30	5.60
3 uses	"	1.00	4.05	5.05

FORMWORK	UNIT	MAT.	INST.	TOTAL
03110.53 PILE CAP FORMWORK				
4 uses	S.F.	0.89	3.85	4.74
5 uses	"	0.74	3.65	4.39
Triangular				
1 use	S.F.	2.35	5.20	7.55
2 uses	"	1.55	4.85	6.40
3 uses	"	1.25	4.55	5.80
4 uses	"	1.00	4.30	5.30
5 uses	"	0.80	4.05	4.85
03110.55 SLAB/MAT FORMWORK				
Mat foundations, job built				
1 use	S.F.	1.90	4.55	6.45
2 uses	"	1.10	4.30	5.40
3 uses	"	0.80	4.05	4.85
4 uses	"	0.68	3.85	4.53
5 uses	"	0.54	3.65	4.19
Edge forms				
6" high				
1 use	L.F.	1.90	3.30	5.20
2 uses	"	1.10	3.15	4.25
3 uses	"	0.80	3.05	3.85
4 uses	"	0.68	2.90	3.58
5 uses	"	0.54	2.80	3.34
12" high				
1 use	L.F.	1.75	3.65	5.40
2 uses	"	1.00	3.45	4.45
3 uses	"	0.75	3.30	4.05
4 uses	"	0.62	3.15	3.77
5 uses	"	0.51	3.05	3.56
Formwork for openings				
1 use	S.F.	2.55	7.30	9.85
2 uses	"	1.45	6.65	8.10
3 uses	"	1.25	6.10	7.35
4 uses	"	0.95	5.60	6.55
5 uses	"	0.80	5.20	6.00
03110.60 STAIR FORMWORK				
Stairway forms, job built				
1 use	S.F.	3.40	7.30	10.70
2 uses	"	1.90	6.65	8.55
3 uses	"	1.45	6.10	7.55
4 uses	"	1.35	5.60	6.95
5 uses	"	1.15	5.20	6.35
Stairs, elevated				
1 use	S.F.	4.05	7.30	11.35
2 uses	"	2.15	6.10	8.25
3 uses	"	1.90	5.20	7.10
4 uses	"	1.65	4.85	6.50
5 uses	"	1.35	4.55	5.90

FORMWORK	UNIT	MAT.	INST.	TOTAL
03110.65 WALL FORMWORK				
Wall forms, exterior, job built				
Up to 8' high wall				
1 use	S.F.	2.25	3.65	5.90
2 uses	"	1.25	3.45	4.70
3 uses	"	1.10	3.30	4.40
4 uses	"	0.94	3.15	4.09
5 uses	"	0.81	3.05	3.86
Over 8' high wall				
1 use	S.F.	2.45	4.55	7.00
2 uses	"	1.40	4.30	5.70
3 uses	"	1.30	4.05	5.35
4 uses	"	1.15	3.85	5.00
5 uses	"	1.00	3.65	4.65
Over 16' high wall				
1 use	S.F.	2.55	5.20	7.75
2 uses	"	1.55	4.85	6.40
3 uses	"	1.40	4.55	5.95
4 uses	"	1.30	4.30	5.60
5 uses	"	1.15	4.05	5.20
Radial wall forms				
1 use	S.F.	2.40	5.60	8.00
2 uses	"	1.45	5.20	6.65
3 uses	"	1.35	4.85	6.20
4 uses	"	1.20	4.55	5.75
5 uses	"	1.10	4.30	5.40
Retaining wall forms				
1 use	S.F.	2.10	4.05	6.15
2 uses	"	1.10	3.85	4.95
3 uses	"	0.96	3.65	4.61
4 uses	"	0.83	3.45	4.28
5 uses	"	0.70	3.30	4.00
Radial retaining wall forms				
1 use	S.F.	2.30	6.10	8.40
2 uses	"	1.40	5.60	7.00
3 uses	"	1.20	5.20	6.40
4 uses	"	1.15	4.85	6.00
5 uses	"	0.99	4.55	5.54
Column pier and pilaster				
1 use	S.F.	2.45	7.30	9.75
2 uses	"	1.45	6.65	8.10
3 uses	"	1.35	6.10	7.45
4 uses	"	1.25	5.60	6.85
5 uses	"	1.10	5.20	6.30
Interior wall forms				
Up to 8' high				
1 use	S.F.	2.25	3.30	5.55
2 uses	"	1.25	3.15	4.40
3 uses	"	1.10	3.05	4.15
4 uses	"	0.95	2.90	3.85
5 uses	"	0.80	2.80	3.60
Over 8' high				
1 use	S.F.	2.45	4.05	6.50

FORMWORK	UNIT	MAT.	INST.	TOTAL
03110.65 WALL FORMWORK				
2 uses	S.F.	1.40	3.85	5.25
3 uses	"	1.30	3.65	4.95
4 uses	"	1.15	3.45	4.60
5 uses	"	1.00	3.30	4.30
Over 16' high				
1 use	S.F.	2.55	4.55	7.10
2 uses	"	1.55	4.30	5.85
3 uses	"	1.40	4.05	5.45
4 uses	"	1.30	3.85	5.15
5 uses	"	1.15	3.65	4.80
Radial wall forms				
1 use	S.F.	2.40	4.85	7.25
2 uses	"	1.45	4.55	6.00
3 uses	"	1.35	4.30	5.65
4 uses	"	1.20	4.05	5.25
5 uses	"	1.10	3.85	4.95
Curved wall forms, 24" sections				
1 use	S.F.	2.40	7.30	9.70
2 uses	"	1.45	6.65	8.10
3 uses	"	1.35	6.10	7.45
4 uses	"	1.20	5.60	6.80
5 uses	"	1.10	5.20	6.30
PVC form liner, per side, smooth finish				
1 use	S.F.	3.85	3.05	6.90
2 uses	"	2.10	2.90	5.00
3 uses	"	1.75	2.80	4.55
4 uses	"	1.35	2.60	3.95
5 uses	"	1.10	2.45	3.55
03110.90 MISCELLANEOUS FORMWORK				
Keyway forms (5 uses)				
2 x 4	L.F.	0.14	1.80	1.94
2 x 6	"	0.22	2.05	2.27
Bulkheads				
Walls, with keyways				
2 piece	L.F.	2.45	3.30	5.75
3 piece	"	3.05	3.65	6.70
Elevated slab, with keyway				
2 piece	L.F.	2.80	3.05	5.85
3 piece	"	4.05	3.30	7.35
Ground slab, with keyway				
2 piece	L.F.	3.05	2.60	5.65
3 piece	"	4.05	2.80	6.85
Chamfer strips				
Wood				
1/2" wide	L.F.	0.19	0.81	1.00
3/4" wide	"	0.24	0.81	1.05
1" wide	"	0.31	0.81	1.12
PVC				
1/2" wide	L.F.	0.56	0.81	1.37
3/4" wide	"	0.62	0.81	1.43
1" wide	"	0.79	0.81	1.60

FORMWORK

03110.90 MISCELLANEOUS FORMWORK	UNIT	MAT.	INST.	TOTAL
Radius				
1"	L.F.	0.85	0.87	1.72
1-1/2"	"	1.50	0.87	2.37
Reglets				
Galvanized steel, 24 ga.	L.F.	0.90	1.45	2.35
Metal formwork				
Straight edge forms				
4" high	L.F.	0.12	2.30	2.42
6" high	"	0.13	2.45	2.58
8" high	"	0.15	2.60	2.75
12" high	"	0.20	2.80	3.00
16" high	"	0.24	3.05	3.29
Curb form, S-shape				
12" x				
1'-6"	L.F.	0.24	5.20	5.44
2'	"	0.30	4.85	5.15
2'-6"	"	0.33	4.55	4.88
3'	"	0.37	4.05	4.42

REINFORCEMENT

03210.05 BEAM REINFORCING	UNIT	MAT.	INST.	TOTAL
Beam-girders				
#3 - #4	TON	730.00	1,010	1,740
#5 - #6	"	640.00	810.00	1,450
#7 - #8	"	610.00	680.00	1,290
#9 - #10	"	610.00	580.00	1,190
#11 - #12	"	610.00	540.00	1,150
#13 - #14	"	610.00	510.00	1,120
Galvanized				
#3 - #4	TON	1,210	1,010	2,220
#5 - #6	"	1,140	810.00	1,950
#7 - #8	"	1,100	680.00	1,780
#9 - #10	"	1,100	580.00	1,680
#11 - #12	"	1,100	540.00	1,640
#13 - #14	"	1,100	510.00	1,610
Epoxy coated				
#3 - #4	TON	1,080	1,160	2,240
#5 - #6	"	1,010	900.00	1,910
#7 - #8	"	980.00	740.00	1,720
#9 - #10	"	980.00	620.00	1,600
#11 - #12	"	980.00	580.00	1,560
#13 - #14	"	980.00	540.00	1,520
Bond Beams				

03 CONCRETE

REINFORCEMENT

REINFORCEMENT	UNIT	MAT.	INST.	TOTAL
03210.05 BEAM REINFORCING				
#3 - #4	TON	730.00	1,350	2,080
#5 - #6	"	640.00	1,010	1,650
#7 - #8	"	610.00	900.00	1,510
Galvanized				
#3 - #4	TON	1,210	1,350	2,560
#5 - #6	"	1,140	1,010	2,150
#7 - #8	"	1,100	900.00	2,000
Epoxy coated				
#3 - #4	TON	1,080	1,620	2,700
#5 - #6	"	1,010	1,160	2,170
#7 - #8	"	980.00	1,010	1,990
03210.10 BOX CULVERT REINFORCING				
Box culverts				
#3 - #4	TON	730.00	510.00	1,240
#5 - #6	"	640.00	450.00	1,090
#7 - #8	"	610.00	410.00	1,020
#9 - #10	"	610.00	370.00	980.00
#11 - #12	"	610.00	340.00	950.00
Galvanized				
#3 - #4	TON	1,210	510.00	1,720
#5 - #6	"	1,140	450.00	1,590
#7 - #8	"	1,100	410.00	1,510
#9 - #10	"	1,100	370.00	1,470
#11 - #12	"	1,100	340.00	1,440
Epoxy coated				
#3 - #4	TON	1,080	540.00	1,620
#5 - #6	"	1,010	480.00	1,490
#7 - #8	"	980.00	430.00	1,410
#9 - #10	"	980.00	390.00	1,370
#11 - #12	"	980.00	350.00	1,330
03210.15 COLUMN REINFORCING				
Columns				
#3 - #4	TON	730.00	1,160	1,890
#5 - #6	"	640.00	900.00	1,540
#7 - #8	"	610.00	810.00	1,420
#9 - #10	"	610.00	740.00	1,350
#11 - #12	"	610.00	680.00	1,290
#13 - #14	"	610.00	620.00	1,230
#15 - #16	"	610.00	580.00	1,190
Galvanized				
#3 - #4	TON	1,210	1,160	2,370
#5 - #6	"	1,140	900.00	2,040
#7 - #8	"	1,100	810.00	1,910
#9 - #10	"	1,100	740.00	1,840
#11 - #12	"	1,100	680.00	1,780
#13 - #14	"	1,100	620.00	1,720
#15 - #16	"	1,100	580.00	1,680
Epoxy coated				
#3 - #4	TON	1,080	1,350	2,430

REINFORCEMENT	UNIT	MAT.	INST.	TOTAL
03210.15 COLUMN REINFORCING				
#5 - #6	TON	1,010	1,010	2,020
#7 - #8	"	980.00	900.00	1,880
#9 - #10	"	980.00	810.00	1,790
#11 - #12	"	980.00	740.00	1,720
#13 - #14	"	980.00	680.00	1,660
#15 - #16	"	980.00	620.00	1,600
Spirals				
8" to 24" dia.	TON	1,120	1,010	2,130
24" to 48" dia.	"	1,120	900.00	2,020
48" to 84" dia.	"	1,230	810.00	2,040
03210.20 ELEVATED SLAB REINFORCING				
Elevated slab				
#3 - #4	TON	730.00	510.00	1,240
#5 - #6	"	640.00	450.00	1,090
#7 - #8	"	610.00	410.00	1,020
#9 - #10	"	610.00	370.00	980.00
#11 - #12	"	610.00	340.00	950.00
Galvanized				
#3 - #4	TON	1,210	510.00	1,720
#5 - #6	"	1,140	450.00	1,590
#7 - #8	"	1,100	410.00	1,510
#9 - #10	"	1,100	370.00	1,470
#11 - #12	"	1,100	340.00	1,440
Epoxy coated				
#3 - #4	TON	1,080	540.00	1,620
#5 - #6	"	1,010	480.00	1,490
#7 - #8	"	980.00	430.00	1,410
#9 - #10	"	980.00	390.00	1,370
#11 - #12	"	980.00	350.00	1,330
03210.25 EQUIP. PAD REINFORCING				
Equipment pad				
#3 - #4	TON	730.00	810.00	1,540
#5 - #6	"	640.00	740.00	1,380
#7 - #8	"	610.00	680.00	1,290
#9 - #10	"	610.00	620.00	1,230
#11 - #12	"	610.00	580.00	1,190
03210.35 FOOTING REINFORCING				
Footings				
Grade 50				
#3 - #4	TON	730.00	680.00	1,410
#5 - #6	"	640.00	580.00	1,220
#7 - #8	"	610.00	510.00	1,120
#9 - #10	"	610.00	450.00	1,060
Grade 60				
#3 - #4	TON	730.00	680.00	1,410
#5 - #6	"	640.00	580.00	1,220
#7 - #8	"	610.00	510.00	1,120
#9 - #10	"	610.00	450.00	1,060
Grade 70				

REINFORCEMENT	UNIT	MAT.	INST.	TOTAL
03210.35 FOOTING REINFORCING				
#3 - #4	TON	730.00	680.00	1,410
#5 - #6	"	640.00	580.00	1,220
#7 - #8	"	610.00	510.00	1,120
#9 - #10	"	610.00	450.00	1,060
#11- #12	"	610.00	410.00	1,020
Straight dowels, 24" long				
1" dia. (#8)	EA.	2.25	4.05	6.30
3/4" dia. (#6)	"	2.05	4.05	6.10
5/8" dia. (#5)	"	1.75	3.40	5.15
1/2" dia. (#4)	"	1.30	2.90	4.20
03210.45 FOUNDATION REINFORCING				
Foundations				
#3 - #4	TON	730.00	680.00	1,410
#5 - #6	"	640.00	580.00	1,220
#7 - #8	"	610.00	510.00	1,120
#9 - #10	"	610.00	450.00	1,060
#11 - #12	"	610.00	410.00	1,020
Galvanized				
#3 - #4	TON	1,210	680.00	1,890
#5 - #6	"	1,140	580.00	1,720
#7 - #8	"	1,100	510.00	1,610
#9 - #10	"	1,100	450.00	1,550
#11 - #12	"	1,100	410.00	1,510
Epoxy Coated				
#3 - #4	TON	1,080	740.00	1,820
#5 - #6	"	1,010	620.00	1,630
#7 - #8	"	980.00	540.00	1,520
#9 - #10	"	980.00	480.00	1,460
#11 - #12	"	980.00	430.00	1,410
03210.50 GRADE BEAM REINFORCING				
Grade beams				
#3 - #4	TON	730.00	620.00	1,350
#5 - #6	"	640.00	540.00	1,180
#7 - #8	"	610.00	480.00	1,090
#9 - #10	"	610.00	430.00	1,040
#11 - #12	"	610.00	390.00	1,000
Galvanized				
#3 - #4	TON	1,210	620.00	1,830
#5 - #6	"	1,140	540.00	1,680
#7 - #8	"	1,100	480.00	1,580
#9 - #10	"	1,100	430.00	1,530
#11 - #12	"	1,100	390.00	1,490
Epoxy coated				
#3 - #4	TON	1,080	680.00	1,760
#5 - #6	"	1,010	580.00	1,590
#7 - #8	"	980.00	510.00	1,490
#9 - #10	"	980.00	450.00	1,430
#11 - #12	"	980.00	410.00	1,390

REINFORCEMENT

REINFORCEMENT	UNIT	MAT.	INST.	TOTAL
03210.53 PILE CAP REINFORCING				
Pile caps				
#3 - #4	TON	730.00	1,010	1,740
#5 - #6	"	640.00	900.00	1,540
#7 - #8	"	610.00	810.00	1,420
#9 - #10	"	610.00	740.00	1,350
#11 - #12	"	610.00	680.00	1,290
Galvanized				
#3 - #4	TON	1,210	1,010	2,220
#5 - #6	"	1,140	900.00	2,040
#7 - #8	"	1,100	810.00	1,910
#9 - #10	"	1,100	740.00	1,840
#11 - #12	"	1,100	680.00	1,780
Epoxy coated				
#3 - #4	TON	1,080	1,160	2,240
#5 - #6	"	1,010	1,010	2,020
#7 - #8	"	980.00	900.00	1,880
#9 - #10	"	980.00	810.00	1,790
#11 - #12	"	980.00	740.00	1,720
03210.55 SLAB/MAT REINFORCING				
Bars, slabs				
#3 - #4	TON	730.00	680.00	1,410
#5 - #6	"	640.00	580.00	1,220
#7 - #8	"	610.00	510.00	1,120
#9 - #10	"	610.00	450.00	1,060
#11 - #12	"	610.00	410.00	1,020
Galvanized				
#3 - #4	TON	1,210	680.00	1,890
#5 - #6	"	1,140	580.00	1,720
#7 - #8	"	1,100	510.00	1,610
#9 - #10	"	1,100	450.00	1,550
#11 - #12	"	1,100	410.00	1,510
Epoxy coated				
#3 - #4	TON	1,080	740.00	1,820
#5 - #6	"	1,010	620.00	1,630
#7 - #8	"	980.00	540.00	1,520
#9 - #10	"	980.00	480.00	1,460
#11 - #12	"	980.00	430.00	1,410
Wire mesh, slabs				
Galvanized				
4x4				
W1.4xW1.4	S.F.	0.17	0.27	0.44
W2.0xW2.0	"	0.22	0.29	0.51
W2.9xW2.9	"	0.32	0.31	0.63
W4.0xW4.0	"	0.47	0.34	0.81
6x6				
W1.4xW1.4	S.F.	0.15	0.20	0.35
W2.0xW2.0	"	0.22	0.23	0.45
W2.9xW2.9	"	0.31	0.24	0.55
W4.0xW4.0	"	0.33	0.27	0.60
Standard				
2x2				

REINFORCEMENT

REINFORCEMENT	UNIT	MAT.	INST.	TOTAL
03210.55 SLAB/MAT REINFORCING				
W.9xW.9	S.F.	0.18	0.27	0.45
4x4				
W1.4xW1.4	S.F.	0.12	0.27	0.39
W2.0xW2.0	"	0.15	0.29	0.44
W2.9xW2.9	"	0.21	0.31	0.52
W4.0xW4.0	"	0.33	0.34	0.67
6x6				
W1.4xW1.4	S.F.	0.08	0.20	0.28
W2.0xW2.0	"	0.11	0.23	0.34
W2.9xW2.9	"	0.15	0.24	0.39
W4.0xW4.0	"	0.22	0.27	0.49
03210.60 STAIR REINFORCING				
Stairs				
#3 - #4	TON	730.00	810.00	1,540
#5 - #6	"	640.00	680.00	1,320
#7 - #8	"	610.00	580.00	1,190
#9 - #10	"	610.00	510.00	1,120
Galvanized				
#3 - #4	TON	1,210	810.00	2,020
#5 - #6	"	1,140	680.00	1,820
#7 - #8	"	1,100	580.00	1,680
#9 - #10	"	1,100	510.00	1,610
Epoxy coated				
#3 - #4	TON	1,080	900.00	1,980
#5 - #6	"	1,010	740.00	1,750
#7 - #8	"	980.00	620.00	1,600
#9 - #10	"	980.00	540.00	1,520
03210.65 WALL REINFORCING				
Walls				
#3 - #4	TON	730.00	580.00	1,310
#5 - #6	"	640.00	510.00	1,150
#7 - #8	"	610.00	450.00	1,060
#9 - #10	"	610.00	410.00	1,020
Galvanized				
#3 - #4	TON	1,210	580.00	1,790
#5 - #6	"	1,140	510.00	1,650
#7 - #8	"	1,100	450.00	1,550
#9 - #10	"	1,100	410.00	1,510
Epoxy coated				
#3 - #4	TON	1,080	620.00	1,700
#5 - #6	"	1,010	540.00	1,550
#7 - #8	"	980.00	480.00	1,460
#9 - #10	"	980.00	430.00	1,410
Masonry wall (horizontal)				
#3 - #4	TON	730.00	1,620	2,350
#5 - #6	"	640.00	1,350	1,990
Galvanized				
#3 - #4	TON	1,210	1,620	2,830

03 CONCRETE

REINFORCEMENT

03210.65 WALL REINFORCING	UNIT	MAT.	INST.	TOTAL
#5 - #6	TON	1,140	1,350	2,490
Masonry wall (vertical)				
#3 - #4	TON	730.00	2,030	2,760
#5 - #6	"	640.00	1,620	2,260
Galvanized				
#3 - #4	TON	1,210	2,030	3,240
#5 - #6	"	1,140	1,620	2,760

ACCESSORIES

03250.40 CONCRETE ACCESSORIES	UNIT	MAT.	INST.	TOTAL
Expansion joint, poured				
Asphalt				
1/2" x 1"	L.F.	0.39	0.57	0.96
1" x 2"	"	1.15	0.62	1.77
Liquid neoprene, cold applied				
1/2" x 1"	L.F.	1.40	0.58	1.98
1" x 2"	"	5.55	0.64	6.19
Polyurethane, 2 parts				
1/2" x 1"	L.F.	1.30	0.95	2.25
1" x 2"	"	5.35	1.05	6.40
Rubberized asphalt, cold				
1/2" x 1"	L.F.	0.32	0.57	0.89
1" x 2"	"	1.00	0.62	1.62
Hot, fuel resistant				
1/2" x 1"	L.F.	0.59	0.57	1.16
1" x 2"	"	2.95	0.62	3.57
Expansion joint, premolded, in slabs				
Asphalt				
1/2" x 6"	L.F.	0.43	0.71	1.15
1" x 12"	"	0.72	0.95	1.67
Cork				
1/2" x 6"	L.F.	0.95	0.71	1.67
1" x 12"	"	3.80	0.95	4.75
Neoprene sponge				
1/2" x 6"	L.F.	1.30	0.71	2.02
1" x 12"	"	4.75	0.95	5.70
Polyethylene foam				
1/2" x 6"	L.F.	0.48	0.71	1.20
1" x 12"	"	2.35	0.95	3.30
Polyurethane foam				
1/2" x 6"	L.F.	0.64	0.71	1.36
1" x 12"	"	1.45	0.95	2.40

03 CONCRETE

ACCESSORIES

03250.40 CONCRETE ACCESSORIES

ACCESSORIES	UNIT	MAT.	INST.	TOTAL
Polyvinyl chloride foam				
1/2" x 6"	L.F.	1.40	0.71	2.12
1" x 12"	"	5.05	0.95	6.00
Rubber, gray sponge				
1/2" x 6"	L.F.	2.25	0.71	2.96
1" x 12"	"	9.55	0.95	10.50
Asphalt felt control joints or bond breaker, screed joints				
4" slab	L.F.	0.59	0.57	1.16
6" slab	"	0.77	0.64	1.41
8" slab	"	1.05	0.71	1.76
10" slab	"	1.45	0.82	2.27
Keyed cold expansion and control joints, 24 ga.				
4" slab	L.F.	0.54	1.80	2.34
5" slab	"	0.66	1.80	2.46
6" slab	"	0.77	1.90	2.67
8" slab	"	0.96	2.05	3.01
10" slab	"	1.10	2.20	3.30
Waterstops				
Polyvinyl chloride				
Ribbed				
3/16" thick x				
4" wide	L.F.	0.90	1.45	2.35
6" wide	"	1.10	1.60	2.70
1/2" thick x				
9" wide	L.F.	3.05	1.80	4.85
Ribbed with center bulb				
3/16" thick x 9" wide	L.F.	2.55	1.80	4.35
3/8" thick x 9" wide	"	3.15	1.80	4.95
Dumbbell type, 3/8" thick x 6" wide	"	3.15	1.60	4.75
Plain, 3/8" thick x 9" wide	"	3.85	1.80	5.65
Center bulb, 3/8" thick x 9" wide	"	5.20	1.80	7.00
Rubber				
Flat dumbbell				
3/8" thick x				
6" wide	L.F.	5.05	1.60	6.65
9" wide	"	7.95	1.80	9.75
Center bulb				
3/8" thick x				
6" wide	L.F.	4.70	1.60	6.30
9" wide	"	9.55	1.80	11.35
Vapor barrier				
4 mil polyethylene	S.F.	0.02	0.10	0.12
6 mil polyethylene	"	0.03	0.10	0.13
Gravel porous fill, under floor slabs, 3/4" stone	C.Y.	14.60	47.70	62.30
Reinforcing accessories				
Beam bolsters				
1-1/2" high, plain	L.F.	0.34	0.41	0.75
Galvanized	"	0.42	0.41	0.83
3" high				
Plain	L.F.	0.79	0.51	1.30
Galvanized	"	0.87	0.51	1.38
Slab bolsters				

ACCESSORIES

03250.40 CONCRETE ACCESSORIES	UNIT	MAT.	INST.	TOTAL
1" high				
Plain	L.F.	0.22	0.20	0.42
Galvanized	"	0.32	0.20	0.52
2" high				
Plain	L.F.	0.32	0.23	0.55
Galvanized	"	0.36	0.23	0.58
Chairs, high chairs				
3" high				
Plain	EA.	0.37	1.00	1.37
Galvanized	"	0.48	1.00	1.48
5" high				
Plain	EA.	0.48	1.05	1.53
Galvanized	"	0.59	1.05	1.64
8" high				
Plain	EA.	0.99	1.15	2.14
Galvanized	"	1.30	1.15	2.45
12" high				
Plain	EA.	2.10	1.35	3.45
Galvanized	"	2.50	1.35	3.85
Continuous, high chair				
3" high				
Plain	L.F.	0.43	0.27	0.70
Galvanized	"	0.53	0.27	0.80
5" high				
Plain	L.F.	0.65	0.29	0.94
Galvanized	"	0.87	0.29	1.16
8" high				
Plain	L.F.	0.88	0.31	1.19
Galvanized	"	1.15	0.31	1.46
12" high				
Plain	L.F.	1.90	0.34	2.24
Galvanized	"	2.25	0.34	2.59

CAST-IN-PLACE CONCRETE

03300.10 CONCRETE ADMIXTURES	UNIT	MAT.	INST.	TOTAL
Concrete admixtures				
Water reducing admixture	GAL			10.43
Set retarder	"			18.70
Air entraining agent	"			6.35

03350.10 CONCRETE FINISHES	UNIT	MAT.	INST.	TOTAL
Floor finishes				
Broom	S.F.	0.00	0.41	0.41

03 CONCRETE

CAST-IN-PLACE CONCRETE	UNIT	MAT.	INST.	TOTAL
03350.10 CONCRETE FINISHES				
Screed	S.F.	0.00	0.36	0.36
Darby	"	0.00	0.36	0.36
Steel float	"	0.00	0.48	0.48
Granolithic topping				
1/2" thick	S.F.	0.21	1.30	1.51
1" thick	"	0.39	1.45	1.84
2" thick	"	0.67	1.60	2.27
Wall finishes				
Burlap rub, with cement paste	S.F.	0.06	0.48	0.54
Float finish	"	0.08	0.71	0.80
Etch with acid	"	0.25	0.48	0.73
Sandblast				
Minimum	S.F.	0.06	0.73	0.79
Maximum	"	0.31	0.73	1.04
Bush hammer				
Green concrete	S.F.	0.00	1.45	1.45
Cured concrete	"	0.00	2.20	2.20
Break ties and patch holes	"	0.00	0.57	0.57
Carborundum				
Dry rub	S.F.	0.00	0.95	0.95
Wet rub	"	0.00	1.45	1.45
Floor hardeners				
Metallic				
Light service	S.F.	0.20	0.36	0.56
Heavy service	"	0.66	0.48	1.14
Non-metallic				
Light service	S.F.	0.10	0.36	0.46
Heavy service	"	0.46	0.48	0.94
Rusticated concrete finish				
Beveled edge	L.F.	0.23	1.60	1.83
Square edge	"	0.30	2.05	2.35
Solid board concrete finish				
Standard	S.F.	0.61	2.40	3.01
Rustic	"	0.54	2.85	3.39
03360.10 PNEUMATIC CONCRETE				
Pneumatic applied concrete (gunite)				
2" thick	S.F.	3.25	1.85	5.10
3" thick	"	4.05	2.50	6.55
4" thick	"	4.85	3.00	7.85
Finish surface				
Minimum	S.F.	0.00	1.80	1.80
Maximum	"	0.00	3.65	3.65
03370.10 CURING CONCRETE				
Sprayed membrane				
Slabs	S.F.	0.04	0.06	0.10
Walls	"	0.07	0.07	0.14
Curing paper				
Slabs	S.F.	0.07	0.07	0.14
Walls	"	0.07	0.08	0.15

CAST-IN-PLACE CONCRETE

03370.10 CURING CONCRETE	UNIT	MAT.	INST.	TOTAL
Burlap				
7.5 oz.	S.F.	0.06	0.10	0.16
12 oz.	"	0.08	0.10	0.18

PLACING CONCRETE

03380.05 BEAM CONCRETE	UNIT	MAT.	INST.	TOTAL
Beams and girders				
2500# or 3000# concrete				
By crane	C.Y.	72.50	53.50	126.00
By pump	"	72.50	48.50	121.00
By hand buggy	"	72.50	28.60	101.10
3500# or 4000# concrete				
By crane	C.Y.	77.00	53.50	130.50
By pump	"	77.00	48.50	125.50
By hand buggy	"	77.00	28.60	105.60
5000# concrete				
By crane	C.Y.	81.50	53.50	135.00
By pump	"	81.50	48.50	130.00
By hand buggy	"	81.50	28.60	110.10
Bond beam, 3000# concrete				
By pump				
8" high				
4" wide	L.F.	0.20	1.05	1.25
6" wide	"	0.46	1.20	1.66
8" wide	"	0.61	1.35	1.96
10" wide	"	0.80	1.50	2.30
12" wide	"	1.10	1.65	2.75
16" high				
8" wide	L.F.	1.50	1.65	3.15
10" wide	"	1.95	1.90	3.85
12" wide	"	2.60	2.20	4.80
By crane				
8" high				
4" wide	L.F.	0.23	1.15	1.38
6" wide	"	0.46	1.25	1.71
8" wide	"	0.61	1.35	1.96
10" wide	"	0.80	1.50	2.30
12" wide	"	1.10	1.65	2.75
16" high				
8" wide	L.F.	1.50	1.65	3.15
10" wide	"	1.95	1.80	3.75
12" wide	"	2.60	2.05	4.65

PLACING CONCRETE

PLACING CONCRETE	UNIT	MAT.	INST.	TOTAL
03380.15 COLUMN CONCRETE				
Columns				
2500# or 3000# concrete				
By crane	C.Y.	72.50	48.50	121.00
By pump	"	72.50	44.40	116.90
3500# or 4000# concrete				
By crane	C.Y.	77.00	48.50	125.50
By pump	"	77.00	44.40	121.40
5000# concrete				
By crane	C.Y.	81.50	48.50	130.00
By pump	"	81.50	44.40	125.90
03380.20 ELEVATED SLAB CONCRETE				
Elevated slab				
2500# or 3000# concrete				
By crane	C.Y.	72.50	26.70	99.20
By pump	"	72.50	20.50	93.00
By hand buggy	"	72.50	28.60	101.10
3500# or 4000# concrete				
By crane	C.Y.	77.00	26.70	103.70
By pump	"	77.00	20.50	97.50
By hand buggy	"	77.00	28.60	105.60
5000# concrete				
By crane	C.Y.	81.50	26.70	108.20
By pump	"	81.50	20.50	102.00
By hand buggy	"	81.50	28.60	110.10
Topping				
2500# or 3000# concrete				
By crane	C.Y.	72.50	26.70	99.20
By pump	"	72.50	20.50	93.00
By hand buggy	"	72.50	28.60	101.10
3500# or 4000# concrete				
By crane	C.Y.	77.00	26.70	103.70
By pump	"	77.00	20.50	97.50
By hand buggy	"	77.00	28.60	105.60
5000# concrete				
By crane	C.Y.	81.50	26.70	108.20
By pump	"	81.50	20.50	102.00
By hand buggy	"	81.50	28.60	110.10
03380.25 EQUIPMENT PAD CONCRETE				
Equipment pad				
2500# or 3000# concrete				
By chute	C.Y.	72.50	9.55	82.05
By pump	"	72.50	38.10	110.60
By crane	"	72.50	44.40	116.90
3500# or 4000# concrete				
By chute	C.Y.	77.00	9.55	86.55
By pump	"	77.00	38.10	115.10
By crane	"	77.00	44.40	121.40
5000# concrete				
By chute	C.Y.	81.50	9.55	91.05
By pump	"	81.50	38.10	119.60

03 CONCRETE

PLACING CONCRETE

PLACING CONCRETE	UNIT	MAT.	INST.	TOTAL
03380.25 EQUIPMENT PAD CONCRETE				
By crane	C.Y.	81.50	44.40	125.90
03380.35 FOOTING CONCRETE				
Continuous footing				
2500# or 3000# concrete				
By chute	C.Y.	72.50	9.55	82.05
By pump	"	72.50	33.30	105.80
By crane	"	72.50	38.10	110.60
3500# or 4000# concrete				
By chute	C.Y.	77.00	9.55	86.55
By pump	"	77.00	33.30	110.30
By crane	"	77.00	38.10	115.10
5000# concrete				
By chute	C.Y.	81.50	9.55	91.05
By pump	"	81.50	33.30	114.80
By crane	"	81.50	38.10	119.60
Spread footing				
2500# or 3000# concrete				
Under 5 cy				
By chute	C.Y.	72.50	9.55	82.05
By pump	"	72.50	35.50	108.00
By crane	"	72.50	41.00	113.50
Over 5 cy				
By chute	C.Y.	72.50	7.15	79.65
By pump	"	72.50	31.40	103.90
By crane	"	72.50	35.50	108.00
3500# or 4000# concrete				
Under 5 c.y.				
By chute	C.Y.	77.00	9.55	86.55
By pump	"	77.00	35.50	112.50
By crane	"	77.00	41.00	118.00
Over 5 c.y.				
By chute	C.Y.	77.00	7.15	84.15
By pump	"	77.00	31.40	108.40
By crane	"	77.00	35.50	112.50
5000# concrete				
Under 5 c.y.				
By chute	C.Y.	81.50	9.55	91.05
By pump	"	81.50	35.50	117.00
By crane	"	81.50	41.00	122.50
Over 5 c.y.				
By chute	C.Y.	81.50	7.15	88.65
By pump	"	81.50	31.40	112.90
By crane	"	81.50	35.50	117.00
03380.50 GRADE BEAM CONCRETE				
Grade beam				
2500# or 3000# concrete				
By chute	C.Y.	72.50	9.55	82.05

03 CONCRETE

PLACING CONCRETE	UNIT	MAT.	INST.	TOTAL
03380.50 GRADE BEAM CONCRETE				
By crane	C.Y.	72.50	38.10	110.60
By pump	"	72.50	33.30	105.80
By hand buggy	"	72.50	28.60	101.10
3500# or 4000# concrete				
By chute	C.Y.	77.00	9.55	86.55
By crane	"	77.00	38.10	115.10
By pump	"	77.00	33.30	110.30
By hand buggy	"	77.00	28.60	105.60
5000# concrete				
By chute	C.Y.	81.50	9.55	91.05
By crane	"	81.50	38.10	119.60
By pump	"	81.50	33.30	114.80
By hand buggy	"	81.50	28.60	110.10
03380.53 PILE CAP CONCRETE				
Pile cap				
2500# or 3000 concrete				
By chute	C.Y.	72.50	9.55	82.05
By crane	"	72.50	44.40	116.90
By pump	"	72.50	38.10	110.60
By hand buggy	"	72.50	28.60	101.10
3500# or 4000# concrete				
By chute	C.Y.	77.00	9.55	86.55
By crane	"	77.00	44.40	121.40
By pump	"	77.00	38.10	115.10
By hand buggy	"	77.00	28.60	105.60
5000# concrete				
By chute	C.Y.	81.50	9.55	91.05
By crane	"	81.50	44.40	125.90
By pump	"	81.50	38.10	119.60
By hand buggy	"	81.50	28.60	110.10
03380.55 SLAB/MAT CONCRETE				
Slab on grade				
2500# or 3000# concrete				
By chute	C.Y.	72.50	7.15	79.65
By crane	"	72.50	22.20	94.70
By pump	"	72.50	19.05	91.55
By hand buggy	"	72.50	19.05	91.55
3500# or 4000# concrete				
By chute	C.Y.	77.00	7.15	84.15
By crane	"	77.00	22.20	99.20
By pump	"	77.00	19.05	96.05
By hand buggy	"	77.00	19.05	96.05
5000# concrete				
By chute	C.Y.	81.50	7.15	88.65
By crane	"	81.50	22.20	103.70
By pump	"	81.50	19.05	100.55
By hand buggy	"	81.50	19.05	100.55
Foundation mat				
2500# or 3000# concrete, over 20 cy				

PLACING CONCRETE

PLACING CONCRETE	UNIT	MAT.	INST.	TOTAL
03380.55 SLAB/MAT CONCRETE				
By chute	C.Y.	65.50	5.70	71.20
By crane	"	65.50	19.05	84.55
By pump	"	72.50	16.65	89.15
By hand buggy	"	72.50	14.30	86.80
03380.58 SIDEWALKS				
Walks, cast in place with wire mesh, base not incl.				
4" thick	S.F.	0.99	0.95	1.94
5" thick	"	1.35	1.15	2.50
6" thick	"	1.60	1.45	3.05
03380.60 STAIR CONCRETE				
Stairs				
2500# or 3000# concrete				
By chute	C.Y.	72.50	9.55	82.05
By crane	"	72.50	44.40	116.90
By pump	"	72.50	38.10	110.60
By hand buggy	"	72.50	28.60	101.10
3500# or 4000# concrete				
By chute	C.Y.	77.00	9.55	86.55
By crane	"	77.00	44.40	121.40
By pump	"	77.00	38.10	115.10
By hand buggy	"	77.00	28.60	105.60
5000# concrete				
By chute	C.Y.	81.50	9.55	91.05
By crane	"	81.50	44.40	125.90
By pump	"	81.50	38.10	119.60
By hand buggy	"	81.50	28.60	110.10
03380.65 WALL CONCRETE				
Walls				
2500# or 3000# concrete				
To 4'				
By chute	C.Y.	72.50	8.15	80.65
By crane	"	72.50	44.40	116.90
By pump	"	72.50	41.00	113.50
To 8'				
By crane	C.Y.	72.50	48.50	121.00
By pump	"	72.50	44.40	116.90
To 16'				
By crane	C.Y.	72.50	53.50	126.00
By pump	"	72.50	48.50	121.00
Over 16'				
By crane	C.Y.	72.50	59.00	131.50
By pump	"	72.50	53.50	126.00
3500# or 4000# concrete				
To 4'				
By chute	C.Y.	77.00	8.15	85.15
By crane	"	77.00	44.40	121.40
By pump	"	77.00	41.00	118.00
To 8'				
By crane	C.Y.	77.00	48.50	125.50

03 CONCRETE

PLACING CONCRETE

03380.65 WALL CONCRETE	UNIT	MAT.	INST.	TOTAL
By pump	C.Y.	77.00	44.40	121.40
To 16'				
By crane	C.Y.	77.00	53.50	130.50
By pump	"	77.00	48.50	125.50
Over 16'				
By crane	C.Y.	77.00	59.00	136.00
By pump	"	77.00	53.50	130.50
5000# concrete				
To 4'				
By chute	C.Y.	81.50	8.15	89.65
By crane	"	81.50	44.40	125.90
By pump	"	81.50	41.00	122.50
To 8'				
By crane	C.Y.	81.50	48.50	130.00
By pump	"	81.50	44.40	125.90
To 16'				
By crane	C.Y.	81.50	53.50	135.00
By pump	"	81.50	48.50	130.00
Filled block (CMU)				
3000# concrete, by pump				
4" wide	S.F.	0.26	1.90	2.16
6" wide	"	0.61	2.20	2.81
8" wide	"	0.94	2.65	3.59
10" wide	"	1.25	3.15	4.40
12" wide	"	1.60	3.80	5.40
Pilasters, 3000# concrete	C.F.	3.70	53.50	57.20
Wall cavity, 2" thick, 3000# concrete	S.F.	0.68	1.80	2.48

PRECAST CONCRETE

03400.10 PRECAST BEAMS	UNIT	MAT.	INST.	TOTAL
Prestressed, double tee, 24" deep, 8' wide				
35' span				
115 psf	S.F.	6.75	0.73	7.48
140 psf	"	7.15	0.73	7.88
40' span				
80 psf	S.F.	6.45	0.78	7.23
143 psf	"	6.90	0.78	7.68
45' span				
50 psf	S.F.	6.20	0.68	6.88
70 psf	"	6.60	0.68	7.28
100 psf	"	6.75	0.68	7.42
130 psf	"	7.45	0.68	8.13

PRECAST CONCRETE	UNIT	MAT.	INST.	TOTAL
03400.10 PRECAST BEAMS				
50' span				
75 psf	S.F.	6.20	0.61	6.81
100 psf	"	6.75	0.61	7.36
Precast beams, girders and joists				
1000 lb/lf live load				
10' span	L.F.	57.00	14.65	71.65
20' span	"	60.00	8.80	68.80
30' span	"	75.50	7.30	82.80
3000 lb/lf live load				
10' span	L.F.	59.50	14.65	74.15
20' span	"	67.50	8.80	76.30
30' span	"	89.50	7.30	96.80
5000 lb/lf live load				
10' span	L.F.	61.00	14.65	75.65
20' span	"	79.50	8.80	88.30
30' span	"	100.00	7.30	107.30
03400.20 PRECAST COLUMNS				
Prestressed concrete columns				
10" x 10"				
10' long	EA.	160.00	88.00	248.00
15' long	"	250.00	91.50	341.50
20' long	"	330.00	97.50	427.50
25' long	"	430.00	100.00	530.00
30' long	"	510.00	110.00	620.00
12" x 12"				
20' long	EA.	450.00	110.00	560.00
25' long	"	570.00	120.00	690.00
30' long	"	700.00	130.00	830.00
16" x 16"				
20' long	EA.	700.00	110.00	810.00
25' long	"	1,030	120.00	1,150
30' long	"	1,220	130.00	1,350
20" x 20"				
20' long	EA.	1,260	120.00	1,380
25' long	"	1,720	120.00	1,840
30' long	"	2,020	130.00	2,150
24" x 24"				
20' long	EA.	1,920	120.00	2,040
25' long	"	2,330	130.00	2,460
30' long	"	2,880	140.00	3,020
28" x 28"				
20' long	EA.	2,610	140.00	2,750
25' long	"	3,160	150.00	3,310
30' long	"	3,910	160.00	4,070
32" x 32"				
20' long	EA.	3,290	150.00	3,440
25' long	"	4,260	160.00	4,420
30' long	"	4,870	170.00	5,040
36" x 36"				
20' long	EA.	4,120	160.00	4,280
25' long	"	5,150	170.00	5,320

PRECAST CONCRETE	UNIT	MAT.	INST.	TOTAL
03400.20 PRECAST COLUMNS				
30' long	EA.	6,180	180.00	6,360
03400.30 PRECAST SLABS				
Prestressed flat slab				
6" thick, 4' wide				
20' span				
80 psf	S.F.	9.80	1.85	11.65
110 psf	"	9.90	1.85	11.75
25' span				
80 psf	S.F.	10.30	1.75	12.05
Cored slab				
6" thick, 4' wide				
20' span				
80 psf	S.F.	4.65	1.85	6.50
100 psf	"	4.65	1.85	6.50
130 psf	"	4.75	1.85	6.60
8" thick, 4' wide				
25' span				
70 psf	S.F.	5.15	1.75	6.90
125 psf	"	5.25	1.75	7.00
170 psf	"	5.35	1.75	7.10
30' span				
70 psf	S.F.	5.15	1.45	6.60
90 psf	"	5.50	1.45	6.95
35' span				
70 psf	S.F.	5.40	1.35	6.75
10" thick, 4' wide				
30' span				
75 psf	S.F.	5.40	1.45	6.85
100 psf	"	5.55	1.45	7.00
130 psf	"	5.70	1.45	7.15
35' span				
60 psf	S.F.	5.55	1.35	6.90
80 psf	"	5.70	1.35	7.05
120 psf	"	5.95	1.35	7.30
40' span				
65 psf	S.F.	5.95	1.10	7.05
Slabs, roof and floor members, 4' wide				
6" thick, 25' span	S.F.	4.75	1.75	6.50
8" thick, 30' span	"	5.50	1.35	6.85
10" thick, 40' span	"	6.75	1.20	7.95
Tee members				
Multiple tee, roof and floor				
Minimum	S.F.	6.20	1.10	7.30
Maximum	"	7.80	2.20	10.00
Double tee wall member				
Minimum	S.F.	5.70	1.25	6.95
Maximum	"	7.20	2.45	9.65
Single tee				
Short span, roof members				

PRECAST CONCRETE	UNIT	MAT.	INST.	TOTAL
03400.30 PRECAST SLABS				
Minimum	S.F.	6.40	1.35	7.75
Maximum	"	7.95	2.75	10.70
Long span, roof members				
Minimum	S.F.	8.15	1.10	9.25
Maximum	"	9.65	2.20	11.85
03400.40 PRECAST WALLS				
Wall panel, 8' x 20'				
Gray cement				
Liner finish				
4" wall	S.F.	7.50	1.25	8.75
5" wall	"	8.10	1.30	9.40
6" wall	"	9.25	1.35	10.60
8" wall	"	9.60	1.35	10.95
Sandblast finish				
4" wall	S.F.	8.65	1.25	9.90
5" wall	"	9.45	1.30	10.75
6" wall	"	10.40	1.35	11.75
8" wall	"	10.85	1.35	12.20
White cement				
Liner finish				
4" wall	S.F.	9.15	1.25	10.40
5" wall	"	9.70	1.30	11.00
6" wall	"	10.65	1.35	12.00
8" wall	"	11.30	1.35	12.65
Sandblast finish				
4" wall	S.F.	9.80	1.25	11.05
5" wall	"	10.40	1.30	11.70
6" wall	"	10.65	1.35	12.00
8" wall	"	11.80	1.35	13.15
Double tee wall panel, 24" deep				
Gray cement				
Liner finish	S.F.	5.40	1.45	6.85
Sandblast finish	"	6.80	1.45	8.25
White cement				
Form liner finish	S.F.	7.60	1.45	9.05
Sandblast finish	"	9.80	1.45	11.25
Partition panels				
4" wall	S.F.	8.15	1.45	9.60
5" wall	"	8.80	1.45	10.25
6" wall	"	9.65	1.45	11.10
8" wall	"	10.45	1.45	11.90
Cladding panels				
4" wall	S.F.	8.05	1.55	9.60
5" wall	"	8.80	1.55	10.35
6" wall	"	9.75	1.55	11.30
8" wall	"	10.35	1.55	11.90
Sandwich panel, 2.5" cladding panel, 2" insulation				
5" wall	S.F.	11.90	1.55	13.45
6" wall	"	12.50	1.55	14.05
8" wall	"	13.20	1.55	14.75

03 CONCRETE

PRECAST CONCRETE

PRECAST CONCRETE	UNIT	MAT.	INST.	TOTAL
03400.40 PRECAST WALLS				
Adjustable tilt-up brace	EA.	0.00	7.15	7.15
03400.90 PRECAST SPECIALTIES				
Precast concrete, coping, 4' to 8' long				
12" wide	L.F.	5.45	3.75	9.20
10" wide	"	4.85	4.25	9.10
Splash block, 30"x12"x4"	EA.	8.05	24.90	32.95
Stair unit, per riser	"	53.00	24.90	77.90
Sun screen and trellis, 8' long, 12" high				
4" thick blades	EA.	61.00	18.70	79.70
5" thick blades	"	71.50	18.70	90.20
6" thick blades	"	88.50	19.95	108.45
8" thick blades	"	120.00	19.95	139.95
Bearing pads for precast members, 2" wide strips				
1/8" thick	L.F.	0.19	0.11	0.30
1/4" thick	"	0.25	0.11	0.36
1/2" thick	"	0.28	0.11	0.39
3/4" thick	"	0.56	0.13	0.69
1" thick	"	0.59	0.14	0.73
1-1/2" thick	"	0.73	0.14	0.87

CEMENTITOUS TOPPINGS

CEMENTITOUS TOPPINGS	UNIT	MAT.	INST.	TOTAL
03550.10 CONCRETE TOPPINGS				
Gypsum fill				
2" thick	S.F.	1.05	0.28	1.33
2-1/2" thick	"	1.20	0.28	1.48
3" thick	"	1.45	0.29	1.74
3-1/2" thick	"	1.65	0.30	1.95
4" thick	"	1.95	0.33	2.28
Formboard				
Mineral fiber board				
1" thick	S.F.	0.99	0.71	1.71
1-1/2" thick	"	2.60	0.82	3.42
Cement fiber board				
1" thick	S.F.	0.77	0.95	1.72
1-1/2" thick	"	0.99	1.10	2.09
Glass fiber board				
1" thick	S.F.	1.20	0.71	1.92
1-1/2" thick	"	1.60	0.82	2.42
Poured deck				

CEMENTITOUS TOPPINGS

03550.10 CONCRETE TOPPINGS

CEMENTITOUS TOPPINGS	UNIT	MAT.	INST.	TOTAL
Vermiculite or perlite				
1 to 4 mix	C.Y.	110.00	44.40	154.40
1 to 6 mix	"	97.00	41.00	138.00
Vermiculite or perlite				
2" thick				
1 to 4 mix	S.F.	1.00	0.28	1.28
1 to 6 mix	"	0.74	0.25	0.99
3" thick				
1 to 4 mix	S.F.	1.35	0.41	1.76
1 to 6 mix	"	1.10	0.38	1.48
Concrete plank, lightweight				
2" thick	S.F.	5.25	2.20	7.45
2-1/2" thick	"	5.40	2.20	7.60
3-1/2" thick	"	5.60	2.45	8.05
4" thick	"	5.85	2.45	8.30
Channel slab, lightweight, straight				
2-3/4" thick	S.F.	4.25	2.20	6.45
3-1/2" thick	"	4.40	2.20	6.60
3-3/4" thick	"	4.70	2.20	6.90
4-3/4" thick	"	5.95	2.45	8.40
Gypsum plank				
2" thick	S.F.	2.00	2.20	4.20
3" thick	"	2.05	2.20	4.25
Cement fiber, T and G planks				
1" thick	S.F.	1.10	2.00	3.10
1-1/2" thick	"	1.15	2.00	3.15
2" thick	"	1.35	2.20	3.55
2-1/2" thick	"	1.40	2.20	3.60
3" thick	"	1.85	2.20	4.05
3-1/2" thick	"	2.10	2.45	4.55
4" thick	"	2.40	2.45	4.85

GROUT

03600.10 GROUTING

GROUT	UNIT	MAT.	INST.	TOTAL
Grouting for bases				
Nonshrink				
Metallic grout				
1" deep	S.F.	4.50	7.30	11.80
2" deep	"	8.55	8.10	16.65
Non-metallic grout				
1" deep	S.F.	3.35	7.30	10.65
2" deep	"	6.45	8.10	14.55

GROUT

03600.10 GROUTING	UNIT	MAT.	INST.	TOTAL
Fluid type				
Non-metallic				
1" deep	S.F.	3.35	7.30	10.65
2" deep	"	6.25	8.10	14.35
Grouting for joints				
Portland cement grout (1 cement to 3 sand, by volume)				
1/2" joint thickness				
6" wide joints	L.F.	0.08	1.20	1.28
8" wide joints	"	0.12	1.45	1.57
1" joint thickness				
4" wide joints	L.F.	0.12	1.15	1.27
6" wide joints	"	0.19	1.25	1.44
8" wide joints	"	0.23	1.50	1.73
Nonshrink, nonmetallic grout				
1/2" joint thickness				
4" wide joint	L.F.	0.55	1.05	1.60
6" wide joint	"	0.79	1.20	1.99
8" wide joint	"	1.05	1.45	2.50
1" joint thickness				
4" wide joint	L.F.	1.05	1.15	2.20
6" wide joint	"	1.60	1.25	2.85
8" wide joint	"	2.15	1.50	3.65

CONCRETE RESTORATION

03730.10 CONCRETE REPAIR	UNIT	MAT.	INST.	TOTAL
Epoxy grout floor patch, 1/4" thick	S.F.	3.85	2.85	6.70
Grout, epoxy, 2 component system	C.F.			194.19
Epoxy sand	BAG			13.06
Epoxy modifier	GAL			83.73
Epoxy gel grout	S.F.	1.95	28.60	30.55
Injection valve, 1 way, threaded plastic	EA.	5.35	5.70	11.05
Grout crack seal, 2 component	C.F.	450.00	28.60	478.60
Grout, non shrink	"	46.10	28.60	74.70
Concrete, epoxy modified				
Sand mix	C.F.	73.00	11.45	84.45
Gravel mix	"	56.00	10.60	66.60
Concrete repair				
Soffit repair				
16" wide	L.F.	2.30	5.70	8.00
18" wide	"	2.40	5.95	8.35
24" wide	"	2.85	6.35	9.20
30" wide	"	3.30	6.80	10.10
32" wide	"	3.50	7.15	10.65

CONCRETE RESTORATION

03730.10 CONCRETE REPAIR

CONCRETE RESTORATION	UNIT	MAT.	INST.	TOTAL
Edge repair				
2" spall	L.F.	1.10	7.15	8.25
3" spall	"	1.10	7.55	8.65
4" spall	"	1.15	7.75	8.90
6" spall	"	1.20	7.95	9.15
8" spall	"	1.25	8.40	9.65
9" spall	"	1.30	9.55	10.85
Crack repair, 1/8" crack	"	2.15	2.85	5.00
Reinforcing steel repair				
1 bar, 4 ft				
#4 bar	L.F.	0.31	5.05	5.36
#5 bar	"	0.43	5.05	5.48
#6 bar	"	0.54	5.40	5.94
#8 bar	"	0.96	5.40	6.36
#9 bar	"	1.20	5.80	7.00
#11 bar	"	1.90	5.80	7.70
Form fabric, nylon				
18" diameter	L.F.			8.13
20" diameter	"			8.26
24" diameter	"			13.59
30" diameter	"			13.96
36" diameter	"			16.00
Pile repairs				
Polyethylene wrap				
30 mil thick				
60" wide	S.F.	8.75	9.55	18.30
72" wide	"	9.45	11.45	20.90
60 mil thick				
60" wide	S.F.	10.50	9.55	20.05
80" wide	"	12.10	13.00	25.10
Pile spall, average repair 3'				
18" x 18"	EA.	27.30	23.85	51.15
20" x 20"	"	36.40	28.60	65.00

04 MASONRY

MORTAR AND GROUT

04100.10 MASONRY GROUT

MORTAR AND GROUT	UNIT	MAT.	INST.	TOTAL
Grout, non shrink, non-metallic, trowelable	C.F.	8.85	1.00	9.85
Grout door frame, hollow metal				
Single	EA.	12.85	37.40	50.25
Double	"	19.25	39.30	58.55
Grout-filled concrete block (CMU)				
4" wide	S.F.	0.37	1.25	1.62
6" wide	"	0.77	1.35	2.12
8" wide	"	1.35	1.50	2.85
12" wide	"	1.95	1.55	3.50
Grout-filled individual CMU cells				
4" wide	L.F.	0.19	0.75	0.94
6" wide	"	0.41	0.75	1.16
8" wide	"	0.53	0.75	1.28
10" wide	"	0.70	0.85	1.55
12" wide	"	0.79	0.85	1.64
Bond beams or lintels, 8" deep				
6" thick	L.F.	0.81	1.20	2.01
8" thick	"	1.05	1.35	2.40
10" thick	"	1.30	1.50	2.80
12" thick	"	1.50	1.65	3.15
Cavity walls				
2" thick	S.F.	0.89	1.80	2.69
3" thick	"	1.30	1.80	3.10
4" thick	"	1.75	1.90	3.65
6" thick	"	2.65	2.20	4.85

04150.10 MASONRY ACCESSORIES

MORTAR AND GROUT	UNIT	MAT.	INST.	TOTAL
Foundation vents	EA.	18.70	14.15	32.85
Bar reinforcing				
Horizontal				
#3 - #4	Lb.	0.39	1.40	1.79
#5 - #6	"	0.36	1.20	1.56
Vertical				
#3 - #4	Lb.	0.39	1.75	2.14
#5 - #6	"	0.36	1.40	1.76
Horizontal joint reinforcing				
Truss type				
4" wide, 6" wall	L.F.	0.14	0.14	0.28
6" wide, 8" wall	"	0.14	0.15	0.29
8" wide, 10" wall	"	0.18	0.15	0.33
10" wide, 12" wall	"	0.18	0.16	0.34
12" wide, 14" wall	"	0.22	0.17	0.39
Ladder type				
4" wide, 6" wall	L.F.	0.11	0.14	0.25
6" wide, 8" wall	"	0.12	0.15	0.27
8" wide, 10" wall	"	0.13	0.15	0.28
10" wide, 12" wall	"	0.14	0.15	0.29
Rectangular wall ties				
3/16" dia., galvanized				
2" x 6"	EA.	0.15	0.59	0.74
2" x 8"	"	0.17	0.59	0.76

MORTAR AND GROUT

04150.10 MASONRY ACCESSORIES	UNIT	MAT.	INST.	TOTAL
2" x 10"	EA.	0.19	0.59	0.78
2" x 12"	"	0.22	0.59	0.81
4" x 6"	"	0.18	0.71	0.89
4" x 8"	"	0.20	0.71	0.91
4" x 10"	"	0.26	0.71	0.97
4" x 12"	"	0.31	0.71	1.02
1/4" dia., galvanized				
2" x 6"	EA.	0.30	0.59	0.89
2" x 8"	"	0.33	0.59	0.92
2" x 10"	"	0.37	0.59	0.96
2" x 12"	"	0.42	0.59	1.01
4" x 6"	"	0.34	0.71	1.05
4" x 8"	"	0.37	0.71	1.08
4" x 10"	"	0.42	0.71	1.13
4" x 12"	"	0.44	0.71	1.15
"Z" type wall ties, galvanized				
6" long				
1/8" dia.	EA.	0.15	0.59	0.74
3/16" dia.	"	0.17	0.59	0.76
1/4" dia.	"	0.18	0.59	0.77
8" long				
1/8" dia.	EA.	0.17	0.59	0.76
3/16" dia.	"	0.18	0.59	0.77
1/4" dia.	"	0.19	0.59	0.78
10" long				
1/8" dia.	EA.	0.18	0.59	0.77
3/16" dia.	"	0.20	0.59	0.79
1/4" dia.	"	0.23	0.59	0.82
Dovetail anchor slots				
Galvanized steel, filled				
24 ga.	L.F.	0.42	0.88	1.30
20 ga.	"	0.53	0.88	1.41
16 oz. copper, foam filled	"	1.05	0.88	1.93
Dovetail anchors				
16 ga.				
3-1/2" long	EA.	0.12	0.59	0.71
5-1/2" long	"	0.14	0.59	0.73
12 ga.				
3-1/2" long	EA.	0.15	0.59	0.74
5-1/2" long	"	0.33	0.59	0.92
Dovetail, triangular galvanized ties, 12 ga.				
3" x 3"	EA.	0.30	0.59	0.89
5" x 5"	"	0.32	0.59	0.91
7" x 7"	"	0.36	0.59	0.95
7" x 9"	"	0.39	0.59	0.98
Brick anchors				
Corrugated, 3-1/2" long				
16 ga.	EA.	0.12	0.59	0.71
12 ga.	"	0.20	0.59	0.79
Non-corrugated, 3-1/2" long				
16 ga.	EA.	0.15	0.59	0.74
12 ga.	"	0.29	0.59	0.88

04 MASONRY

MORTAR AND GROUT

04150.20 MASONRY CONTROL JOINTS	UNIT	MAT.	INST.	TOTAL
Control joint, cross shaped PVC	L.F.	2.80	0.88	3.68
Closed cell joint filler				
1/2"	L.F.	0.42	0.88	1.30
3/4"	"	0.76	0.88	1.64
Rubber, for				
4" wall	L.F.	3.05	0.88	3.93
6" wall	"	4.75	0.93	5.68
8" wall	"	5.60	0.98	6.58
PVC, for				
4" wall	L.F.	2.35	0.88	3.23
6" wall	"	2.90	0.93	3.83
8" wall	"	3.30	0.98	4.28

04150.50 MASONRY FLASHING	UNIT	MAT.	INST.	TOTAL
Through-wall flashing				
5 oz. coated copper	S.F.	2.70	2.95	5.65
0.030" elastomeric	"	0.59	2.35	2.94

UNIT MASONRY

04210.10 BRICK MASONRY	UNIT	MAT.	INST.	TOTAL
Standard size brick, running bond				
Face brick, red (6.4/sf)				
Veneer	S.F.	2.70	5.90	8.60
Cavity wall	"	2.70	5.05	7.75
9" solid wall	"	5.15	10.10	15.25
Back-up				
4" thick	S.F.	2.75	4.40	7.15
8" thick	"	5.45	7.05	12.50
Firewall				
12" thick	S.F.	8.30	11.80	20.10
16" thick	"	11.00	16.05	27.05
Glazed brick (7.4/sf)				
Veneer	S.F.	8.85	6.45	15.30
Buff or gray face brick (6.4/sf)				
Veneer	S.F.	3.00	5.90	8.90
Cavity wall	"	3.00	5.05	8.05
Jumbo or oversize brick (3/sf)				
4" veneer	S.F.	3.45	3.55	7.00
4" back-up	"	3.45	2.95	6.40
8" back-up	"	8.10	5.05	13.15
12" firewall	"	11.55	8.85	20.40
16" firewall	"	17.10	11.80	28.90

04 MASONRY

UNIT MASONRY	UNIT	MAT.	INST.	TOTAL
04210.10 BRICK MASONRY				
Norman brick, red face, (4.5/sf)				
4" veneer	S.F.	4.05	4.40	8.45
Cavity wall	"	4.05	3.95	8.00
Chimney, standard brick, including flue				
16" x 16"	L.F.	16.15	35.40	51.55
16" x 20"	"	19.10	35.40	54.50
16" x 24"	"	22.60	35.40	58.00
20" x 20"	"	24.75	44.20	68.95
20" x 24"	"	31.80	44.20	76.00
20" x 32"	"	40.10	50.50	90.60
Window sill, face brick on edge	"	2.40	8.85	11.25
04210.20 STRUCTURAL TILE				
Structural glazed tile				
6T series, 5-1/2" x 12"				
Glazed on one side				
2" thick	S.F.	6.00	3.55	9.55
4" thick	"	7.50	3.55	11.05
6" thick	"	11.00	3.95	14.95
8" thick	"	14.45	4.40	18.85
Glazed on two sides				
4" thick	S.F.	11.55	4.40	15.95
6" thick	"	17.35	5.05	22.40
04210.60 PAVERS, MASONRY				
Brick walk laid on sand, sand joints				
Laid flat, (4.5 per sf)	S.F.	2.70	3.95	6.65
Laid on edge, (7.2 per sf)	"	4.25	5.90	10.15
Precast concrete patio blocks				
2" thick				
Natural	S.F.	2.10	1.20	3.30
Colors	"	2.90	1.20	4.10
Exposed aggregates, local aggregate				
Natural	S.F.	2.45	1.20	3.65
Colors	"	3.25	1.20	4.45
Granite or limestone aggregate	"	4.45	1.20	5.65
White tumblestone aggregate	"	3.45	1.20	4.65
Stone pavers, set in mortar				
Bluestone				
1" thick				
Irregular	S.F.	2.20	8.85	11.05
Snapped rectangular	"	3.35	7.05	10.40
1-1/2" thick, random rectangular	"	3.80	8.85	12.65
2" thick, random rectangular	"	4.50	10.10	14.60
Slate				
Natural cleft				
Irregular, 3/4" thick	S.F.	2.10	10.10	12.20
Random rectangular				
1-1/4" thick	S.F.	4.50	8.85	13.35
1-1/2" thick	"	5.10	9.80	14.90
Granite blocks				

UNIT MASONRY

UNIT MASONRY	UNIT	MAT.	INST.	TOTAL
04210.60 PAVERS, MASONRY				
3" thick, 3" to 6" wide				
4" to 12" long	S.F.	4.85	11.80	16.65
6" to 15" long	"	2.85	10.10	12.95
Crushed stone, white marble, 3" thick	"	1.15	0.57	1.72
04220.10 CONCRETE MASONRY UNITS				
Hollow, load bearing				
4"	S.F.	1.25	2.60	3.85
6"	"	1.55	2.70	4.25
8"	"	2.05	2.95	5.00
10"	"	2.70	3.20	5.90
12"	"	2.85	3.55	6.40
Solid, load bearing				
4"	S.F.	1.65	2.60	4.25
6"	"	2.00	2.70	4.70
8"	"	2.30	2.95	5.25
10"	"	3.15	3.20	6.35
12"	"	3.50	3.55	7.05
Back-up block, 8" x 16"				
2"	S.F.	0.83	2.00	2.83
4"	"	1.20	2.10	3.30
6"	"	1.50	2.20	3.70
8"	"	1.75	2.35	4.10
10"	"	2.35	2.55	4.90
12"	"	2.45	2.70	5.15
Foundation wall, 8" x 16"				
6"	S.F.	1.60	2.55	4.15
8"	"	1.80	2.70	4.50
10"	"	2.70	2.95	5.65
12"	"	3.00	3.20	6.20
Solid				
6"	S.F.	2.25	2.70	4.95
8"	"	2.75	2.95	5.70
10"	"	3.40	3.20	6.60
12"	"	4.10	3.55	7.65
Exterior, styrofoam inserts, standard weight, 8" x 16"				
6"	S.F.	2.40	2.70	5.10
8"	"	2.65	2.95	5.60
10"	"	3.75	3.20	6.95
12"	"	3.95	3.55	7.50
Acoustical slotted block				
4"	S.F.	2.65	3.20	5.85
6"	"	3.25	3.20	6.45
8"	"	4.85	3.55	8.40
Filled cavities				
4"	S.F.	3.35	3.95	7.30
6"	"	4.15	4.15	8.30
8"	"	5.55	4.40	9.95
Hollow, split face				
4"	S.F.	2.55	2.60	5.15
6"	"	2.95	2.70	5.65
8"	"	2.65	2.95	5.60

UNIT MASONRY	UNIT	MAT.	INST.	TOTAL
04220.10 CONCRETE MASONRY UNITS				
10"	S.F.	3.45	3.20	6.65
12"	"	3.70	3.55	7.25
Split rib profile				
4"	S.F.	1.85	3.20	5.05
6"	"	2.15	3.20	5.35
8"	"	2.30	3.55	5.85
10"	"	2.55	3.55	6.10
12"	"	2.75	3.55	6.30
High strength block, 3500 psi				
2"	S.F.	1.20	2.60	3.80
4"	"	1.50	2.70	4.20
6"	"	1.80	2.70	4.50
8"	"	2.05	2.95	5.00
10"	"	2.35	3.20	5.55
12"	"	2.80	3.55	6.35
Solar screen concrete block				
4" thick				
6" x 6"	S.F.	3.60	7.85	11.45
8" x 8"	"	4.30	7.05	11.35
12" x 12"	"	2.50	5.45	7.95
8" thick				
8" x 16"	S.F.	2.65	5.05	7.70
Glazed block				
Cove base, glazed 1 side, 2"	L.F.	8.10	3.95	12.05
4"	"	8.10	3.95	12.05
6"	"	8.55	4.40	12.95
8"	"	9.15	4.40	13.55
Single face				
2"	S.F.	6.00	2.95	8.95
4"	"	6.50	2.95	9.45
6"	"	6.60	3.20	9.80
8"	"	6.80	3.55	10.35
10"	"	7.05	3.95	11.00
12"	"	7.30	4.15	11.45
Double face				
4"	S.F.	9.60	3.70	13.30
6"	"	10.40	3.95	14.35
8"	"	10.70	4.40	15.10
Corner or bullnose				
2"	EA.	8.10	4.40	12.50
4"	"	10.40	5.05	15.45
6"	"	12.70	5.05	17.75
8"	"	13.85	5.90	19.75
10"	"	15.00	6.45	21.45
12"	"	16.15	7.05	23.20
04220.90 BOND BEAMS & LINTELS				
Bond beam, no grout or reinforcement				
8" x 16" x				
4" thick	L.F.	1.60	2.70	4.30
6" thick	"	1.95	2.85	4.80

UNIT MASONRY

UNIT MASONRY	UNIT	MAT.	INST.	TOTAL
04220.90 BOND BEAMS & LINTELS				
8" thick	L.F.	2.35	2.95	5.30
10" thick	"	2.50	3.05	5.55
12" thick	"	2.50	3.20	5.70
Beam lintel, no grout or reinforcement				
8" x 16" x				
10" thick	L.F.	4.35	3.55	7.90
12" thick	"	5.55	3.95	9.50
Precast masonry lintel				
6 lf, 8" high x				
4" thick	L.F.	4.15	5.90	10.05
6" thick	"	5.30	5.90	11.20
8" thick	"	6.00	6.45	12.45
10" thick	"	7.15	6.45	13.60
10 lf, 8" high x				
4" thick	L.F.	5.25	3.55	8.80
6" thick	"	6.45	3.55	10.00
8" thick	"	7.15	3.95	11.10
10" thick	"	9.70	3.95	13.65
Steel angles and plates				
Minimum	Lb.	0.39	0.51	0.90
Maximum	"	0.52	0.88	1.40
Various size angle lintels				
1/4" stock				
3" x 3"	L.F.	2.65	2.20	4.85
3" x 3-1/2"	"	2.75	2.20	4.95
3/8" stock				
3" x 4"	L.F.	4.30	2.20	6.50
3-1/2" x 4"	"	4.50	2.20	6.70
4" x 4"	"	4.85	2.20	7.05
5" x 3-1/2"	"	5.10	2.20	7.30
6" x 3-1/2"	"	5.20	2.20	7.40
1/2" stock				
6" x 4"	L.F.	6.45	2.20	8.65
04270.10 GLASS BLOCK				
Glass block, 4" thick				
6" x 6"	S.F.	24.95	11.80	36.75
8" x 8"	"	21.55	8.85	30.40
12" x 12"	"	21.55	7.05	28.60
Replacement glass blocks, 4" x 8" x 8"				
Minimum	S.F.	26.20	35.40	61.60
Maximum	"	31.00	70.50	101.50
04295.10 PARGING/MASONRY PLASTER				
Parging				
1/2" thick	S.F.	0.45	2.35	2.80
3/4" thick	"	0.57	2.95	3.52
1" thick	"	0.79	3.55	4.34

04 MASONRY

STONE

04400.10 STONE	UNIT	MAT.	INST.	TOTAL
Rubble stone				
Walls set in mortar				
8" thick	S.F.	9.90	8.85	18.75
12" thick	"	11.95	14.15	26.10
18" thick	"	15.85	17.70	33.55
24" thick	"	19.85	23.55	43.40
Dry set wall				
8" thick	S.F.	11.15	5.90	17.05
12" thick	"	12.60	8.85	21.45
18" thick	"	17.40	11.80	29.20
24" thick	"	21.20	14.15	35.35
Cut stone				
Facing panels				
3/4" thick	S.F.	26.40	14.15	40.55
1-1/2" thick	"	39.70	16.05	55.75

MASONRY RESTORATION

04520.10 RESTORATION AND CLEANING	UNIT	MAT.	INST.	TOTAL
Masonry cleaning				
Washing brick				
Smooth surface	S.F.	0.30	0.59	0.89
Rough surface	"	0.40	0.79	1.19
Steam clean masonry				
Smooth face				
Minimum	S.F.	0.00	0.46	0.46
Maximum	"	0.00	0.66	0.66
Rough face				
Minimum	S.F.	0.00	0.61	0.61
Maximum	"	0.00	0.91	0.91
Sandblast masonry				
Minimum	S.F.	0.35	0.73	1.08
Maximum	"	0.53	1.20	1.73
Pointing masonry				
Brick	S.F.	0.91	1.40	2.31
Concrete block	"	0.41	1.00	1.41
Cut and repoint				
Brick				
Minimum	S.F.	0.25	1.75	2.00
Maximum	"	0.43	3.55	3.98
Stone work	L.F.	1.15	2.70	3.85
Cut and recaulk				
Oil base caulks	L.F.	0.89	2.35	3.24

MASONRY RESTORATION	UNIT	MAT.	INST.	TOTAL
04520.10 RESTORATION AND CLEANING				
Butyl caulks	L.F.	0.78	2.35	3.13
Polysulfides and acrylics	"	1.50	2.35	3.85
Silicones	"	1.75	2.35	4.10
Cement and sand grout on walls, to 1/8" thick				
Minimum	S.F.	0.44	1.40	1.84
Maximum	"	1.10	1.75	2.85
Brick removal and replacement				
Minimum	EA.	0.46	4.40	4.86
Average	"	0.58	5.90	6.48
Maximum	"	1.15	17.70	18.85

05 METALS

METAL FASTENING

METAL FASTENING	UNIT	MAT.	INST.	TOTAL
05050.10 STRUCTURAL WELDING				
Welding				
Single pass				
1/8"	L.F.	0.41	2.00	2.41
3/16"	"	0.77	2.65	3.42
1/4"	"	1.05	3.30	4.35
Miscellaneous steel shapes				
Plain	Lb.	0.77	0.08	0.85
Galvanized	"	1.20	0.13	1.33
Plates				
Plain	Lb.	0.72	0.10	0.82
Galvanized	"	1.10	0.16	1.26
05050.90 METAL ANCHORS				
Anchor bolts				
1/2" x				
8" long	EA.			3.37
10" long	"			3.59
12" long	"			3.93
18" long	"			4.27
3/4" x				
8" long	EA.			4.49
12" long	"			5.05
18" long	"			6.95
24" long	"			9.20
1" x				
12" long	EA.			8.98
18" long	"			11.22
24" long	"			13.46
36" long	"			20.20
Expansion shield				
1/4"	EA.			0.67
3/8"	"			1.12
1/2"	"			2.24
5/8"	"			3.26
3/4"	"			3.93
1"	"			5.39
Non-drilling anchor				
1/4"	EA.			0.42
3/8"	"			0.53
1/2"	"			0.80
5/8"	"			1.33
3/4"	"			2.28
Self-drilling anchor				
1/4"	EA.			1.12
5/16"	"			1.41
3/8"	"			1.68
1/2"	"			2.24
5/8"	"			4.27
3/4"	"			5.61
7/8"	"			7.85
Add 25% for galvanized anchor bolts				

05 METALS

METAL FASTENING	UNIT	MAT.	INST.	TOTAL
05050.90 METAL ANCHORS				
Channel door frame, with anchors	Lb.	1.10	0.44	1.54
Corner guard angle, with anchors	"	1.00	0.66	1.66
05050.95 METAL LINTELS				
Lintels, steel				
Plain	Lb.	0.75	0.99	1.74
Galvanized	"	1.10	0.99	2.09
05120.10 STRUCTURAL STEEL				
Beams and girders, A-36				
Welded	TON	1,250	440.00	1,690
Bolted	"	1,220	400.00	1,620
Columns				
Pipe				
6" dia.	Lb.	0.84	0.44	1.28
12" dia.	"	0.72	0.37	1.09
Purlins and girts				
Welded	TON	1,560	730.00	2,290
Bolted	"	1,540	630.00	2,170
Column base plates				
Up to 150 lb each	Lb.	0.90	0.27	1.17
Over 150 lb each	"	0.74	0.33	1.07
Structural pipe				
3" to 5" o.d.	TON	1,680	880.00	2,560
6" to 12" o.d.	"	1,560	630.00	2,190
Structural tube				
6" square				
Light sections	TON	1,950	880.00	2,830
Heavy sections	"	1,830	630.00	2,460
6" wide rectangular				
Light sections	TON	1,950	730.00	2,680
Heavy sections	"	1,830	550.00	2,380
Greater than 6" wide rectangular				
Light sections	TON	2,140	730.00	2,870
Heavy sections	"	2,020	550.00	2,570
Miscellaneous structural shapes				
Steel angle	TON	1,390	1,100	2,490
Steel plate	"	1,580	730.00	2,310
Trusses, field welded				
60 lb/lf	TON	2,020	550.00	2,570
100 lb/lf	"	1,760	440.00	2,200
150 lb/lf	"	1,660	370.00	2,030
Bolted				
60 lb/lf	TON	1,990	490.00	2,480
100 lb/lf	"	1,740	400.00	2,140
150 lb/lf	"	1,650	340.00	1,990
Add for galvanizing	"			396.00

05 METALS

COLD FORMED FRAMING

05410.10 METAL FRAMING

COLD FORMED FRAMING	UNIT	MAT.	INST.	TOTAL
Furring channel, galvanized				
Beams and columns, 3/4"				
12" o.c.	S.F.	0.34	3.95	4.29
16" o.c.	"	0.29	3.60	3.89
Walls, 3/4"				
12" o.c.	S.F.	0.33	2.00	2.33
16" o.c.	"	0.26	1.65	1.91
24" o.c.	"	0.19	1.30	1.49
1-1/2"				
12" o.c.	S.F.	0.43	2.00	2.43
16" o.c.	"	0.34	1.65	1.99
24" o.c.	"	0.26	1.30	1.56
Stud, load bearing				
16" o.c.				
16 ga.				
2-1/2"	S.F.	0.91	1.75	2.66
3-5/8"	"	1.15	1.75	2.90
4"	"	1.25	1.75	3.00
6"	"	1.80	2.00	3.80
24" o.c.				
16 ga.				
2-1/2"	S.F.	0.73	1.55	2.28
3-5/8"	"	0.97	1.55	2.52
4"	"	1.10	1.55	2.65
6"	"	1.55	1.65	3.20
8"	"	1.70	1.65	3.35

METAL FABRICATIONS

05510.10 STAIRS

METAL FABRICATIONS	UNIT	MAT.	INST.	TOTAL
Stock unit, steel, complete, per riser				
Tread				
3'-6" wide	EA.	81.50	49.60	131.10
4' wide	"	94.50	56.50	151.00
5' wide	"	110.00	66.00	176.00
Metal pan stair, cement filled, per riser				
3'-6" wide	EA.	87.50	39.70	127.20
4' wide	"	100.00	44.10	144.10
5' wide	"	110.00	49.60	159.60
Landing, steel pan	S.F.	35.00	9.90	44.90
Cast iron tread, steel stringers, stock units, per riser				
Tread				
3'-6" wide	EA.	160.00	49.60	209.60
4' wide	"	180.00	56.50	236.50

05 METALS

METAL FABRICATIONS

METAL FABRICATIONS	UNIT	MAT.	INST.	TOTAL
05510.10 STAIRS				
5' wide	EA.	210.00	66.00	276.00
Stair treads, abrasive, 12" x 3'-6"				
Cast iron				
3/8"	EA.	85.50	19.85	105.35
1/2"	"	110.00	19.85	129.85
Cast aluminum				
5/16"	EA.	100.00	19.85	119.85
3/8"	"	110.00	19.85	129.85
1/2"	"	130.00	19.85	149.85
05515.10 LADDERS				
Ladder, 18" wide				
With cage	L.F.	61.50	26.50	88.00
Without cage	"	39.30	19.85	59.15
05520.10 RAILINGS				
Railing, pipe				
1-1/4" diameter, welded steel				
2-rail				
Primed	L.F.	12.35	7.95	20.30
Galvanized	"	15.70	7.95	23.65
3-rail				
Primed	L.F.	15.70	9.90	25.60
Galvanized	"	20.20	9.90	30.10
Wall mounted, single rail, welded steel				
Primed	L.F.	5.25	6.10	11.35
Galvanized	"	7.85	6.10	13.95
1-1/2" diameter, welded steel				
2-rail				
Primed	L.F.	13.45	7.95	21.40
Galvanized	"	16.85	7.95	24.80
3-rail				
Primed	L.F.	16.85	9.90	26.75
Galvanized	"	21.30	9.90	31.20
Wall mounted, single rail, welded steel				
Primed	L.F.	6.15	6.10	12.25
Galvanized	"	8.40	6.10	14.50
2" diameter, welded steel				
2-rail				
Primed	L.F.	16.85	8.80	25.65
Galvanized	"	21.30	8.80	30.10
3-rail				
Primed	L.F.	21.30	11.35	32.65
Galvanized	"	25.80	11.35	37.15
Wall mounted, single rail, welded steel				
Primed	L.F.	6.75	6.60	13.35
Galvanized	"	8.40	6.60	15.00

METAL FABRICATIONS

05530.10 METAL GRATING

	UNIT	MAT.	INST.	TOTAL
Floor plate, checkered, steel				
1/4"				
Primed	S.F.	5.25	0.57	5.82
Galvanized	"	7.00	0.57	7.57
3/8"				
Primed	S.F.	7.00	0.61	7.61
Galvanized	"	9.35	0.61	9.96
Aluminum grating, pressure-locked bearing bars				
3/4" x 1/8"	S.F.	11.20	0.99	12.19
1" x 1/8"	"	12.35	0.99	13.34
1-1/4" x 1/8"	"	15.70	0.99	16.69
1-1/4" x 3/16"	"	16.85	0.99	17.84
1-1/2" x 1/8"	"	17.95	0.99	18.94
1-3/4" x 3/16"	"	20.20	0.99	21.19
Miscellaneous expenses				
Cutting				
Minimum	L.F.	0.00	2.65	2.65
Maximum	"	0.00	3.95	3.95
Banding				
Minimum	L.F.	0.00	6.60	6.60
Maximum	"	0.00	7.95	7.95
Toe plates				
Minimum	L.F.	0.00	7.95	7.95
Maximum	"	0.00	9.90	9.90
Steel grating, primed				
3/4" x 1/8"	S.F.	4.95	1.30	6.25
1" x 1/8"	"	5.15	1.30	6.45
1-1/4" x 1/8"	"	5.50	1.30	6.80
1-1/4" x 3/16"	"	6.70	1.30	8.00
1-1/2" x 1/8"	"	6.15	1.30	7.45
1-3/4" x 3/16"	"	8.75	1.30	10.05
Galvanized				
3/4" x 1/8"	S.F.	6.15	1.30	7.45
1" x 1/8"	"	6.55	1.30	7.85
1-1/4" x 1/8"	"	7.10	1.30	8.40
1-1/4" x 3/16"	"	8.45	1.30	9.75
1-1/2" x 1/8"	"	7.60	1.30	8.90
1-3/4" x 3/16"	"	11.10	1.30	12.40
Miscellaneous expenses				
Cutting				
Minimum	L.F.	0.00	2.85	2.85
Maximum	"	0.00	4.40	4.40
Banding				
Minimum	L.F.	0.00	7.20	7.20
Maximum	"	0.00	8.80	8.80
Toe plates				
Minimum	L.F.	0.00	8.80	8.80
Maximum	"	0.00	11.35	11.35

05540.10 CASTINGS

	UNIT	MAT.	INST.	TOTAL
Miscellaneous castings				
Light sections	Lb.	1.10	0.79	1.89

05 METALS

METAL FABRICATIONS

05540.10 CASTINGS

	UNIT	MAT.	INST.	TOTAL
Heavy sections	Lb.	0.91	0.57	1.48
Manhole covers and frames				
Regular, city type				
18" dia.				
100 lb	EA.	220.00	79.50	299.50
24" dia.				
200 lb	EA.	230.00	79.50	309.50
300 lb	"	230.00	88.00	318.00
400 lb	"	250.00	88.00	338.00
26" dia., 475 lb	"	300.00	100.00	400.00
30" dia., 600 lb	"	350.00	110.00	460.00
8" square, 75 lb	"	98.50	15.85	114.35
24" square				
126 lb	EA.	220.00	79.50	299.50
500 lb	"	350.00	100.00	450.00
Watertight type				
20" dia., 200 lb	EA.	200.00	100.00	300.00
24" dia., 350 lb	"	340.00	130.00	470.00
Steps, cast iron				
7" x 9"	EA.	9.50	7.95	17.45
8" x 9"	"	11.35	8.80	20.15
Manhole covers and frames, aluminum				
12" x 12"	EA.	45.40	15.85	61.25
18" x 18"	"	47.20	15.85	63.05
24" x 24"	"	51.50	19.85	71.35
Corner protection				
Steel angle guard with anchors				
2" x 2" x 3/16"	L.F.	9.85	5.65	15.50
2" x 3" x 1/4"	"	11.20	5.65	16.85
3" x 3" x 5/16"	"	13.00	5.65	18.65
3" x 4" x 5/16"	"	15.60	6.10	21.70
4" x 4" x 5/16"	"	16.25	6.10	22.35

MISC. FABRICATIONS

05580.10 METAL SPECIALTIES

	UNIT	MAT.	INST.	TOTAL
Kick plate				
4" high x 1/4" thick				
Primed	L.F.	4.70	7.95	12.65
Galvanized	"	5.35	7.95	13.30
6" high x 1/4" thick				
Primed	L.F.	5.15	8.80	13.95
Galvanized	"	6.05	8.80	14.85

MISC. FABRICATIONS

MISC. FABRICATIONS	UNIT	MAT.	INST.	TOTAL
05700.10 ORNAMENTAL METAL				
Railings, vertical square bars, 6" o.c., with shaped top rails				
Steel	L.F.	50.50	19.85	70.35
Aluminum	"	69.00	19.85	88.85
Bronze	"	110.00	26.50	136.50
Stainless steel	"	100.00	26.50	126.50
Laminated metal or wood handrails with metal supports				
2-1/2" round or oval shape	L.F.	95.50	19.85	115.35
05800.10 EXPANSION CONTROL				
Expansion joints with covers, floor assembly type				
With 1" space				
Aluminum	L.F.	16.40	6.60	23.00
Bronze	"	34.20	6.60	40.80
Stainless steel	"	26.50	6.60	33.10
With 2" space				
Aluminum	L.F.	20.25	6.60	26.85
Bronze	"	35.90	6.60	42.50
Stainless steel	"	31.60	6.60	38.20
Ceiling and wall assembly type				
With 1" space				
Aluminum	L.F.	10.50	7.95	18.45
Bronze	"	35.90	7.95	43.85
Stainless steel	"	32.40	7.95	40.35
With 2" space				
Aluminum	L.F.	11.55	7.95	19.50
Bronze	"	38.90	7.95	46.85
Stainless steel	"	34.20	7.95	42.15
Exterior roof and wall, aluminum				
Roof to roof				
With 1" space	L.F.	30.60	6.60	37.20
With 2" space	"	33.10	6.60	39.70
Roof to wall				
With 1" space	L.F.	23.80	7.20	31.00
With 2" space	"	28.30	7.20	35.50

FASTENERS AND ADHESIVES

06050.10 ACCESSORIES	UNIT	MAT.	INST.	TOTAL
Column/post base, cast aluminum				
4" x 4"	EA.	2.00	9.10	11.10
6" x 6"	"	4.40	9.10	13.50
Bridging, metal, per pair				
12" o.c.	EA.	0.74	3.65	4.39
16" o.c.	"	0.74	3.30	4.04
Anchors				
Bolts, threaded two ends, with nuts and washers				
1/2" dia.				
4" long	EA.	0.99	2.30	3.29
7-1/2" long	"	1.20	2.30	3.50
15" long	"	2.50	2.30	4.80
Bolts, carriage				
1/4 x 4	EA.	1.00	3.65	4.65
5/16 x 6	"	1.05	3.85	4.90
Joist and beam hangers				
18 ga.				
2 x 4	EA.	0.48	3.65	4.13
2 x 6	"	0.63	3.65	4.28
2 x 8	"	0.74	3.65	4.39
2 x 10	"	0.85	4.05	4.90
2 x 12	"	0.99	4.55	5.54
16 ga.				
3 x 6	EA.	1.75	4.05	5.80
3 x 8	"	1.80	4.05	5.85
3 x 10	"	2.10	4.30	6.40
3 x 12	"	2.30	4.85	7.15
3 x 14	"	2.45	5.20	7.65
4 x 6	"	2.00	4.05	6.05
4 x 8	"	2.30	4.05	6.35
4 x 10	"	2.85	4.30	7.15
4 x 12	"	3.20	4.85	8.05
4 x 14	"	3.50	5.20	8.70
Rafter anchors, 18 ga., 1-1/2" wide				
5-1/4" long	EA.	0.54	3.05	3.59
10-3/4" long	"	0.83	3.05	3.88
Sill anchors				
Embedded in concrete	EA.	1.35	3.65	5.00
Strap ties, 14 ga., 1-3/8" wide				
12" long	EA.	1.10	3.05	4.15

06 WOOD AND PLASTICS

ROUGH CARPENTRY	UNIT	MAT.	INST.	TOTAL
06110.10 BLOCKING				
Steel construction				
Walls				
2x4	L.F.	0.47	2.45	2.92
2x6	"	0.67	2.80	3.47
2x8	"	0.96	3.05	4.01
2x10	"	1.30	3.30	4.60
2x12	"	1.75	3.65	5.40
Ceilings				
2x4	L.F.	0.47	2.80	3.27
2x6	"	0.67	3.30	3.97
2x8	"	0.96	3.65	4.61
2x10	"	1.30	4.05	5.35
2x12	"	1.75	4.55	6.30
Wood construction				
Walls				
2x4	L.F.	0.47	2.05	2.52
2x6	"	0.67	2.30	2.97
2x8	"	0.96	2.45	3.41
2x10	"	1.30	2.60	3.90
2x12	"	1.75	2.80	4.55
Ceilings				
2x4	L.F.	0.47	2.30	2.77
2x6	"	0.67	2.60	3.27
2x8	"	0.96	2.80	3.76
2x10	"	1.30	3.05	4.35
2x12	"	1.75	3.30	5.05
06110.20 CEILING FRAMING				
Ceiling joists				
16" o.c.				
2x4	S.F.	0.47	0.70	1.17
2x6	"	0.72	0.73	1.45
2x8	"	1.00	0.76	1.76
2x10	"	1.45	0.79	2.24
2x12	"	1.90	0.83	2.73
Headers and nailers				
2x4	L.F.	0.47	1.20	1.67
2x6	"	0.67	1.20	1.87
2x8	"	0.96	1.30	2.26
2x10	"	1.30	1.40	2.70
2x12	"	1.75	1.50	3.25
Sister joists for ceilings				
2x4	L.F.	0.47	2.60	3.07
2x6	"	0.67	3.05	3.72
2x8	"	0.96	3.65	4.61
2x10	"	1.30	4.55	5.85
2x12	"	1.75	6.10	7.85
06110.30 FLOOR FRAMING				
Floor joists				
16" o.c.				

06 WOOD AND PLASTICS

ROUGH CARPENTRY	UNIT	MAT.	INST.	TOTAL
06110.30 FLOOR FRAMING				
2x6	S.F.	0.72	0.61	1.33
2x8	"	1.00	0.62	1.62
2x10	"	1.45	0.63	2.08
2x12	"	1.90	0.65	2.55
2x14	"	2.35	0.68	3.03
3x6	"	1.60	0.63	2.23
3x8	"	2.10	0.65	2.75
3x10	"	2.60	0.68	3.28
3x12	"	3.10	0.70	3.80
3x14	"	3.65	0.73	4.38
4x6	"	2.10	0.63	2.73
4x8	"	2.80	0.65	3.45
4x10	"	3.45	0.68	4.13
4x12	"	4.10	0.70	4.80
4x14	"	4.85	0.73	5.58
Sister joists for floors				
2x4	L.F.	0.47	2.30	2.77
2x6	"	0.67	2.60	3.27
2x8	"	0.96	3.05	4.01
2x10	"	1.30	3.65	4.95
2x12	"	1.75	4.55	6.30
06110.40 FURRING				
Furring, wood strips				
Walls				
On masonry or concrete walls				
1x2 furring				
12" o.c.	S.F.	0.22	1.15	1.37
16" o.c.	"	0.19	1.05	1.24
24" o.c.	"	0.15	0.96	1.11
1x3 furring				
12" o.c.	S.F.	0.32	1.15	1.47
16" o.c.	"	0.28	1.05	1.33
24" o.c.	"	0.22	0.96	1.18
On wood walls				
1x2 furring				
12" o.c.	S.F.	0.22	0.81	1.03
16" o.c.	"	0.19	0.73	0.92
24" o.c.	"	0.15	0.66	0.81
1x3 furring				
12" o.c.	S.F.	0.32	0.81	1.13
16" o.c.	"	0.28	0.73	1.01
24" o.c.	"	0.22	0.66	0.88
Ceilings				
On masonry or concrete ceilings				
1x2 furring				
12" o.c.	S.F.	0.22	2.05	2.27
16" o.c.	"	0.19	1.80	1.99
24" o.c.	"	0.15	1.65	1.80
1x3 furring				
12" o.c.	S.F.	0.32	2.05	2.37

ROUGH CARPENTRY

ROUGH CARPENTRY	UNIT	MAT.	INST.	TOTAL
06110.40 FURRING				
16" o.c.	S.F.	0.28	1.80	2.08
24" o.c.	"	0.22	1.65	1.87
On wood ceilings				
1x2 furring				
12" o.c.	S.F.	0.22	1.35	1.57
16" o.c.	"	0.19	1.20	1.39
24" o.c.	"	0.15	1.10	1.25
1x3				
12" o.c.	S.F.	0.32	1.35	1.67
16" o.c.	"	0.28	1.20	1.48
24" o.c.	"	0.22	1.10	1.32
06110.50 ROOF FRAMING				
Roof framing				
Rafters, gable end				
4-6 pitch (4-in-12 to 6-in-12)				
16" o.c.				
2x6	S.F.	0.75	0.68	1.43
2x8	"	1.10	0.70	1.80
2x10	"	1.60	0.73	2.33
2x12	"	2.10	0.76	2.86
24" o.c.				
2x6	S.F.	0.61	0.57	1.18
2x8	"	0.84	0.59	1.43
2x10	"	1.25	0.63	1.88
2x12	"	1.70	0.70	2.40
Ridge boards				
2x6	L.F.	0.67	1.80	2.47
2x8	"	0.96	2.05	3.01
2x10	"	1.30	2.30	3.60
2x12	"	1.75	2.60	4.35
Hip rafters				
2x6	L.F.	0.67	1.30	1.97
2x8	"	0.96	1.35	2.31
2x10	"	1.30	1.40	2.70
2x12	"	1.75	1.45	3.20
Jack rafters				
4-6 pitch (4-in-12 to 6-in-12)				
16" o.c.				
2x6	S.F.	0.76	1.05	1.81
2x8	"	1.10	1.10	2.20
2x10	"	1.60	1.20	2.80
2x12	"	2.15	1.20	3.35
24" o.c.				
2x6	S.F.	0.61	0.83	1.44
2x8	"	0.84	0.85	1.69
2x10	"	1.25	0.89	2.14
2x12	"	1.70	0.91	2.61
Sister rafters				
2x4	L.F.	0.47	2.60	3.07
2x6	"	0.67	3.05	3.72

06 WOOD AND PLASTICS

ROUGH CARPENTRY	UNIT	MAT.	INST.	TOTAL
06110.50 ROOF FRAMING				
2x8	L.F.	0.96	3.65	4.61
2x10	"	1.30	4.55	5.85
2x12	"	1.75	6.10	7.85
Fascia boards				
2x4	L.F.	0.47	1.80	2.27
2x6	"	0.67	1.80	2.47
2x8	"	0.96	2.05	3.01
2x10	"	1.30	2.05	3.35
2x12	"	1.75	2.30	4.05
Cant strips				
Fiber				
3x3	L.F.	0.23	1.05	1.28
4x4	"	0.31	1.10	1.41
Wood				
3x3	L.F.	1.30	1.10	2.40
06110.60 SLEEPERS				
Sleepers, over concrete				
12" o.c.				
1x2	S.F.	0.20	0.83	1.03
1x3	"	0.28	0.87	1.15
2x4	"	0.57	1.00	1.57
2x6	"	0.84	1.05	1.89
16" o.c.				
1x2	S.F.	0.17	0.73	0.90
1x3	"	0.23	0.73	0.96
2x4	"	0.47	0.87	1.34
2x6	"	0.72	0.91	1.63
06110.65 SOFFITS				
Soffit framing				
2x3	L.F.	0.37	2.60	2.97
2x4	"	0.47	2.80	3.27
2x6	"	0.67	3.05	3.72
2x8	"	0.96	3.30	4.26
06110.70 WALL FRAMING				
Framing wall, studs				
12" o.c.				
2x3	S.F.	0.41	0.68	1.09
2x4	"	0.57	0.68	1.25
2x6	"	0.84	0.73	1.57
2x8	"	1.20	0.76	1.96
16" o.c.				
2x3	S.F.	0.35	0.57	0.92
2x4	"	0.47	0.57	1.04
2x6	"	0.67	0.61	1.28
2x8	"	1.00	0.63	1.63
24" o.c.				

06 WOOD AND PLASTICS

ROUGH CARPENTRY	UNIT	MAT.	INST.	TOTAL
06110.70 WALL FRAMING				
2x3	S.F.	0.26	0.49	0.75
2x4	"	0.39	0.49	0.88
2x6	"	0.57	0.52	1.09
2x8	"	0.77	0.54	1.31
Plates, top or bottom				
2x3	L.F.	0.37	1.05	1.42
2x4	"	0.47	1.15	1.62
2x6	"	0.67	1.20	1.87
2x8	"	0.96	1.30	2.26
Headers, door or window				
2x8				
Single				
4' long	EA.	4.00	22.80	26.80
8' long	"	8.00	28.10	36.10
Double				
4' long	EA.	8.00	26.10	34.10
8' long	"	15.95	33.20	49.15
2x12				
Single				
6' long	EA.	11.50	28.10	39.60
12' long	"	23.00	36.50	59.50
Double				
6' long	EA.	23.00	33.20	56.20
12' long	"	46.00	40.50	86.50
06115.10 FLOOR SHEATHING				
Sub-flooring, plywood, CDX				
1/2" thick	S.F.	0.50	0.46	0.96
5/8" thick	"	0.59	0.52	1.11
3/4" thick	"	0.70	0.61	1.31
Structural plywood				
1/2" thick	S.F.	0.55	0.46	1.01
5/8" thick	"	0.65	0.52	1.17
3/4" thick	"	0.77	0.56	1.33
Underlayment				
Hardboard, 1/4" tempered	S.F.	0.40	0.46	0.86
Plywood, CDX				
3/8" thick	S.F.	0.44	0.46	0.90
1/2" thick	"	0.51	0.49	1.00
5/8" thick	"	0.61	0.52	1.13
3/4" thick	"	0.72	0.56	1.28
06115.20 ROOF SHEATHING				
Sheathing				
Plywood, CDX				
3/8" thick	S.F.	0.44	0.47	0.91
1/2" thick	"	0.51	0.49	1.00
5/8" thick	"	0.61	0.52	1.13
3/4" thick	"	0.72	0.56	1.28

06 WOOD AND PLASTICS

ROUGH CARPENTRY	UNIT	MAT.	INST.	TOTAL
06115.30 WALL SHEATHING				
Sheathing				
Plywood, CDX				
3/8" thick	S.F.	0.44	0.54	0.98
1/2" thick	"	0.51	0.56	1.07
5/8" thick	"	0.61	0.61	1.22
3/4" thick	"	0.72	0.66	1.38
06125.10 WOOD DECKING				
Decking, T&G solid				
Fir				
3" thick	S.F.	2.50	0.91	3.41
4" thick	"	3.05	0.97	4.02
Southern yellow pine				
3" thick	S.F.	2.45	1.05	3.50
4" thick	"	2.65	1.10	3.75
White pine				
3" thick	S.F.	3.10	0.91	4.01
4" thick	"	4.10	0.97	5.07
06130.10 HEAVY TIMBER				
Mill framed structures				
Beams to 20' long				
Douglas fir				
6x8	L.F.	4.40	4.45	8.85
6x10	"	5.20	4.60	9.80
Southern yellow pine				
6x8	L.F.	3.50	4.45	7.95
6x10	"	4.25	4.60	8.85
Columns to 12' high				
6x6	L.F.	3.15	6.65	9.80
10x10	"	9.50	7.40	16.90
06190.20 WOOD TRUSSES				
Truss, fink, 2x4 members				
3-in-12 slope				
24' span	EA.	56.50	38.10	94.60
30' span	"	72.00	40.40	112.40

FINISH CARPENTRY	UNIT	MAT.	INST.	TOTAL
06200.10 FINISH CARPENTRY				
Casing				
11/16 x 2-1/2	L.F.	1.10	1.65	2.75

FINISH CARPENTRY

FINISH CARPENTRY	UNIT	MAT.	INST.	TOTAL
06200.10 FINISH CARPENTRY				
11/16 x 3-1/2	L.F.	1.40	1.75	3.15
Half round				
1/2	L.F.	0.43	1.45	1.88
5/8	"	0.57	1.45	2.02
Railings, balusters				
1-1/8 x 1-1/8	L.F.	1.90	3.65	5.55
1-1/2 x 1-1/2	"	2.20	3.30	5.50
Stop				
5/8 x 1-5/8				
Colonial	L.F.	0.44	2.30	2.74
Ranch	"	0.43	2.30	2.73
Exterior trim, casing, select pine, 1x3	"	1.50	1.80	3.30
Douglas fir				
1x3	L.F.	0.79	1.80	2.59
1x4	"	1.10	1.80	2.90
1x6	"	1.30	2.05	3.35
Cornices, white pine, #2 or better				
1x2	L.F.	0.48	1.80	2.28
1x4	"	0.74	1.80	2.54
1x8	"	1.30	2.15	3.45
Shelving, pine				
1x8	L.F.	1.55	2.80	4.35
1x10	"	1.90	2.90	4.80
1x12	"	2.45	3.05	5.50
06220.10 MILLWORK				
Countertop, laminated plastic				
25" x 7/8" thick				
Minimum	L.F.	12.35	9.10	21.45
Average	"	22.20	12.15	34.35
Maximum	"	37.00	14.60	51.60
Base cabinets, 34-1/2" high, 24" deep, hardwood, no tops				
Minimum	L.F.	43.20	14.60	57.80
Average	"	74.00	18.25	92.25
Maximum	"	120.00	24.30	144.30
Wall cabinets				
Minimum	L.F.	34.60	12.15	46.75
Average	"	55.50	14.60	70.10
Maximum	"	110.00	18.25	128.25

06 WOOD AND PLASTICS

WOOD TREATMENT

WOOD TREATMENT	UNIT	MAT.	INST.	TOTAL
06300.10 WOOD TREATMENT				
Creosote preservative treatment				
8 lb/cf	B.F.			0.26
10 lb/cf	"			0.34
Salt preservative treatment				
Oil borne				
Minimum	B.F.			0.20
Maximum	"			0.32
Water borne				
Minimum	B.F.			0.15
Maximum	"			0.24
Fire retardant treatment				
Minimum	B.F.			0.33
Maximum	"			0.40

ARCHITECTURAL WOODWORK

ARCHITECTURAL WOODWORK	UNIT	MAT.	INST.	TOTAL
06420.10 PANEL WORK				
Plywood unfinished, 1/4" thick				
Birch				
Natural	S.F.	1.05	1.20	2.25
Select	"	1.35	1.20	2.55
Knotty pine	"	1.35	1.20	2.55
Plywood, prefinished, 1/4" thick, premium grade				
Birch veneer	S.F.	1.85	1.45	3.30
Cherry veneer	"	2.10	1.45	3.55
06430.10 STAIRWORK				
Risers, 1x8, 42" wide				
White oak	EA.	15.15	18.25	33.40
Pine	"	10.70	18.25	28.95
Treads, 1-1/16" x 9-1/2" x 42"				
White oak	EA.	21.45	22.80	44.25
06440.10 COLUMNS				
Column, hollow, round wood				
12" diameter				
10' high	EA.	420.00	49.80	469.80
12' high	"	520.00	53.50	573.50
24" diameter				
16' high	EA.	1,770	74.50	1,845
18' high	"	2,000	78.50	2,079

07 THERMAL AND MOISTURE

MOISTURE PROTECTION	UNIT	MAT.	INST.	TOTAL
07100.10 WATERPROOFING				
Membrane waterproofing, elastomeric				
Butyl				
1/32" thick	S.F.	0.59	1.15	1.74
1/16" thick	"	0.77	1.20	1.97
Butyl with nylon				
1/32" thick	S.F.	0.70	1.15	1.85
1/16" thick	"	0.83	1.20	2.03
Neoprene				
1/32" thick	S.F.	0.98	1.15	2.13
1/16" thick	"	1.60	1.20	2.80
Neoprene with nylon				
1/32" thick	S.F.	1.05	1.15	2.20
1/16" thick	"	1.70	1.20	2.90
Plastic vapor barrier (polyethylene)				
4 mil	S.F.	0.01	0.11	0.12
6 mil	"	0.02	0.11	0.13
10 mil	"	0.04	0.14	0.18
Bituminous membrane waterproofing, asphalt felt, 15 lb.				
One ply	S.F.	0.35	0.71	1.06
Two ply	"	0.41	0.87	1.28
Three ply	"	0.52	1.00	1.52
Four ply	"	0.58	1.20	1.78
Five ply	"	0.64	1.50	2.14
Modified asphalt membrane waterproofing, fibrous asphalt				
One ply	S.F.	0.80	1.20	2.00
Two ply	"	1.20	1.45	2.65
Three ply	"	1.35	1.60	2.95
Four ply	"	1.65	1.90	3.55
Five ply	"	1.95	2.30	4.25
Asphalt coated protective board				
1/8" thick	S.F.	0.45	0.71	1.17
1/4" thick	"	0.64	0.71	1.36
3/8" thick	"	0.70	0.71	1.42
1/2" thick	"	0.84	0.75	1.59
Cement protective board				
3/8" thick	S.F.	1.55	0.95	2.50
1/2" thick	"	2.20	0.95	3.15
Fluid applied, neoprene				
50 mil	S.F.	1.90	0.95	2.85
90 mil	"	2.60	0.95	3.55
Tab extended polyurethane				
.050" thick	S.F.	1.00	0.71	1.72
Fluid applied rubber based polyurethane				
6 mil	S.F.	0.54	0.89	1.43
15 mil	"	1.00	0.71	1.72
Bentonite waterproofing, panels				
3/16" thick	S.F.	0.91	0.71	1.63
1/4" thick	"	1.05	0.71	1.76
5/8" thick	"	1.55	0.75	2.30
Granular admixtures, trowel on, 3/8" thick	"	1.25	0.71	1.97
Metallic oxide waterproofing, iron compound, troweled				
5/8" thick	S.F.	0.94	0.71	1.66

07 THERMAL AND MOISTURE

MOISTURE PROTECTION	UNIT	MAT.	INST.	TOTAL
07100.10 WATERPROOFING				
3/4" thick	S.F.	1.15	0.82	1.97
07150.10 DAMPPROOFING				
Silicone dampproofing, sprayed on				
Concrete surface				
1 coat	S.F.	0.37	0.16	0.53
2 coats	"	0.62	0.22	0.84
Concrete block				
1 coat	S.F.	0.37	0.19	0.56
2 coats	"	0.62	0.26	0.88
Brick				
1 coat	S.F.	0.43	0.22	0.65
2 coats	"	0.67	0.29	0.96
07160.10 BITUMINOUS DAMPPROOFING				
Building paper, asphalt felt				
15 lb	S.F.	0.08	1.15	1.23
30 lb	"	0.14	1.20	1.34
Asphalt dampproofing, troweled, cold, primer plus				
1 coat	S.F.	0.51	0.95	1.46
2 coats	"	0.76	1.45	2.21
3 coats	"	1.15	1.80	2.95
Fibrous asphalt dampproofing, hot troweled, primer plus				
1 coat	S.F.	0.79	1.15	1.94
2 coats	"	0.99	1.60	2.59
3 coats	"	1.05	2.05	3.10
Asphaltic paint dampproofing, per coat				
Brush on	S.F.	0.14	0.41	0.55
Spray on	"	0.25	0.32	0.57
07190.10 VAPOR BARRIERS				
Vapor barrier, polyethylene				
2 mil	S.F.	0.01	0.14	0.15
6 mil	"	0.02	0.14	0.16
8 mil	"	0.03	0.16	0.19
10 mil	"	0.04	0.16	0.20

INSULATION

07210.10 BATT INSULATION

INSULATION	UNIT	MAT.	INST.	TOTAL
Ceiling, fiberglass, unfaced				
3-1/2" thick, R11	S.F.	0.26	0.34	0.60
6" thick, R19	"	0.34	0.38	0.72
9" thick, R30	"	0.68	0.44	1.12
Suspended ceiling, unfaced				
3-1/2" thick, R11	S.F.	0.26	0.32	0.58
6" thick, R19	"	0.34	0.36	0.70
9" thick, R30	"	0.68	0.41	1.09
Wall, fiberglass				
Paper backed				
2" thick, R7	S.F.	0.17	0.30	0.47
3" thick, R8	"	0.18	0.32	0.50
4" thick, R11	"	0.31	0.34	0.65
6" thick, R19	"	0.45	0.36	0.81
Foil backed, 1 side				
2" thick, R7	S.F.	0.39	0.30	0.69
3" thick, R11	"	0.41	0.32	0.73
4" thick, R14	"	0.43	0.34	0.77
Foil backed, 2 sides				
Unfaced				
2" thick, R7	S.F.	0.25	0.30	0.55
3" thick, R9	"	0.29	0.32	0.61
4" thick, R11	"	0.31	0.34	0.65
6" thick, R19	"	0.39	0.36	0.75

07210.20 BOARD INSULATION

INSULATION	UNIT	MAT.	INST.	TOTAL
Insulation, rigid				
0.75" thick, R2.78	S.F.	0.24	0.26	0.50
1.06" thick, R4.17	"	0.37	0.27	0.64
1.31" thick, R5.26	"	0.50	0.29	0.79
1.63" thick, R6.67	"	0.62	0.30	0.92
2.25" thick, R8.33	"	0.68	0.32	1.00
Perlite board, roof				
1.00" thick, R2.78	S.F.	0.52	0.24	0.76
1.50" thick, R4.17	"	0.80	0.25	1.05
2.00" thick, R5.92	"	0.99	0.26	1.25
2.50" thick, R6.67	"	1.20	0.27	1.47
Rigid urethane				
Roof				
1" thick, R6.67	S.F.	0.70	0.24	0.94
1.20" thick, R8.33	"	0.79	0.24	1.03
1.50" thick, R11.11	"	0.96	0.25	1.21
2" thick, R14.29	"	1.25	0.26	1.51
2.25" thick, R16.67	"	1.60	0.27	1.87
Polystyrene				
Roof				
1.0" thick, R4.17	S.F.	0.66	0.24	0.90
1.5" thick, R6.26	"	1.10	0.25	1.35
2.0" thick, R8.33	"	1.20	0.26	1.46
Wall				
1.0" thick, R4.17	S.F.	0.66	0.30	0.96
1.5" thick, R6.26	"	1.10	0.32	1.42

INSULATION	UNIT	MAT.	INST.	TOTAL
07210.20 BOARD INSULATION				
2.0" thick, R8.33	S.F.	1.20	0.34	1.54
07210.60 LOOSE FILL INSULATION				
Blown-in type				
Fiberglass				
5" thick, R11	S.F.	0.24	0.24	0.48
6" thick, R13	"	0.29	0.29	0.58
9" thick, R19	"	0.34	0.41	0.75
Poured type				
Fiberglass				
1" thick, R4	S.F.	0.22	0.18	0.40
2" thick, R8	"	0.41	0.20	0.61
3" thick, R12	"	0.61	0.24	0.85
4" thick, R16	"	0.79	0.29	1.08
Vermiculite or perlite				
2" thick, R4.8	S.F.	0.52	0.20	0.72
3" thick, R7.2	"	0.74	0.24	0.98
4" thick, R9.6	"	0.97	0.29	1.26
Masonry, poured vermiculite or perlite				
4" block	S.F.	0.39	0.14	0.53
6" block	"	0.50	0.18	0.68
8" block	"	0.75	0.20	0.95
10" block	"	0.89	0.22	1.11
12" block	"	1.00	0.24	1.24
07210.70 SPRAYED INSULATION				
Foam, sprayed on				
Polystyrene				
1" thick, R4	S.F.	0.50	0.29	0.79
2" thick, R8	"	0.99	0.38	1.37
Urethane				
1" thick, R4	S.F.	0.44	0.29	0.73
2" thick, R8	"	0.92	0.38	1.30
07250.10 FIREPROOFING				
Sprayed on				
1" thick				
On beams	S.F.	0.47	0.64	1.11
On columns	"	0.48	0.57	1.05
On decks				
Flat surface	S.F.	0.48	0.29	0.77
Fluted surface	"	0.61	0.36	0.97
1-1/2" thick				
On beams	S.F.	0.85	0.82	1.67
On columns	"	0.97	0.71	1.69
On decks				
Flat surface	S.F.	0.73	0.36	1.09
Fluted surface	"	0.85	0.48	1.33

SHINGLES AND TILES	UNIT	MAT.	INST.	TOTAL
07310.10 ASPHALT SHINGLES				
Standard asphalt shingles, strip shingles				
210 lb/square	SQ.	38.80	35.20	74.00
240 lb/square	"	42.40	44.00	86.40
Roll roofing, mineral surface				
90 lb	SQ.	21.80	25.10	46.90
140 lb	"	37.60	35.20	72.80
07310.30 METAL SHINGLES				
Aluminum, .020" thick				
Plain	SQ.	180.00	70.50	250.50
Steel, galvanized				
Plain	SQ.	180.00	70.50	250.50
07310.60 SLATE SHINGLES				
Slate shingles				
Ribbon	SQ.	430.00	180.00	610.00
Clear	"	560.00	180.00	740.00
Replacement shingles				
Small jobs	EA.	8.35	11.75	20.10
Large jobs	S.F.	6.55	5.85	12.40
07310.70 WOOD SHINGLES				
Wood shingles, on roofs				
White cedar, #1 shingles				
4" exposure	SQ.	170.00	120.00	290.00
5" exposure	"	150.00	88.00	238.00
#2 shingles				
4" exposure	SQ.	120.00	120.00	240.00
5" exposure	"	100.00	88.00	188.00
Resquared and rebutted				
4" exposure	SQ.	150.00	120.00	270.00
5" exposure	"	130.00	88.00	218.00
Add for fire retarding	"			73.80
07310.80 WOOD SHAKES				
Shakes, hand split, 24" red cedar, on roofs				
5" exposure	SQ.	180.00	180.00	360.00
7" exposure	"	170.00	140.00	310.00
9" exposure	"	150.00	120.00	270.00
Add for fire retarding	"			53.00

07 THERMAL AND MOISTURE

ROOFING AND SIDING	UNIT	MAT.	INST.	TOTAL
07410.10 MANUFACTURED ROOFS				
Aluminum roof panels, for structural steel framing				
Corrugated				
Unpainted finish				
.024"	S.F.	1.25	0.88	2.13
.030"	"	1.45	0.88	2.33
Painted finish				
.024"	S.F.	1.60	0.88	2.48
.030"	"	1.95	0.88	2.83
Steel roof panels, for structural steel framing				
Corrugated, painted				
18 ga.	S.F.	3.70	0.88	4.58
20 ga.	"	3.45	0.88	4.33
07460.10 METAL SIDING PANELS				
Aluminum siding panels				
Corrugated				
Plain finish				
.024"	S.F.	1.10	1.60	2.70
.032"	"	1.35	1.60	2.95
Painted finish				
.024"	S.F.	1.40	1.60	3.00
.032"	"	1.60	1.60	3.20
Steel siding panels				
Corrugated				
22 ga.	S.F.	1.30	2.65	3.95
24 ga.	"	1.15	2.65	3.80
07460.70 STEEL SIDING				
Ribbed, sheets, galvanized				
22 ga.	S.F.	1.30	1.60	2.90
24 ga.	"	1.15	1.60	2.75
Primed				
24 ga.	S.F.	1.50	1.60	3.10
26 ga.	"	1.10	1.60	2.70

MEMBRANE ROOFING	UNIT	MAT.	INST.	TOTAL
07510.10 BUILT-UP ASPHALT ROOFING				
Built-up roofing, asphalt felt, including gravel				
2 ply	SQ.	28.20	88.00	116.20
3 ply	"	38.60	120.00	158.60
4 ply	"	55.00	140.00	195.00
Walkway, for built-up roofs				
3' x 3' x				

MEMBRANE ROOFING

07510.10 BUILT-UP ASPHALT ROOFING

	UNIT	MAT.	INST.	TOTAL
1/2" thick	S.F.	2.70	1.15	3.85
3/4" thick	"	3.55	1.15	4.70
1" thick	"	3.90	1.15	5.05
Roof bonds				
Asphalt felt				
10 yrs	SQ.			21.41
20 yrs	"			22.63
Cant strip, 4" x 4"				
Treated wood	L.F.	1.20	1.00	2.20
Foamglass	"	0.64	0.88	1.52
Mineral fiber	"	0.28	0.88	1.16
New gravel for built-up roofing, 400 lb/sq	SQ.	19.90	70.50	90.40
Roof gravel (ballast)	C.Y.	13.35	180.00	193.35
Aluminum coating, top surfacing, for built-up roofing	SQ.	21.05	58.50	79.55
Remove 4-ply built-up roof (includes gravel)	"	0.00	180.00	180.00
Remove & replace gravel, includes flood coat	"	28.10	120.00	148.10

07530.10 SINGLE-PLY ROOFING

	UNIT	MAT.	INST.	TOTAL
Elastic sheet roofing				
Neoprene, 1/16" thick	S.F.	1.55	0.44	1.99
EPDM rubber				
45 mil	S.F.	0.84	0.44	1.28
PVC				
45 mil	S.F.	1.15	0.44	1.59
Flashing				
Pipe flashing, 90 mil thick				
1" pipe	EA.	16.55	8.80	25.35
Neoprene flashing, 60 mil thick strip				
6" wide	L.F.	0.99	2.95	3.94
12" wide	"	1.95	4.40	6.35
18" wide	"	2.90	5.85	8.75
24" wide	"	3.80	8.80	12.60
Adhesives				
Mastic sealer, applied at joints only				
1/4" bead	L.F.	0.06	0.18	0.24
Fluid applied roofing				
Urethane, 2 components, elastomeric top membrane				
1" thick	S.F.	1.70	0.59	2.29
Vinyl liquid roofing, 2 coats, 2 mils per coat	"	2.70	0.50	3.20
Silicone roofing, 2 coats sprayed, 16 mil per coat	"	2.00	0.59	2.59
Inverted roof system				
Insulated membrane with coarse gravel ballast				
3 ply with 2" polystyrene	S.F.	3.65	0.59	4.24
Ballast, 3/4" through 1-1/2" dia. river gravel, 100lb/sf	"	0.24	35.20	35.44
Walkway for membrane roofs, 1/2" thick	"	1.35	1.15	2.50

FLASHING AND SHEET METAL

FLASHING AND SHEET METAL	UNIT	MAT.	INST.	TOTAL
07610.10 METAL ROOFING				
Sheet metal roofing, copper, 16 oz, batten seam	SQ.	480.00	230.00	710.00
Standing seam	"	470.00	220.00	690.00
Aluminum roofing, natural finish				
Corrugated, on steel frame				
.0175" thick	SQ.	86.50	100.00	186.50
.0215" thick	"	120.00	100.00	220.00
.024" thick	"	140.00	100.00	240.00
.032" thick	"	170.00	100.00	270.00
Ridge cap				
.019" thick	L.F.	2.60	1.15	3.75
Corrugated galvanized steel roofing, on steel frame				
28 ga.	SQ.	84.00	100.00	184.00
26 ga.	"	96.00	100.00	196.00
24 ga.	"	110.00	100.00	210.00
22 ga.	"	120.00	100.00	220.00
07620.10 FLASHING AND TRIM				
Counter flashing				
Aluminum, .032"	S.F.	0.99	3.50	4.49
Stainless steel, .015"	"	2.85	3.50	6.35
Copper				
16 oz.	S.F.	3.00	3.50	6.50
20 oz.	"	3.85	3.50	7.35
24 oz.	"	4.70	3.50	8.20
32 oz.	"	6.05	3.50	9.55
Valley flashing				
Aluminum, .032"	S.F.	0.99	2.20	3.19
Stainless steel, .015	"	2.85	2.20	5.05
Copper				
16 oz.	S.F.	3.00	2.20	5.20
20 oz.	"	3.85	2.95	6.80
24 oz.	"	4.70	2.20	6.90
32 oz.	"	6.05	2.20	8.25
Base flashing				
Aluminum, .040"	S.F.	1.70	2.95	4.65
Stainless steel, .018"	"	3.40	2.95	6.35
Copper				
16 oz.	S.F.	3.00	2.95	5.95
20 oz.	"	3.85	2.20	6.05
24 oz.	"	4.70	2.95	7.65
32 oz.	"	6.05	2.95	9.00
Flashing and trim, aluminum				
.019" thick	S.F.	0.80	2.50	3.30
.032" thick	"	0.99	2.50	3.49
.040" thick	"	1.70	2.70	4.40
Reglets, copper 10 oz.	L.F.	2.10	2.35	4.45
Stainless steel, .020"	"	1.60	2.35	3.95
Gravel stop				
Aluminum, .032"				
4"	L.F.	3.25	1.15	4.40
10"	"	6.80	1.35	8.15

07 THERMAL AND MOISTURE

FLASHING AND SHEET METAL

FLASHING AND SHEET METAL	UNIT	MAT.	INST.	TOTAL
07620.10 FLASHING AND TRIM				
Copper, 16 oz.				
4"	L.F.	4.75	1.15	5.90
10"	"	8.60	1.35	9.95
07700.10 MANUFACTURED SPECIALTIES				
Smoke vent, 48" x 48"				
Aluminum	EA.	620.00	88.00	708.00
Galvanized steel	"	600.00	88.00	688.00
Heat/smoke vent, 48" x 96"				
Aluminum	EA.	1,920	120.00	2,040
Galvanized steel	"	1,800	120.00	1,920
Ridge vent strips				
Mill finish	L.F.	2.55	2.35	4.90
Soffit vents				
Mill finish				
2-1/2" wide	L.F.	0.32	1.40	1.72
Roof hatches				
Steel, plain, primed				
2'6" x 3'0"	EA.	420.00	88.00	508.00
Galvanized steel				
2'6" x 3'0"	EA.	430.00	88.00	518.00
Aluminum				
2'6" x 3'0"	EA.	450.00	88.00	538.00
Gravity ventilators, with curb, base, damper and screen				
Wind driven spinner				
6" dia.	EA.	37.00	23.45	60.45
12" dia.	"	49.70	23.45	73.15

SKYLIGHTS

SKYLIGHTS	UNIT	MAT.	INST.	TOTAL
07810.10 PLASTIC SKYLIGHTS				
Single thickness, not including mounting curb				
2' x 4'	EA.	230.00	44.00	274.00
4' x 4'	"	300.00	58.50	358.50
Double thickness, not including mounting curb				
2' x 4'	EA.	300.00	44.00	344.00
4' x 4'	"	370.00	58.50	428.50

08 DOORS AND WINDOWS

METAL	UNIT	MAT.	INST.	TOTAL
08110.10 METAL DOORS				
Flush hollow metal, standard duty, 20 ga., 1-3/8" thick				
2-6 x 6-8	EA.	150.00	40.50	190.50
2-8 x 6-8	"	180.00	40.50	220.50
3-0 x 6-8	"	190.00	40.50	230.50
1-3/4" thick				
2-6 x 6-8	EA.	180.00	40.50	220.50
2-8 x 6-8	"	190.00	40.50	230.50
3-0 x 6-8	"	200.00	40.50	240.50
Heavy duty, 20 ga., unrated, 1-3/4"				
2-8 x 6-8	EA.	190.00	40.50	230.50
3-0 x 6-8	"	200.00	40.50	240.50
08110.40 METAL DOOR FRAMES				
Hollow metal, stock, 18 ga., 4-3/4" x 1-3/4"				
2-0 x 7-0	EA.	90.00	45.60	135.60
2-4 x 7-0	"	95.50	45.60	141.10
2-6 x 7-0	"	98.50	45.60	144.10
3-0 x 7-0	"	100.00	45.60	145.60
08120.10 ALUMINUM DOORS				
Aluminum doors, commercial				
Narrow stile				
2-6 x 7-0	EA.	450.00	200.00	650.00
3-0 x 7-0	"	470.00	200.00	670.00
3-6 x 7-0	"	490.00	200.00	690.00
Wide stile				
2-6 x 7-0	EA.	690.00	200.00	890.00
3-0 x 7-0	"	720.00	200.00	920.00
3-6 x 7-0	"	740.00	200.00	940.00
08300.10 SPECIAL DOORS				
Overhead door, coiling insulated				
Chain gear, no frame, 12' x 12'	EA.	1,850	460.00	2,310
Sliding metal fire doors, motorized, fusible link, 3 hr.				
3-0 x 6-8	EA.	2,370	730.00	3,100
3-8 x 6-8	"	2,370	730.00	3,100
4-0 x 8-0	"	2,550	730.00	3,280
5-0 x 8-0	"	2,610	730.00	3,340
Counter doors, (roll-up shutters), standard, manual				
Opening, 4' high				
4' wide	EA.	840.00	300.00	1,140
6' wide	"	1,150	300.00	1,450
8' wide	"	1,340	330.00	1,670
10' wide	"	1,460	460.00	1,920
14' wide	"	1,820	460.00	2,280
6' high				
4' wide	EA.	1,040	300.00	1,340
6' wide	"	1,340	330.00	1,670
8' wide	"	1,440	360.00	1,800

METAL	UNIT	MAT.	INST.	TOTAL
08300.10 SPECIAL DOORS				
10' wide	EA.	1,620	460.00	2,080
14' wide	"	1,730	520.00	2,250
For stainless steel, add to material, 40%				
For motor operator, add	EA.			924.00
Service doors, (roll up shutters), standard, manual				
Opening				
8' high x 8' wide	EA.	1,080	200.00	1,280
10' high x 10' wide	"	1,270	300.00	1,570
12' high x 12' wide	"	1,500	460.00	1,960
14' high x 14' wide	"	1,960	610.00	2,570
16' high x 14' wide	"	2,660	610.00	3,270
20' high x 14' wide	"	3,760	910.00	4,670
24' high x 16' wide	"	5,100	810.00	5,910
For motor operator				
Up to 12-0 x 12-0, add	EA.			1,044
Over 12-0 x 12-0, add	"			1,335
Roll-up doors				
13-0 high x 14-0 wide	EA.	920.00	520.00	1,440
12-0 high x 14-0 wide	"	1,160	520.00	1,680
Top coiling grilles, manually operated, steel or aluminum				
Opening, 4' high x				
4' wide	EA.	1,090	150.00	1,240
6' wide	"	1,120	150.00	1,270
8' wide	"	1,270	200.00	1,470
12' wide	"	1,500	200.00	1,700
16' wide	"	1,730	300.00	2,030
6' high x				
4' wide	EA.	1,160	300.00	1,460
6' wide	"	1,270	330.00	1,600
8' wide	"	1,330	360.00	1,690
12' wide	"	1,620	410.00	2,030
16' wide	"	2,080	520.00	2,600
Side coiling grilles, manually operated, aluminum				
Opening, 8' high x				
18' wide	EA.	8,090	3,330	11,420
24' wide	"	10,400	3,810	14,210
12' high x				
12' wide	EA.	8,090	3,330	11,420
18' wide	"	10,400	3,810	14,210
24' wide	"	13,860	4,440	18,300

08 DOORS AND WINDOWS

STOREFRONTS

08410.10 STOREFRONTS	UNIT	MAT.	INST.	TOTAL
Storefront, aluminum and glass				
Minimum	S.F.	12.60	4.95	17.55
Average	"	18.05	5.65	23.70
Maximum	"	30.10	6.60	36.70

METAL WINDOWS

08510.10 STEEL WINDOWS	UNIT	MAT.	INST.	TOTAL
Steel windows, primed				
Casements				
Operable				
Minimum	S.F.	23.80	2.35	26.15
Maximum	"	35.70	2.65	38.35
Fixed sash	"	19.05	2.00	21.05
Double hung	"	35.70	2.20	37.90
Industrial windows				
Horizontally pivoted sash	S.F.	33.30	2.65	35.95
Fixed sash	"	26.20	2.20	28.40
Security sash				
Operable	S.F.	41.60	2.65	44.25
Fixed	"	36.90	2.20	39.10
Picture window	"	17.85	2.20	20.05
Projecting sash				
Minimum	S.F.	30.90	2.50	33.40
Maximum	"	38.10	2.50	40.60
Mullions	L.F.	8.15	2.00	10.15

08520.10 ALUMINUM WINDOWS	UNIT	MAT.	INST.	TOTAL
Fixed window				
6 sf to 8 sf	S.F.	10.40	5.65	16.05
12 sf to 16 sf	"	9.25	4.40	13.65
Projecting window				
6 sf to 8 sf	S.F.	23.10	9.90	33.00
12 sf to 16 sf	"	20.80	6.60	27.40
Horizontal sliding				
6 sf to 8 sf	S.F.	15.05	4.95	20.00
12 sf to 16 sf	"	13.85	3.95	17.80
Double hung				
6 sf to 8 sf	S.F.	20.80	7.95	28.75
10 sf to 12 sf	"	18.50	6.60	25.10

08 DOORS AND WINDOWS

HARDWARE	UNIT	MAT.	INST.	TOTAL
08710.10 HINGES				
Hinges				
3 x 3 butts, steel, interior, plain bearing	PAIR			12.25
4 x 4 butts, steel, standard	"			18.39
5 x 4-1/2 butts, bronze/s. steel, heavy duty	"			49.04
08710.20 LOCKSETS				
Latchset, heavy duty				
Cylindrical	EA.	100.00	22.80	122.80
Mortise	"	100.00	36.50	136.50
Lockset, heavy duty				
Cylindrical	EA.	130.00	22.80	152.80
Mortise	"	150.00	36.50	186.50
08710.30 CLOSERS				
Door closers				
Surface mounted, traditional type, parallel arm				
Standard	EA.	110.00	45.60	155.60
Heavy duty	"	120.00	45.60	165.60
08710.40 DOOR TRIM				
Panic device				
Mortise	EA.	530.00	91.00	621.00
Vertical rod	"	610.00	91.00	701.00
Labelled, rim type	"	550.00	91.00	641.00
Mortise	"	660.00	91.00	751.00
Vertical rod	"	740.00	91.00	831.00
Door plates				
Kick plate, aluminum, 3 beveled edges				
10" x 28"	EA.	11.70	18.25	29.95
10" x 38"	"	16.20	18.25	34.45
Push plate, 4" x 16"				
Aluminum	EA.	11.55	7.30	18.85
Bronze	"	58.00	7.30	65.30
Stainless steel	"	46.20	7.30	53.50
08710.60 WEATHERSTRIPPING				
Weatherstrip, head and jamb, metal strip, neoprene bulb				
Standard duty	L.F.	2.65	2.05	4.70
Heavy duty	"	3.35	2.30	5.65
Thresholds				
Bronze	L.F.	33.30	9.10	42.40
Aluminum				
Plain	L.F.	12.50	9.10	21.60
Vinyl insert	"	19.05	9.10	28.15
Aluminum with grit	"	17.85	9.10	26.95
Steel				
Plain	L.F.	12.50	9.10	21.60
Interlocking	"	19.05	30.40	49.45

08 DOORS AND WINDOWS

GLAZING

08810.10 GLAZING

GLAZING	UNIT	MAT.	INST.	TOTAL
Sheet glass, 1/8" thick	S.F.	4.95	2.20	7.15
Plate glass, bronze or grey, 1/4" thick	"	7.40	3.60	11.00
Clear	"	5.70	3.60	9.30
Polished	"	6.30	3.60	9.90
Plexiglass				
1/8" thick	S.F.	2.15	3.60	5.75
1/4" thick	"	3.95	2.20	6.15
Float glass, clear				
1/4" thick	S.F.	4.25	3.60	7.85
1/2" thick	"	14.30	6.60	20.90
3/4" thick	"	17.85	9.90	27.75
1" thick	"	31.50	13.25	44.75
Tinted glass, polished plate, twin ground				
1/4" thick	S.F.	5.65	3.60	9.25
1/2" thick	"	14.80	6.60	21.40
Total, full vision, all glass window system				
To 10' high				
Minimum	S.F.	23.00	9.90	32.90
Average	"	30.30	9.90	40.20
Maximum	"	36.30	9.90	46.20
10' to 20' high				
Minimum	S.F.	27.80	9.90	37.70
Average	"	33.90	9.90	43.80
Maximum	"	42.40	9.90	52.30
Insulated glass, bronze or gray				
1/2" thick	S.F.	10.55	6.60	17.15
1" thick	"	12.55	9.90	22.45
Spandrel glass, polished bronze/grey, 1 side, 1/4" thick	"	8.45	3.60	12.05
Tempered glass (safety)				
Clear sheet glass				
1/8" thick	S.F.	5.80	2.20	8.00
3/16" thick	"	7.00	3.05	10.05
Clear float glass				
1/4" thick	S.F.	6.05	3.30	9.35
1/2" thick	"	18.15	6.60	24.75
3/4" thick	"	25.40	13.25	38.65
Tinted float glass				
3/16" thick	S.F.	7.25	3.05	10.30
1/4" thick	"	8.00	3.30	11.30
3/8" thick	"	14.50	4.95	19.45
1/2" thick	"	19.35	6.60	25.95
Laminated glass				
Float safety glass with polyvinyl plastic interlayer				
1/4", sheet or float				
Two lites, 1/8" thick, clear glass	S.F.	8.45	3.30	11.75
1/2" thick, float glass				
Two lites, 1/4" thick, clear glass	S.F.	12.60	6.60	19.20
Tinted glass	"	15.05	6.60	21.65
Insulating glass, two lites, clear float glass				
1/2" thick	S.F.	8.25	6.60	14.85
3/4" thick	"	10.45	9.90	20.35
1" thick	"	11.75	13.25	25.00

GLAZING

08810.10 GLAZING	UNIT	MAT.	INST.	TOTAL
Glass seal edge				
3/8" thick	S.F.	6.95	6.60	13.55
Tinted glass				
1/2" thick	S.F.	14.15	6.60	20.75
1" thick	"	15.20	13.25	28.45
Tempered, clear				
1" thick	S.F.	27.70	13.25	40.95
Wire reinforced	"	35.20	13.25	48.45
Plate mirror glass				
1/4" thick				
15 sf	S.F.	7.25	3.95	11.20
Over 15 sf	"	6.65	3.60	10.25
Door type, 1/4" thick	"	7.60	3.95	11.55
Transparent, one way vision, 1/4" thick	"	16.15	3.95	20.10
Sheet mirror glass				
3/16" thick	S.F.	7.00	3.95	10.95
1/4" thick	"	7.40	3.30	10.70
Wall tiles, 12" x 12"				
Clear glass	S.F.	2.40	2.20	4.60
Veined glass	"	2.80	2.20	5.00
Wire glass, 1/4" thick				
Clear	S.F.	9.80	13.25	23.05
Hammered	"	9.95	13.25	23.20
Obscure	"	11.90	13.25	25.15
Glazing accessories				
Neoprene glazing gaskets				
1/4" glass	L.F.	0.90	1.60	2.50
1/2" glass	"	1.40	1.75	3.15
3/4" glass	"	1.90	1.80	3.70
1" glass	"	2.15	2.00	4.15
Mullion section				
1/4" glass	L.F.	0.40	0.79	1.19
3/8" glass	"	0.51	0.99	1.50
1/2" glass	"	0.75	1.15	1.90
3/4" glass	"	1.15	1.30	2.45
1" glass	"	1.55	1.60	3.15
Molded corners	EA.	1.65	26.50	28.15

GLAZED CURTAIN WALLS

08910.10 GLAZED CURTAIN WALLS	UNIT	MAT.	INST.	TOTAL
Curtain wall, aluminum system, framing sections				
2" x 3"				

GLAZED CURTAIN WALLS

08910.10 GLAZED CURTAIN WALLS	UNIT	MAT.	INST.	TOTAL
Jamb	L.F.	7.25	3.30	10.55
Horizontal	"	7.35	3.30	10.65
Mullion	"	9.80	3.30	13.10
2" x 4"				
Jamb	L.F.	9.80	4.95	14.75
Horizontal	"	10.10	4.95	15.05
Mullion	"	9.80	4.95	14.75
3" x 5-1/2"				
Jamb	L.F.	12.95	4.95	17.90
Horizontal	"	14.45	4.95	19.40
Mullion	"	13.10	4.95	18.05
4" corner mullion	"	17.30	6.60	23.90
Coping sections				
1/8" x 8"	L.F.	18.15	6.60	24.75
1/8" x 9"	"	18.25	6.60	24.85
1/8" x 12-1/2"	"	18.70	7.95	26.65
Sill section				
1/8" x 6"	L.F.	18.75	3.95	22.70
1/8" x 7"	"	18.05	3.95	22.00
1/8" x 8-1/2"	"	18.45	3.95	22.40
Column covers, aluminum				
1/8" x 26"	L.F.	18.05	9.90	27.95
1/8" x 34"	"	18.05	10.45	28.50
1/8" x 38"	"	18.25	10.45	28.70
Doors				
Aluminum framed, standard hardware				
Narrow stile				
2-6 x 7-0	EA.	400.00	200.00	600.00
3-0 x 7-0	"	410.00	200.00	610.00
3-6 x 7-0	"	420.00	200.00	620.00
Wide stile				
2-6 x 7-0	EA.	690.00	200.00	890.00
3-0 x 7-0	"	750.00	200.00	950.00
3-6 x 7-0	"	800.00	200.00	1,000
Window wall system, complete				
Minimum	S.F.	13.00	3.95	16.95
Average	"	20.75	4.40	25.15
Maximum	"	48.00	5.65	53.65
Added costs				
For bronze, add 20% to material				
For stainless steel, add 50% to material				

SUPPORT SYSTEMS

09110.10 METAL STUDS

SUPPORT SYSTEMS	UNIT	MAT.	INST.	TOTAL
Studs, non load bearing, galvanized				
2-1/2", 20 ga.				
12" o.c.	S.F.	0.64	0.76	1.40
16" o.c.	"	0.54	0.61	1.15
25 ga.				
12" o.c.	S.F.	0.45	0.76	1.21
16" o.c.	"	0.32	0.61	0.93
24" o.c.	"	0.23	0.51	0.74
3-5/8", 20 ga.				
12" o.c.	S.F.	0.77	0.91	1.68
16" o.c.	"	0.65	0.73	1.38
24" o.c.	"	0.53	0.61	1.14
25 ga.				
12" o.c.	S.F.	0.44	0.91	1.35
16" o.c.	"	0.37	0.73	1.10
24" o.c.	"	0.31	0.61	0.92
4", 20 ga.				
12" o.c.	S.F.	0.89	0.91	1.80
16" o.c.	"	0.72	0.73	1.45
24" o.c.	"	0.58	0.61	1.19
25 ga.				
12" o.c.	S.F.	0.53	0.91	1.44
16" o.c.	"	0.44	0.73	1.17
24" o.c.	"	0.36	0.61	0.97
6", 20 ga.				
12" o.c.	S.F.	1.10	1.15	2.25
16" o.c.	"	0.89	0.91	1.80
24" o.c.	"	0.70	0.76	1.46
25 ga.				
12" o.c.	S.F.	0.66	1.15	1.81
16" o.c.	"	0.54	0.91	1.45
24" o.c.	"	0.45	0.76	1.21
Load bearing studs, galvanized				
3-5/8", 16 ga.				
12" o.c.	S.F.	1.30	0.91	2.21
16" o.c.	"	1.05	0.73	1.78
18 ga.				
12" o.c.	S.F.	1.15	0.61	1.76
16" o.c.	"	0.92	0.73	1.65
4", 16 ga.				
12" o.c.	S.F.	1.40	0.91	2.31
16" o.c.	"	1.15	0.73	1.88
6", 16 ga.				
12" o.c.	S.F.	1.70	1.15	2.85
16" o.c.	"	1.40	0.91	2.31
Furring				
On beams and columns				
7/8" channel	L.F.	0.33	2.45	2.78
1-1/2" channel	"	0.40	2.80	3.20
On ceilings				
3/4" furring channels				
12" o.c.	S.F.	0.32	1.50	1.82

09 FINISHES

SUPPORT SYSTEMS

09110.10 METAL STUDS	UNIT	MAT.	INST.	TOTAL
16" o.c.	S.F.	0.28	1.45	1.73
24" o.c.	"	0.20	1.30	1.50
1-1/2" furring channels				
12" o.c.	S.F.	0.46	1.65	2.11
16" o.c.	"	0.40	1.50	1.90
24" o.c.	"	0.33	1.40	1.73
On walls				
3/4" furring channels				
12" o.c.	S.F.	0.32	1.20	1.52
16" o.c.	"	0.28	1.15	1.43
24" o.c.	"	0.20	1.05	1.25
1-1/2" furring channels				
12" o.c.	S.F.	0.46	1.30	1.76
16" o.c.	"	0.40	1.20	1.60

LATH AND PLASTER

	UNIT	MAT.	INST.	TOTAL
09205.10 GYPSUM LATH				
Gypsum lath, 1/2" thick				
Clipped	S.Y.	4.85	2.05	6.90
Nailed	"	4.70	2.30	7.00
09205.20 METAL LATH				
Stucco lath				
1.8 lb.	S.Y.	3.65	4.55	8.20
3.6 lb.	"	4.10	4.55	8.65
Paper backed				
Minimum	S.Y.	2.85	3.65	6.50
Maximum	"	4.55	5.20	9.75
09210.10 PLASTER				
Gypsum plaster, trowel finish, 2 coats				
Ceilings	S.Y.	4.10	11.30	15.40
Walls	"	4.10	10.60	14.70
3 coats				
Ceilings	S.Y.	5.65	15.70	21.35
Walls	"	5.65	13.90	19.55
Patch holes, average size holes				
1 sf to 5 sf				
Minimum	S.F.	1.65	6.00	7.65
Average	"	2.20	7.20	9.40
Maximum	"	2.95	9.00	11.95
Over 5 sf				

09 FINISHES

LATH AND PLASTER

LATH AND PLASTER	UNIT	MAT.	INST.	TOTAL
09210.10 PLASTER				
Minimum	S.F.	1.45	3.60	5.05
Average	"	1.95	5.15	7.10
Maximum	"	2.40	6.00	8.40
Patch cracks				
Minimum	S.F.	0.85	1.20	2.05
average	"	0.97	1.80	2.77
Maximum	"	1.20	3.60	4.80
09220.10 PORTLAND CEMENT PLASTER				
Stucco, portland, gray, 3 coat, 1" thick				
Sand finish	S.Y.	5.25	15.70	20.95
Trowel finish	"	5.25	16.40	21.65
White cement				
Sand finish	S.Y.	6.00	16.40	22.40
Trowel finish	"	6.00	18.05	24.05
Scratch coat				
For ceramic tile	S.Y.	1.90	3.60	5.50
For quarry tile	"	1.90	3.60	5.50
Portland cement plaster				
2 coats, 1/2"	S.Y.	3.80	7.20	11.00
3 coats, 7/8"	"	4.50	9.00	13.50
09250.10 GYPSUM BOARD				
Drywall, plasterboard, 3/8" clipped to				
Metal furred ceiling	S.F.	0.24	0.41	0.65
Columns and beams	"	0.24	0.91	1.15
Walls	"	0.24	0.36	0.60
Nailed or screwed to				
Wood framed ceiling	S.F.	0.24	0.36	0.60
Columns and beams	"	0.24	0.81	1.05
Walls	"	0.24	0.33	0.57
1/2", clipped to				
Metal furred ceiling	S.F.	0.25	0.41	0.66
Columns and beams	"	0.25	0.91	1.16
Walls	"	0.25	0.36	0.61
Nailed or screwed to				
Wood framed ceiling	S.F.	0.25	0.36	0.61
Columns and beams	"	0.25	0.81	1.06
Walls	"	0.25	0.33	0.58
5/8", clipped to				
Metal furred ceiling	S.F.	0.28	0.46	0.74
Columns and beams	"	0.28	1.00	1.28
Walls	"	0.28	0.41	0.69
Nailed or screwed to				
Wood framed ceiling	S.F.	0.28	0.46	0.74
Columns and beams	"	0.28	1.00	1.28
Walls	"	0.28	0.41	0.69
Taping and finishing joints				
Minimum	S.F.	0.03	0.24	0.27
Average	"	0.04	0.30	0.34
Maximum	"	0.07	0.36	0.44

LATH AND PLASTER

09250.10 GYPSUM BOARD	UNIT	MAT.	INST.	TOTAL
Casing bead				
Minimum	L.F.	0.11	1.05	1.16
Average	"	0.12	1.20	1.32
Maximum	"	0.15	1.80	1.95
Corner bead				
Minimum	L.F.	0.12	1.05	1.17
Average	"	0.15	1.20	1.35
Maximum	"	0.19	1.80	1.99

TILE

09310.10 CERAMIC TILE	UNIT	MAT.	INST.	TOTAL
Glazed wall tile, 4-1/4" x 4-1/4"				
Minimum	S.F.	1.65	2.55	4.20
Average	"	2.60	2.95	5.55
Maximum	"	7.15	3.55	10.70
Unglazed floor tile				
Portland cement bed, cushion edge, face mounted				
1" x 1"	S.F.	4.95	3.20	8.15
1" x 2"	"	8.00	3.05	11.05
2" x 2"	"	5.25	2.95	8.20
Adhesive bed, with white grout				
1" x 1"	S.F.	4.40	3.20	7.60
1" x 2"	"	4.60	3.05	7.65
2" x 2"	"	4.65	2.95	7.60

09330.10 QUARRY TILE	UNIT	MAT.	INST.	TOTAL
Floor				
4 x 4 x 1/2"	S.F.	3.80	4.70	8.50
6 x 6 x 1/2"	"	3.85	4.40	8.25
6 x 6 x 3/4"	"	4.55	4.40	8.95
Wall, applied to 3/4" portland cement bed				
4 x 4 x 1/2"	S.F.	3.45	7.05	10.50
6 x 6 x 3/4"	"	4.35	5.90	10.25
Cove base				
5 x 6 x 1/2" straight top	L.F.	3.40	5.90	9.30
6 x 6 x 3/4" round top	"	3.60	5.90	9.50
Stair treads 6 x 6 x 3/4"	"	6.00	8.85	14.85
Window sill 6 x 8 x 3/4"	"	4.85	7.05	11.90
For abrasive surface, add to material, 25%				

TILE	UNIT	MAT.	INST.	TOTAL
09410.10 TERRAZZO				
Floors on concrete, 1-3/4" thick, 5/8" topping				
Gray cement	S.F.	3.20	5.15	8.35
White cement	"	3.45	5.15	8.60
Sand cushion, 3" thick, 5/8" top, 1/4"				
Gray cement	S.F.	3.75	6.00	9.75
White cement	"	4.20	6.00	10.20
Monolithic terrazzo, 3-1/2" base slab, 5/8" topping	"	2.70	4.50	7.20
Terrazzo wainscot, cast-in-place, 1/2" thick	"	5.60	9.00	14.60
Base, cast in place, terrazzo cove type, 6" high	L.F.	6.00	5.15	11.15
Curb, cast in place, 6" wide x 6" high, polished top	"	6.65	18.05	24.70
For venetian type terrazzo, add to material, 10%				
For abrasive heavy duty terrazzo, add to material, 15%				
Divider strips				
Zinc	L.F.			1.01
Brass	"			1.89
Stairs, cast-in-place, topping on concrete or metal				
1-1/2" thick treads, 12" wide	L.F.	4.00	18.05	22.05
Combined tread and riser	"	6.00	45.10	51.10
Precast terrazzo, thin set				
Terrazzo tiles, non-slip surface				
9" x 9" x 1" thick	S.F.	12.70	5.15	17.85
12" x 12"				
1" thick	S.F.	13.70	4.80	18.50
1-1/2" thick	"	14.30	5.15	19.45
18" x 18" x 1-1/2" thick	"	18.70	5.15	23.85
24" x 24" x 1-1/2" thick	"	24.00	4.25	28.25
For white cement, add to material, 10%				
For venetian type terrazzo, add to material, 25%				
Terrazzo wainscot				
12" x 12" x 1" thick	S.F.	14.00	9.00	23.00
18" x 18" x 1-1/2" thick	"	22.00	10.30	32.30
Base				
6" high				
Straight	L.F.	9.05	2.80	11.85
Coved	"	10.70	2.80	13.50
8" high				
Straight	L.F.	10.15	3.00	13.15
Coved	"	11.90	3.00	14.90
Terrazzo curbs				
8" wide x 8" high	L.F.	21.35	14.45	35.80
6" wide x 6" high	"	19.35	12.05	31.40
Precast terrazzo stair treads, 12" wide				
1-1/2" thick				
Diamond pattern	L.F.	22.70	6.55	29.25
Non-slip surface	"	25.40	6.55	31.95
2" thick				
Diamond pattern	L.F.	25.40	6.55	31.95
Non-slip surface	"	28.00	7.20	35.20
Stair risers, 1" thick to 6" high				
Straight sections	L.F.	9.55	3.60	13.15
Cove sections	"	11.20	3.60	14.80
Combined tread and riser				

09 FINISHES

TILE

09410.10 TERRAZZO	UNIT	MAT.	INST.	TOTAL
Straight sections				
1-1/2" tread, 3/4" riser	L.F.	46.50	10.30	56.80
3" tread, 1" riser	"	49.00	10.30	59.30
Curved sections				
2" tread, 1" riser	L.F.	52.50	12.05	64.55
3" tread, 1" riser	"	54.00	12.05	66.05
Stair stringers, notched for treads and risers				
1" thick	L.F.	24.70	9.00	33.70
2" thick	"	25.60	12.05	37.65
Landings, structural, nonslip				
1-1/2" thick	S.F.	21.20	6.00	27.20
3" thick	"	20.70	7.20	27.90
Conductive terrazzo, spark proof industrial floor				
Epoxy terrazzo				
Floor	S.F.	8.95	2.25	11.20
Base	"	8.95	3.00	11.95
Polyacrylate				
Floor	S.F.	7.20	2.25	9.45
Base	"	7.20	3.00	10.20
Polyester				
Floor	S.F.	2.55	1.45	4.00
Base	"	2.55	1.80	4.35
Synthetic latex mastic				
Floor	S.F.	4.15	2.25	6.40
Base	"	4.15	3.00	7.15

ACOUSTICAL TREATMENT

09510.10 CEILINGS AND WALLS	UNIT	MAT.	INST.	TOTAL
Acoustical panels, suspension system not included				
Fiberglass panels				
5/8" thick				
2' x 2'	S.F.	0.63	0.52	1.15
2' x 4'	"	0.63	0.41	1.03
3/4" thick				
2' x 2'	S.F.	0.76	0.52	1.28
2' x 4'	"	0.76	0.41	1.17
Mineral fiber panels				
5/8" thick				
2' x 2'	S.F.	0.76	0.52	1.28
2' x 4'	"	0.76	0.41	1.17
3/4" thick				
2' x 2'	S.F.	1.00	0.52	1.52

09 FINISHES

ACOUSTICAL TREATMENT

09510.10 CEILINGS AND WALLS	UNIT	MAT.	INST.	TOTAL
2' x 4'	S.F.	1.00	0.41	1.41
Ceiling suspension systems				
T bar system				
2' x 4'	S.F.	0.74	0.36	1.11
2' x 2'	"	0.79	0.41	1.20

FLOORING

09550.10 WOOD FLOORING	UNIT	MAT.	INST.	TOTAL
Wood block industrial flooring				
Creosoted				
2" thick	S.F.	2.90	0.96	3.86
2-1/2" thick	"	3.05	1.15	4.20
3" thick	"	3.15	1.20	4.35
Gym floor, 2 ply felt, 25/32" maple, finished, in mastic	"	5.40	2.05	7.45
Over wood sleepers	"	6.05	2.30	8.35
Finishing, sand, fill, finish, and wax	"	0.51	0.91	1.42
Refinish sand, seal, and 2 coats of polyurethane	"	0.85	1.20	2.05
Clean and wax floors	"	0.14	0.18	0.32

09630.10 UNIT MASONRY FLOORING	UNIT	MAT.	INST.	TOTAL
Clay brick				
9 x 4-1/2 x 3" thick				
Glazed	S.F.	5.80	3.05	8.85
Unglazed	"	5.50	3.05	8.55
8 x 4 x 3/4" thick				
Glazed	S.F.	5.20	3.15	8.35
Unglazed	"	5.10	3.15	8.25
For herringbone pattern, add to labor, 15%				

09660.10 RESILIENT TILE FLOORING	UNIT	MAT.	INST.	TOTAL
Solid vinyl tile, 1/8" thick, 12" x 12"				
Marble patterns	S.F.	1.70	0.91	2.61
Solid colors	"	2.20	0.91	3.11
Travertine patterns	"	2.50	0.91	3.41

09665.10 RESILIENT SHEET FLOORING	UNIT	MAT.	INST.	TOTAL
Vinyl sheet flooring				
Minimum	S.F.	1.50	0.36	1.87
Average	"	2.15	0.44	2.59
Maximum	"	4.55	0.61	5.16
Cove, to 6"	L.F.	0.91	0.73	1.64
Fluid applied resilient flooring				

09 FINISHES

FLOORING

FLOORING	UNIT	MAT.	INST.	TOTAL
09665.10 RESILIENT SHEET FLOORING				
Polyurethane, poured in place, 3/8" thick	S.F.	6.15	3.05	9.20
09678.10 RESILIENT BASE AND ACCESSORIES				
Wall base, vinyl				
4" high	L.F.	0.61	1.20	1.81
6" high	"	0.84	1.20	2.04
Stair accessories				
Treads, 1/4" x 12", rubber diamond surface				
Marbled	L.F.	4.85	3.05	7.90
Plain	"	4.75	3.05	7.80
Grit strip safety tread, 12" wide, colors				
3/16" thick	L.F.	6.65	3.05	9.70
5/16" thick	"	9.35	3.05	12.40
Risers, 7" high, 1/8" thick, colors				
Flat	L.F.	1.40	1.80	3.20
Coved	"	2.35	1.80	4.15

CARPET

CARPET	UNIT	MAT.	INST.	TOTAL
09680.10 FLOOR LEVELING				
Repair and level floors to receive new flooring				
Minimum	S.Y.	0.62	1.20	1.82
Average	"	2.40	3.05	5.45
Maximum	"	3.20	3.65	6.85
09682.10 CARPET PADDING				
Carpet padding				
Jute padding				
Minimum	S.Y.	2.95	1.65	4.60
Average	"	3.80	1.80	5.60
Maximum	"	5.80	2.05	7.85
Sponge rubber cushion				
Minimum	S.Y.	3.45	1.65	5.10
Average	"	4.60	1.80	6.40
Maximum	"	5.80	2.05	7.85
Urethane cushion, 3/8" thick				
Minimum	S.Y.	3.45	1.65	5.10
Average	"	4.05	1.80	5.85
Maximum	"	4.60	2.05	6.65

CARPET

09685.10 CARPET	UNIT	MAT.	INST.	TOTAL
Carpet, acrylic				
24 oz., light traffic	S.Y.	19.50	2.05	21.55
28 oz., medium traffic	"	24.70	2.05	26.75
Commercial				
Nylon				
28 oz., medium traffic	S.Y.	17.95	2.05	20.00
35 oz., heavy traffic	"	21.45	2.05	23.50
Wool				
30 oz., medium traffic	S.Y.	29.20	2.05	31.25
36 oz., medium traffic	"	30.90	2.05	32.95
42 oz., heavy traffic	"	40.80	2.05	42.85
Carpet tile				
Foam backed				
Clean and vacuum carpet				
Minimum	S.Y.	0.20	0.14	0.34
Average	"	0.32	0.24	0.56
Maximum	"	0.44	0.36	0.81

09700.10 SPECIAL FLOORING	UNIT	MAT.	INST.	TOTAL
Epoxy flooring, marble chips				
Epoxy with colored quartz chips in 1/4" base	S.F.	7.50	2.05	9.55
Heavy duty epoxy topping, 3/16" thick	"	6.55	2.05	8.60
Epoxy terrazzo				
1/4" thick chemical resistant	S.F.	7.65	2.30	9.95

PAINTING

09910.10 EXTERIOR PAINTING	UNIT	MAT.	INST.	TOTAL
Exterior painting				
Wood surfaces, 1 coat primer, two coats paint				
Door and frame	EA.	3.80	61.00	64.80
Windows	S.F.	0.08	0.91	0.99
Wood trim	"	0.20	0.91	1.11
Wood siding	"	0.20	0.46	0.66
Hardboard surfaces				
One coat primer, two coats paint	S.F.	0.20	0.46	0.66
Asbestos cement surfaces				
One coat primer, two coats paint	S.F.	0.22	0.46	0.68
Galvanized surfaces, galvanized primer				
One coat primer, two coats paint	S.F.	0.21	0.43	0.64
Stucco surfaces, acrylic primer, acrylic latex paint				
One coat primer, two coats paint	S.F.	0.21	0.61	0.82

09 FINISHES

PAINTING	UNIT	MAT.	INST.	TOTAL
09910.10 EXTERIOR PAINTING				
Concrete masonry unit surfaces, brush work				
One coat filler, one coat paint	S.F.	0.23	0.46	0.69
Two coats epoxy	"	0.33	0.61	0.94
Texture coating	"	0.31	0.36	0.68
Concrete surfaces				
One coat filler, one coat paint	S.F.	0.20	0.46	0.66
Two coats paint	"	0.26	0.61	0.87
Structural steel				
One field coat paint, brush work				
Light framing	S.F.	0.10	0.30	0.40
Heavy framing	"	0.10	0.18	0.28
One field coat paint, spray work				
Light framing	S.F.	0.08	0.09	0.17
Heavy framing	"	0.08	0.07	0.15
Miscellaneous steel items, spray work, one coat				
Exposed decking	S.F.	0.10	0.30	0.40
Joist	"	0.10	0.46	0.56
Columns	"	0.10	0.46	0.56
Pipes, one coat primer, one coat paint				
4" dia.	L.F.	0.08	0.46	0.54
8" dia.	"	0.20	0.61	0.81
12" dia.	"	0.34	0.91	1.25
Paint letters on pipe with brush	"	0.39	0.91	1.30
Paint pipe insulation cloth cover	S.F.	0.23	0.46	0.69
Sprinkler system piping	L.F.	0.24	0.30	0.54
Miscellaneous surfaces				
Stair pipe rails				
Two rails	L.F.	0.17	1.20	1.37
One rail	"	0.09	0.73	0.82
Stair to 4' wide, including rails, per riser	EA.	0.64	5.20	5.84
Gratings and frames	S.F.	0.17	1.20	1.37
Ladders	L.F.	0.20	1.05	1.25
Miscellaneous exposed metal	S.F.	0.17	0.36	0.54
09920.10 INTERIOR PAINTING				
Walls, concrete and masonry, brush, primer, acrylic				
One coat primer, one coat paint	S.F.	0.10	0.46	0.56
Two coats paint	"	0.17	0.61	0.78
Plywood, paint	"	0.06	0.20	0.26
Natural finish	"	0.07	0.23	0.30
Wood, paint	"	0.07	0.23	0.30
Natural finish	"	0.09	0.26	0.35
Metal				
One coat filler	S.F.	0.17	0.23	0.40
One coat primer, one coat paint	"	0.20	0.46	0.66
Two coats paint	"	0.24	0.61	0.85
Plaster or gypsum board, paint	"	0.09	0.20	0.29
Epoxy	"	0.12	0.23	0.35
Ceilings, one coat paint, wood	"	0.08	0.26	0.34
Concrete	"	0.12	0.23	0.35
Plaster	"	0.08	0.20	0.28

09 FINISHES

PAINTING	UNIT	MAT.	INST.	TOTAL
09920.10 INTERIOR PAINTING				
Miscellaneous metal, brushwork	S.F.	0.17	0.46	0.63
09920.30 DOORS AND MILLWORK				
Painting, doors				
Minimum	S.F.	0.13	1.20	1.33
Average	"	0.24	1.80	2.04
Maximum	"	0.40	2.45	2.85
Cabinets, shelves, and millwork				
Minimum	S.F.	0.09	0.61	0.70
Average	"	0.20	1.05	1.25
Maximum	"	0.30	1.80	2.10
09920.60 WINDOWS				
Painting, windows				
Minimum	S.F.	0.06	0.73	0.79
Average	"	0.09	0.91	1.00
Maximum	"	0.14	1.45	1.59
09955.10 WALL COVERING				
Vinyl wall covering				
Medium duty	S.F.	0.95	0.52	1.47
Heavy duty	"	1.40	0.61	2.01
Over pipes and irregular shapes				
Flexible gypsum coated wall fabric, fire resistant	S.F.	1.05	0.36	1.42
Vinyl corner guards				
3/4" x 3/4" x 8'	EA.	4.90	4.55	9.45
2-3/4" x 2-3/4" x 4'	"	2.90	4.55	7.45
09980.10 PAINTING PREPARATION				
Cleaning, light				
Wood	S.F.	0.03	0.09	0.12
Plaster or gypsum wallboard	"	0.03	0.08	0.11
Prepare sprinkler piping for painting	L.F.	0.03	0.07	0.10
Prepare insulated pipe for painting	S.F.	0.03	0.09	0.12
Normal painting prep, masonry and concrete				
Unpainted	S.F.	0.03	0.06	0.09
Painted	"	0.03	0.09	0.12
Plaster or gypsum				
Unpainted	S.F.	0.03	0.06	0.09
Painted	"	0.03	0.09	0.12
Wood				
Unpainted	S.F.	0.03	0.06	0.09
Painted	"	0.03	0.09	0.12
Painted steel, light rusting	"	0.04	0.12	0.16
Sandblasting				
Brush off blast	S.F.	0.12	0.24	0.36
Commercial blast	"	0.24	0.61	0.85
Near white metal blast	"	0.35	1.05	1.40
White metal blast	"	0.47	1.20	1.67

09 FINISHES

PAINTING

09980.15 PAINT	UNIT	MAT.	INST.	TOTAL
Paint, enamel				
600 sf per gal.	GAL			27.64
550 sf per gal.	"			24.93
500 sf per gal.	"			18.13
450 sf per gal.	"			17.00
350 sf per gal.	"			15.86
Filler, 60 sf per gal.	"			19.26
Latex, 400 sf per gal.	"			15.86
Aluminum				
400 sf per gal.	GAL			22.66
500 sf per gal.	"			40.79
Red lead, 350 sf per gal.	"			33.99
Primer				
400 sf per gal.	GAL			21.15
300 sf per gal.	"			21.02
Latex base, interior, white	"			17.59
Sealer and varnish				
400 sf per gal.	GAL			16.25
425 sf per gal.	"			23.77
600 sf per gal.	"			31.84

10 SPECIALTIES

SPECIALTIES	UNIT	MAT.	INST.	TOTAL
10110.10 CHALKBOARDS				
Chalkboard, metal frame, 1/4" thick				
48"x60"	EA.	200.00	36.50	236.50
48"x96"	"	330.00	40.50	370.50
48"x144"	"	500.00	45.60	545.60
48"x192"	"	640.00	52.00	692.00
Liquid chalkboard				
48"x60"	EA.	240.00	36.50	276.50
48"x96"	"	360.00	40.50	400.50
48"x144"	"	560.00	45.60	605.60
48"x192"	"	710.00	52.00	762.00
Map rail, deluxe	L.F.	6.10	1.80	7.90
Average	PCT.			3.10
10165.10 TOILET PARTITIONS				
Toilet partition, plastic laminate				
Ceiling mounted	EA.	630.00	120.00	750.00
Floor mounted	"	590.00	91.00	681.00
Metal				
Ceiling mounted	EA.	460.00	120.00	580.00
Floor mounted	"	430.00	91.00	521.00
Wheel chair partition, plastic laminate				
Ceiling mounted	EA.	840.00	120.00	960.00
Floor mounted	"	780.00	91.00	871.00
Painted metal				
Ceiling mounted	EA.	660.00	120.00	780.00
Floor mounted	"	590.00	91.00	681.00
Urinal screen, plastic laminate				
Wall hung	EA.	300.00	45.60	345.60
Floor mounted	"	250.00	45.60	295.60
Porcelain enameled steel, floor mounted	"	350.00	45.60	395.60
Painted metal, floor mounted	"	220.00	45.60	265.60
Stainless steel, floor mounted	"	460.00	45.60	505.60
Metal toilet partitions				
Toilet partitions, front door and side divider, floor mounted				
Porcelain enameled steel	EA.	780.00	91.00	871.00
Painted steel	"	430.00	91.00	521.00
Stainless steel	"	1,030	91.00	1,121
10185.10 SHOWER STALLS				
Shower receptors				
Precast, terrazzo				
32" x 32"	EA.	340.00	33.50	373.50
32" x 48"	"	430.00	40.20	470.20
Concrete				
32" x 32"	EA.	170.00	33.50	203.50
48" x 48"	"	210.00	44.70	254.70
Shower door, trim and hardware				
Porcelain enameled steel, flush	EA.	330.00	40.20	370.20
Baked enameled steel, flush	"	190.00	40.20	230.20
Aluminum frame, tempered glass, 48" wide, sliding	"	420.00	50.50	470.50
Folding	"	400.00	50.50	450.50

SPECIALTIES	UNIT	MAT.	INST.	TOTAL
10185.10 SHOWER STALLS				
Shower compartment, precast concrete receptor				
Single entry type				
Porcelain enameled steel	EA.	1,560	400.00	1,960
Baked enameled steel	"	900.00	400.00	1,300
Stainless steel	"	1,800	400.00	2,200
Double entry type				
Porcelain enameled steel	EA.	4,450	500.00	4,950
Baked enameled steel	"	2,880	500.00	3,380
Stainless steel	"	4,510	500.00	5,010
10210.10 VENTS AND WALL LOUVERS				
Vents w/screen, 4" deep, 8" wide, 5" high				
Modular	EA.	62.00	12.40	74.40
Aluminum gable louvers	S.F.	11.05	6.60	17.65
Vent screen aluminum, 4" wide, continuous	L.F.	3.05	1.30	4.35
Wall louver, aluminum mill finish				
Under, 2 sf	S.F.	23.30	4.95	28.25
2 to 4 sf	"	19.60	4.40	24.00
5 to 10 sf	"	19.00	4.40	23.40
Galvanized steel				
Under 2 sf	S.F.	22.05	4.95	27.00
2 to 4 sf	"	15.95	4.40	20.35
5 to 10 sf	"	14.70	4.40	19.10
10225.10 DOOR LOUVERS				
Fixed, 1" thick, enameled steel				
8"x8"	EA.	35.60	4.55	40.15
12"x12"	"	45.40	5.20	50.60
20"x20"	"	87.00	14.60	101.60
24"x24"	"	93.00	16.60	109.60
10290.10 PEST CONTROL				
Termite control				
Under slab spraying				
Minimum	S.F.	0.13	0.07	0.20
Average	"	0.17	0.14	0.31
Maximum	"	0.23	0.29	0.52
10350.10 FLAGPOLES				
Installed in concrete base				
Fiberglass				
25' high	EA.	920.00	240.00	1,160
50' high	"	4,120	610.00	4,730
Aluminum				
25' high	EA.	1,140	240.00	1,380
50' high	"	3,780	610.00	4,390
Bonderized steel				
25' high	EA.	1,720	280.00	2,000

SPECIALTIES	UNIT	MAT.	INST.	TOTAL
10350.10 FLAGPOLES				
50' high	EA.	4,000	730.00	4,730
Freestanding tapered, fiberglass				
30' high	EA.	1,030	260.00	1,290
40' high	"	1,830	330.00	2,160
50' high	"	3,660	360.00	4,020
60' high	"	4,800	430.00	5,230
Wall mounted, with collar, brushed aluminum finish				
15' long	EA.	800.00	180.00	980.00
18' long	"	1,030	180.00	1,210
20' long	"	1,140	190.00	1,330
24' long	"	1,260	210.00	1,470
Outrigger, wall, including base				
10' long	EA.	690.00	240.00	930.00
20' long	"	1,030	300.00	1,330
10400.10 IDENTIFYING DEVICES				
Directory and bulletin boards				
Open face boards				
Chrome plated steel frame	S.F.	20.40	18.25	38.65
Aluminum framed	"	17.95	18.25	36.20
Bronze framed	"	19.80	18.25	38.05
Stainless steel framed	"	38.40	18.25	56.65
Tack board, aluminum framed	"	11.50	18.25	29.75
Visual aid board, aluminum framed	"	11.50	18.25	29.75
Glass encased boards, hinged and keyed				
Aluminum framed	S.F.	44.60	45.60	90.20
Bronze framed	"	48.30	45.60	93.90
Stainless steel framed	"	110.00	45.60	155.60
Chrome plated steel framed	"	110.00	45.60	155.60
Metal plaque				
Cast bronze	S.F.	290.00	30.40	320.40
Aluminum	"	270.00	30.40	300.40
Metal engraved plaque				
Porcelain steel	S.F.	370.00	30.40	400.40
Stainless steel	"	330.00	30.40	360.40
Brass	"	370.00	30.40	400.40
Aluminum	"	170.00	30.40	200.40
Metal built-up plaque				
Bronze	S.F.	280.00	36.50	316.50
Copper and bronze	"	290.00	36.50	326.50
Copper and aluminum	"	320.00	36.50	356.50
Metal nameplate plaques				
Cast bronze	S.F.	280.00	22.80	302.80
Cast aluminum	"	260.00	22.80	282.80
Engraved, 1-1/2" x 6"				
Bronze	EA.	89.00	22.80	111.80
Aluminum	"	89.00	22.80	111.80
Letters, on masonry or concrete, aluminum, satin finish				
1/2" thick				
2" high	EA.	10.55	14.60	25.15
4" high	"	11.75	18.25	30.00
6" high	"	15.50	20.25	35.75

10 SPECIALTIES

SPECIALTIES	UNIT	MAT.	INST.	TOTAL
10400.10 IDENTIFYING DEVICES				
3/4" thick				
8" high	EA.	20.40	22.80	43.20
10" high	"	26.00	26.10	52.10
1" thick				
12" high	EA.	34.00	30.40	64.40
14" high	"	38.40	36.50	74.90
16" high	"	53.00	45.60	98.60
For polished aluminum add, 15%				
For clear anodized aluminum add, 15%				
For colored anodic aluminum add, 30%				
For profiled and color enameled letters add, 50%				
Cast bronze, satin finish letters				
3/8" thick				
2" high	EA.	19.20	14.60	33.80
4" high	"	20.80	18.25	39.05
1/2" thick, 6" high	"	27.00	20.25	47.25
5/8" thick, 8" high	"	34.90	22.80	57.70
1" thick				
10" high	EA.	52.50	26.10	78.60
12" high	"	67.50	30.40	97.90
14" high	"	75.50	36.50	112.00
16" high	"	95.50	45.60	141.10
Interior door signs, adhesive, flexible				
2" x 8"	EA.	13.00	7.15	20.15
4" x 4"	"	13.00	7.15	20.15
6" x 7"	"	16.95	7.15	24.10
6" x 9"	"	22.30	7.15	29.45
10" x 9"	"	29.70	7.15	36.85
10" x 12"	"	38.40	7.15	45.55
Hard plastic type, no frame				
3" x 8"	EA.	27.20	7.15	34.35
4" x 4"	"	26.00	7.15	33.15
4" x 12"	"	32.20	7.15	39.35
Hard plastic type, with frame				
3" x 8"	EA.	86.50	7.15	93.65
4" x 4"	"	70.50	7.15	77.65
4" x 12"	"	110.00	7.15	117.15
10450.10 CONTROL				
Access control, 7' high, indoor or outdoor impenetrability				
Remote or card control, type B	EA.	2,950	500.00	3,450
Free passage, type B	"	2,720	500.00	3,220
Remote or card control, type AA	"	3,850	500.00	4,350
Free passage, type AA	"	3,510	500.00	4,010
10500.10 LOCKERS				
Locker bench, floor mounted, laminated maple				
4'	EA.	93.50	30.40	123.90
6'	"	120.00	30.40	150.40
Wardrobe locker, 12" x 60" x 15", baked on enamel				

SPECIALTIES	UNIT	MAT.	INST.	TOTAL
10500.10 LOCKERS				
1-tier	EA.	120.00	18.25	138.25
2-tier	"	130.00	18.25	148.25
3-tier	"	130.00	19.20	149.20
4-tier	"	160.00	19.20	179.20
12" x 72" x 15", baked on enamel				
1-tier	EA.	120.00	18.25	138.25
2-tier	"	140.00	18.25	158.25
4-tier	"	180.00	19.20	199.20
5-tier	"	180.00	19.20	199.20
15" x 60" x 15", baked on enamel				
1-tier	EA.	160.00	18.25	178.25
4-tier	"	170.00	19.20	189.20
Wardrobe locker, single tier type				
12" x 15" x 72"	EA.	120.00	36.50	156.50
18" x 15" x 72"	"	150.00	38.40	188.40
12" x 18" x 72"	"	140.00	40.50	180.50
18" x 18" x 72"	"	160.00	42.90	202.90
Double tier type				
12" x 15" x 36"	EA.	84.50	18.25	102.75
18" x 15" x 36"	"	87.00	18.25	105.25
12" x 18" x 36"	"	78.50	18.25	96.75
18" x 18" x 36"	"	89.50	18.25	107.75
Two person unit				
18" x 15" x 72"	EA.	240.00	61.00	301.00
18" x 18" x 72"	"	140.00	73.00	213.00
Duplex unit				
15" x 15" x 72"	EA.	230.00	36.50	266.50
15" x 21" x 72"	"	240.00	36.50	276.50
Basket lockers, basket sets with baskets				
24 basket set	SET	490.00	180.00	670.00
30 basket set	"	590.00	230.00	820.00
36 basket set	"	660.00	300.00	960.00
42 basket set	"	730.00	360.00	1,090
10520.10 FIRE PROTECTION				
Portable fire extinguishers				
Water pump tank type				
2.5 gal.				
Red enameled galvanized	EA.	81.00	19.05	100.05
Red enameled copper	"	93.00	19.05	112.05
Polished copper	"	140.00	19.05	159.05
Carbon dioxide type, red enamel steel				
Squeeze grip with hose and horn				
2.5 lb	EA.	46.70	19.05	65.75
5 lb	"	110.00	22.00	132.00
10 lb	"	160.00	28.60	188.60
15 lb	"	190.00	35.80	225.80
20 lb	"	220.00	35.80	255.80
Wheeled type				
125 lb	EA.	1,180	57.00	1,237
250 lb	"	1,210	57.00	1,267

10 SPECIALTIES

SPECIALTIES	UNIT	MAT.	INST.	TOTAL
10520.10 FIRE PROTECTION				
500 lb	EA.	2,270	57.00	2,327
Dry chemical, pressurized type				
Red enameled steel				
2.5 lb	EA.	28.00	19.05	47.05
5 lb	"	46.70	22.00	68.70
10 lb	"	58.50	28.60	87.10
20 lb	"	93.50	35.80	129.30
30 lb	"	250.00	35.80	285.80
Chrome plated steel, 2.5 lb	"	58.50	19.05	77.55
Other type extinguishers				
2.5 gal, stainless steel, pressurized water tanks	EA.	58.50	19.05	77.55
Soda and acid type	"	110.00	19.05	129.05
Cartridge operated, water type	"	180.00	19.05	199.05
Loaded stream, water type	"	82.00	19.05	101.05
Foam type	"	67.50	19.05	86.55
40 gal, wheeled foam type	"	3,560	57.00	3,617
Fire extinguisher cabinets				
Enameled steel				
8" x 12" x 27"	EA.	130.00	57.00	187.00
8" x 16" x 38"	"	170.00	57.00	227.00
Aluminum				
8" x 12" x 27"	EA.	170.00	57.00	227.00
8" x 16" x 38"	"	200.00	57.00	257.00
8" x 12" x 27"	"	290.00	57.00	347.00
Stainless steel				
8" x 16" x 38"	EA.	370.00	57.00	427.00
10550.10 POSTAL SPECIALTIES				
Mail chutes				
Single mail chute				
Finished aluminum	L.F.	460.00	91.00	551.00
Bronze	"	670.00	91.00	761.00
Single mail chute receiving box				
Finished aluminum	EA.	720.00	180.00	900.00
Bronze	"	1,350	180.00	1,530
Twin mail chute, double parallel				
Finished aluminum	FLR	860.00	180.00	1,040
Bronze	"	1,350	180.00	1,530
Receiving box, 36" x 20" x 12"				
Finished aluminum	EA.	1,500	300.00	1,800
Bronze	"	2,150	300.00	2,450
Locked receiving mail box				
Finished aluminum	EA.	720.00	180.00	900.00
Bronze	"	1,230	180.00	1,410
Commercial postal accessories for mail chutes				
Letter slot, brass	EA.	240.00	61.00	301.00
Bulk mail slot, brass	"	690.00	61.00	751.00
Mail boxes				
Residential postal accessories				
Letter slot	EA.	71.00	18.25	89.25
Rural letter box	"	50.50	45.60	96.10

SPECIALTIES	UNIT	MAT.	INST.	TOTAL
10670.10 SHELVING				
Shelving, enamel, closed side and back, 12" x 36"				
5 shelves	EA.	120.00	61.00	181.00
8 shelves	"	150.00	81.00	231.00
Open				
5 shelves	EA.	80.00	61.00	141.00
8 shelves	"	80.00	81.00	161.00
Metal storage shelving, baked enamel				
7 shelf unit, 72" or 84" high				
12" shelf	L.F.	34.00	38.40	72.40
24" shelf	"	44.90	45.60	90.50
36" shelf	"	58.50	52.00	110.50
4 shelf unit, 40" high				
12" shelf	L.F.	29.10	33.20	62.30
24" shelf	"	41.30	40.50	81.80
3 shelf unit, 32" high				
12" shelf	L.F.	23.05	19.20	42.25
24" shelf	"	30.40	22.80	53.20
Single shelf unit, attached to masonry				
12" shelf	L.F.	9.10	6.65	15.75
24" shelf	"	12.75	7.95	20.70
For stainless steel, add to material, 120%				
For attachment to gypsum board, add to labor, 50%				
10800.10 BATH ACCESSORIES				
Ash receiver, wall mounted, aluminum	EA.	130.00	18.25	148.25
Grab bar, 1-1/2" dia., stainless steel, wall mounted				
24" long	EA.	35.60	18.25	53.85
36" long	"	51.50	19.20	70.70
48" long	"	71.00	21.45	92.45
1" dia., stainless steel				
12" long	EA.	28.20	15.85	44.05
24" long	"	33.10	18.25	51.35
36" long	"	41.70	20.25	61.95
48" long	"	42.90	21.45	64.35
Hand dryer, surface mounted, 110 volt	"	400.00	45.60	445.60
Medicine cabinet, 16 x 22, baked enamel, steel, lighted	"	63.50	14.60	78.10
With mirror, lighted	"	110.00	24.30	134.30
Mirror, 1/4" plate glass, up to 10 sf	S.F.	6.35	3.65	10.00
Mirror, stainless steel frame				
18"x24"	EA.	100.00	12.15	112.15
24"x30"	"	130.00	18.25	148.25
24"x48"	"	250.00	30.40	280.40
30"x30"	"	140.00	36.50	176.50
48"x72"	"	420.00	61.00	481.00
With shelf, 18"x24"	"	140.00	14.60	154.60
Sanitary napkin dispenser, stainless steel, wall mounted	"	350.00	24.30	374.30
Shower rod, 1" diameter				
Chrome finish over brass	EA.	57.50	18.25	75.75
Stainless steel	"	55.00	18.25	73.25
Soap dish, stainless steel, wall mounted	"	70.00	24.30	94.30
Toilet tissue dispenser, stainless, wall mounted				
Single roll	EA.	34.30	9.10	43.40

SPECIALTIES	UNIT	MAT.	INST.	TOTAL
10800.10 BATH ACCESSORIES				
Double roll	EA.	54.00	10.40	64.40
Towel dispenser, stainless steel				
Flush mounted	EA.	100.00	20.25	120.25
Surface mounted	"	93.00	18.25	111.25
Combination towel dispenser and waste receptacle	"	310.00	24.30	334.30
Towel bar, stainless steel				
18" long	EA.	29.40	14.60	44.00
24" long	"	34.30	16.60	50.90
30" long	"	36.80	18.25	55.05
36" long	"	40.40	20.25	60.65
Waste receptacle, stainless steel, wall mounted	"	210.00	30.40	240.40

ARCHITECTURAL EQUIPMENT

11020.10 SECURITY EQUIPMENT

ARCHITECTURAL EQUIPMENT	UNIT	MAT.	INST.	TOTAL
Office safes, 30" x 20" x 20", 1 hr rating	EA.	1,810	91.00	1,901
30" x 16" x 15", 2 hr rating	"	1,700	73.00	1,773
30" x 28" x 20", H&G rating	"	7,930	45.60	7,976
Surveillance system				
Minimum	EA.	4,230	730.00	4,960
Maximum	"	32,730	3,650	36,380
Insulated file room door				
1 hr rating				
32" wide	EA.	2,860	360.00	3,220
40" wide	"	3,040	410.00	3,450

11161.10 LOADING DOCK EQUIPMENT

ARCHITECTURAL EQUIPMENT	UNIT	MAT.	INST.	TOTAL
Dock leveler, 10 ton capacity				
6' x 8'	EA.	3,400	360.00	3,760
7' x 8'	"	3,510	360.00	3,870
Bumpers, laminated rubber				
4-1/2" thick				
6" x 14"	EA.	57.50	7.30	64.80
10" x 14"	"	62.00	9.10	71.10
10" x 36"	"	80.00	12.15	92.15
12" x 14"	"	68.00	9.60	77.60
12" x 36"	"	90.50	13.50	104.00
6" thick				
10" x 14"	EA.	72.50	10.40	82.90
10" x 24"	"	95.00	12.60	107.60
10" x 36"	"	99.00	18.25	117.25
Extruded rubber bumpers				
T-section, 22" x 22" x 3"	EA.	91.50	7.30	98.80
Molded rubber bumpers				
24" x 12" x 3" thick	EA.	56.00	18.25	74.25
Door seal, 12" x 12", vinyl covered	L.F.	32.10	9.10	41.20
Dock boards, heavy duty, 5' x 5'				
5000 lb				
Minimum	EA.	900.00	300.00	1,200
Maximum	"	1,550	300.00	1,850
9000 lb				
Minimum	EA.	1,040	300.00	1,340
Maximum	"	1,730	330.00	2,060
15,000 lb	"	1,360	330.00	1,690
Truck shelters				
Minimum	EA.	1,080	280.00	1,360
Maximum	"	1,790	520.00	2,310

11170.10 WASTE HANDLING

ARCHITECTURAL EQUIPMENT	UNIT	MAT.	INST.	TOTAL
Incinerator, electric				
100 lb/hr				
Minimum	EA.	16,430	370.00	16,800
Maximum	"	28,330	370.00	28,700
400 lb/hr				
Minimum	EA.	29,160	750.00	29,910
Maximum	"	56,650	750.00	57,400
1000 lb/hr				

11 EQUIPMENT

ARCHITECTURAL EQUIPMENT	UNIT	MAT.	INST.	TOTAL
11170.10 WASTE HANDLING				
Minimum	EA.	56,650	1,130	57,780
Maximum	"	90,640	1,130	91,770
11480.10 ATHLETIC EQUIPMENT				
Basketball backboard				
Fixed	EA.	620.00	460.00	1,080
Swing-up	"	1,440	730.00	2,170
Portable, hydraulic	"	7,990	180.00	8,170
Suspended type, standard	"	2,820	730.00	3,550
For glass backboard, add	"			865.33
For electrically operated, add	"			1,130
Bleacher, telescoping, manual				
15 tier, minimum	SEAT	79.50	7.30	86.80
Maximum	"	91.50	7.30	98.80
20 tier, minimum	"	80.50	8.10	88.60
Maximum	"	92.50	8.10	100.60
30 tier, minimum	"	81.50	12.15	93.65
Maximum	"	94.00	12.15	106.15
Boxing ring elevated, complete, 22' x 22'	EA.	110.00	3,650	3,760
Gym divider curtain				
Minimum	S.F.	3.25	0.49	3.74
Maximum	"	4.45	0.49	4.94
Scoreboards, single face				
Minimum	EA.	2,940	360.00	3,300
Maximum	"	18,990	1,820	20,810
Parallel bars				
Minimum	EA.	990.00	360.00	1,350
Maximum	"	2,760	610.00	3,370
11500.10 INDUSTRIAL EQUIPMENT				
Vehicular paint spray booth, solid back, 14'4" x 9'6"				
24' deep	EA.	10,150	360.00	10,510
26'6" deep	"	10,270	360.00	10,630
28'6" deep	"	10,460	360.00	10,820
Drive through, 14'9" x 9'6"				
24' deep	EA.	10,460	360.00	10,820
26'6" deep	"	10,710	360.00	11,070
28'6" deep	"	10,830	360.00	11,190
Water wash, paint spray booth				
5' x 11'2" x 10'8"	EA.	5,260	360.00	5,620
6' x 11'2" x 10'8"	"	5,380	360.00	5,740
8' x 11'2" x 10'8"	"	6,810	360.00	7,170
10' x 11'2" x 11'2"	"	7,550	360.00	7,910
12' x 12'2" x 11'2"	"	9,220	360.00	9,580
14' x 12'2" x 11'2"	"	10,830	360.00	11,190
16' x 12'2" x 11'2"	"	12,190	360.00	12,550
20' x 12'2" x 11'2"	"	14,610	360.00	14,970
Dry type spray booth, with paint arrestors				
5'4" x 7'2" x 6'8"	EA.	2,290	360.00	2,650
6'4" x 7'2" x 6'8"	"	2,790	360.00	3,150
8'4" x 7'2" x 9'2"	"	3,160	360.00	3,520
10'4" x 7'2" x 9'2"	"	3,650	360.00	4,010

ARCHITECTURAL EQUIPMENT

11500.10 INDUSTRIAL EQUIPMENT

ARCHITECTURAL EQUIPMENT	UNIT	MAT.	INST.	TOTAL
12'4" x 7'6" x 9'2"	EA.	3,590	360.00	3,950
14'4" x 7'6" x 9'8"	"	4,890	360.00	5,250
16'4" x 7'7" x 9'8"	"	5,510	360.00	5,870
20'4" x 7'7" x 10'8"	"	6,190	360.00	6,550
Air compressor, electric				
1 hp				
115 volt	EA.	2,790	240.00	3,030
5 hp				
115 volt	EA.	5,760	360.00	6,120
230 volt	"	4,890	360.00	5,250
Hydraulic lifts				
8,000 lb capacity	EA.	430.00	910.00	1,340
11,000 lb capacity	"	6,000	1,460	7,460
24,000 lb capacity	"	8,480	2,430	10,910
Power tools				
Band saws				
10"	EA.	580.00	30.40	610.40
14"	"	1,020	36.50	1,057
Motorized shaper	"	540.00	28.10	568.10
Motorized lathe	"	640.00	30.40	670.40
Bench saws				
9" saw	EA.	1,160	24.30	1,184
10" saw	"	1,790	26.10	1,816
12" saw	"	2,290	30.40	2,320
Electric grinders				
1/3 hp	EA.	220.00	14.60	234.60
1/2 hp	"	320.00	15.85	335.85
3/4 hp	"	380.00	15.85	395.85

12 FURNISHINGS

INTERIOR	UNIT	MAT.	INST.	TOTAL
12690.40 FLOOR MATS				
Recessed entrance mat, 3/8" thick, aluminum link	S.F.	17.50	18.25	35.75
Steel, flexible	"	7.70	18.25	25.95

CONSTRUCTION	UNIT	MAT.	INST.	TOTAL
13121.10 PRE-ENGINEERED BUILDINGS				
Pre-engineered metal building, 40'x100'				
14' eave height	S.F.	8.15	2.95	11.10
16' eave height	"	8.60	3.40	12.00
20' eave height	"	9.85	4.40	14.25
60'x100'				
14' eave height	S.F.	7.05	2.95	10.00
16' eave height	"	7.40	3.40	10.80
20' eave height	"	8.25	4.40	12.65
80'x100'				
14' eave height	S.F.	7.05	2.95	10.00
16' eave height	"	7.30	3.40	10.70
20' eave height	"	8.35	4.40	12.75
100'x100'				
14' eave height	S.F.	10.35	2.95	13.30
16' eave height	"	7.40	3.40	10.80
20' eave height	"	8.10	4.40	12.50
100'x150'				
14' eave height	S.F.	6.45	2.95	9.40
16' eave height	"	6.90	3.40	10.30
20' eave height	"	7.65	4.40	12.05
120'x150'				
14' eave height	S.F.	6.45	2.95	9.40
16' eave height	"	6.80	3.40	10.20
20' eave height	"	7.40	4.40	11.80
140'x150'				
14' eave height	S.F.	6.05	2.95	9.00
16' eave height	"	6.55	3.40	9.95
20' eave height	"	7.30	4.40	11.70
160'x200'				
14' eave height	S.F.	5.95	2.95	8.90
16' eave height	"	6.00	3.40	9.40
20' eave height	"	6.80	4.40	11.20
200'x200'				
14' eave height	S.F.	5.95	2.95	8.90
16' eave height	"	6.20	3.40	9.60
20' eave height	"	6.70	4.40	11.10
Hollow metal door and frame, 6' x 7'	EA.			630.63
Sectional steel overhead door, manually operated				
8' x 8'	EA.			681.45
12' x 12'	"			1,022
Roll-up steel door, manually operated				
10' x 10'	EA.			1,262
12' x 12'	"			1,515
For gravity ridge ventilator with birdscreen	"			466.92
9" throat x 10'	"			504.75
12" throat x 10'	"			530.09
For 20" rotary vent with damper	"			214.53
For 4' x 3' fixed louver	"			227.15
For 4' x 3' aluminum sliding window	"			233.51
For 3' x 9' fiberglass panels	"			93.39
Liner panel, 26 ga, painted steel	S.F.	1.45	0.99	2.44
Wall panel insulated, 26 ga. steel, foam core	"	4.65	0.99	5.64

13 SPECIAL

CONSTRUCTION

13121.10 PRE-ENGINEERED BUILDINGS	UNIT	MAT.	INST.	TOTAL
Roof panel, 26 ga. painted steel	S.F.	1.20	0.57	1.77
Plastic (sky light)	"	3.45	0.57	4.02
Insulation, 3-1/2" thick blanket, R11	"	1.10	0.27	1.37

14 CONVEYING

ELEVATORS

14210.10 ELEVATORS	UNIT	MAT.	INST.	TOTAL
Passenger elevators, electric, geared				
Based on a shaft of 6 stops and 6 openings				
50 fpm, 2000 lb	EA.	72,080	1,490	73,570
100 fpm, 2000 lb	"	74,750	1,660	76,410
150 fpm				
2000 lb	EA.	82,760	1,870	84,630
3000 lb	"	104,110	2,140	106,250
4000 lb	"	108,120	2,490	110,610

LIFTS

14410.10 PERSONNEL LIFTS	UNIT	MAT.	INST.	TOTAL
Electrically operated, 1 or 2 person lift				
With attached foot platforms				
3 stops	EA.			15,169
5 stops	"			19,416
7 stops	"			23,055
For each additional stop, add $1250				
Residential stair climber, per story	EA.	4,490	310.00	4,800

14450.10 VEHICLE LIFTS	UNIT	MAT.	INST.	TOTAL
Automotive hoist, one post, semi-hydraulic, 8,000 lb	EA.	3,660	1,490	5,150
Full hydraulic, 8,000 lb	"	3,470	1,490	4,960
2 post, semi-hydraulic, 10,000 lb	"	3,790	2,140	5,930
Full hydraulic				
10,000 lb	EA.	4,040	2,140	6,180
13,000 lb	"	4,980	3,740	8,720
18,500 lb	"	6,500	3,740	10,240
24,000 lb	"	8,770	3,740	12,510
26,000 lb	"	10,160	3,740	13,900
Pneumatic hoist, fully hydraulic				
11,000 lb	EA.	4,860	4,980	9,840
24,000 lb	"	8,520	4,980	13,500

HOISTS AND CRANES

14600.10 INDUSTRIAL HOISTS

HOISTS AND CRANES	UNIT	MAT.	INST.	TOTAL
Industrial hoists, electric, light to medium duty				
500 lb	EA.	5,070	190.00	5,260
1000 lb	"	5,320	200.00	5,520
2000 lb	"	5,460	210.00	5,670
5000 lb	"	8,890	250.00	9,140
10,000 lb	"	16,570	280.00	16,850
20,000 lb	"	29,070	310.00	29,380
30,000 lb	"	31,250	370.00	31,620
Heavy duty				
500 lb	EA.	7,480	190.00	7,670
1000 lb	"	7,710	200.00	7,910
2000 lb	"	8,580	210.00	8,790
5000 lb	"	9,940	250.00	10,190
10,000 lb	"	14,660	280.00	14,940
20,000 lb	"	20,020	310.00	20,330
30,000 lb	"	30,340	370.00	30,710
Air powered hoists				
500 lb	EA.	2,800	190.00	2,990
1000 lb	"	3,120	190.00	3,310
2000 lb	"	3,410	200.00	3,610
4000 lb	"	3,980	220.00	4,200
6000 lb	"	4,750	290.00	5,040
Overhead traveling bridge crane				
Single girder, 20' span				
3 ton	EA.	15,940	750.00	16,690
5 ton	"	17,530	750.00	18,280
7.5 ton	"	22,180	750.00	22,930
10 ton	"	22,180	930.00	23,110
15 ton	"	27,410	930.00	28,340
30' span				
3 ton	EA.	16,960	750.00	17,710
5 ton	"	18,740	750.00	19,490
10 ton	"	23,970	930.00	24,900
15 ton	"	28,680	930.00	29,610
Double girder, 40' span				
3 ton	EA.	32,890	1,660	34,550
5 ton	"	33,780	1,660	35,440
7.5 ton	"	34,930	1,660	36,590
10 ton	"	36,720	2,140	38,860
15 ton	"	41,300	2,140	43,440
25 ton	"	61,190	2,140	63,330
50' span				
3 ton	EA.	35,890	1,660	37,550
5 ton	"	37,160	1,660	38,820
7.5 ton	"	38,500	1,660	40,160
10 ton	"	41,810	2,140	43,950
15 ton	"	48,440	2,140	50,580
25 ton	"	66,290	2,140	68,430
Rail for bridge crane, including splice bars	Lb.			

14 CONVEYING

HOISTS AND CRANES

14650.10 JIB CRANES

HOISTS AND CRANES	UNIT	MAT.	INST.	TOTAL
Self supporting, swinging 8' boom, 200 deg rotation				
2000 lb	EA.	1,530	330.00	1,860
4000 lb	"	2,040	660.00	2,700
10,000 lb	"	3,020	660.00	3,680
Wall mounted, 180 deg rotation				
2000 lb	EA.	820.00	330.00	1,150
4000 lb	"	1,080	660.00	1,740
10,000 lb	"	2,420	660.00	3,080

BASIC MATERIALS	UNIT	MAT.	INST.	TOTAL
15100.10 SPECIALTIES				
Wall penetration				
Concrete wall, 6" thick				
2" dia.	EA.	0.00	9.55	9.55
4" dia.	"	0.00	14.30	14.30
12" thick				
2" dia.	EA.	0.00	13.00	13.00
4" dia.	"	0.00	20.45	20.45
15120.10 BACKFLOW PREVENTERS				
Backflow preventer, flanged, cast iron, with valves				
3" pipe	EA.	2,490	200.00	2,690
4" pipe	"	3,400	220.00	3,620
Threaded				
3/4" pipe	EA.	570.00	25.20	595.20
2" pipe	"	1,020	40.20	1,060
15140.11 PIPE HANGERS, LIGHT				
A band, black iron				
1/2"	EA.	0.45	2.85	3.30
1"	"	0.57	3.00	3.57
1-1/4"	"	0.63	3.10	3.73
1-1/2"	"	0.72	3.35	4.07
2"	"	0.79	3.65	4.44
2-1/2"	"	1.10	4.00	5.10
3"	"	1.10	4.45	5.55
4"	"	1.65	5.05	6.70
Copper				
1/2"	EA.	0.58	2.85	3.43
3/4"	"	0.58	3.00	3.58
1"	"	0.63	3.00	3.63
1-1/4"	"	0.64	3.10	3.74
1-1/2"	"	0.73	3.35	4.08
2"	"	0.84	3.65	4.49
2-1/2"	"	1.10	4.00	5.10
3"	"	1.25	4.45	5.70
4"	"	1.80	5.05	6.85
2 hole clips, galvanized				
3/4"	EA.	0.14	2.70	2.84
1"	"	0.17	2.80	2.97
1-1/4"	"	0.22	2.85	3.07
1-1/2"	"	0.28	3.00	3.28
2"	"	0.36	3.10	3.46
2-1/2"	"	0.66	3.20	3.86
3"	"	0.97	3.35	4.32
4"	"	2.05	3.65	5.70
Perforated strap				
3/4"				
Galvanized, 20 ga.	L.F.	0.23	2.00	2.23
Copper, 22 ga.	"	0.28	2.00	2.28
J-Hooks				
1/2"	EA.	0.25	1.85	2.10

15 MECHANICAL

BASIC MATERIALS

BASIC MATERIALS	UNIT	MAT.	INST.	TOTAL
15140.11 PIPE HANGERS, LIGHT				
3/4"	EA.	0.28	1.85	2.13
1"	"	0.29	1.90	2.19
1-1/4"	"	0.30	1.95	2.25
1-1/2"	"	0.31	2.00	2.31
2"	"	0.32	2.00	2.32
3"	"	0.37	2.10	2.47
4"	"	0.37	2.10	2.47
PVC coated hangers, galvanized, 28 ga.				
1-1/2" x 12"	EA.	0.75	2.70	3.45
2" x 12"	"	0.83	2.85	3.68
3" x 12"	"	0.91	3.10	4.01
4" x 12"	"	1.00	3.35	4.35
Copper, 30 ga.				
1-1/2" x 12"	EA.	0.91	2.70	3.61
2" x 12"	"	1.10	2.85	3.95
3" x 12"	"	1.20	3.10	4.30
4" x 12"	"	1.30	3.35	4.65
Wire hook hangers				
Black wire, 1/2" x				
4"	EA.	0.26	2.00	2.26
6"	"	0.31	2.10	2.41
Copper wire hooks				
1/2" x				
4"	EA.	0.32	2.00	2.32
6"	"	0.36	2.10	2.46
8"	"	0.41	2.25	2.66
10"	"	0.51	2.35	2.86
12"	"	0.58	2.50	3.08
15240.10 VIBRATION CONTROL				
Vibration isolator, in-line, stainless connector, screwed				
3/4"	EA.	68.50	23.65	92.15
1"	"	73.00	25.20	98.20
2"	"	120.00	31.00	151.00
3"	"	310.00	36.60	346.60
4"	"	410.00	40.20	450.20

INSULATION

INSULATION	UNIT	MAT.	INST.	TOTAL
15260.10 FIBERGLASS PIPE INSULATION				
Fiberglass insulation on 1/2" pipe				
1" thick	L.F.	1.10	1.35	2.45
1-1/2" thick	"	2.75	1.70	4.45
3/4" pipe				

INSULATION	UNIT	MAT.	INST.	TOTAL
15260.10 FIBERGLASS PIPE INSULATION				
1" thick	L.F.	2.45	1.35	3.80
1-1/2" thick	"	2.95	1.70	4.65
1" pipe				
1" thick	L.F.	2.50	1.35	3.85
1-1/2" thick	"	3.20	1.70	4.90
2" thick	"	4.40	2.00	6.40
1-1/4" dia. pipe				
1" thick	L.F.	3.20	1.70	4.90
1-1/2" thick	"	3.80	1.85	5.65
1-1/2" pipe				
1" thick	L.F.	3.30	1.70	5.00
1-1/2" thick	"	3.95	1.85	5.80
2" pipe				
1" thick	L.F.	3.55	1.70	5.25
1-1/2" thick	"	4.20	1.85	6.05
2-1/2" pipe				
1" thick	L.F.	3.95	1.70	5.65
1-1/2" thick	"	4.55	1.85	6.40
3" pipe				
1" thick	L.F.	4.05	1.90	5.95
1-1/2" thick	"	4.90	2.00	6.90
4" pipe				
1" thick	L.F.	5.05	1.90	6.95
1-1/2" thick	"	6.25	2.00	8.25
6" pipe				
1" thick	L.F.	6.25	2.10	8.35
2" thick	"	9.80	2.25	12.05
10" pipe				
2" thick	L.F.	14.60	2.10	16.70
3" thick	"	20.25	2.25	22.50
15260.20 CALCIUM SILICATE				
Calcium silicate insulation, 6" pipe				
2" thick	L.F.	7.60	2.85	10.45
3" thick	"	16.75	3.35	20.10
6" thick	"	36.70	4.00	40.70
12" pipe				
2" thick	L.F.	12.65	3.10	15.75
3" thick	"	27.90	3.65	31.55
6" thick	"	62.00	4.45	66.45
15260.60 EXTERIOR PIPE INSULATION				
Fiberglass insulation, aluminum jacket				
1/2" pipe				
1" thick	L.F.	3.10	3.10	6.20
1-1/2" thick	"	3.80	3.35	7.15
1" pipe				
1" thick	L.F.	3.50	3.10	6.60
1-1/2" thick	"	4.25	3.35	7.60
2" pipe				
1" thick	L.F.	4.40	3.65	8.05
1-1/2" thick	"	5.30	3.85	9.15

INSULATION	UNIT	MAT.	INST.	TOTAL
15260.60 EXTERIOR PIPE INSULATION				
3" pipe				
1" thick	L.F.	4.90	4.00	8.90
1-1/2" thick	"	5.80	4.25	10.05
4" pipe				
1" thick	L.F.	5.80	4.00	9.80
1-1/2" thick	"	6.95	4.25	11.20
6" pipe				
1" thick	L.F.	6.40	4.45	10.85
2" thick	"	10.55	4.75	15.30
10" pipe				
2" thick	L.F.	16.15	4.45	20.60
3" thick	"	23.20	4.75	27.95
15260.90 PIPE INSULATION FITTINGS				
Insulation protection saddle				
1" thick covering				
1/2" pipe	EA.	3.65	16.10	19.75
3/4" pipe	"	3.65	16.10	19.75
1" pipe	"	3.65	16.10	19.75
2" pipe	"	5.20	16.10	21.30
3" pipe	"	5.90	18.30	24.20
6" pipe	"	7.35	25.20	32.55
1-1/2" thick covering				
3/4" pipe	EA.	5.95	16.10	22.05
1" pipe	"	5.95	16.10	22.05
2" pipe	"	6.65	16.10	22.75
3" pipe	"	7.50	16.10	23.60
6" pipe	"	11.05	25.20	36.25
10" pipe	"	13.95	33.50	47.45
15280.10 EQUIPMENT INSULATION				
Equipment insulation, 2" thick, cellular glass	S.F.	2.10	2.50	4.60
Urethane, rigid, field applied jacket, plastered finish	"	2.20	5.05	7.25
Fiberglass, rigid, with vapor barrier	"	2.10	2.25	4.35
15290.10 DUCTWORK INSULATION				
Fiberglass duct insulation, plain blanket				
1-1/2" thick	S.F.	0.65	0.50	1.15
2" thick	"	0.77	0.67	1.44
With vapor barrier				
1-1/2" thick	S.F.	0.40	0.50	0.90
2" thick	"	0.45	0.67	1.12
Rigid with vapor barrier				
2" thick	S.F.	1.30	1.35	2.65

FIRE PROTECTION

15330.10 WET SPRINKLER SYSTEM

FIRE PROTECTION	UNIT	MAT.	INST.	TOTAL
Sprinkler head, 212 deg, brass, exposed piping	EA.	5.50	16.10	21.60
Chrome, concealed piping	"	5.25	22.35	27.60
Water motor alarm	"	73.00	67.00	140.00
Fire department inlet connection	"	110.00	80.50	190.50
Wall plate for fire dept connection	"	54.00	33.50	87.50
Swing check valve flanged iron body, 4"	"	110.00	130.00	240.00
Check valve, 6"	"	510.00	200.00	710.00
Wet pipe valve, flange to groove, 4"	"	450.00	44.70	494.70
Flange to flange				
6"	EA.	570.00	67.00	637.00
8"	"	800.00	130.00	930.00
Alarm valve, flange to flange, (wet valve)				
4"	EA.	540.00	44.70	584.70
8"	"	820.00	340.00	1,160
Inspector's test connection	"	34.30	33.50	67.80
Wall hydrant, polished brass, 2-1/2" x 2-1/2", single	"	270.00	28.70	298.70
2-way	"	620.00	28.70	648.70
3-way	"	1,260	28.70	1,289
Wet valve trim, includes retard chamber & gauges, 4"-6"	"	430.00	33.50	463.50
Retard pressure switch for wet systems	"	760.00	80.50	840.50
Air maintenance device	"	230.00	33.50	263.50
Wall hydrant non-freeze, 8" thick wall, vacuum breaker	"	25.20	20.10	45.30
12" thick wall	"	27.50	20.10	47.60

PLUMBING

15410.05 C.I. PIPE, ABOVE GROUND

PLUMBING	UNIT	MAT.	INST.	TOTAL
No hub pipe				
1-1/2" pipe	L.F.	4.35	2.85	7.20
2" pipe	"	5.00	3.35	8.35
3" pipe	"	7.05	4.00	11.05
4" pipe	"	9.60	6.70	16.30
No hub fittings, 1-1/2" pipe				
1/4 bend	EA.	5.65	13.40	19.05
1/8 bend	"	4.05	13.40	17.45
Sanitary tee	"	6.90	20.10	27.00
Sanitary cross	"	8.55	20.10	28.65
Plug	"			1.25
Coupling	"			7.55
Wye	"	7.55	20.10	27.65
2" pipe				
1/4 bend	EA.	5.80	16.10	21.90
1/8 bend	"	4.15	16.10	20.25

PLUMBING	UNIT	MAT.	INST.	TOTAL
15410.05 C.I. PIPE, ABOVE GROUND				
Sanitary tee	EA.	7.15	26.80	33.95
Coupling	"			7.55
Wye	"	8.05	33.50	41.55
3" pipe				
1/4 bend	EA.	7.30	20.10	27.40
1/8 bend	"	6.55	20.10	26.65
Sanitary tee	"	8.55	25.20	33.75
Coupling	"			8.81
Wye	"	8.80	33.50	42.30
4" pipe				
1/4 bend	EA.	10.10	20.10	30.20
1/8 bend	"	8.55	20.10	28.65
Sanitary tee	"	12.60	33.50	46.10
Coupling	"			9.83
Wye	"	12.60	33.50	46.10
15410.06 C.I. PIPE, BELOW GROUND				
No hub pipe				
1-1/2" pipe	L.F.	4.55	2.00	6.55
2" pipe	"	5.00	2.25	7.25
3" pipe	"	7.05	2.50	9.55
4" pipe	"	9.60	3.35	12.95
Fittings, 1-1/2"				
1/4 bend	EA.	5.65	11.50	17.15
1/8 bend	"	4.40	11.50	15.90
Plug	"			1.25
Wye	"	7.55	16.10	23.65
Wye & 1/8 bend	"	7.55	11.50	19.05
P-trap	"	10.95	11.50	22.45
2"				
1/4 bend	EA.	5.80	13.40	19.20
1/8 bend	"	4.15	13.40	17.55
Plug	"			1.51
Double wye	"	10.30	25.20	35.50
Wye & 1/8 bend	"	10.20	20.10	30.30
Double wye & 1/8 bend	"	15.10	25.20	40.30
P-trap	"	10.95	13.40	24.35
3"				
1/4 bend	EA.	7.30	16.10	23.40
1/8 bend	"	6.55	16.10	22.65
Plug	"			2.27
Wye	"	8.80	25.20	34.00
3x2" wye	"	9.05	25.20	34.25
Wye & 1/8 bend	"	11.35	25.20	36.55
Double wye & 1/8 bend	"	20.75	25.20	45.95
3x2" double wye & 1/8 bend	"	15.75	25.20	40.95
3x2" reducer	"	4.80	16.10	20.90
P-trap	"	13.20	16.10	29.30
4"				
1/4 bend	EA.	10.10	16.10	26.20
1/8 bend	"	8.55	16.10	24.65

PLUMBING	UNIT	MAT.	INST.	TOTAL
15410.06 C.I. PIPE, BELOW GROUND				
Wye	EA.	12.60	25.20	37.80
15410.09 SERVICE WEIGHT PIPE				
Service weight pipe, single hub				
3" x 5'	EA.	28.80	8.55	37.35
4" x 5'	"	34.80	8.95	43.75
6" x 5'	"	55.50	10.05	65.55
1/8 bend				
3"	EA.	6.25	16.10	22.35
4"	"	8.65	18.30	26.95
6"	"	13.85	20.10	33.95
1/4 bend				
3"	EA.	7.80	16.10	23.90
4"	"	10.90	18.30	29.20
6"	"	18.60	20.10	38.70
Sweep				
3"	EA.	12.00	16.10	28.10
4"	"	16.80	18.30	35.10
6"	"	30.10	20.10	50.20
Sanitary T				
3"	EA.	12.65	28.70	41.35
4"	"	15.60	33.50	49.10
6"	"	31.30	36.60	67.90
Wye				
3"	EA.	13.85	22.35	36.20
4"	"	17.40	23.65	41.05
6"	"	36.00	28.70	64.70
15410.10 COPPER PIPE				
Type "K" copper				
1/2"	L.F.	1.80	1.25	3.05
3/4"	"	3.30	1.35	4.65
1"	"	4.15	1.45	5.60
DWV, copper				
1-1/4"	L.F.	3.30	1.70	5.00
1-1/2"	"	4.25	1.85	6.10
2"	"	4.80	2.00	6.80
3"	"	9.05	2.25	11.30
4"	"	14.95	2.50	17.45
6"	"	48.00	2.85	50.85
Type "L" copper				
1/4"	L.F.	1.05	1.20	2.25
3/8"	"	1.15	1.20	2.35
1/2"	"	1.80	1.25	3.05
3/4"	"	2.00	1.35	3.35
1"	"	2.75	1.45	4.20
Type "M" copper				
1/2"	L.F.	1.30	1.25	2.55
3/4"	"	1.80	1.35	3.15
1"	"	2.25	1.45	3.70

PLUMBING	UNIT	MAT.	INST.	TOTAL
15410.10 COPPER PIPE				
Type "K" tube, coil				
1/4" x 60'	EA.			49.03
1/2" x 60'	"			91.66
3/4" x 60'	"			149.23
1" x 60'	"			223.84
Type "L" tube, coil				
1/4" x 60'	EA.			46.90
3/8" x 60'	"			70.34
1/2" x 60'	"			93.80
3/4" x 60'	"			140.70
1" x 60'	"			239.83
15410.11 COPPER FITTINGS				
DWV fittings, coupling with stop				
1-1/4"	EA.	3.75	23.65	27.40
1-1/2"	"	4.25	25.20	29.45
1-1/2" x 1-1/4"	"	4.90	25.20	30.10
2"	"	4.80	26.80	31.60
2" x 1-1/4"	"	6.40	26.80	33.20
2" x 1-1/2"	"	6.40	26.80	33.20
3"	"	8.10	33.50	41.60
3" x 1-1/2"	"	16.00	33.50	49.50
3" x 2"	"	16.00	33.50	49.50
4"	"	17.05	40.20	57.25
Slip coupling				
1-1/2"	EA.	4.80	25.20	30.00
2"	"	5.25	26.80	32.05
3"	"	7.45	33.50	40.95
90 ells				
1-1/2"	EA.	8.95	25.20	34.15
1-1/2" x 1-1/4"	"	11.30	25.20	36.50
2"	"	9.60	26.80	36.40
2" x 1-1/2"	"	13.85	26.80	40.65
3"	"	27.70	33.50	61.20
4"	"	110.00	40.20	150.20
Street, 90 elbows				
1-1/2"	EA.	7.45	25.20	32.65
2"	"	13.30	26.80	40.10
3"	"	26.70	33.50	60.20
4"	"	120.00	40.20	160.20
45 ells				
1-1/4"	EA.	5.35	23.65	29.00
1-1/2"	"	4.90	25.20	30.10
2"	"	9.15	26.80	35.95
3"	"	17.05	33.50	50.55
4"	"	49.00	40.20	89.20
Street, 45 ell				
1-1/2"	EA.	5.95	25.20	31.15
2"	"	10.65	26.80	37.45
3"	"	25.60	33.50	59.10
Wye				

PLUMBING	UNIT	MAT.	INST.	TOTAL
15410.11 COPPER FITTINGS				
1-1/4"	EA.	17.05	23.65	40.70
1-1/2"	"	14.95	25.20	40.15
2"	"	24.50	26.80	51.30
3"	"	57.50	33.50	91.00
4"	"	130.00	40.20	170.20
Sanitary tee				
1-1/4"	EA.	12.25	23.65	35.90
1-1/2"	"	11.75	25.20	36.95
2"	"	17.05	26.80	43.85
3"	"	40.50	33.50	74.00
4"	"	100.00	40.20	140.20
No-hub adapters				
1-1/2" x 2"	EA.	11.75	25.20	36.95
2"	"	11.75	26.80	38.55
2" x 3"	"	32.00	26.80	58.80
3"	"	19.70	33.50	53.20
3" x 4"	"	33.00	33.50	66.50
4"	"	35.20	40.20	75.40
Fitting reducers				
1-1/2" x 1-1/4"	EA.	4.90	25.20	30.10
2" x 1-1/2"	"	4.80	26.80	31.60
3" x 1-1/2"	"	15.45	33.50	48.95
3" x 2"	"	15.80	33.50	49.30
Copper caps				
1-1/2"	EA.	8.30	25.20	33.50
2"	"	13.30	26.80	40.10
Copper pipe fittings				
1/2"				
90 deg ell	EA.	0.76	8.95	9.71
45 deg ell	"	0.96	8.95	9.91
Tee	"	1.30	11.50	12.80
Cap	"	0.52	4.45	4.97
Coupling	"	0.55	8.95	9.50
Union	"	3.85	10.05	13.90
3/4"				
90 deg ell	EA.	1.65	10.05	11.70
45 deg ell	"	1.95	10.05	12.00
Tee	"	2.75	13.40	16.15
Cap	"	1.00	4.75	5.75
Coupling	"	1.10	10.05	11.15
Union	"	5.65	11.50	17.15
1"				
90 deg ell	EA.	3.85	13.40	17.25
45 deg ell	"	5.00	13.40	18.40
Tee	"	6.30	16.10	22.40
Cap	"	1.85	6.70	8.55
Coupling	"	2.75	13.40	16.15
Union	"	7.40	13.40	20.80
1-1/4"				
90 deg ell	EA.	5.25	11.50	16.75
45 deg ell	"	6.50	11.50	18.00
Tee	"	8.55	20.10	28.65

PLUMBING	UNIT	MAT.	INST.	TOTAL
15410.11 COPPER FITTINGS				
Cap	EA.	1.50	6.70	8.20
Union	"	12.25	14.35	26.60
1-1/2"				
90 deg ell	EA.	6.80	14.35	21.15
45 deg ell	"	8.10	14.35	22.45
Tee	"	11.20	22.35	33.55
Cap	"	1.50	6.70	8.20
Coupling	"	5.00	13.40	18.40
Union	"	18.65	18.30	36.95
2"				
90 deg ell	EA.	13.30	16.10	29.40
45 deg ell	"	12.25	25.20	37.45
Tee	"	19.20	25.20	44.40
Cap	"	3.10	8.05	11.15
Coupling	"	8.10	16.10	24.20
Union	"	20.25	20.10	40.35
2-1/2"				
90 deg ell	EA.	25.60	20.10	45.70
45 deg ell	"	22.40	20.10	42.50
Tee	"	25.60	28.70	54.30
Cap	"	6.30	10.05	16.35
Coupling	"	12.25	20.10	32.35
Union	"	37.30	22.35	59.65
15410.15 BRASS FITTINGS				
Compression fittings, union				
3/8"	EA.	2.65	6.70	9.35
1/2"	"	3.95	6.70	10.65
5/8"	"	4.85	6.70	11.55
Union elbow				
3/8"	EA.	4.40	6.70	11.10
1/2"	"	6.95	6.70	13.65
5/8"	"	8.65	6.70	15.35
Union tee				
3/8"	EA.	5.80	6.70	12.50
1/2"	"	9.25	6.70	15.95
5/8"	"	13.30	6.70	20.00
Male connector				
3/8"	EA.	1.85	6.70	8.55
1/2"	"	1.25	6.70	7.95
5/8"	"	2.90	6.70	9.60
Female connector				
3/8"	EA.	3.45	6.70	10.15
1/2"	"	4.40	6.70	11.10
5/8"	"	4.40	6.70	11.10
15410.30 PVC/CPVC PIPE				
PVC schedule 40				
1/2" pipe	L.F.	1.45	1.70	3.15
3/4" pipe	"	1.70	1.85	3.55
1" pipe	"	1.95	2.00	3.95

PLUMBING	UNIT	MAT.	INST.	TOTAL
15410.30 PVC/CPVC PIPE				
1-1/4" pipe	L.F.	2.20	2.25	4.45
1-1/2" pipe	"	2.30	2.50	4.80
2" pipe	"	2.65	2.85	5.50
2-1/2" pipe	"	3.65	3.35	7.00
3" pipe	"	3.85	4.00	7.85
4" pipe	"	5.70	5.05	10.75
6" pipe	"	10.50	10.05	20.55
8" pipe	"	16.05	13.40	29.45
Fittings, 1/2"				
90 deg ell	EA.	0.35	5.05	5.40
45 deg ell	"	0.55	5.05	5.60
Tee	"	0.47	5.75	6.22
Polypropylene, acid resistant, DWV pipe				
Schedule 40				
1-1/2" pipe	L.F.	3.75	2.85	6.60
2" pipe	"	5.00	3.35	8.35
3" pipe	"	9.25	4.00	13.25
4" pipe	"	13.10	5.05	18.15
6" pipe	"	23.50	10.05	33.55
Polyethylene pipe and fittings				
SDR-21				
3" pipe	L.F.	2.45	5.05	7.50
4" pipe	"	3.70	6.70	10.40
6" pipe	"	6.15	10.05	16.20
8" pipe	"	9.90	11.50	21.40
10" pipe	"	11.10	13.40	24.50
12" pipe	"	17.30	16.10	33.40
14" pipe	"	22.20	20.10	42.30
16" pipe	"	27.10	25.20	52.30
18" pipe	"	29.60	31.00	60.60
20" pipe	"	37.00	40.20	77.20
22" pipe	"	43.20	44.70	87.90
24" pipe	"	52.00	50.50	102.50
Fittings, 3"				
90 deg elbow	EA.	66.00	20.10	86.10
45 deg elbow	"	41.70	20.10	61.80
Tee	"	38.00	33.50	71.50
45 deg wye	"	98.00	33.50	131.50
Reducer	"	21.80	25.20	47.00
Flange assembly	"	15.10	20.10	35.20
4"				
90 deg elbow	EA.	97.00	25.20	122.20
45 deg elbow	"	59.00	25.20	84.20
Tee	"	83.50	40.20	123.70
45 deg wye	"	150.00	40.20	190.20
Reducer	"	52.50	33.50	86.00
Flange assembly	"	52.50	25.20	77.70
8"				
90 deg elbow	EA.	260.00	50.50	310.50
45 deg elbow	"	150.00	50.50	200.50
Tee	"	250.00	80.50	330.50
45 deg wye	"	380.00	80.50	460.50

PLUMBING	UNIT	MAT.	INST.	TOTAL
15410.30 PVC/CPVC PIPE				
Reducer	EA.	130.00	67.00	197.00
Flange assembly	"	140.00	50.50	190.50
10"				
90 deg elbow	EA.	360.00	67.00	427.00
45 deg elbow	"	200.00	67.00	267.00
Tee	"	330.00	100.00	430.00
45 deg wye	"	520.00	100.00	620.00
Reducer	"	170.00	80.50	250.50
Flange assembly	"	170.00	67.00	237.00
12"				
90 deg elbow	EA.	590.00	80.50	670.50
45 deg elbow	"	360.00	80.50	440.50
Tee	"	460.00	130.00	590.00
45 deg wye	"	720.00	130.00	850.00
Reducer	"	280.00	100.00	380.00
Flange assembly	"	220.00	80.50	300.50
14"				
90 deg elbow	EA.	800.00	100.00	900.00
45 deg elbow	"	460.00	100.00	560.00
Tee	"	590.00	160.00	750.00
45 deg wye	"	1,080	160.00	1,240
Reducer	"	230.00	130.00	360.00
Flange assembly	"	270.00	100.00	370.00
16"				
90 deg elbow	EA.	1,000	100.00	1,100
45 deg elbow	"	590.00	100.00	690.00
Tee	"	730.00	160.00	890.00
45 deg wye	"	1,210	160.00	1,370
Reducer	"	470.00	130.00	600.00
Flange assembly	"	340.00	100.00	440.00
18"				
90 deg elbow	EA.	1,530	130.00	1,660
45 deg elbow	"	990.00	130.00	1,120
Tee	"	1,180	200.00	1,380
45 deg wye	"	2,040	200.00	2,240
Reducer	"	440.00	130.00	570.00
Flange assembly	"	600.00	130.00	730.00
20"				
90 deg elbow	EA.	1,230	130.00	1,360
45 deg elbow	"	730.00	130.00	860.00
15410.33 ABS DWV PIPE				
Schedule 40 ABS				
1-1/2" pipe	L.F.	1.80	2.00	3.80
2" pipe	"	2.40	2.25	4.65
3" pipe	"	3.65	2.85	6.50
4" pipe	"	5.45	4.00	9.45
6" pipe	"	8.45	5.05	13.50
15410.35 PLASTIC PIPE				
Fiberglass reinforced pipe				
2" pipe	L.F.	9.90	3.10	13.00

PLUMBING	UNIT	MAT.	INST.	TOTAL
15410.35 PLASTIC PIPE				
3" pipe	L.F.	13.30	3.35	16.65
4" pipe	"	16.95	3.65	20.60
6" pipe	"	26.60	4.00	30.60
8" pipe	"	38.70	6.70	45.40
10" pipe	"	48.40	8.05	56.45
12" pipe	"	60.50	10.05	70.55
Fittings				
90 deg elbow, flanged				
2"	EA.	79.00	40.20	119.20
3"	"	94.50	44.70	139.20
4"	"	120.00	50.50	170.50
6"	"	220.00	67.00	287.00
8"	"	400.00	80.50	480.50
10"	"	530.00	100.00	630.00
12"	"	710.00	130.00	840.00
45 deg elbow, flanged				
2"	EA.	79.00	33.50	112.50
3"	"	94.50	40.20	134.70
4"	"	120.00	50.50	170.50
6"	"	220.00	67.00	287.00
8"	"	340.00	80.50	420.50
10"	"	450.00	100.00	550.00
12"	"	570.00	130.00	700.00
Tee, flanged				
2"	EA.	100.00	50.50	150.50
3"	"	140.00	57.50	197.50
4"	"	160.00	67.00	227.00
6"	"	270.00	80.50	350.50
8"	"	480.00	100.00	580.00
10"	"	780.00	130.00	910.00
12"	"	1,070	200.00	1,270
Wye, flanged				
2"	EA.	200.00	50.50	250.50
3"	"	270.00	57.50	327.50
4"	"	360.00	67.00	427.00
6"	"	450.00	80.50	530.50
8"	"	610.00	100.00	710.00
10"	"	1,040	130.00	1,170
12"	"	1,330	200.00	1,530
Concentric reducer, flanged				
2"	EA.	76.50	33.50	110.00
4"	"	110.00	40.20	150.20
6"	"	150.00	57.50	207.50
8"	"	250.00	80.50	330.50
10"	"	380.00	100.00	480.00
12"	"	550.00	130.00	680.00
Adapter, bell x male or female				
2"	EA.	9.80	33.50	43.30
3"	"	19.60	36.60	56.20
4"	"	20.70	40.20	60.90
6"	"	44.20	57.50	101.70
8"	"	64.50	80.50	145.00

PLUMBING	UNIT	MAT.	INST.	TOTAL
15410.35 PLASTIC PIPE				
10"	EA.	92.50	100.00	192.50
12"	"	160.00	130.00	290.00
Nipples				
2" x 6"	EA.	4.15	4.00	8.15
2" x 12"	"	6.25	5.05	11.30
3" x 8"	"	6.25	6.20	12.45
3" x 12"	"	6.25	6.70	12.95
4" x 8"	"	6.25	6.70	12.95
4" x 12"	"	7.00	8.05	15.05
6" x 12"	"	16.00	10.05	26.05
8" x 18"	"	45.80	10.05	55.85
8" x 24"	"	53.50	11.50	65.00
10" x 18"	"	53.50	13.40	66.90
10" x 24"	"	72.50	16.10	88.60
12" x 18"	"	72.50	18.30	90.80
12" x 24"	"	85.00	20.10	105.10
Sleeve coupling				
2"	EA.	9.05	33.50	42.55
3"	"	9.90	40.20	50.10
4"	"	13.70	57.50	71.20
6"	"	33.10	80.50	113.60
8"	"	54.50	100.00	154.50
10"	"	81.50	130.00	211.50
Flanges				
2"	EA.	13.50	33.50	47.00
3"	"	18.20	40.20	58.40
4"	"	24.15	57.50	81.65
6"	"	41.30	80.50	121.80
8"	"	72.50	100.00	172.50
10"	"	100.00	130.00	230.00
12"	"	130.00	130.00	260.00
15410.70 STAINLESS STEEL PIPE				
Stainless steel, schedule 40, threaded				
1/2" pipe	L.F.	4.95	5.75	10.70
1" pipe	"	7.70	6.20	13.90
1-1/2" pipe	"	10.45	6.70	17.15
2" pipe	"	15.70	7.30	23.00
2-1/2" pipe	"	22.00	8.05	30.05
3" pipe	"	30.80	8.95	39.75
4" pipe	"	39.60	10.05	49.65
15410.80 STEEL PIPE				
Black steel, extra heavy pipe, threaded				
1/2" pipe	L.F.	1.45	1.60	3.05
3/4" pipe	"	1.75	1.60	3.35
1" pipe	"	2.40	2.00	4.40
1-1/2" pipe	"	3.65	2.25	5.90
2-1/2" pipe	"	7.25	5.05	12.30
3" pipe	"	9.70	6.70	16.40
4" pipe	"	14.70	8.05	22.75

15 MECHANICAL

PLUMBING	UNIT	MAT.	INST.	TOTAL
15410.80 STEEL PIPE				
5" pipe	L.F.	19.75	10.05	29.80
6" pipe	"	24.70	10.05	34.75
8" pipe	"	36.40	13.40	49.80
10" pipe	"	50.50	16.10	66.60
12" pipe	"	64.00	20.10	84.10
Fittings, malleable iron, threaded, 1/2" pipe				
90 deg ell	EA.	1.45	13.40	14.85
45 deg ell	"	2.35	13.40	15.75
Tee	"	1.95	20.10	22.05
3/4" pipe				
90 deg ell	EA.	1.95	13.40	15.35
45 deg ell	"	3.20	20.10	23.30
Tee	"	3.25	20.10	23.35
1-1/2" pipe				
90 deg ell	EA.	6.10	20.10	26.20
45 deg ell	"	7.15	20.10	27.25
Tee	"	8.80	28.70	37.50
2-1/2" pipe				
90 deg ell	EA.	24.70	50.50	75.20
45 deg ell	"	30.90	50.50	81.40
Tee	"	33.30	67.00	100.30
3" pipe				
90 deg ell	EA.	35.80	67.00	102.80
45 deg ell	"	40.70	67.00	107.70
Tee	"	45.70	100.00	145.70
4" pipe				
90 deg ell	EA.	71.50	80.50	152.00
45 deg ell	"	73.00	80.50	153.50
Tee	"	100.00	130.00	230.00
6" pipe				
90 deg ell	EA.	200.00	80.50	280.50
45 deg ell	"	250.00	80.50	330.50
Tee	"	280.00	130.00	410.00
8" pipe				
90 deg ell	EA.	210.00	160.00	370.00
45 deg ell	"	260.00	160.00	420.00
Tee	"	310.00	250.00	560.00
10" pipe				
90 deg ell	EA.	240.00	200.00	440.00
45 deg ell	"	280.00	200.00	480.00
Tee	"	350.00	250.00	600.00
12" pipe				
90 deg ell	EA.	280.00	250.00	530.00
45 deg ell	"	320.00	250.00	570.00
Tee	"	410.00	340.00	750.00
Butt welded, 1/2" pipe				
90 deg ell	EA.	9.75	13.40	23.15
45 deg ell	"	14.20	13.40	27.60
Tee	"	39.50	20.10	59.60
3/4" pipe				
90 deg ell	EA.	9.75	13.40	23.15
45 deg. ell	"	14.80	13.40	28.20

PLUMBING

15410.80 STEEL PIPE	UNIT	MAT.	INST.	TOTAL
Tee	EA.	40.70	20.10	60.80
1" pipe				
90 deg ell	EA.	9.75	16.10	25.85
45 deg ell	"	14.80	16.10	30.90
Tee	"	40.70	22.35	63.05
1-1/2" pipe				
90 deg ell	EA.	13.80	20.10	33.90
45 deg. ell	"	14.70	20.10	34.80
Tee	"	36.40	28.70	65.10
Reducing tee	"	24.30	28.70	53.00
Cap	"	12.95	16.10	29.05
2-1/2" pipe				
90 deg. ell	EA.	28.40	40.20	68.60
45 deg. ell	"	28.40	40.20	68.60
Tee	"	110.00	57.50	167.50
Reducing tee	"	23.45	57.50	80.95
Cap	"	29.60	20.10	49.70
3" pipe				
90 deg ell	EA.	27.10	50.50	77.60
45 deg. ell	"	66.50	50.50	117.00
Tee	"	27.10	67.00	94.10
Reducing tee	"	28.40	67.00	95.40
Cap	"	14.80	33.50	48.30
4" pipe				
90 deg ell	EA.	54.50	67.00	121.50
45 deg. ell	"	44.40	67.00	111.40
Tee	"	97.50	100.00	197.50
Reducing tee	"	35.80	100.00	135.80
Cap	"	16.05	33.50	49.55
6" pipe				
90 deg. ell	EA.	170.00	80.50	250.50
45 deg. ell	"	120.00	80.50	200.50
Tee	"	270.00	130.00	400.00
Reducing tee	"	240.00	130.00	370.00
Cap	"	28.40	40.20	68.60
8" pipe				
90 deg. ell	EA.	270.00	130.00	400.00
45 deg. ell	"	200.00	130.00	330.00
Tee	"	600.00	200.00	800.00
Reducing tee	"	500.00	200.00	700.00
Cap	"	43.20	80.50	123.70
10" pipe				
90 deg ell	EA.	580.00	130.00	710.00
45 deg. ell	"	400.00	130.00	530.00
Tee	"	1,070	200.00	1,270
Reducing tee	"	210.00	200.00	410.00
Cap	"	68.00	100.00	168.00
12" pipe				
90 deg. ell	EA.	980.00	160.00	1,140
45 deg. ell	"	640.00	160.00	800.00
Tee	"	1,360	290.00	1,650
Reducing tee	"	480.00	290.00	770.00

PLUMBING	UNIT	MAT.	INST.	TOTAL
15410.80 STEEL PIPE				
Cap	EA.	82.50	100.00	182.50
Cast iron fittings				
1/2" pipe				
90 deg. ell	EA.	2.45	13.40	15.85
45 deg. ell	"	4.95	13.40	18.35
Tee	"	3.20	20.10	23.30
Reducing tee	"	6.05	20.10	26.15
3/4" pipe				
90 deg. ell	EA.	2.60	13.40	16.00
45 deg. ell	"	3.20	13.40	16.60
Tee	"	4.00	20.10	24.10
Reducing tee	"	5.25	20.10	25.35
1" pipe				
90 deg. ell	EA.	3.10	16.10	19.20
45 deg. ell	"	4.30	16.10	20.40
Tee	"	3.95	22.35	26.30
Reducing tee	"	5.20	22.35	27.55
1-1/2" pipe				
90 deg. ell	EA.	6.15	20.10	26.25
45 deg. ell	"	8.70	20.10	28.80
Tee	"	8.95	28.70	37.65
Reducing tee	"	12.10	28.70	40.80
2-1/2" pipe				
90 deg. ell	EA.	19.55	40.20	59.75
45 deg. ell	"	23.20	40.20	63.40
Tee	"	27.80	57.50	85.30
Reducing tee	"	32.30	57.50	89.80
3" pipe				
90 deg. ell	EA.	31.60	50.50	82.10
45 deg. ell	"	36.30	50.50	86.80
Tee	"	42.00	80.50	122.50
Reducing tee	"	48.60	80.50	129.10
4" pipe				
90 deg. ell	EA.	57.00	67.00	124.00
45 deg. ell	"	70.50	67.00	137.50
Tee	"	81.00	100.00	181.00
Reducing tee	"	97.00	100.00	197.00
6" pipe				
90 deg. ell	EA.	130.00	67.00	197.00
45 deg. ell	"	150.00	67.00	217.00
Tee	"	190.00	100.00	290.00
Reducing tee	"	220.00	100.00	320.00
8" pipe				
90 deg. ell	EA.	270.00	130.00	400.00
45 deg. ell	"	300.00	130.00	430.00
Tee	"	390.00	200.00	590.00
Reducing tee	"	420.00	200.00	620.00
15410.82 GALVANIZED STEEL PIPE				
Galvanized pipe				
1/2" pipe	L.F.	1.55	4.00	5.55

15 MECHANICAL

PLUMBING

PLUMBING	UNIT	MAT.	INST.	TOTAL
15410.82 GALVANIZED STEEL PIPE				
3/4" pipe	L.F.	2.05	5.05	7.10
1" pipe	"	3.15	5.75	8.90
1-1/4" pipe	"	3.65	6.70	10.35
1-1/2" pipe	"	3.85	8.05	11.90
2" pipe	"	5.55	10.05	15.60
2-1/2" pipe	"	8.00	13.40	21.40
3" pipe	"	11.10	14.35	25.45
4" pipe	"	15.15	16.75	31.90
6" pipe	"	30.30	33.50	63.80
15430.23 CLEANOUTS				
Cleanout, wall				
2"	EA.	56.50	26.80	83.30
3"	"	62.50	26.80	89.30
4"	"	81.50	33.50	115.00
6"	"	140.00	40.20	180.20
8"	"	170.00	50.50	220.50
Floor				
2"	EA.	58.50	33.50	92.00
3"	"	74.00	33.50	107.50
4"	"	81.50	40.20	121.70
6"	"	110.00	50.50	160.50
8"	"	210.00	57.50	267.50
15430.24 GREASE TRAPS				
Grease traps, cast iron, 3" pipe				
35 gpm, 70 lb capacity	EA.	1,910	400.00	2,310
50 gpm, 100 lb capacity	"	2,430	500.00	2,930
15430.25 HOSE BIBBS				
Hose bibb				
1/2"	EA.	5.30	13.40	18.70
3/4"	"	5.65	13.40	19.05
15430.60 VALVES				
Gate valve, 125 lb, bronze, soldered				
1/2"	EA.	13.50	10.05	23.55
3/4"	"	15.95	10.05	26.00
1"	"	19.60	13.40	33.00
1-1/2"	"	34.30	16.10	50.40
2"	"	47.80	20.10	67.90
2-1/2"	"	98.00	25.20	123.20
Threaded				
1/4", 125 lb	EA.	13.20	16.10	29.30
1/2"				
125 lb	EA.	14.70	16.10	30.80
150 lb	"	19.60	16.10	35.70
300 lb	"	34.30	16.10	50.40

PLUMBING	UNIT	MAT.	INST.	TOTAL
15430.60 VALVES				
3/4"				
125 lb	EA.	17.15	16.10	33.25
150 lb	"	23.30	16.10	39.40
300 lb	"	44.10	16.10	60.20
1"				
125 lb	EA.	22.05	16.10	38.15
150 lb	"	30.60	16.10	46.70
300 lb	"	61.50	20.10	81.60
1-1/2"				
125 lb	EA.	67.50	20.10	87.60
150 lb	"	51.50	20.10	71.60
300 lb	"	120.00	22.35	142.35
2"				
125 lb	EA.	46.60	28.70	75.30
150 lb	"	71.00	28.70	99.70
300 lb	"	170.00	33.50	203.50
Cast iron, flanged				
2", 150 lb	EA.	200.00	33.50	233.50
2-1/2"				
125 lb	EA.	200.00	33.50	233.50
150 lb	"	310.00	33.50	343.50
250 lb	"	390.00	33.50	423.50
3"				
125 lb	EA.	230.00	40.20	270.20
150 lb	"	320.00	40.20	360.20
250 lb	"	490.00	40.20	530.20
4"				
125 lb	EA.	280.00	57.50	337.50
150 lb	"	380.00	57.50	437.50
250 lb	"	660.00	57.50	717.50
6"				
125 lb	EA.	430.00	80.50	510.50
250 lb	"	1,100	80.50	1,181
8"				
125 lb	EA.	860.00	100.00	960.00
250 lb	"	2,080	100.00	2,180
OS&Y, flanged				
2"				
125 lb	EA.	150.00	33.50	183.50
250 lb	"	340.00	33.50	373.50
2-1/2"				
125 lb	EA.	160.00	33.50	193.50
250 lb	"	390.00	40.20	430.20
3"				
125 lb	EA.	160.00	40.20	200.20
250 lb	"	430.00	40.20	470.20
4"				
125 lb	EA.	230.00	67.00	297.00
250 lb	"	660.00	67.00	727.00
6"				
125 lb	EA.	360.00	80.50	440.50
250 lb	"	1,070	80.50	1,151

PLUMBING

15430.60 VALVES

PLUMBING	UNIT	MAT.	INST.	TOTAL
Check valve, bronze, soldered, 125 lb				
1/2"	EA.	16.35	10.05	26.40
3/4"	"	18.65	10.05	28.70
1"	"	24.50	13.40	37.90
1-1/4"	"	34.30	16.10	50.40
1-1/2"	"	40.00	16.10	56.10
2"	"	57.00	20.10	77.10
Threaded				
1/2"				
125 lb	EA.	13.50	13.40	26.90
150 lb	"	22.05	13.40	35.45
200 lb	"	22.55	13.40	35.95
3/4"				
125 lb	EA.	17.15	16.10	33.25
150 lb	"	27.00	16.10	43.10
200 lb	"	27.00	16.10	43.10
1"				
125 lb	EA.	22.05	20.10	42.15
150 lb	"	35.50	20.10	55.60
200 lb	"	35.50	20.10	55.60
Flow check valve, cast iron, threaded				
1"	EA.	34.40	16.10	50.50
1-1/4"	"	36.60	20.10	56.70
1-1/2"				
125 lb	EA.	33.50	20.10	53.60
150 lb	"	55.50	20.10	75.60
200 lb	"	55.50	22.35	77.85
2"				
125 lb	EA.	49.10	22.35	71.45
150 lb	"	81.00	22.35	103.35
200 lb	"	81.00	25.20	106.20
2-1/2"				
125 lb	EA.	110.00	33.50	143.50
250 lb	"	360.00	40.20	400.20
3"				
125 lb	EA.	130.00	40.20	170.20
250 lb	"	450.00	50.50	500.50
4"				
125 lb	EA.	200.00	57.50	257.50
250 lb	"	570.00	67.00	637.00
6"				
125 lb	EA.	270.00	80.50	350.50
250 lb	"	950.00	80.50	1,031
Vertical check valve, bronze, 125 lb, threaded				
1/2"	EA.	27.00	16.10	43.10
3/4"	"	35.50	18.30	53.80
1"	"	42.90	20.10	63.00
1-1/4"	"	52.50	22.35	74.85
1-1/2"	"	59.00	25.20	84.20
2"	"	95.50	28.70	124.20
Cast iron, flanged				
2-1/2"	EA.	350.00	40.20	390.20

PLUMBING	UNIT	MAT.	INST.	TOTAL
15430.60 VALVES				
3"	EA.	390.00	50.50	440.50
4"	"	530.00	67.00	597.00
6	"	920.00	80.50	1,001
8"	"	1,800	100.00	1,900
10"	"	2,700	130.00	2,830
12"	"	3,340	160.00	3,500
Globe valve, bronze, soldered, 125 lb				
1/2"	EA.	14.70	11.50	26.20
3/4"	"	17.15	12.55	29.70
1"	"	30.60	13.40	44.00
1-1/4"	"	47.80	14.35	62.15
1-1/2"	"	62.50	16.75	79.25
2"	"	110.00	20.10	130.10
Threaded				
1/2"				
125 lb	EA.	14.70	13.40	28.10
150 lb	"	41.70	13.40	55.10
300 lb	"	51.50	13.40	64.90
3/4"				
125 lb	EA.	17.15	16.10	33.25
150 lb	"	44.10	16.10	60.20
300 lb	"	54.00	16.10	70.10
1"				
125 lb	EA.	30.60	20.10	50.70
150 lb	"	97.00	20.10	117.10
300 lb	"	110.00	20.10	130.10
1-1/4"				
125 lb	EA.	55.00	20.10	75.10
150 lb	"	130.00	20.10	150.10
300 lb	"	150.00	20.10	170.10
1-1/2"				
125 lb	EA.	63.50	22.35	85.85
150 lb	"	140.00	22.35	162.35
300 lb	"	170.00	22.35	192.35
2"				
125 lb	EA.	110.00	26.80	136.80
150 lb	"	280.00	26.80	306.80
300 lb	"	340.00	26.80	366.80
Cast iron flanged				
2-1/2"				
125 lb	EA.	520.00	40.20	560.20
250 lb	"	960.00	40.20	1,000
3"				
125 lb	EA.	620.00	50.50	670.50
250 lb	"	1,030	50.50	1,081
4"				
125 lb	EA.	880.00	67.00	947.00
250 lb	"	1,440	67.00	1,507
6"				
125 lb	EA.	1,580	80.50	1,661
250 lb	"	1,940	80.50	2,021
8"				

PLUMBING	UNIT	MAT.	INST.	TOTAL
15430.60 VALVES				
125 lb	EA.	2,600	100.00	2,700
250 lb	"	3,030	100.00	3,130
Butterfly valve, cast iron, wafer type				
2"				
150 lb	EA.	130.00	28.70	158.70
200 lb	"	170.00	33.50	203.50
2-1/2"				
150 lb	EA.	170.00	33.50	203.50
200 lb	"	180.00	36.60	216.60
3"				
150 lb	EA.	170.00	40.20	210.20
200 lb	"	180.00	44.70	224.70
4"				
150 lb	EA.	190.00	57.50	247.50
200 lb	"	220.00	67.00	287.00
6"				
150 lb	EA.	230.00	80.50	310.50
200 lb	"	290.00	80.50	370.50
8"				
150 lb	EA.	290.00	89.50	379.50
200 lb	"	340.00	100.00	440.00
10"				
150 lb	EA.	340.00	100.00	440.00
200 lb	"	460.00	130.00	590.00
Ball valve, bronze, 250 lb, threaded				
1/2"	EA.	8.15	16.10	24.25
3/4"	"	12.25	16.10	28.35
1"	"	14.00	20.10	34.10
1-1/4"	"	22.75	22.35	45.10
1-1/2"	"	36.20	25.20	61.40
2"	"	40.80	28.70	69.50
Angle valve, bronze, 150 lb, threaded				
1/2"	EA.	44.30	14.35	58.65
3/4"	"	58.50	16.10	74.60
1"	"	84.00	16.10	100.10
1-1/4"	"	100.00	20.10	120.10
1-1/2"	"	130.00	22.35	152.35
Balancing valve, with meter connections, circuit setter				
1/2"	EA.	42.90	16.10	59.00
3/4"	"	45.30	18.30	63.60
1"	"	59.00	20.10	79.10
1-1/4"	"	81.00	22.35	103.35
1-1/2"	"	95.50	26.80	122.30
2"	"	140.00	33.50	173.50
2-1/2"	"	270.00	40.20	310.20
3"	"	400.00	50.50	450.50
4"	"	560.00	67.00	627.00
Balancing valve, straight type				
1/2"	EA.	11.65	16.10	27.75
3/4"	"	14.10	16.10	30.20
Angle type				
1/2"	EA.	15.70	16.10	31.80

PLUMBING

PLUMBING	UNIT	MAT.	INST.	TOTAL
15430.60 VALVES				
3/4"	EA.	21.80	16.10	37.90
Square head cock, 125 lb, bronze body				
1/2"	EA.	9.20	13.40	22.60
3/4"	"	11.05	16.10	27.15
1"	"	15.30	18.30	33.60
1-1/4"	"	20.85	20.10	40.95
Pressure regulating valve, bronze, class 300				
1"	EA.	540.00	25.20	565.20
1-1/2"	"	720.00	31.00	751.00
2"	"	820.00	40.20	860.20
3"	"	920.00	57.50	977.50
4"	"	1,160	80.50	1,241
5"	"	1,750	100.00	1,850
6"	"	1,770	130.00	1,900
15430.68 STRAINERS				
Strainer, Y pattern, 125 psi, cast iron body, threaded				
3/4"	EA.	11.55	14.35	25.90
1"	"	13.85	16.10	29.95
1-1/4"	"	18.50	20.10	38.60
1-1/2"	"	23.10	20.10	43.20
2"	"	34.70	25.20	59.90
250 psi, brass body, threaded				
3/4"	EA.	26.60	16.10	42.70
1"	"	37.00	16.10	53.10
1-1/4"	"	46.20	20.10	66.30
1-1/2"	"	67.00	20.10	87.10
2"	"	110.00	25.20	135.20
Cast iron body, threaded				
3/4"	EA.	16.85	16.10	32.95
1"	"	21.50	16.10	37.60
1-1/4"	"	28.60	20.10	48.70
1-1/2"	"	38.00	20.10	58.10
2"	"	48.20	25.20	73.40
15430.70 DRAINS, ROOF & FLOOR				
Floor drain, cast iron, with cast iron top				
2"	EA.	47.60	33.50	81.10
3"	"	72.50	33.50	106.00
4"	"	95.00	33.50	128.50
6"	"	170.00	40.20	210.20
Roof drain, cast iron				
2"	EA.	120.00	33.50	153.50
3"	"	130.00	33.50	163.50
4"	"	140.00	33.50	173.50
5"	"	160.00	40.20	200.20
6"	"	190.00	40.20	230.20

PLUMBING FIXTURES

PLUMBING FIXTURES	UNIT	MAT.	INST.	TOTAL
15440.15 FAUCETS				
Washroom				
Minimum	EA.	140.00	67.00	207.00
Average	"	240.00	80.50	320.50
Maximum	"	350.00	100.00	450.00
Handicapped				
Minimum	EA.	180.00	80.50	260.50
Average	"	260.00	100.00	360.00
Maximum	"	390.00	130.00	520.00
For trim and rough-in				
Minimum	EA.	52.50	80.50	133.00
Average	"	78.50	100.00	178.50
Maximum	"	120.00	200.00	320.00
15440.18 HYDRANTS				
Wall hydrant				
8" thick	EA.	110.00	67.00	177.00
12" thick	"	120.00	80.50	200.50
18" thick	"	140.00	89.50	229.50
24" thick	"	160.00	100.00	260.00
Ground hydrant				
2' deep	EA.	210.00	50.50	260.50
4' deep	"	220.00	57.50	277.50
6' deep	"	250.00	67.00	317.00
8' deep	"	300.00	100.00	400.00
15440.20 LAVATORIES				
Lavatory, counter top, porcelain enamel on cast iron				
Minimum	EA.	130.00	80.50	210.50
Average	"	200.00	100.00	300.00
Maximum	"	360.00	130.00	490.00
Wall hung, china				
Minimum	EA.	190.00	80.50	270.50
Average	"	230.00	100.00	330.00
Maximum	"	520.00	130.00	650.00
Handicapped				
Minimum	EA.	310.00	100.00	410.00
Average	"	360.00	130.00	490.00
Maximum	"	570.00	200.00	770.00
For trim and rough-in				
Minimum	EA.	120.00	100.00	220.00
Average	"	150.00	130.00	280.00
Maximum	"	260.00	200.00	460.00
15440.30 SHOWERS				
Shower, fiberglass, 36"x34"x84"				
Minimum	EA.	510.00	290.00	800.00
Average	"	710.00	400.00	1,110
Maximum	"	1,020	400.00	1,420
Steel, 1 piece, 36"x36"				
Minimum	EA.	470.00	290.00	760.00
Average	"	710.00	400.00	1,110
Maximum	"	840.00	400.00	1,240

PLUMBING FIXTURES	UNIT	MAT.	INST.	TOTAL
15440.30 SHOWERS				
Receptor, molded stone, 36"x36"				
Minimum	EA.	160.00	130.00	290.00
Average	"	270.00	200.00	470.00
Maximum	"	410.00	340.00	750.00
For trim and rough-in				
Minimum	EA.	120.00	180.00	300.00
Average	"	160.00	220.00	380.00
Maximum	"	210.00	400.00	610.00
15440.40 SINKS				
Service sink, 24"x29"				
Minimum	EA.	470.00	100.00	570.00
Average	"	590.00	130.00	720.00
Maximum	"	790.00	200.00	990.00
Mop sink, 24"x36"x10"				
Minimum	EA.	350.00	80.50	430.50
Average	"	420.00	100.00	520.00
Maximum	"	560.00	130.00	690.00
For trim and rough-in				
Minimum	EA.	190.00	130.00	320.00
Average	"	290.00	200.00	490.00
Maximum	"	360.00	270.00	630.00
15440.50 URINALS				
Urinal, flush valve, floor mounted				
Minimum	EA.	590.00	100.00	690.00
Average	"	710.00	130.00	840.00
Maximum	"	830.00	200.00	1,030
Wall mounted				
Minimum	EA.	300.00	100.00	400.00
Average	"	420.00	130.00	550.00
Maximum	"	540.00	200.00	740.00
For trim and rough-in				
Minimum	EA.	100.00	100.00	200.00
Average	"	150.00	200.00	350.00
Maximum	"	200.00	270.00	470.00
15440.60 WATER CLOSETS				
Water closet flush tank, floor mounted				
Minimum	EA.	240.00	100.00	340.00
Average	"	480.00	130.00	610.00
Maximum	"	940.00	200.00	1,140
Handicapped				
Minimum	EA.	280.00	130.00	410.00
Average	"	490.00	200.00	690.00
Maximum	"	750.00	400.00	1,150
Bowl, with flush valve, floor mounted				
Minimum	EA.	340.00	100.00	440.00
Average	"	370.00	130.00	500.00
Maximum	"	720.00	200.00	920.00
Wall mounted				
Minimum	EA.	340.00	100.00	440.00

PLUMBING FIXTURES	UNIT	MAT.	INST.	TOTAL
15440.60 WATER CLOSETS				
Average	EA.	400.00	130.00	530.00
Maximum	"	760.00	200.00	960.00
For trim and rough-in				
Minimum	EA.	120.00	100.00	220.00
Average	"	140.00	130.00	270.00
Maximum	"	190.00	200.00	390.00
15440.70 WATER HEATERS				
Water heater, electric				
6 gal	EA.	210.00	67.00	277.00
10 gal	"	220.00	67.00	287.00
20 gal	"	270.00	80.50	350.50
40 gal	"	300.00	80.50	380.50
80 gal	"	550.00	100.00	650.00
100 gal	"	670.00	130.00	800.00
120 gal	"	860.00	130.00	990.00
15440.90 MISCELLANEOUS FIXTURES				
Electric water cooler				
Floor mounted	EA.	520.00	130.00	650.00
Wall mounted	"	650.00	130.00	780.00
Wash fountain				
Wall mounted	EA.	910.00	200.00	1,110
Circular, floor supported	"	1,810	400.00	2,210
Deluge shower and eye wash	"	660.00	200.00	860.00
15440.95 FIXTURE CARRIERS				
Water fountain, wall carrier				
Minimum	EA.	71.50	40.20	111.70
Average	"	95.00	50.50	145.50
Maximum	"	120.00	67.00	187.00
Lavatory, wall carrier				
Minimum	EA.	97.00	40.20	137.20
Average	"	140.00	50.50	190.50
Maximum	"	180.00	67.00	247.00
Sink, industrial, wall carrier				
Minimum	EA.	100.00	40.20	140.20
Average	"	130.00	50.50	180.50
Maximum	"	150.00	67.00	217.00
Toilets, water closets, wall carrier				
Minimum	EA.	130.00	40.20	170.20
Average	"	150.00	50.50	200.50
Maximum	"	190.00	67.00	257.00
Floor support				
Minimum	EA.	60.50	33.50	94.00
Average	"	73.00	40.20	113.20
Maximum	"	85.00	50.50	135.50
Urinals, wall carrier				
Minimum	EA.	79.00	40.20	119.20
Average	"	100.00	50.50	150.50
Maximum	"	130.00	67.00	197.00

15 MECHANICAL

PLUMBING FIXTURES

PLUMBING FIXTURES	UNIT	MAT.	INST.	TOTAL
15440.95 FIXTURE CARRIERS				
Floor support				
Minimum	EA.	66.50	33.50	100.00
Average	"	97.00	40.20	137.20
Maximum	"	120.00	50.50	170.50
15450.30 PUMPS				
In-line pump, bronze, centrifugal				
5 gpm, 20' head	EA.	490.00	25.20	515.20
20 gpm, 40' head	"	910.00	25.20	935.20
50 gpm				
50' head	EA.	1,190	50.50	1,241
100' head	"	1,820	50.50	1,871
70 gpm, 100' head	"	1,820	67.00	1,887
100 gpm, 80' head	"	1,820	67.00	1,887
250 gpm, 150' head	"	5,710	100.00	5,810
Cast iron, centrifugal				
50 gpm, 200' head	EA.	2,510	50.50	2,561
100 gpm				
100' head	EA.	2,320	67.00	2,387
200' head	"	2,700	67.00	2,767
200 gpm				
100' head	EA.	2,520	100.00	2,620
200' head	"	3,410	100.00	3,510
Centrifugal, close coupled, c.i., single stage				
50 gpm, 100' head	EA.	1,020	50.50	1,071
100 gpm, 100' head	"	1,240	67.00	1,307
Base mounted				
50 gpm, 100' head	EA.	1,660	50.50	1,711
100 gpm, 50' head	"	1,880	67.00	1,947
200 gpm, 100' head	"	2,410	100.00	2,510
300 gpm, 175' head	"	3,130	100.00	3,230
Suction diffuser, flanged, strainer				
3" inlet, 2-1/2" outlet	EA.	270.00	50.50	320.50
3" outlet	"	280.00	50.50	330.50
4" inlet				
3" outlet	EA.	320.00	67.00	387.00
4" outlet	"	380.00	67.00	447.00
6" inlet				
4" outlet	EA.	440.00	80.50	520.50
5" outlet	"	530.00	80.50	610.50
6" Outlet	"	550.00	80.50	630.50
8" inlet				
6" outlet	EA.	600.00	100.00	700.00
8" outlet	"	1,030	100.00	1,130
10" inlet				
8" outlet	EA.	1,390	130.00	1,520
Vertical turbine				
Single stage, C.I., 3550 rpm, 200 gpm, 50'head	EA.	3,540	130.00	3,670
Multi stage, 3550 rpm				
50 gpm, 100' head	EA.	3,540	100.00	3,640
100 gpm				
100' head	EA.	3,710	100.00	3,810

PLUMBING FIXTURES

PLUMBING FIXTURES	UNIT	MAT.	INST.	TOTAL
15450.30 PUMPS				
200 gpm				
50' head	EA.	3,960	130.00	4,090
100' head	"	4,020	130.00	4,150
Bronze				
Single stage, 3550 rpm, 100 gpm, 50' head	EA.	3,670	100.00	3,770
Multi stage, 3550 rpm, 50 gpm, 100' head	"	3,800	100.00	3,900
100 gpm				
100' head	EA.	3,800	100.00	3,900
200 gpm				
50' head	EA.	3,800	130.00	3,930
100' head	"	4,060	130.00	4,190
Sump pump, bronze, 1750 rpm, 25 gpm				
20' head	EA.	3,330	500.00	3,830
150' head	"	4,640	670.00	5,310
50 gpm				
100' head	EA.	3,970	500.00	4,470
100 gpm				
50' head	EA.	3,450	500.00	3,950
15480.10 SPECIAL SYSTEMS				
Air compressor, air cooled, two stage				
5.0 cfm, 175 psi	EA.	2,560	800.00	3,360
10 cfm, 175 psi	"	3,130	890.00	4,020
20 cfm, 175 psi	"	4,310	960.00	5,270
50 cfm, 125 psi	"	6,280	1,060	7,340
80 cfm, 125 psi	"	9,000	1,150	10,150
Single stage, 125 psi				
1.0 cfm	EA.	2,360	570.00	2,930
1.5 cfm	"	2,400	570.00	2,970
2.0 cfm	"	2,450	570.00	3,020
Automotive compressor, hose reel, air and water, 50' hose	"	1,100	340.00	1,440
Lube equipment, 3 reel, with pumps	"	5,750	1,610	7,360
Tire changer				
Truck	EA.	9,000	570.00	9,570
Passenger car	"	2,120	310.00	2,430
Air hose reel, includes, 50' hose	"	560.00	310.00	870.00
Hose reel, 5 reel, motor oil, gear oil, lube, air & water	"	5,250	1,610	6,860
Water hose reel, 50' hose	"	560.00	310.00	870.00
Pump, air operated, for motor or gear oil, fits 55 gal drum	"	740.00	40.20	780.20
For chassis lube	"	1,200	40.20	1,240
Fuel dispensing pump, lighted dial, one product				
One hose	EA.	2,620	340.00	2,960
Two hose	"	4,620	340.00	4,960
Two products, two hose	"	4,870	340.00	5,210

15 MECHANICAL

HEATING & VENTILATING

HEATING & VENTILATING	UNIT	MAT.	INST.	TOTAL
15610.10 FURNACES				
Electric, hot air				
40 mbh	EA.	560.00	200.00	760.00
80 mbh	"	600.00	220.00	820.00
100 mbh	"	970.00	240.00	1,210
160 mbh	"	1,560	250.00	1,810
200 mbh	"	2,870	260.00	3,130
400 mbh	"	5,120	270.00	5,390
Gas fired hot air				
40 mbh	EA.	560.00	200.00	760.00
80 mbh	"	690.00	220.00	910.00
100 mbh	"	720.00	240.00	960.00
160 mbh	"	940.00	250.00	1,190
200 mbh	"	1,670	260.00	1,930
400 mbh	"	5,950	270.00	6,220
Oil fired hot air				
40 mbh	EA.	750.00	200.00	950.00
80 mbh	"	980.00	220.00	1,200
100 mbh	"	1,150	240.00	1,390
160 mbh	"	1,380	250.00	1,630
200 mbh	"	2,950	260.00	3,210
400 mbh	"	6,250	270.00	6,520
15780.20 ROOFTOP UNITS				
Packaged, single zone rooftop unit, with roof curb				
2 ton	EA.	3,270	400.00	3,670
3 ton	"	3,780	400.00	4,180
4 ton	"	4,530	500.00	5,030
5 ton	"	4,580	670.00	5,250
7.5 ton	"	7,420	800.00	8,220
15830.70 UNIT HEATERS				
Steam unit heater, horizontal				
12,500 btuh, 200 cfm	EA.	400.00	67.00	467.00
17,000 btuh, 300 cfm	"	450.00	67.00	517.00
40,000 btuh, 500 cfm	"	570.00	67.00	637.00
60,000 btuh, 700 cfm	"	630.00	67.00	697.00
70,000 btuh, 1000 cfm	"	790.00	100.00	890.00
Vertical				
12,500 btuh, 200 cfm	EA.	340.00	67.00	407.00
17,000 btuh, 300 cfm	"	570.00	67.00	637.00
40,000 btuh, 500 cfm	"	680.00	67.00	747.00
60,000 btuh, 700 cfm	"	790.00	67.00	857.00
70,000 btuh, 1000 cfm	"	910.00	67.00	977.00
Gas unit heater, horizontal				
27,400 btuh	EA.	570.00	160.00	730.00
38,000 btuh	"	620.00	160.00	780.00
56,000 btuh	"	680.00	160.00	840.00
82,200 btuh	"	750.00	160.00	910.00
103,900 btuh	"	780.00	250.00	1,030
125,700 btuh	"	800.00	250.00	1,050
133,200 btuh	"	850.00	250.00	1,100

15 MECHANICAL

HEATING & VENTILATING

15830.70 UNIT HEATERS	UNIT	MAT.	INST.	TOTAL
149,000 btuh	EA.	990.00	250.00	1,240
172,000 btuh	"	1,080	250.00	1,330
190,000 btuh	"	1,130	250.00	1,380
225,000 btuh	"	1,250	250.00	1,500
Hot water unit heater, horizontal				
12,500 btuh, 200 cfm	EA.	340.00	67.00	407.00
17,000 btuh, 300 cfm	"	400.00	67.00	467.00
25,000 btuh, 500 cfm	"	420.00	67.00	487.00
30,000 btuh, 700 cfm	"	510.00	67.00	577.00
50,000 btuh, 1000 cfm	"	570.00	100.00	670.00
60,000 btuh, 1300 cfm	"	790.00	100.00	890.00
Vertical				
12,500 btuh, 200 cfm	EA.	420.00	67.00	487.00
17,000 btuh, 300 cfm	"	450.00	67.00	517.00
25,000 btuh, 500 cfm	"	480.00	67.00	547.00
30,000 btuh, 700 cfm	"	520.00	67.00	587.00
50,000 btuh, 1000 cfm	"	570.00	67.00	637.00
60,000 btuh, 1300 cfm	"	620.00	67.00	687.00
Cabinet unit heaters, ceiling, exposed, hot water				
200 cfm	EA.	790.00	130.00	920.00
300 cfm	"	850.00	160.00	1,010
400 cfm	"	880.00	190.00	1,070
600 cfm	"	910.00	210.00	1,120
800 cfm	"	1,130	250.00	1,380
1000 cfm	"	1,470	290.00	1,760
1200 cfm	"	1,590	340.00	1,930
2000 cfm	"	2,720	450.00	3,170

AIR HANDLING

15855.10 AIR HANDLING UNITS	UNIT	MAT.	INST.	TOTAL
Air handling unit, medium pressure, single zone				
1500 cfm	EA.	2,570	250.00	2,820
3000 cfm	"	3,210	450.00	3,660
4000 cfm	"	3,590	500.00	4,090
5000 cfm	"	4,110	540.00	4,650
6000 cfm	"	4,620	570.00	5,190
7000 cfm	"	5,130	620.00	5,750
8500 cfm	"	6,030	670.00	6,700
Rooftop air handling units				
4950 cfm	EA.	7,720	450.00	8,170
7370 cfm	"	8,360	570.00	8,930

AIR HANDLING	UNIT	MAT.	INST.	TOTAL
15870.20 EXHAUST FANS				
Belt drive roof exhaust fans				
640 cfm, 2618 fpm	EA.	700.00	50.50	750.50
940 cfm, 2604 fpm	"	900.00	50.50	950.50
1050 cfm, 3325 fpm	"	810.00	50.50	860.50
1170 cfm, 2373 fpm	"	1,170	50.50	1,221
2440 cfm, 4501 fpm	"	920.00	50.50	970.50

AIR DISTRIBUTION	UNIT	MAT.	INST.	TOTAL
15890.10 METAL DUCTWORK				
Rectangular duct				
Galvanized steel				
Minimum	Lb.	1.20	3.65	4.85
Average	"	1.50	4.45	5.95
Maximum	"	2.30	6.70	9.00
Aluminum				
Minimum	Lb.	3.05	8.05	11.10
Average	"	4.05	10.05	14.10
Maximum	"	5.05	13.40	18.45
Fittings				
Minimum	EA.	10.10	13.40	23.50
Average	"	15.15	20.10	35.25
Maximum	"	22.20	40.20	62.40
For work				
10-20' high, add per pound, $.30				
30-50', add per pound, $.50				
15890.30 FLEXIBLE DUCTWORK				
Flexible duct, 1.25" fiberglass				
6" dia.	L.F.	2.15	2.25	4.40
8" dia.	"	2.75	2.50	5.25
12" dia.	"	3.80	3.10	6.90
16" dia.	"	5.70	3.65	9.35
Flexible duct connector, 3" wide fabric	"	1.60	6.70	8.30
15910.10 DAMPERS				
Horizontal parallel aluminum backdraft damper				
12" x 12"	EA.	43.90	10.05	53.95
24" x 24"	"	71.50	20.10	91.60
36" x 36"	"	160.00	28.70	188.70
15940.10 DIFFUSERS				
Ceiling diffusers, round, baked enamel finish				
6" dia.	EA.	31.90	13.40	45.30

15 MECHANICAL

AIR DISTRIBUTION	UNIT	MAT.	INST.	TOTAL
15940.10 DIFFUSERS				
8" dia.	EA.	38.20	16.75	54.95
12" dia.	"	57.50	16.75	74.25
16" dia.	"	83.00	18.30	101.30
20" dia.	"	120.00	20.10	140.10
Rectangular				
6x6"	EA.	35.20	13.40	48.60
12x12"	"	62.00	20.10	82.10
18x18"	"	96.00	20.10	116.10
24x24"	"	180.00	25.20	205.20
15940.40 REGISTERS AND GRILLES				
Lay in flush mounted, perforated face, return				
6x6/24x24	EA.	29.40	16.10	45.50
8x8/24x24	"	29.40	16.10	45.50
9x9/24x24	"	31.90	16.10	48.00
10x10/24x24	"	34.30	16.10	50.40
12x12/24x24	"	34.30	16.10	50.40
Rectangular, ceiling return, single deflection				
10x10	EA.	14.70	20.10	34.80
12x12	"	17.15	20.10	37.25
16x16	"	17.15	20.10	37.25
20x20	"	31.90	20.10	52.00
24x18	"	60.00	20.10	80.10
36x24	"	120.00	22.35	142.35
36x30	"	170.00	22.35	192.35
Wall, return air register				
12x12	EA.	24.45	10.05	34.50
16x16	"	36.00	10.05	46.05
18x18	"	42.40	10.05	52.45
20x20	"	50.00	10.05	60.05
24x24	"	69.50	10.05	79.55

BASIC MATERIALS

16050.30 BUS DUCT

BASIC MATERIALS	UNIT	MAT.	INST.	TOTAL
Bus duct, 100a, plug-in				
10', 600v	EA.	170.00	130.00	300.00
With ground	"	220.00	200.00	420.00
Circuit breakers, with enclosure				
1 pole				
15a-60a	EA.	170.00	46.80	216.80
70a-100a	"	190.00	58.50	248.50
2 pole				
15a-60a	EA.	250.00	51.50	301.50
70a-100a	"	300.00	61.00	361.00
Circuit breaker, adapter cubicle				
225a	EA.	2,130	70.50	2,201
200a	"	2,520	75.00	2,595
Fusible switches, 240v, 3 phase				
30a	EA.	170.00	46.80	216.80
60a	"	210.00	58.50	268.50
100a	"	280.00	70.50	350.50
200a	"	490.00	98.50	588.50

16110.12 CABLE TRAY

BASIC MATERIALS	UNIT	MAT.	INST.	TOTAL
Cable tray, 6"	L.F.	11.95	2.75	14.70
Ventilated cover	"	4.85	1.40	6.25
Solid cover	"	3.75	1.40	5.15

16110.20 CONDUIT SPECIALTIES

BASIC MATERIALS	UNIT	MAT.	INST.	TOTAL
Rod beam clamp, 1/2"	EA.	3.80	2.35	6.15
Hanger rod				
3/8"	L.F.	0.79	1.85	2.64
1/2"	"	1.30	2.35	3.65
Hanger channel, 1-1/2"				
No holes	EA.	3.10	1.40	4.50
Holes	"	3.45	1.40	4.85
Channel strap				
1/2"	EA.	0.79	2.35	3.14
1"	"	1.05	2.35	3.40
2"	"	1.50	3.75	5.25
3"	"	1.80	5.75	7.55
4"	"	2.50	6.80	9.30
5"	"	3.50	6.80	10.30
6"	"	4.20	6.80	11.00
Conduit penetrations, roof and wall, 8" thick				
1/2"	EA.	0.00	28.80	28.80
1"	"	0.00	37.40	37.40
2"	"	0.00	75.00	75.00
3"	"	0.00	75.00	75.00
4"	"	0.00	93.50	93.50
Fireproofing, for conduit penetrations				
1/2"	EA.	1.90	23.35	25.25
1"	"	2.00	23.35	25.35
2"	"	2.85	34.00	36.85
3"	"	5.60	45.20	50.80

BASIC MATERIALS

BASIC MATERIALS	UNIT	MAT.	INST.	TOTAL
16110.20 CONDUIT SPECIALTIES				
4"	EA.	7.90	70.50	78.40
16110.21 ALUMINUM CONDUIT				
Aluminum conduit				
1/2"	L.F.	0.90	1.40	2.30
3/4"	"	1.15	1.85	3.00
1"	"	1.20	2.35	3.55
1-1/4"	"	2.20	2.75	4.95
1-1/2"	"	2.75	3.75	6.50
2"	"	3.65	4.15	7.80
2-1/2"	"	5.80	4.65	10.45
3"	"	7.65	5.00	12.65
3-1/2"	"	9.20	5.75	14.95
4"	"	10.85	6.80	17.65
5"	"	15.55	8.50	24.05
6"	"	20.50	9.35	29.85
16110.22 EMT CONDUIT				
EMT conduit				
1/2"	L.F.	0.29	1.40	1.69
3/4"	"	0.40	1.85	2.25
1"	"	0.55	2.35	2.90
1-1/4"	"	0.80	2.75	3.55
1-1/2"	"	1.05	3.75	4.80
2"	"	1.15	4.15	5.30
2-1/2"	"	2.55	4.65	7.20
3"	"	3.65	5.75	9.40
3-1/2"	"	4.80	6.80	11.60
4"	"	5.40	8.50	13.90
16110.23 FLEXIBLE CONDUIT				
Flexible conduit, steel				
3/8"	L.F.	0.22	1.40	1.62
1/2	"	0.23	1.40	1.63
3/4"	"	0.33	1.85	2.18
1"	"	0.65	1.85	2.50
1-1/4"	"	0.81	2.35	3.16
1-1/2"	"	1.10	2.75	3.85
2"	"	1.40	3.75	5.15
2-1/2"	"	1.65	4.15	5.80
3"	"	2.05	5.00	7.05
16110.24 GALVANIZED CONDUIT				
Galvanized rigid steel conduit				
1/2"	L.F.	0.92	1.85	2.77
3/4"	"	1.15	2.35	3.50
1"	"	1.50	2.75	4.25

BASIC MATERIALS	UNIT	MAT.	INST.	TOTAL
16110.24 GALVANIZED CONDUIT				
1-1/4"	L.F.	2.20	3.75	5.95
1-1/2"	"	2.50	4.15	6.65
2"	"	3.15	4.65	7.80
2-1/2"	"	5.15	6.80	11.95
3"	"	6.75	8.50	15.25
3-1/2"	"	6.95	8.90	15.85
4"	"	10.10	9.85	19.95
5"	"	20.60	13.35	33.95
6"	"	29.80	17.80	47.60
16110.25 PLASTIC CONDUIT				
PVC conduit, schedule 40				
1/2"	L.F.	0.31	1.40	1.71
3/4"	"	0.37	1.40	1.77
1"	"	0.51	1.85	2.36
1-1/4"	"	0.66	1.85	2.51
1-1/2"	"	0.85	2.35	3.20
2"	"	1.10	2.35	3.45
2-1/2"	"	1.75	2.75	4.50
3"	"	2.00	2.75	4.75
3-1/2"	"	2.65	3.75	6.40
4"	"	2.90	3.75	6.65
5"	"	4.10	4.15	8.25
6"	"	5.45	4.65	10.10
16110.27 PLASTIC COATED CONDUIT				
Rigid steel conduit, plastic coated				
1/2"	L.F.	2.45	2.35	4.80
3/4"	"	2.85	2.75	5.60
1"	"	3.75	3.75	7.50
1-1/4"	"	4.65	4.65	9.30
1-1/2"	"	5.70	5.75	11.45
2"	"	7.45	6.80	14.25
2-1/2"	"	11.35	8.90	20.25
3"	"	14.30	10.40	24.70
3-1/2"	"	17.40	11.70	29.10
4"	"	21.25	14.40	35.65
5"	"	36.40	17.80	54.20
90 degree elbows				
1/2"	EA.	9.40	14.40	23.80
3/4"	"	9.60	17.80	27.40
1"	"	11.15	20.80	31.95
1-1/4"	"	13.95	23.35	37.30
1-1/2"	"	16.90	28.80	45.70
2"	"	23.50	37.40	60.90
2-1/2"	"	45.50	53.50	99.00
3"	"	46.00	62.50	108.50
3-1/2"	"	98.50	76.50	175.00
4"	"	100.00	93.50	193.50
5"	"	240.00	120.00	360.00

BASIC MATERIALS	UNIT	MAT.	INST.	TOTAL
16110.27 PLASTIC COATED CONDUIT				
Couplings				
1/2"	EA.	2.75	2.75	5.50
3/4"	"	2.90	3.75	6.65
1"	"	3.90	4.15	8.05
1-1/4"	"	4.55	5.00	9.55
1-1/2"	"	6.30	5.75	12.05
2"	"	7.90	6.80	14.70
2-1/2"	"	19.80	8.50	28.30
3"	"	23.10	8.90	32.00
3-1/2"	"	31.80	9.35	41.15
4"	"	39.00	10.40	49.40
5"	"	110.00	11.70	121.70
1 hole conduit straps				
3/4"	EA.	3.15	2.35	5.50
1"	"	4.55	2.35	6.90
1-1/4"	"	5.75	2.75	8.50
1-1/2"	"	6.95	2.75	9.70
2"	"	8.00	2.75	10.75
3"	"	12.40	3.75	16.15
3-1/2"	"	13.85	3.75	17.60
4"	"	18.90	4.65	23.55
16110.28 STEEL CONDUIT				
Intermediate metal conduit (IMC)				
1/2"	L.F.	0.90	1.40	2.30
3/4"	"	1.05	1.85	2.90
1"	"	1.45	2.35	3.80
1-1/4"	"	1.85	2.75	4.60
1-1/2"	"	2.20	3.75	5.95
2"	"	2.95	4.15	7.10
2-1/2"	"	4.50	5.60	10.10
3"	"	6.40	6.80	13.20
3-1/2"	"	8.35	8.50	16.85
4"	"	9.75	8.90	18.65
90 degree ell				
1/2"	EA.	2.60	11.70	14.30
3/4"	"	3.45	14.40	17.85
1"	"	5.00	17.80	22.80
1-1/4"	"	6.95	20.80	27.75
1-1/2"	"	8.85	23.35	32.20
2"	"	12.75	26.70	39.45
2-1/2"	"	22.05	31.20	53.25
3"	"	33.80	41.60	75.40
3-1/2"	"	59.50	53.50	113.00
4"	"	70.50	62.50	133.00
Couplings				
1/2"	EA.	1.05	2.35	3.40
3/4"	"	1.25	2.75	4.00
1"	"	1.85	3.75	5.60
1-1/4"	"	2.30	4.15	6.45
1-1/2"	"	2.90	4.65	7.55

BASIC MATERIALS

BASIC MATERIALS	UNIT	MAT.	INST.	TOTAL
16110.28 STEEL CONDUIT				
2"	EA.	3.85	5.00	8.85
2-1/2"	"	8.55	5.75	14.30
3"	"	11.70	6.80	18.50
3-1/2"	"	15.65	6.80	22.45
4"	"	16.50	7.50	24.00
16110.35 WIREMOLD				
Wiremold raceway with fittings, surface mounted				
#200	L.F.	0.51	1.40	1.91
#500	"	0.62	1.40	2.02
#700	"	0.67	1.85	2.52
#800	"	1.10	1.85	2.95
Fittings, #200, 90 degree flat elbow	EA.	0.99	2.35	3.34
Internal elbow	"	1.95	2.35	4.30
Extension adapter	"	3.45	2.75	6.20
#200, #500, #700				
Single pole switch and box	EA.	6.55	18.70	25.25
Duplex receptacle with box	"	7.60	14.40	22.00
#500, #700				
90 deg. flat elbow	EA.	0.85	3.75	4.60
Internal elbow	"	1.10	3.75	4.85
Junction box	"	5.55	6.25	11.80
Fixture box	"	5.65	6.25	11.90
Shallow switch and receptacle box	"	4.60	3.75	8.35
#800				
90 deg. flat elbow	EA.	1.05	3.75	4.80
Internal elbow	"	1.05	3.75	4.80
Junction box	"	3.30	5.75	9.05
16110.60 TRENCH DUCT				
Trench duct, with cover				
9"	L.F.	71.00	7.95	78.95
12"	"	79.50	9.35	88.85
18"	"	110.00	12.45	122.45
Tees				
9"	EA.	260.00	81.50	341.50
12"	"	290.00	93.50	383.50
18"	"	370.00	100.00	470.00
Vertical elbows				
9"	EA.	88.50	37.40	125.90
12"	"	93.00	51.00	144.00
18"	"	120.00	63.50	183.50
Cabinet connectors				
9"	EA.	110.00	93.50	203.50
12"	"	130.00	98.50	228.50
18"	"	150.00	110.00	260.00
End closers				
9"	EA.	27.50	28.80	56.30
12"	"	28.60	31.20	59.80
18"	"	42.70	37.40	80.10

BASIC MATERIALS	UNIT	MAT.	INST.	TOTAL
16110.60 TRENCH DUCT				
Horizontal elbows				
9"	EA.	260.00	70.50	330.50
12"	"	290.00	81.50	371.50
18"	"	370.00	98.50	468.50
Crosses				
9"	EA.	410.00	93.50	503.50
12"	"	430.00	100.00	530.00
18"	"	520.00	120.00	640.00
16110.80 WIREWAYS				
Wireway, hinge cover type				
2-1/2" x 2-1/2"				
1' section	EA.	8.95	7.20	16.15
2'	"	12.70	8.90	21.60
3'	"	17.25	11.70	28.95
16120.41 ALUMINUM CONDUCTORS				
Type XHHW, stranded aluminum, 600v				
#8	L.F.	0.14	0.23	0.37
#6	"	0.20	0.28	0.48
#4	"	0.26	0.37	0.63
#2	"	0.32	0.42	0.74
1/0	"	0.51	0.52	1.03
2/0	"	0.61	0.56	1.17
3/0	"	0.73	0.66	1.39
4/0	"	0.89	0.70	1.59
THW, stranded				
#8	L.F.	0.12	0.23	0.35
#6	"	0.13	0.28	0.41
#4	"	0.17	0.37	0.54
#3	"	0.22	0.42	0.64
#1	"	0.32	0.47	0.79
1/0	"	0.40	0.52	0.92
2/0	"	0.45	0.55	1.00
3/0	"	0.56	0.55	1.11
4/0	"	0.66	0.70	1.36
16120.43 COPPER CONDUCTORS				
Copper conductors, type THW, solid				
#14	L.F.	0.07	0.19	0.26
#12	"	0.08	0.23	0.31
#10	"	0.09	0.28	0.37
Stranded				
#14	L.F.	0.09	0.19	0.28
#12	"	0.10	0.23	0.33
#10	"	0.12	0.28	0.40
#8	"	0.19	0.37	0.56
#6	"	0.25	0.42	0.67
#4	"	0.36	0.47	0.83
#3	"	0.51	0.47	0.98

BASIC MATERIALS

16120.43 COPPER CONDUCTORS

BASIC MATERIALS	UNIT	MAT.	INST.	TOTAL
#2	L.F.	0.63	0.56	1.19
#1	"	0.81	0.66	1.47
1/0	"	0.98	0.75	1.73
2/0	"	1.15	0.94	2.09
3/0	"	1.45	1.15	2.60
4/0	"	1.75	1.30	3.05
THHN-THWN, solid				
#14	L.F.	0.07	0.19	0.26
#12	"	0.09	0.23	0.32
#10	"	0.11	0.28	0.39
Stranded				
#14	L.F.	0.09	0.19	0.28
#12	"	0.11	0.23	0.34
#10	"	0.12	0.28	0.40
#8	"	0.19	0.37	0.56
#6	"	0.25	0.42	0.67
#4	"	0.36	0.47	0.83
#2	"	0.63	0.56	1.19
#1	"	0.80	0.66	1.46
1/0	"	0.97	0.75	1.72
2/0	"	1.15	0.94	2.09
3/0	"	1.45	1.15	2.60
4/0	"	1.75	1.30	3.05
XLP, 600v				
#12	L.F.	0.12	0.23	0.35
#10	"	0.18	0.28	0.46
#8	"	0.25	0.37	0.62
#6	"	0.30	0.42	0.72
#4	"	0.42	0.47	0.89
#3	"	0.52	0.52	1.04
#2	"	0.64	0.56	1.20
#1	"	0.86	0.66	1.52
1/0	"	1.05	0.75	1.80
2/0	"	1.25	0.94	2.19
3/0	"	1.55	1.20	2.75
4/0	"	1.85	1.30	3.15
Bare solid wire				
#14	L.F.	0.08	0.19	0.27
#12	"	0.10	0.23	0.33
#10	"	0.11	0.28	0.39
#8	"	0.19	0.37	0.56
#6	"	0.25	0.42	0.67
#4	"	0.36	0.47	0.83
#2	"	0.58	0.56	1.14
Bare stranded wire				
#8	L.F.	0.19	0.37	0.56
#6	"	0.26	0.47	0.73
#4	"	0.36	0.47	0.83
#2	"	0.61	0.52	1.13
#1	"	0.79	0.66	1.45
1/0	"	0.97	0.84	1.81
2/0	"	1.15	0.94	2.09

BASIC MATERIALS	UNIT	MAT.	INST.	TOTAL
16120.43 COPPER CONDUCTORS				
3/0	L.F.	1.40	1.15	2.55
4/0	"	1.75	1.30	3.05
Type "BX" solid armored cable				
#14/2	L.F.	0.33	1.15	1.48
#14/3	"	0.41	1.30	1.71
#14/4	"	0.55	1.45	2.00
#12/2	"	0.35	1.30	1.65
#12/3	"	0.52	1.45	1.97
#12/4	"	0.74	1.65	2.39
#10/2	"	0.59	1.45	2.04
#10/3	"	0.78	1.65	2.43
#10/4	"	1.20	1.85	3.05
#8/2	"	0.97	1.65	2.62
#8/3	"	1.25	1.85	3.10
Steel type, metal clad cable, solid, with ground				
#14/2	L.F.	0.35	0.84	1.19
#14/3	"	0.48	0.94	1.42
#14/4	"	0.66	1.05	1.71
#12/2	"	0.42	0.94	1.36
#12/3	"	0.58	1.15	1.73
#12/4	"	0.79	1.40	2.19
#10/2	"	0.69	1.05	1.74
#10/3	"	1.05	1.30	2.35
#10/4	"	1.40	1.55	2.95
Metal clad cable, stranded, with ground				
#8/2	L.F.	1.15	1.30	2.45
#8/3	"	1.65	1.65	3.30
#8/4	"	2.15	1.95	4.10
#6/2	"	1.75	1.40	3.15
#6/3	"	2.10	1.80	3.90
#6/4	"	2.50	2.10	4.60
#4/2	"	2.25	1.85	4.10
#4/3	"	2.95	2.10	5.05
#4/4	"	3.50	2.60	6.10
#3/3	"	3.65	2.35	6.00
#3/4	"	4.30	2.75	7.05
#2/3	"	4.30	2.65	6.95
#2/4	"	5.25	3.10	8.35
#1/3	"	5.20	3.55	8.75
#1/4	"	6.15	3.95	10.10
16120.47 SHEATHED CABLE				
Non-metallic sheathed cable				
Type NM cable with ground				
#14/2	L.F.	0.17	0.70	0.87
#12/2	"	0.22	0.75	0.97
#10/2	"	0.35	0.83	1.18
#8/2	"	0.68	0.94	1.62
#6/2	"	1.00	1.15	2.15
#14/3	"	0.26	1.20	1.46
#12/3	"	0.36	1.25	1.61
#10/3	"	0.53	1.25	1.78

BASIC MATERIALS	UNIT	MAT.	INST.	TOTAL
16120.47 SHEATHED CABLE				
#8/3	L.F.	1.25	1.30	2.55
#6/3	"	1.45	1.30	2.75
#4/3	"	2.20	1.50	3.70
#2/3	"	3.15	1.65	4.80
Type U.F. cable with ground				
#14/2	L.F.	0.21	0.75	0.96
#12/2	"	0.26	0.89	1.15
#10/2	"	0.33	0.94	1.26
#8/2	"	1.05	1.05	2.10
#6/2	"	1.25	1.25	2.50
#14/3	"	0.28	0.94	1.22
#12/3	"	0.37	1.00	1.37
#10/3	"	0.53	1.15	1.68
#8/3	"	1.35	1.30	2.65
#6/3	"	2.00	1.50	3.50
Type S.F.U. cable, 3 conductor				
#8	L.F.	0.84	1.30	2.14
#6	"	1.15	1.45	2.60
Type SER cable, 4 conductor				
#6	L.F.	1.55	1.70	3.25
#4	"	2.35	1.80	4.15
Flexible cord, type STO cord				
#18/2	L.F.	0.64	0.19	0.83
#18/3	"	0.76	0.23	0.99
#18/4	"	1.05	0.28	1.33
#16/2	"	0.74	0.19	0.93
#16/3	"	0.89	0.21	1.10
#16/4	"	1.20	0.23	1.43
#14/2	"	1.20	0.23	1.43
#14/3	"	1.50	0.29	1.79
#14/4	"	1.80	0.33	2.13
#12/2	"	1.60	0.28	1.88
#12/3	"	1.90	0.31	2.21
#12/4	"	2.20	0.37	2.57
#10/2	"	2.00	0.33	2.33
#10/3	"	2.50	0.37	2.87
#10/4	"	2.80	0.42	3.22
#8/2	"	3.35	0.37	3.72
#8/3	"	3.70	0.42	4.12
#8/4	"	5.20	0.47	5.67
16130.40 BOXES				
Round cast box, type SEH				
1/2"	EA.	15.45	16.25	31.70
3/4"	"	16.85	19.70	36.55
SEHC				
1/2"	EA.	19.40	16.25	35.65
3/4"	"	20.35	19.70	40.05
SEHL				
1/2"	EA.	23.95	16.25	40.20
3/4"	"	25.30	20.80	46.10
SEHT				

BASIC MATERIALS

16130.40 BOXES

BASIC MATERIALS	UNIT	MAT.	INST.	TOTAL
1/2"	EA.	27.20	19.70	46.90
3/4"	"	28.10	23.35	51.45
SEHX				
1/2"	EA.	29.50	23.35	52.85
3/4"	"	30.60	28.80	59.40
Blank cover	"	3.85	6.80	10.65
1/2", hub cover	"	2.90	6.80	9.70
Cover with gasket	"	3.20	8.30	11.50
Rectangle, type FS boxes				
1/2"	EA.	7.70	16.25	23.95
3/4"	"	8.15	18.70	26.85
1"	"	8.70	23.35	32.05
FSA				
1/2"	EA.	7.85	16.25	24.10
3/4"	"	8.85	18.70	27.55
FSC				
1/2"	EA.	8.55	16.25	24.80
3/4"	"	9.40	19.70	29.10
1"	"	11.70	23.35	35.05
FSL				
1/2"	EA.	9.45	16.25	25.70
3/4"	"	9.75	18.70	28.45
FSR				
1/2"	EA.	9.30	16.25	25.55
3/4"	"	11.80	18.70	30.50
FSS				
1/2"	EA.	8.25	16.25	24.50
3/4"	"	9.05	18.70	27.75
FSLA				
1/2"	EA.	9.45	16.25	25.70
3/4"	"	10.70	18.70	29.40
FSCA				
1/2"	EA.	12.55	16.25	28.80
3/4"	"	14.40	18.70	33.10
FSCC				
1/2"	EA.	10.10	18.70	28.80
3/4"	"	13.35	23.35	36.70
FSCT				
1/2"	EA.	10.50	18.70	29.20
3/4"	"	13.10	23.35	36.45
1"	"	16.50	26.70	43.20
FST				
1/2"	EA.	13.60	23.35	36.95
3/4"	"	15.30	26.70	42.00
FSX				
1/2"	EA.	13.85	28.80	42.65
3/4"	"	16.05	34.00	50.05
FSCD boxes				
1/2"	EA.	15.25	28.80	44.05
3/4"	"	16.15	34.00	50.15
Rectangle, type FS, 2 gang boxes				
1/2"	EA.	14.05	16.25	30.30

BASIC MATERIALS	UNIT	MAT.	INST.	TOTAL
16130.40 BOXES				
3/4"	EA.	15.10	18.70	33.80
1"	"	16.55	23.35	39.90
Weatherproof cast aluminum boxes, 1 gang, 3 outlets				
1/2"	EA.	4.15	18.70	22.85
3/4"	"	4.50	23.35	27.85
2 gang, 3 outlets				
1/2"	EA.	7.95	23.35	31.30
3/4"	"	8.30	24.95	33.25
1 gang, 4 outlets				
1/2"	EA.	5.10	28.80	33.90
3/4"	"	5.45	34.00	39.45
2 gang, 4 outlets				
1/2"	EA.	8.65	28.80	37.45
3/4"	"	9.05	34.00	43.05
Weatherproof and type FS box covers, blank, 1 gang	"	1.80	6.80	8.60
Tumbler switch, 1 gang	"	2.20	6.80	9.00
1 gang, single recept	"	2.30	6.80	9.10
Duplex recept	"	2.20	6.80	9.00
Despard	"	2.10	6.80	8.90
Red pilot light	"	14.00	6.80	20.80
SW and				
Single recept	EA.	6.55	9.35	15.90
Duplex recept	"	5.40	9.35	14.75
2 gang				
Blank	EA.	2.40	8.50	10.90
Tumbler switch	"	3.35	8.50	11.85
Single recept	"	3.75	8.50	12.25
Duplex recept	"	4.05	8.50	12.55
Box covers				
Surface	EA.	9.90	9.35	19.25
Sealing	"	10.80	9.35	20.15
Dome	"	14.95	9.35	24.30
1/2" nipple	"	19.05	9.35	28.40
3/4" nipple	"	19.60	9.35	28.95
16130.60 PULL AND JUNCTION BOXES				
4"				
Octagon box	EA.	1.35	5.35	6.70
Box extension	"	1.95	2.75	4.70
Plaster ring	"	1.05	2.75	3.80
Cover blank	"	0.43	2.75	3.18
Square box	"	1.60	5.35	6.95
Box extension	"	1.75	2.75	4.50
Plaster ring	"	0.79	2.75	3.54
Cover blank	"	0.48	2.75	3.23
4-11/16"				
Square box	EA.	2.90	5.35	8.25
Box extension	"	3.85	2.75	6.60
Plaster ring	"	2.25	2.75	5.00
Cover blank	"	0.87	2.75	3.62
Switch and device boxes				
2 gang	EA.	8.40	5.35	13.75

BASIC MATERIALS	UNIT	MAT.	INST.	TOTAL
16130.60 PULL AND JUNCTION BOXES				
3 gang	EA.	10.35	5.35	15.70
4 gang	"	14.65	7.50	22.15
Device covers				
2 gang	EA.	4.10	2.75	6.85
3 gang	"	4.85	2.75	7.60
4 gang	"	7.05	2.75	9.80
Handy box	"	1.25	5.35	6.60
Extension	"	1.70	2.75	4.45
Switch cover	"	0.48	2.75	3.23
Switch box with knockout	"	1.40	6.80	8.20
Weatherproof cover, spring type	"	4.85	3.75	8.60
Cover plate, dryer receptacle 1 gang plastic	"	2.25	4.65	6.90
For 4" receptacle, 2 gang	"	3.85	4.65	8.50
Duplex receptacle cover plate, plastic	"	0.43	2.75	3.18
4", vertical bracket box, 1-1/2" with				
RMX clamps	EA.	2.05	6.80	8.85
BX clamps	"	3.10	6.80	9.90
4", octagon device cover				
1 switch	EA.	1.40	2.75	4.15
1 duplex recept	"	1.30	2.75	4.05
4" octagon adjustable bar hangers				
18-1/2"	EA.	1.60	2.35	3.95
26-1/2"	"	1.95	2.35	4.30
With clip				
18-1/2"	EA.	1.75	2.35	4.10
26-1/2"	"	1.70	2.35	4.05
4" square to round plaster rings	"	1.35	2.75	4.10
2 gang device plaster rings	"	1.60	2.75	4.35
Surface covers				
1 gang switch	EA.	1.55	2.75	4.30
2 gang switch	"	1.60	2.75	4.35
1 single recept	"	1.55	2.75	4.30
1 20a twist lock recept	"	1.95	2.75	4.70
1 30a twist lock recept	"	2.00	2.75	4.75
1 duplex recept	"	1.20	2.75	3.95
2 duplex recept	"	1.40	2.75	4.15
Switch and duplex recept	"	1.70	2.75	4.45
4" plastic round boxes, ground straps				
Box only	EA.	0.99	6.80	7.79
Box w/clamps	"	1.20	9.35	10.55
Box w/16" bar	"	1.75	10.70	12.45
Box w/24" bar	"	1.90	11.70	13.60
4" plastic round box covers				
Blank cover	EA.	0.57	2.75	3.32
Plaster ring	"	0.64	2.75	3.39
4" plastic square boxes				
Box only	EA.	0.73	6.80	7.53
Box w/clamps	"	0.90	9.35	10.25
Box w/hanger	"	1.10	11.70	12.80
Box w/nails and clamp	"	1.60	11.70	13.30
4" plastic square box covers				
Blank cover	EA.	0.64	2.75	3.39

16 ELECTRICAL

BASIC MATERIALS

BASIC MATERIALS	UNIT	MAT.	INST.	TOTAL
16130.60 PULL AND JUNCTION BOXES				
1 gang ring	EA.	0.61	2.75	3.36
2 gang ring	"	0.64	2.75	3.39
Round ring	"	0.73	2.75	3.48
16130.65 PULL BOXES AND CABINETS				
Galvanized pull boxes, screw cover				
4x4x4	EA.	4.95	8.90	13.85
4x6x4	"	6.15	8.90	15.05
16130.80 RECEPTACLES				
Contractor grade duplex receptacles, 15 a 120v				
Duplex	EA.	2.15	9.35	11.50
125 volt, 20a, duplex, grounding type, standard grade	"	5.60	9.35	14.95
Ground fault interrupter type	"	41.30	13.85	55.15
250 volt, 20a, 2 pole, single receptacle, ground type	"	6.30	9.35	15.65
120/208v, 4 pole, single receptacle, twist lock				
20a	EA.	13.55	16.25	29.80
50a	"	25.80	16.25	42.05
125/250v, 3 pole, flush receptacle				
30a	EA.	8.60	13.85	22.45
50a	"	9.90	13.85	23.75
60a	"	40.10	16.25	56.35
Clock receptacle, 2 pole, grounding type	"	6.95	9.35	16.30
125/250v, 3 pole, 3 wire surface recepts				
30a	EA.	16.60	13.85	30.45
50a	"	18.45	13.85	32.30
60a	"	40.30	16.25	56.55
Cord set, 3 wire, 6' cord				
30a	EA.	17.95	13.85	31.80
50a	"	25.10	13.85	38.95
125/250v, 3 pole, 3 wire cap				
30a	EA.	14.70	18.70	33.40
50a	"	26.70	18.70	45.40
60a	"	34.30	20.80	55.10
16198.10 ELECTRIC MANHOLES				
Precast, handhole, 4' deep				
2'x2'	EA.	350.00	160.00	510.00
3'x3'	"	470.00	260.00	730.00
4'x4'	"	1,010	480.00	1,490
Power manhole, complete, precast, 8' deep				
4'x4'	EA.	1,440	660.00	2,100
6'x6'	"	1,920	940.00	2,860
8'x8'	"	2,280	980.00	3,260
6' deep, 9' x 12'	"	2,520	1,170	3,690
Cast in place, power manhole, 8' deep				
4'x4'	EA.	1,680	660.00	2,340
6'x6'	"	2,160	940.00	3,100
8'x8'	"	2,400	980.00	3,380

16 ELECTRICAL

BASIC MATERIALS	UNIT	MAT.	INST.	TOTAL
16199.10 UTILITY POLES & FITTINGS				
Wood pole, creosoted				
25'	EA.	300.00	110.00	410.00
30'	"	360.00	140.00	500.00
35'	"	480.00	160.00	640.00
40'	"	580.00	180.00	760.00
45'	"	670.00	330.00	1,000
50'	"	790.00	340.00	1,130
55'	"	910.00	350.00	1,260
Treated, wood preservative, 6"x6"				
8'	EA.	30.30	23.35	53.65
10'	"	44.80	37.40	82.20
12'	"	47.20	41.60	88.80
14'	"	59.50	62.50	122.00
16'	"	71.50	75.00	146.50
18'	"	84.50	93.50	178.00
20'	"	100.00	93.50	193.50
Aluminum, brushed, no base				
8'	EA.	400.00	93.50	493.50
10'	"	450.00	120.00	570.00
15'	"	510.00	130.00	640.00
20'	"	680.00	150.00	830.00
25'	"	910.00	180.00	1,090
30'	"	1,360	210.00	1,570
35'	"	1,590	230.00	1,820
40'	"	2,040	290.00	2,330
Steel, no base				
10'	EA.	460.00	120.00	580.00
15'	"	510.00	140.00	650.00
20'	"	570.00	180.00	750.00
25'	"	740.00	210.00	950.00
30'	"	990.00	240.00	1,230
35'	"	1,360	290.00	1,650
Concrete, no base				
13'	EA.	570.00	260.00	830.00
16'	"	840.00	340.00	1,180
18'	"	1,010	410.00	1,420
25'	"	1,240	470.00	1,710
30'	"	1,650	570.00	2,220
35'	"	2,120	660.00	2,780
40'	"	2,470	750.00	3,220
45'	"	2,950	800.00	3,750
50'	"	3,650	850.00	4,500
55'	"	4,070	890.00	4,960
60'	"	4,650	940.00	5,590
Pole line hardware				
Wood crossarm				
4'	EA.	31.20	62.50	93.70
8'	"	62.50	78.00	140.50
10'	"	120.00	96.00	216.00
Angle steel brace				
1 piece	EA.	6.95	11.70	18.65
2 piece	"	14.85	16.25	31.10

16 ELECTRICAL

BASIC MATERIALS

16199.10 UTILITY POLES & FITTINGS

BASIC MATERIALS	UNIT	MAT.	INST.	TOTAL
Eye nut, 5/8"	EA.	3.45	2.35	5.80
Bolt (14-16"), 5/8"	"	9.10	9.35	18.45
Transformer, ground connection	"	4.05	11.70	15.75
Stirrup	"	12.40	14.40	26.80
Secondary lead support	"	13.85	18.70	32.55
Spool insulator	"	10.20	9.35	19.55
Guy grip, preformed				
7/16"	EA.	2.20	6.80	9.00
1/2"	"	3.30	6.80	10.10
Hook	"	2.20	11.70	13.90
Strain insulator	"	17.80	17.00	34.80
Wire				
5/16"	L.F.	0.36	0.23	0.59
7/16"	"	0.72	0.28	1.00
1/2"	"	1.20	0.37	1.57
Soft drawn ground, copper, #8	"	0.30	0.37	0.67
Ground clamp	EA.	4.75	14.40	19.15
Perforated strapping for conduit, 1-1/2"	L.F.	1.80	6.80	8.60
Hot line clamp	EA.	13.85	37.40	51.25
Lightning arrester				
3kv	EA.	76.50	46.80	123.30
10kv	"	120.00	75.00	195.00
30kv	"	230.00	93.50	323.50
36kv	"	460.00	120.00	580.00
Fittings				
Plastic molding	L.F.	2.10	6.80	8.90
Molding staples	EA.	0.50	2.35	2.85
Ground wires staples	"	0.21	1.40	1.61
Copper butt plate	"	0.59	13.85	14.44
Anchor bond clamp	"	2.60	6.80	9.40
Guy wire				
1/4"	L.F.	0.29	1.40	1.69
3/8"	"	0.45	2.35	2.80
Guy grip				
1/4"	EA.	1.45	2.35	3.80
3/8"	"	2.55	2.35	4.90

POWER GENERATION

16210.10 GENERATORS

POWER GENERATION	UNIT	MAT.	INST.	TOTAL
Diesel generator, with auto transfer switch				
50kw	EA.	22,480	1,440	23,920
125kw	"	33,840	2,340	36,180
300kw	"	55,430	4,680	60,110

POWER GENERATION

POWER GENERATION	UNIT	MAT.	INST.	TOTAL
16210.10 GENERATORS				
750kw	EA.	161,630	9,350	170,980
16320.10 TRANSFORMERS				
Floor mounted, single phase, int. dry, 480v-120/240v				
3 kva	EA.	250.00	85.00	335.00
5 kva	"	330.00	140.00	470.00
7.5 kva	"	500.00	160.00	660.00
10 kva	"	580.00	180.00	760.00
15 kva	"	780.00	200.00	980.00
100 kva	"	2,720	540.00	3,260
Three phase, 480v-120/208v				
15 kva	EA.	930.00	280.00	1,210
30 kva	"	1,200	440.00	1,640
45 kva	"	1,610	510.00	2,120
225 kva	"	5,280	720.00	6,000
16350.10 CIRCUIT BREAKERS				
Molded case, 240v, 15-60a, bolt-on				
1 pole	EA.	10.40	11.70	22.10
2 pole	"	20.80	16.25	37.05
70-100a, 2 pole	"	58.00	24.95	82.95
15-60a, 3 pole	"	69.50	18.70	88.20
70-100a, 3 pole	"	92.50	28.80	121.30
480v, 2 pole				
15-60a	EA.	150.00	13.85	163.85
70-100a	"	180.00	18.70	198.70
3 pole				
15-60a	EA.	180.00	18.70	198.70
70-100a	"	220.00	20.80	240.80
70-225a	"	460.00	28.80	488.80
Load center circuit breakers, 240v				
1 pole, 10-60a	EA.	10.40	11.70	22.10
2 pole				
10-60a	EA.	20.80	18.70	39.50
70-100a	"	34.70	31.20	65.90
110-150a	"	130.00	34.00	164.00
3 pole				
10-60a	EA.	69.50	23.35	92.85
70-100a	"	100.00	34.00	134.00
Load center, G.F.I. breakers, 240v				
1 pole, 15-30a	EA.	69.50	13.85	83.35
2 pole, 15-30a	"	120.00	18.70	138.70
Key operated breakers, 240v, 1 pole, 10-30a	"	11.55	13.85	25.40
Tandem breakers, 240v				
1 pole, 15-30a	EA.	19.65	18.70	38.35
2 pole, 15-30a	"	35.80	24.95	60.75
Bolt-on, G.F.I. breakers, 240v, 1 pole, 15-30a	"	84.50	16.25	100.75

POWER GENERATION

POWER GENERATION	UNIT	MAT.	INST.	TOTAL
16360.10 SAFETY SWITCHES				
Fused, 3 phase, 30 amp, 600v, heavy duty				
NEMA 1	EA.	130.00	53.50	183.50
NEMA 3r	"	210.00	53.50	263.50
NEMA 4	"	570.00	75.00	645.00
NEMA 12	"	210.00	81.50	291.50
60a				
NEMA 1	EA.	170.00	53.50	223.50
NEMA 3r	"	260.00	53.50	313.50
NEMA 4	"	690.00	75.00	765.00
NEMA 12	"	220.00	81.50	301.50
100a				
NEMA 1	EA.	320.00	81.50	401.50
NEMA 3r	"	450.00	81.50	531.50
NEMA 4	"	1,370	93.50	1,464
NEMA 12	"	380.00	120.00	500.00
200a				
NEMA 1	EA.	460.00	120.00	580.00
NEMA 3r	"	600.00	120.00	720.00
NEMA 4	"	1,940	130.00	2,070
NEMA 12	"	580.00	160.00	740.00
Non-fused, 240-600v, heavy duty, 3 phase, 30 amp				
NEMA 1	EA.	91.50	53.50	145.00
NEMA 3r	"	150.00	53.50	203.50
NEMA 4	"	570.00	81.50	651.50
NEMA 12	"	170.00	81.50	251.50
60a				
NEMA1	EA.	150.00	53.50	203.50
NEMA 3r	"	240.00	53.50	293.50
NEMA 4	"	690.00	81.50	771.50
NEMA 12	"	220.00	81.50	301.50
100a				
NEMA 1	EA.	230.00	81.50	311.50
NEMA 3r	"	340.00	81.50	421.50
NEMA 4	"	1,370	120.00	1,490
NEMA 12	"	300.00	120.00	420.00
200a, NEMA 1	"	350.00	120.00	470.00
600a, NEMA 12	"	1,540	580.00	2,120
16365.10 FUSES				
Fuse, one-time, 250v				
30a	EA.	1.15	2.35	3.50
60a	"	1.70	2.35	4.05
100a	"	6.85	2.35	9.20
200a	"	17.15	2.35	19.50
400a	"	29.70	2.35	32.05
600a	"	57.00	2.35	59.35
600v				
30a	EA.	5.15	2.35	7.50
60a	"	5.70	2.35	8.05
100a	"	16.00	2.35	18.35
200a	"	34.30	2.35	36.65

16 ELECTRICAL

POWER GENERATION

16365.10 FUSES	UNIT	MAT.	INST.	TOTAL
400a	EA.	68.50	2.35	70.85

16395.10 GROUNDING	UNIT	MAT.	INST.	TOTAL
Ground rods, copper clad, 1/2" x				
6'	EA.	8.55	31.20	39.75
8'	"	12.80	34.00	46.80
10'	"	16.00	46.80	62.80
5/8" x				
5'	EA.	11.75	28.80	40.55
6'	"	12.80	34.00	46.80
8'	"	16.00	46.80	62.80
10'	"	18.15	58.50	76.65
3/4" x				
8'	EA.	23.45	34.00	57.45
10'	"	32.00	37.40	69.40
Ground rod clamp				
5/8"	EA.	3.85	5.75	9.60
3/4"	"	4.95	5.75	10.70
Ground rod couplings				
1/2"	EA.	5.50	4.65	10.15
5/8"	"	7.70	4.65	12.35
Ground rod, driving stud				
1/2"	EA.	4.40	4.65	9.05
5/8"	"	4.95	4.65	9.60
3/4"	"	5.50	4.65	10.15
Ground rod clamps, #8-2 to				
1" pipe	EA.	4.40	9.35	13.75
2" pipe	"	5.50	11.70	17.20
3" pipe	"	22.00	13.85	35.85
5" pipe	"	35.20	16.25	51.45
6" pipe	"	48.40	20.80	69.20

SERVICE AND DISTRIBUTION

16425.10 SWITCHBOARDS	UNIT	MAT.	INST.	TOTAL
Switchboard, 90" high, no main disconnect, 208/120v				
400a	EA.	1,640	370.00	2,010
600a	"	2,540	370.00	2,910
1000a	"	3,200	370.00	3,570
1200a	"	3,420	470.00	3,890
1600a	"	3,760	560.00	4,320
2000a	"	4,030	660.00	4,690

SERVICE AND DISTRIBUTION

SERVICE AND DISTRIBUTION	UNIT	MAT.	INST.	TOTAL
16425.10 SWITCHBOARDS				
2500a	EA.	3,310	750.00	4,060
277/480v				
600a	EA.	2,920	380.00	3,300
800a	"	3,200	380.00	3,580
1600a	"	4,030	560.00	4,590
2000a	"	4,310	660.00	4,970
2500a	"	4,590	750.00	5,340
3000a	"	5,290	1,290	6,580
4000a	"	6,400	1,390	7,790
16430.20 METERING				
Outdoor wp meter sockets, 1 gang, 240v, 1 phase				
Includes sealing ring, 100a	EA.	27.60	70.50	98.10
150a	"	36.80	83.00	119.80
200a	"	46.00	93.50	139.50
Die cast hubs, 1-1/4"	"	4.25	14.95	19.20
1-1/2"	"	4.90	14.95	19.85
2"	"	5.90	14.95	20.85
16470.10 PANELBOARDS				
Indoor load center, 1 phase 240v main lug only				
30a - 2 spaces	EA.	12.50	93.50	106.00
100a - 8 spaces	"	42.30	110.00	152.30
150a - 16 spaces	"	100.00	140.00	240.00
200a - 24 spaces	"	130.00	160.00	290.00
200a - 42 spaces	"	250.00	190.00	440.00
Main circuit breaker				
100a - 8 spaces	EA.	100.00	110.00	210.00
100a - 16 spaces	"	130.00	130.00	260.00
150a - 16 spaces	"	210.00	140.00	350.00
150a - 24 spaces	"	250.00	150.00	400.00
200a - 24 spaces	"	250.00	160.00	410.00
200a - 42 spaces	"	350.00	170.00	520.00
3 phase, 480/277v, main lugs only, 120a, 30 circuits	"	690.00	160.00	850.00
277/480v, 4 wire, flush surface				
225a, 30 circuits	EA.	730.00	190.00	920.00
400a, 30 circuits	"	840.00	230.00	1,070
600a, 42 circuits	"	1,050	280.00	1,330
208/120v, main circuit breaker, 3 phase, 4 wire				
100a				
12 circuits	EA.	600.00	240.00	840.00
20 circuits	"	740.00	290.00	1,030
30 circuits	"	1,090	330.00	1,420
400a				
30 circuits	EA.	2,290	690.00	2,980
42 circuits	"	2,750	750.00	3,500
600a, 42 circuits	"	5,350	850.00	6,200
120/208v, flush, 3 ph., 4 wire, main only				
100a				
12 circuits	EA.	420.00	240.00	660.00

SERVICE AND DISTRIBUTION

SERVICE AND DISTRIBUTION	UNIT	MAT.	INST.	TOTAL
16470.10 PANELBOARDS				
20 circuits	EA.	580.00	290.00	870.00
30 circuits	"	870.00	330.00	1,200
225a				
30 circuits	EA.	880.00	360.00	1,240
42 circuits	"	1,110	450.00	1,560
400a				
30 circuits	EA.	1,680	690.00	2,370
42 circuits	"	2,460	750.00	3,210
600a, 42 circuits	"	3,830	850.00	4,680
16480.10 MOTOR CONTROLS				
Motor generator set, 3 phase, 480/277v, w/controls				
10kw	EA.	7,560	1,290	8,850
15kw	"	9,860	1,440	11,300
20kw	"	10,940	1,500	12,440
40kw	"	15,430	1,780	17,210
100kw	"	25,240	2,880	28,120
200kw	"	57,690	3,400	61,090
300kw	"	72,110	3,740	75,850
2 pole, 230 volt starter, w/NEMA-1				
1 hp, 9 amp, size 00	EA.	110.00	46.80	156.80
2 hp, 18amp, size 0	"	140.00	46.80	186.80
3 hp, 27amp, size 1	"	150.00	46.80	196.80
5 hp, 45amp, size 1p	"	200.00	46.80	246.80
7-1/2 hp, 45a, size 2	"	340.00	46.80	386.80
15 hp, 90a, size 3	"	520.00	46.80	566.80
16490.10 SWITCHES				
Fused interrupter load, 35kv				
20A				
1 pole	EA.	15,140	750.00	15,890
2 pole	"	16,410	800.00	17,210
3 way	"	17,670	800.00	18,470
4 way	"	18,930	850.00	19,780
30a, 1 pole	"	15,140	750.00	15,890
3 way	"	17,670	800.00	18,470
4 way	"	18,930	850.00	19,780
Weatherproof switch, including box & cover, 20a				
1 pole	EA.	16,410	750.00	17,160
2 pole	"	17,670	800.00	18,470
3 way	"	18,930	850.00	19,780
4 way	"	20,190	850.00	21,040
Photo electric switches				
1000 watt				
105-135v	EA.	21.50	34.00	55.50
Dimmer switch and switch plate				
600 w	EA.	11.90	14.40	26.30
Contractor grade wall switch 15a, 120v				
Single pole	EA.	1.70	7.50	9.20
Three way	"	2.90	9.35	12.25

SERVICE AND DISTRIBUTION

SERVICE AND DISTRIBUTION	UNIT	MAT.	INST.	TOTAL
16490.10 SWITCHES				
Four way	EA.	14.40	12.45	26.85
Specification grade toggle switches, 20a, 120-277v				
Single pole	EA.	10.85	9.35	20.20
Double pole	"	12.45	13.85	26.30
3 way	"	11.45	11.70	23.15
4 way	"	29.20	13.85	43.05
Switch plates, plastic ivory				
1 gang	EA.	0.41	3.75	4.16
2 gang	"	0.81	4.65	5.46
3 gang	"	1.25	5.60	6.85
4 gang	"	2.05	6.80	8.85
5 gang	"	4.70	7.50	12.20
6 gang	"	5.55	8.50	14.05
Stainless steel				
1 gang	EA.	1.40	3.75	5.15
2 gang	"	2.70	4.65	7.35
3 gang	"	4.70	5.75	10.45
4 gang	"	6.50	6.80	13.30
5 gang	"	8.95	7.50	16.45
6 gang	"	11.10	8.50	19.60
Brass				
1 gang	EA.	3.00	3.75	6.75
2 gang	"	5.95	4.65	10.60
3 gang	"	10.85	5.75	16.60
4 gang	"	14.60	6.80	21.40
5 gang	"	18.10	7.50	25.60
6 gang	"	21.55	8.50	30.05
16490.20 TRANSFER SWITCHES				
Automatic transfer switch 600v, 3 pole				
30a	EA.	1,520	160.00	1,680
100a	"	3,160	220.00	3,380
400a	"	6,940	470.00	7,410
800a	"	11,620	850.00	12,470
1200a	"	18,940	1,070	20,010
2600a	"	44,200	1,970	46,170
16490.80 SAFETY SWITCHES				
Safety switch, 600v, 3 pole, heavy duty, NEMA-1				
30a	EA.	130.00	46.80	176.80
60a	"	170.00	53.50	223.50
100a	"	320.00	75.00	395.00
200a	"	450.00	120.00	570.00
400a	"	1,130	260.00	1,390
600a	"	2,040	370.00	2,410
800a	"	3,140	490.00	3,630
1200a	"	3,890	670.00	4,560

LIGHTING

16510.05 INTERIOR LIGHTING

LIGHTING	UNIT	MAT.	INST.	TOTAL
Recessed fluorescent fixtures, 2'x2'				
2 lamp	EA.	54.00	34.00	88.00
4 lamp	"	72.50	34.00	106.50
2 lamp w/flange	"	68.00	46.80	114.80
4 lamp w/flange	"	85.50	46.80	132.30
1'x4'				
2 lamp	EA.	55.00	31.20	86.20
3 lamp	"	76.00	31.20	107.20
2 lamp w/flange	"	68.00	34.00	102.00
3 lamp w/flange	"	89.00	34.00	123.00
2'x4'				
2 lamp	EA.	68.00	34.00	102.00
3 lamp	"	79.00	34.00	113.00
4 lamp	"	77.50	34.00	111.50
2 lamp w/flange	"	80.50	46.80	127.30
3 lamp w/flange	"	92.00	46.80	138.80
4 lamp w/flange	"	90.50	46.80	137.30
4'x4'				
4 lamp	EA.	160.00	46.80	206.80
6 lamp	"	190.00	46.80	236.80
8 lamp	"	200.00	46.80	246.80
4 lamp w/flange	"	190.00	70.50	260.50
6 lamp w/flange	"	250.00	70.50	320.50
8 lamp, w/flange	"	270.00	70.50	340.50
Surface mounted incandescent fixtures				
40w	EA.	49.10	31.20	80.30
75w	"	53.50	31.20	84.70
100w	"	63.00	31.20	94.20
150w	"	74.50	31.20	105.70
Pendant				
40w	EA.	53.50	37.40	90.90
75w	"	59.00	37.40	96.40
100w	"	67.50	37.40	104.90
150w	"	77.00	37.40	114.40
Recessed incandescent fixtures				
40w	EA.	90.50	70.50	161.00
75w	"	94.50	70.50	165.00
100w	"	97.00	70.50	167.50
150w	"	110.00	70.50	180.50
Exit lights, 120v				
Recessed	EA.	70.00	58.50	128.50
Back mount	"	49.10	34.00	83.10
Universal mount	"	56.00	34.00	90.00
Emergency battery units, 6v-120v, 50 unit	"	100.00	70.50	170.50
With 1 head	"	120.00	70.50	190.50
With 2 heads	"	140.00	70.50	210.50
Mounting bucket	"	21.05	34.00	55.05
Light track single circuit				
2'	EA.	24.95	23.35	48.30
4'	"	46.60	23.35	69.95
8'	"	83.00	46.80	129.80
12'	"	130.00	70.50	200.50

LIGHTING	UNIT	MAT.	INST.	TOTAL
16510.05 INTERIOR LIGHTING				
Fixtures, square				
R-20	EA.	66.50	6.80	73.30
R-30	"	69.50	6.80	76.30
Mini spot	"	92.00	6.80	98.80
16510.10 LIGHTING INDUSTRIAL				
Surface mounted fluorescent, wrap around lens				
1 lamp	EA.	50.00	37.40	87.40
2 lamps	"	54.50	41.60	96.10
4 lamps	"	86.00	46.80	132.80
Wall mounted fluorescent				
2-20w lamps	EA.	40.10	23.35	63.45
2-30w lamps	"	43.00	23.35	66.35
2-40w lamps	"	50.00	31.20	81.20
Indirect, with wood shielding, 2049w lamps				
4'	EA.	86.00	46.80	132.80
8'	"	120.00	75.00	195.00
Industrial fluorescent, 2 lamp				
4'	EA.	86.00	34.00	120.00
8'	"	140.00	62.50	202.50
Strip fluorescent				
4'				
1 lamp	EA.	28.70	31.20	59.90
2 lamps	"	35.80	31.20	67.00
8'				
1 lamp	EA.	57.50	34.00	91.50
2 lamps	"	64.50	41.60	106.10
Wire guard for strip fixture, 4' long	"	14.30	16.25	30.55
Strip fluorescent, 8' long, two 4' lamps	"	86.00	62.50	148.50
With four 4' lamps	"	110.00	75.00	185.00
Wet location fluorescent, plastic housing				
4' long				
1 lamp	EA.	100.00	46.80	146.80
2 lamps	"	110.00	62.50	172.50
8' long				
2 lamps	EA.	180.00	75.00	255.00
4 lamps	"	240.00	81.50	321.50
Parabolic troffer, 2'x2'				
With 2 "U" lamps	EA.	120.00	46.80	166.80
With 3 "U" lamps	"	140.00	53.50	193.50
2'x4'				
With 2 40w lamps	EA.	160.00	53.50	213.50
With 3 40w lamps	"	180.00	62.50	242.50
With 4 40w lamps	"	200.00	62.50	262.50
1'x4'				
With 1 T-12 lamp, 9 cell	EA.	120.00	34.00	154.00
With 2 T-12 lamps	"	140.00	41.60	181.60
With 1 T-12 lamp, 20 cell	"	140.00	34.00	174.00
With 2 T-12 lamps	"	150.00	41.60	191.60
Steel sided surface fluorescent, 2'x4'				
3 lamps	EA.	110.00	62.50	172.50
4 lamps	"	110.00	62.50	172.50

LIGHTING

16510.10 LIGHTING INDUSTRIAL

LIGHTING	UNIT	MAT.	INST.	TOTAL
Outdoor sign fluor., 1 lamp, remote ballast				
4' long	EA.	2,050	280.00	2,330
6' long	"	2,460	370.00	2,830
Recess mounted, commercial, 2'x2', 13" high				
100w	EA.	680.00	190.00	870.00
250w	"	750.00	210.00	960.00
High pressure sodium, hi-bay open				
400w	EA.	360.00	81.50	441.50
1000w	"	630.00	110.00	740.00
Enclosed				
400w	EA.	560.00	110.00	670.00
1000w	"	830.00	140.00	970.00
Metal halide hi-bay, open				
400w	EA.	330.00	81.50	411.50
1000w	"	560.00	110.00	670.00
Enclosed				
400w	EA.	530.00	110.00	640.00
1000w	"	790.00	140.00	930.00
High pressure sodium, low bay, surface mounted				
100w	EA.	230.00	46.80	276.80
150w	"	250.00	53.50	303.50
250w	"	280.00	62.50	342.50
400w	"	360.00	75.00	435.00
Metal halide, low bay, pendant mounted				
175w	EA.	250.00	62.50	312.50
250w	"	330.00	75.00	405.00
400w	"	410.00	100.00	510.00
Indirect luminare, square, metal halide, freestanding				
175w	EA.	660.00	46.80	706.80
250w	"	670.00	46.80	716.80
400w	"	730.00	46.80	776.80
High pressure sodium				
150w	EA.	670.00	46.80	716.80
250w	"	710.00	46.80	756.80
400w	"	770.00	46.80	816.80
Round, metal halide				
175w	EA.	620.00	46.80	666.80
250w	"	660.00	46.80	706.80
400w	"	700.00	46.80	746.80
High pressure sodium				
150w	EA.	630.00	46.80	676.80
250w	"	690.00	46.80	736.80
400w	"	710.00	46.80	756.80
Wall mounted, metal halide				
175w	EA.	520.00	120.00	640.00
250w	"	550.00	120.00	670.00
400w	"	910.00	150.00	1,060
High pressure sodium				
150w	EA.	590.00	120.00	710.00
250w	"	730.00	120.00	850.00
400w	"	940.00	150.00	1,090
Wall pack lithonia, high pressure sodium				

16 ELECTRICAL

LIGHTING

16510.10 LIGHTING INDUSTRIAL

LIGHTING	UNIT	MAT.	INST.	TOTAL
35w	EA.	160.00	41.60	201.60
55w	"	180.00	46.80	226.80
150w	"	200.00	75.00	275.00
250w	"	250.00	81.50	331.50
Low pressure sodium				
35w	EA.	220.00	81.50	301.50
55w	"	340.00	93.50	433.50
Wall pack hubbell, high pressure sodium				
35w	EA.	170.00	41.60	211.60
150w	"	220.00	75.00	295.00
250w	"	280.00	81.50	361.50
Compact fluorescent				
2-7w	EA.	110.00	46.80	156.80
2-13w	"	120.00	62.50	182.50
1-18w	"	140.00	62.50	202.50
Handball & racquet ball court, 2'x2', metal halide				
250w	EA.	370.00	120.00	490.00
400w	"	460.00	130.00	590.00
High pressure sodium				
250w	EA.	410.00	120.00	530.00
400w	"	450.00	130.00	580.00
Bollard light, 42" w/found., high pressure sodium				
70w	EA.	630.00	120.00	750.00
100w	"	640.00	120.00	760.00
150w	"	650.00	120.00	770.00
Light fixture lamps				
Lamp				
20w med. bipin base, cool white, 24"	EA.	4.65	6.80	11.45
30w cool white, rapid start, 36"	"	6.00	6.80	12.80
40w cool white "U", 3"	"	11.50	6.80	18.30
40w cool white, rapid start, 48"	"	2.75	6.80	9.55
70w high pressure sodium, mogul base	"	59.50	9.35	68.85
75w slimline, 96"	"	7.70	9.35	17.05
100w				
Incandescent, 100a, inside frost	EA.	2.35	4.65	7.00
Mercury vapor, clear, mogul base	"	29.10	9.35	38.45
High pressure sodium, mogul base	"	63.50	9.35	72.85
150w				
Par 38 flood or spot, incandescent	EA.	7.15	4.65	11.80
High pressure sodium, 1/2 mogul base	"	66.00	9.35	75.35
175w				
Mercury vapor, clear, mogul base	EA.	50.50	9.35	59.85
Metal halide, clear, mogul base	"	22.50	9.35	31.85
250w				
High pressure sodium, mogul base	EA.	70.00	9.35	79.35
Mercury vapor, clear, mogul base	"	39.70	9.35	49.05
Metal halide, clear, mogul base	"	63.50	9.35	72.85
High pressure sodium, mogul base	"	70.00	9.35	79.35
400w				
Mercury vapor, clear, mogul base	EA.	31.80	9.35	41.15
Metal halide, clear, mogul base	"	59.50	9.35	68.85
High pressure sodium, mogul base	"	75.50	9.35	84.85

LIGHTING

LIGHTING	UNIT	MAT.	INST.	TOTAL
16510.10 LIGHTING INDUSTRIAL				
1000w				
Mercury vapor, clear, mogul base	EA.	69.00	11.70	80.70
High pressure sodium, mogul base	"	180.00	11.70	191.70
16510.30 EXTERIOR LIGHTING				
Exterior light fixtures				
Rectangle, high pressure sodium				
70w	EA.	400.00	120.00	520.00
100w	"	420.00	120.00	540.00
150w	"	450.00	120.00	570.00
250w	"	610.00	130.00	740.00
400w	"	680.00	160.00	840.00
Flood, rectangular, high pressure sodium				
70w	EA.	420.00	120.00	540.00
100w	"	440.00	120.00	560.00
150w	"	480.00	120.00	600.00
400w	"	720.00	160.00	880.00
1000w	"	1,090	210.00	1,300
Round				
400w	EA.	750.00	160.00	910.00
1000w	"	1,170	210.00	1,380
Round, metal halide				
400w	EA.	820.00	160.00	980.00
1000w	"	1,220	210.00	1,430
Light fixture arms, cobra head, 6', high press. sodium				
100w	EA.	450.00	93.50	543.50
150w	"	710.00	120.00	830.00
250w	"	740.00	120.00	860.00
400w	"	760.00	140.00	900.00
Flood, metal halide				
400w	EA.	780.00	160.00	940.00
1000w	"	1,190	210.00	1,400
1500w	"	1,610	280.00	1,890
Mercury vapor				
250w	EA.	460.00	130.00	590.00
400w	"	800.00	160.00	960.00
Incandescent				
300w	EA.	83.50	81.50	165.00
500w	"	140.00	93.50	233.50
1000w	"	150.00	150.00	300.00
16510.90 POWER LINE FILTERS				
Heavy duty power line filter, 240v				
100a	EA.	2,890	470.00	3,360
300a	"	9,590	750.00	10,340
600a	"	13,290	1,130	14,420

16 ELECTRICAL

LIGHTING	UNIT	MAT.	INST.	TOTAL
16610.30 UNINTERRUPTIBLE POWER				
Uninterruptible power systems, (U.P.S.), 3kva	"	10,300	370.00	10,670
5 kva	"	11,440	510.00	11,950
7.5 kva	"	13,730	750.00	14,480
10 kva	"	17,160	1,030	18,190
15 kva	"	20,590	1,070	21,660
20 kva	"	28,600	1,120	29,720
25 kva	"	36,610	1,170	37,780
30 kva	"	37,750	1,210	38,960
35 kva	"	40,040	1,260	41,300
40 kva	"	43,470	1,310	44,780
45 kva	"	45,760	1,360	47,120
50 kva	"	49,190	1,400	50,590
62.5 kva	"	58,340	1,500	59,840
75 kva	"	67,500	1,630	69,130
100 kva	"	90,380	1,680	92,060
150 kva	"	137,280	2,340	139,620
200 kva	"	183,040	2,580	185,620
300 kva	"	274,560	3,500	278,060
400 kva	"	410,740	4,200	414,940
500 kva	"	513,430	5,120	518,550
16670.10 LIGHTNING PROTECTION				
Lightning protection				
Copper point, nickel plated, 12'				
1/2" dia.	EA.	33.00	46.80	79.80
5/8" dia.	"	39.60	46.80	86.40

COMMUNICATIONS	UNIT	MAT.	INST.	TOTAL
16720.10 FIRE ALARM SYSTEMS				
Master fire alarm box, pedestal mounted	EA.	4,640	750.00	5,390
Master fire alarm box	"	2,390	280.00	2,670
Box light	"	130.00	23.35	153.35
Ground assembly for box	"	66.50	31.20	97.70
Bracket for pole type box	"	86.00	34.00	120.00
Pull station				
Waterproof	EA.	66.50	23.35	89.85
Manual	"	39.80	18.70	58.50
Horn, waterproof	"	59.50	46.80	106.30
Interior alarm	"	39.80	34.00	73.80
Coded transmitter, automatic	"	700.00	93.50	793.50
Control panel, 8 zone	"	1,400	370.00	1,770
Battery charger and cabinet	"	470.00	93.50	563.50
Batteries, nickel cadmium or lead calcium	"	350.00	230.00	580.00

COMMUNICATIONS

16720.10 FIRE ALARM SYSTEMS

COMMUNICATIONS	UNIT	MAT.	INST.	TOTAL
CO2 pressure switch connection	EA.	66.50	34.00	100.50
Annunciator panels				
Fire detection annunciator, remote type, 8 zone	EA.	240.00	85.00	325.00
12 zone	"	310.00	93.50	403.50
16 zone	"	380.00	120.00	500.00
Fire alarm systems				
Bell	EA.	73.00	28.80	101.80
Weatherproof bell	"	90.00	31.20	121.20
Horn	"	42.40	34.00	76.40
Siren	"	430.00	93.50	523.50
Chime	"	53.00	28.80	81.80
Audio/visual	"	77.00	34.00	111.00
Strobe light	"	71.00	34.00	105.00
Smoke detector	"	110.00	31.20	141.20
Heat detection	"	19.90	23.35	43.25
Thermal detector	"	18.55	23.35	41.90
Ionization detector	"	93.00	24.95	117.95
Duct detector	"	310.00	130.00	440.00
Test switch	"	53.00	23.35	76.35
Remote indicator	"	26.80	26.70	53.50
Door holder	"	120.00	34.00	154.00
Telephone jack	"	19.90	13.85	33.75
Fireman phone	"	280.00	46.80	326.80
Speaker	"	56.50	37.40	93.90
Remote fire alarm annunciator panel				
24 zone	EA.	1,460	310.00	1,770
48 zone	"	2,920	610.00	3,530
Control panel				
12 zone	EA.	990.00	140.00	1,130
16 zone	"	1,300	210.00	1,510
24 zone	"	1,990	310.00	2,300
48 zone	"	3,710	750.00	4,460
Power supply	"	230.00	70.50	300.50
Status command	"	6,300	230.00	6,530
Printer	"	2,450	70.50	2,521
Transponder	"	200.00	42.00	242.00
Transformer	"	150.00	31.20	181.20
Transceiver	"	200.00	34.00	234.00
Relays	"	79.50	23.35	102.85
Flow switch	"	260.00	93.50	353.50
Tamper switch	"	160.00	140.00	300.00
End of line resistor	"	11.25	16.25	27.50
Printed ckt. card	"	100.00	23.35	123.35
Central processing unit	"	12,980	290.00	13,270
UPS backup to c.p.u.	"	12,600	420.00	13,020
Smoke detector, fixed temp. & rate of rise comb.	"	200.00	75.00	275.00

16720.50 SECURITY SYSTEMS

COMMUNICATIONS	UNIT	MAT.	INST.	TOTAL
Sensors	EA.			
Balanced magnetic door switch, surface mounted	"	95.50	23.35	118.85
With remote test	"	130.00	46.80	176.80
Flush mounted	"	91.50	87.00	178.50

16 ELECTRICAL

COMMUNICATIONS	UNIT	MAT.	INST.	TOTAL
16720.50 SECURITY SYSTEMS				
Mounted bracket	EA.	7.00	16.25	23.25
Mounted bracket spacer	"	6.30	16.25	22.55
Photoelectric sensor, for fence	"			
6 beam	"	10,290	130.00	10,420
9 beam	"	12,570	200.00	12,770
Photoelectric sensor, 12 volt dc	"			
500' range	"	310.00	75.00	385.00
800' range	"	340.00	93.50	433.50
Monitor cabinet, wall mounted				
1 zone	EA.	490.00	46.80	536.80
5 zone	"	1,760	75.00	1,835
10 zone	"	770.00	81.50	851.50
20 zone	"	2,440	93.50	2,534
16730.20 CLOCK SYSTEMS				
Clock systems				
Single face	EA.	140.00	37.40	177.40
Double face	"	390.00	37.40	427.40
Skeleton	"	300.00	130.00	430.00
Master	"	3,120	230.00	3,350
Signal generator	"	2,690	190.00	2,880
Elapsed time indicator	"	540.00	37.40	577.40
Controller	"	110.00	24.95	134.95
Clock and speaker	"	230.00	51.00	281.00
Bell				
Standard	EA.	78.00	24.95	102.95
Weatherproof	"	100.00	37.40	137.40
Horn				
Standard	EA.	49.70	34.00	83.70
Weatherproof	"	64.00	44.50	108.50
Chime	"	56.50	24.95	81.45
Buzzer	"	21.25	24.95	46.20
Flasher	"	78.00	28.80	106.80
Control Board	"	350.00	160.00	510.00
Program unit	"	370.00	230.00	600.00
Block back box	"	21.25	23.35	44.60
Double clock back box	"	42.80	31.20	74.00
Wire guard	"	14.20	9.35	23.55
16740.10 TELEPHONE SYSTEMS				
Communication cable				
25 pair	L.F.	0.44	1.20	1.64
100 pair	"	2.10	1.35	3.45
150 pair	"	3.20	1.55	4.75
200 pair	"	4.30	1.85	6.15
300 pair	"	5.45	1.95	7.40
400 pair	"	7.45	2.10	9.55
Cable tap in manhole or junction box				
25 pair cable	EA.	3.50	180.00	183.50
50 pair cable	"	7.10	350.00	357.10

16 ELECTRICAL

COMMUNICATIONS

COMMUNICATIONS	UNIT	MAT.	INST.	TOTAL
16740.10 TELEPHONE SYSTEMS				
75 pair cable	EA.	10.60	530.00	540.60
100 pair cable	"	14.15	710.00	724.15
150 pair cable	"	21.25	1,040	1,061
200 pair cable	"	28.30	1,390	1,418
300 pair cable	"	42.50	2,080	2,123
400 pair cable	"	56.50	2,880	2,937
Cable terminations, manhole or junction box				
25 pair cable	EA.	3.50	180.00	183.50
50 pair cable	"	7.10	350.00	357.10
100 pair cable	"	14.15	710.00	724.15
150 pair cable	"	21.25	1,040	1,061
200 pair cable	"	28.30	1,390	1,418
300 pair cable	"	42.50	2,080	2,123
400 pair cable	"	44.10	2,880	2,924
Telephones, standard				
1 button	EA.	73.50	140.00	213.50
2 button	"	110.00	160.00	270.00
6 button	"	170.00	250.00	420.00
12 button	"	430.00	360.00	790.00
18 button	"	450.00	420.00	870.00
Hazardous area				
Desk	EA.	1,290	340.00	1,630
Wall	"	600.00	230.00	830.00
Accessories				
Standard ground	EA.	21.25	75.00	96.25
Push button	"	21.65	75.00	96.65
Buzzer	"	22.65	75.00	97.65
Interface device	"	12.00	37.40	49.40
Long cord	"	12.75	37.40	50.15
Interior jack	"	7.75	18.70	26.45
Exterior jack	"	15.55	28.80	44.35
Hazardous area				
Selector switch	EA.	130.00	150.00	280.00
Bell	"	190.00	150.00	340.00
Horn	"	290.00	200.00	490.00
Horn relay	"	240.00	140.00	380.00
16770.30 SOUND SYSTEMS				
Power amplifiers	EA.	670.00	160.00	830.00
Pre-amplifiers	"	530.00	130.00	660.00
Tuner	"	340.00	68.00	408.00
Horn				
Equilizer	EA.	860.00	75.00	935.00
Mixer	"	360.00	100.00	460.00
Tape recorder	"	1,130	87.00	1,217
Microphone	"	98.50	46.80	145.30
Cassette Player	"	570.00	100.00	670.00
Record player	"	50.00	89.00	139.00
Equipment rack	"	64.50	60.50	125.00
Speaker				
Wall	EA.	360.00	190.00	550.00
Paging	"	130.00	37.40	167.40

16 ELECTRICAL

COMMUNICATIONS

	UNIT	MAT.	INST.	TOTAL
16770.30 SOUND SYSTEMS				
Column	EA.	190.00	24.95	214.95
Single	"	42.90	28.80	71.70
Double	"	130.00	210.00	340.00
Volume control	"	42.90	24.95	67.85
Plug-in	"	130.00	37.40	167.40
Desk	"	110.00	18.70	128.70
Outlet	"	21.40	18.70	40.10
Stand	"	42.90	13.85	56.75
Console	"	10,100	370.00	10,470
Power supply	"	190.00	60.50	250.50
16780.10 ANTENNAS AND TOWERS				
Guy cable, alumaweld				
1x3, 7/32"	L.F.	0.30	2.35	2.65
1x3, 1/4"	"	0.35	2.35	2.70
1x3, 25/64"	"	0.50	2.75	3.25
1x19, 1/2"	"	1.25	3.25	4.50
1x7, 35/64"	"	1.45	3.75	5.20
1x19, 13/16"	"	1.55	4.65	6.20
Preformed alumaweld end grip				
1/4" cable	EA.	2.30	4.65	6.95
3/8" cable	"	3.05	4.65	7.70
1/2" cable	"	3.80	6.80	10.60
9/16" cable	"	4.75	9.35	14.10
5/8" cable	"	5.80	11.70	17.50
Fiberglass guy rod, white epoxy coated				
1/4" dia.	L.F.	1.50	6.80	8.30
3/8" dia	"	2.30	6.80	9.10
1/2" dia	"	3.05	9.35	12.40
5/8" dia	"	3.80	11.70	15.50
Preformed glass grip end grip, guy rod				
1/4" dia.	EA.	8.40	6.80	15.20
3/8" dia.	"	9.70	9.35	19.05
1/2" dia.	"	11.95	11.70	23.65
5/8" dia.	"	13.40	11.70	25.10
Spelter socket end grip, 1/4" dia. guy rod				
Standard strength	EA.	24.90	23.35	48.25
High performance	"	30.60	23.35	53.95
3/8" dia. guy rod				
Standard strength	EA.	24.20	16.25	40.45
High performance	"	30.60	23.35	53.95
Timber pole, Douglas Fir				
80-85 ft	EA.	2,010	910.00	2,920
90-95 ft	"	2,520	1,040	3,560
Southern yellow pine				
35-45 ft	EA.	1,260	510.00	1,770
50-55 ft	"	1,760	660.00	2,420
16780.50 TELEVISION SYSTEMS				
TV outlet, self terminating, w/cover plate	EA.	7.65	14.40	22.05
Thru splitter	"	16.65	75.00	91.65

COMMUNICATIONS

16780.50 TELEVISION SYSTEMS	UNIT	MAT.	INST.	TOTAL
End of line	EA.	13.90	62.50	76.40
In line splitter multitap				
4 way	EA.	27.80	85.00	112.80
2 way	"	20.85	79.50	100.35
Equipment cabinet	"	69.50	75.00	144.50
Antenna				
Broad band uhf	EA.	140.00	160.00	300.00
Lightning arrester	"	37.50	34.00	71.50
TV cable	L.F.	0.32	0.23	0.55
Coaxial cable rg	"	0.21	0.23	0.44
Cable drill, with replacement tip	EA.	6.95	23.35	30.30
Cable blocks for in-line taps	"	13.90	34.00	47.90
In-line taps ptu-series 36 tv system	"	16.65	53.50	70.15
Control receptacles	"	10.90	21.00	31.90
Coupler	"	20.85	110.00	130.85
Head end equipment	"	2,630	310.00	2,940
TV camera	"	1,420	78.00	1,498
TV power bracket	"	130.00	37.40	167.40
TV monitor	"	1,090	68.00	1,158
Video recorder	"	2,120	98.50	2,219
Console	"	4,380	400.00	4,780
Selector switch	"	680.00	64.50	744.50
TV controller	"	330.00	65.50	395.50

RESISTANCE HEATING

16850.10 ELECTRIC HEATING	UNIT	MAT.	INST.	TOTAL
Baseboard heater				
2', 375w	EA.	41.80	46.80	88.60
3', 500w	"	48.50	46.80	95.30
4', 750w	"	53.00	53.50	106.50
5', 935w	"	74.00	62.50	136.50
6', 1125w	"	78.50	75.00	153.50
7', 1310w	"	92.50	85.00	177.50
8', 1500w	"	98.00	93.50	191.50
9', 1680w	"	140.00	100.00	240.00
10', 1875w	"	150.00	110.00	260.00
Unit heater, wall mounted				
750w	EA.	120.00	75.00	195.00
1500w	"	190.00	78.00	268.00
2000w	"	200.00	81.50	281.50
2500w	"	210.00	85.00	295.00
3000w	"	240.00	93.50	333.50

RESISTANCE HEATING

16850.10 ELECTRIC HEATING

	UNIT	MAT.	INST.	TOTAL
4000w	EA.	280.00	110.00	390.00
Thermostat				
Integral	EA.	32.00	23.35	55.35
Line voltage	"	32.40	23.35	55.75
Electric heater connection	"	1.35	11.70	13.05
Fittings				
Inside corner	EA.	20.85	18.70	39.55
Outside corner	"	22.30	18.70	41.00
Receptacle section	"	23.65	18.70	42.35
Blank section	"	29.20	18.70	47.90
Infrared heaters				
600w	EA.	130.00	46.80	176.80
2000w	"	140.00	56.00	196.00
3000w	"	210.00	93.50	303.50
4000w	"	290.00	120.00	410.00
Controller	"	55.50	31.20	86.70
Wall bracket	"	110.00	34.00	144.00
Radiant ceiling heater panels				
500w	EA.	140.00	46.80	186.80
750w	"	170.00	46.80	216.80
Unit heaters, suspended, single phase				
3.0 kw	EA.	290.00	130.00	420.00
5.0 kw	"	300.00	130.00	430.00
7.5 kw	"	460.00	150.00	610.00
10.0 kw	"	510.00	180.00	690.00
Three phase				
5 kw	EA.	370.00	130.00	500.00
7.5 kw	"	470.00	150.00	620.00
10 kw	"	510.00	180.00	690.00
15 kw	"	870.00	200.00	1,070
20 kw	"	1,160	250.00	1,410
25 kw	"	1,390	300.00	1,690
30 kw	"	1,620	370.00	1,990
35 kw	"	1,960	370.00	2,330
Unit heater thermostat	"	37.30	24.95	62.25
Mounting bracket	"	38.40	34.00	72.40
Relay	"	48.70	28.80	77.50
Duct heaters, three phase				
10 kw	EA.	440.00	180.00	620.00
15 kw	"	520.00	180.00	700.00
17.5 kw	"	550.00	190.00	740.00
20 kw	"	590.00	290.00	880.00

CONTROLS

16910.40 CONTROL CABLE

CONTROLS	UNIT	MAT.	INST.	TOTAL
Control cable, 600v, #14 THWN, PVC jacket				
2 wire	L.F.	0.19	0.37	0.56
4 wire	"	0.30	0.47	0.77
6 wire	"	0.51	6.15	6.66
8 wire	"	0.58	6.80	7.38
10 wire	"	0.69	7.50	8.19
12 wire	"	1.05	8.50	9.55
14 wire	"	1.15	9.85	11.00
16 wire	"	1.25	10.40	11.65
18 wire	"	1.45	11.35	12.80
20 wire	"	1.60	11.70	13.30
22 wire	"	1.65	13.35	15.00
Audio cables, shielded, #24 gauge				
3 conductor	L.F.	0.15	0.19	0.34
4 conductor	"	0.19	0.28	0.47
5 conductor	"	0.21	0.33	0.54
6 conductor	"	0.24	0.42	0.66
7 conductor	"	0.28	0.51	0.79
8 conductor	"	0.31	0.56	0.87
9 conductor	"	0.32	0.66	0.98
10 conductor	"	0.35	0.70	1.05
15 conductor	"	0.61	0.84	1.45
20 conductor	"	0.81	1.05	1.86
25 conductor	"	1.00	1.25	2.25
30 conductor	"	1.20	1.40	2.60
40 conductor	"	1.55	1.70	3.25
50 conductor	"	2.00	1.95	3.95
#22 gauge				
3 conductor	L.F.	0.13	0.19	0.32
4 conductor	"	0.32	0.28	0.60
#20 gauge				
3 conductor	L.F.	0.17	0.19	0.36
10 conductor	"	0.54	0.70	1.24
15 conductor	"	0.69	0.84	1.53
#18 gauge				
3 conductor	L.F.	0.21	0.19	0.40
4 conductor	"	0.28	0.28	0.56
Microphone cables, #24 gauge				
2 conductor	L.F.	0.21	0.19	0.40
3 conductor	"	0.24	0.23	0.47
#20 gauge				
1 conductor	L.F.	0.23	0.19	0.42
2 conductor	"	0.36	0.19	0.55
2 conductor	"	0.42	0.19	0.61
3 conductor	"	0.51	0.28	0.79
4 conductor	"	0.70	0.33	1.03
5 conductor	"	0.88	0.42	1.30
7 conductor	"	0.96	0.51	1.47
8 conductor	"	1.05	0.56	1.61
Computer cables shielded, #24 gauge				
1 pair	L.F.	0.14	0.19	0.33
2 pair	"	0.20	0.19	0.39

CONTROLS

16910.40 CONTROL CABLE

	UNIT	MAT.	INST.	TOTAL
3 pair	L.F.	0.23	0.28	0.51
4 pair	"	0.28	0.33	0.61
5 pair	"	0.35	0.42	0.77
6 pair	"	0.42	0.52	0.94
7 pair	"	0.44	0.56	1.00
8 pair	"	0.51	0.66	1.17
50 pair	"	2.95	1.80	4.75
Fire alarm cables, #22 gauge				
6 conductor	L.F.	0.21	0.47	0.68
9 conductor	"	0.28	0.70	0.98
12 conductor	"	0.32	0.75	1.07
#18 gauge				
2 conductor	L.F.	0.21	0.23	0.44
4 conductor	"	0.28	0.33	0.61
#16 gauge				
2 conductor	L.F.	0.21	0.33	0.54
4 conductor	"	0.30	0.37	0.67
#14 gauge				
2 conductor	L.F.	0.32	0.37	0.69
#12 gauge				
2 conductor	L.F.	0.39	0.47	0.86
Plastic jacketed thermostat cable				
2 conductor	L.F.	0.19	0.19	0.38
3 conductor	"	0.24	0.23	0.47
4 conductor	"	0.30	0.28	0.58
5 conductor	"	0.32	0.37	0.69
6 conductor	"	0.37	0.42	0.79
7 conductor	"	0.42	0.56	0.98
8 conductor	"	0.47	0.61	1.08

Man-Hour Tables

The man-hour productivities used to develop the labor costs are listed in the following section of this book. These productivities represent typical installation labor for thousands of construction items. The data takes into account all activities involved in normal construction under commonly experienced working conditions. As with the Costbook pages, these items are listed according to the CSI MASTERFORMAT. In order to best use the information in this book, please review this sample page and read the "Features in this Book" section.

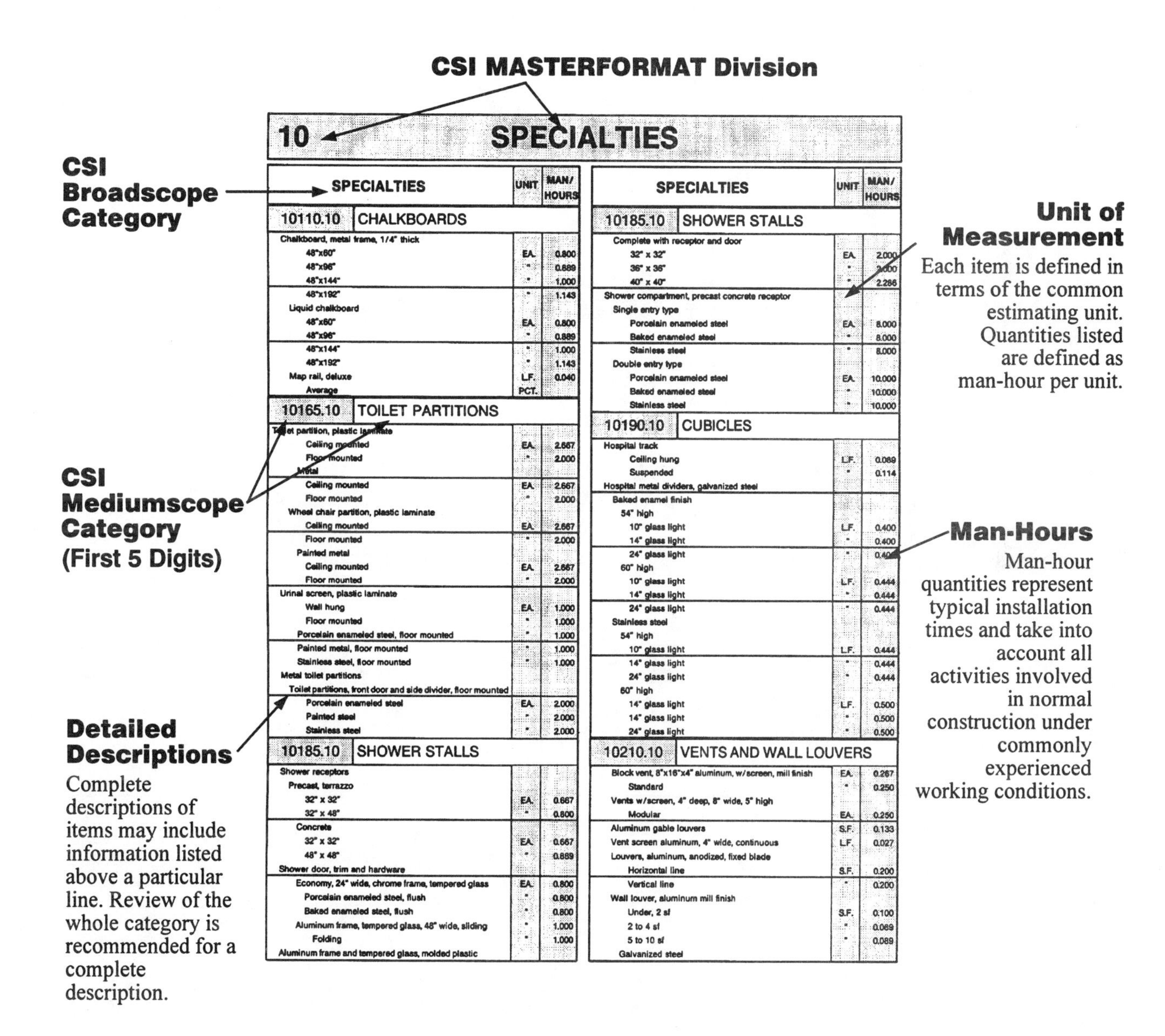

10 SPECIALTIES

SPECIALTIES	UNIT	MAN/ HOURS
10110.10 CHALKBOARDS		
Chalkboard, metal frame, 1/4" thick		
48"x60"	EA.	0.800
48"x96"	"	0.889
48"x144"	"	1.000
48"x192"	"	1.143
Liquid chalkboard		
48"x60"	EA.	0.800
48"x96"	"	0.889
48"x144"	"	1.000
48"x192"	"	1.143
Map rail, deluxe	L.F.	0.040
Average	PCT.	
10165.10 TOILET PARTITIONS		
Toilet partition, plastic laminate		
Ceiling mounted	EA.	2.667
Floor mounted	"	2.000
Metal		
Ceiling mounted	EA.	2.667
Floor mounted	"	2.000
Wheel chair partition, plastic laminate		
Ceiling mounted	EA.	2.667
Floor mounted	"	2.000
Painted metal		
Ceiling mounted	EA.	2.667
Floor mounted	"	2.000
Urinal screen, plastic laminate		
Wall hung	EA.	1.000
Floor mounted	"	1.000
Porcelain enameled steel, floor mounted	"	1.000
Painted metal, floor mounted	"	1.000
Stainless steel, floor mounted	"	1.000
Metal toilet partitions		
Toilet partitions, front door and side divider, floor mounted		
Porcelain enameled steel	EA.	2.000
Painted steel	"	2.000
Stainless steel	"	2.000
10185.10 SHOWER STALLS		
Shower receptors		
Precast, terrazzo		
32" x 32"	EA.	0.667
32" x 48"	"	0.800
Concrete		
32" x 32"	EA.	0.667
48" x 48"	"	0.889
Shower door, trim and hardware		
Economy, 24" wide, chrome frame, tempered glass	EA.	0.800
Porcelain enameled steel, flush	"	0.800
Baked enameled steel, flush	"	0.800
Aluminum frame, tempered glass, 48" wide, sliding	"	1.000
Folding	"	1.000
Aluminum frame and tempered glass, molded plastic		

SPECIALTIES	UNIT	MAN/ HOURS
10185.10 SHOWER STALLS		
Complete with receptor and door		
32" x 32"	EA.	2.000
36" x 36"	"	[illegible]
40" x 40"	"	2.286
Shower compartment, precast concrete receptor		
Single entry type		
Porcelain enameled steel	EA.	8.000
Baked enameled steel	"	8.000
Stainless steel	"	8.000
Double entry type		
Porcelain enameled steel	EA.	10.000
Baked enameled steel	"	10.000
Stainless steel	"	10.000
10190.10 CUBICLES		
Hospital track		
Ceiling hung	L.F.	0.089
Suspended	"	0.114
Hospital metal dividers, galvanized steel		
Baked enamel finish		
54" high		
10" glass light	L.F.	0.400
14" glass light	"	0.400
24" glass light	"	0.40[illegible]
60" high		
10" glass light	L.F.	0.444
14" glass light	"	0.444
24" glass light	"	0.444
Stainless steel		
54" high		
10" glass light	L.F.	0.444
14" glass light	"	0.444
24" glass light	"	0.444
60" high		
14" glass light	L.F.	0.500
14" glass light	"	0.500
24" glass light	"	0.500
10210.10 VENTS AND WALL LOUVERS		
Block vent, 8"x16"x4" aluminum, w/screen, mill finish	EA.	0.267
Standard	"	0.250
Vents w/screen, 4" deep, 8" wide, 5" high		
Modular	EA.	0.250
Aluminum gable louvers	S.F.	0.133
Vent screen aluminum, 4" wide, continuous	L.F.	0.027
Louvers, aluminum, anodized, fixed blade		
Horizontal line	S.F.	0.200
Vertical line	"	0.200
Wall louver, aluminum mill finish		
Under, 2 sf	S.F.	0.100
2 to 4 sf	"	0.069
5 to 10 sf	"	0.089
Galvanized steel		

BNi® Building News

SOIL TESTS

02010.10 SOIL BORING	UNIT	MAN/ HOURS
Borings, uncased, stable earth		
2-1/2" dia.		
Minimum	L.F.	0.200
Average	"	0.300
Maximum	"	0.480
4" dia.		
Minimum	L.F.	0.218
Average	"	0.343
Maximum	"	0.600
Cased, including samples		
2-1/2" dia.		
Minimum	L.F.	0.240
Average	"	0.400
Maximum	"	0.800
4" dia.		
Minumum	L.F.	0.480
Average	"	0.686
Maximum	"	0.960
Drilling in rock		
No sampling		
Minimum	L.F.	0.436
Average	"	0.632
Maximum	"	0.857
With casing and sampling		
Minimum	L.F.	0.600
Average	"	0.800
Maximum	"	1.200
Test pits		
Light soil		
Minimum	EA.	3.000
Average	"	4.000
Maximum	"	8.000
Heavy soil		
Minimum	EA.	4.800
Average	"	6.000
Maximum	"	12.000

DEMOLITION

02060.10 BUILDING DEMOLITION	UNIT	MAN/ HOURS
Building, complete with disposal		
Wood frame	C.F.	0.003
Concrete	"	0.004
Steel frame	"	0.005
Partition removal		
Concrete block partitions		
4" thick	S.F.	0.040
8" thick	S.F.	0.053
12" thick	"	0.073
Brick masonry partitions		
4" thick	S.F.	0.040
8" thick	"	0.050
12" thick	"	0.067
16" thick	"	0.100
Cast in place concrete partitions		
Unreinforced		
6" thick	S.F.	0.160
8" thick	"	0.171
10" thick	"	0.200
12" thick	"	0.240
Reinforced		
6" thick	S.F.	0.185
8" thick	"	0.240
10" thick	"	0.267
12" thick	"	0.320
Terra cotta		
To 6" thick	S.F.	0.040
Stud partitions		
Metal or wood, with drywall both sides	S.F.	0.040
Metal studs, both sides, lath and plaster	"	0.053
Door and frame removal		
Hollow metal in masonry wall		
Single		
2'6"x6'8"	EA.	1.000
3'x7'	"	1.333
Double		
3'x7'	EA.	1.600
4'x8'	"	1.600
Wood in framed wall		
Single		
2'6"x6'8"	EA.	0.571
3'x6'8"	"	0.667
Double		
2'6"x6'8"	EA.	0.800
3'x6'8"	"	0.889
Remove for re-use		
Hollow metal	EA.	2.000
Wood	"	1.333
Floor removal		
Brick flooring	S.F.	0.032
Ceramic or quarry tile	"	0.018
Terrazzo	"	0.036
Heavy wood	"	0.021
Residential wood	"	0.023
Resilient tile or linoleum	"	0.008
Ceiling removal		
Acoustical tile ceiling		
Adhesive fastened	S.F.	0.008
Furred and glued	"	0.007
Suspended grid	"	0.005
Drywall ceiling		

02 SITEWORK

DEMOLITION	UNIT	MAN/ HOURS
02060.10 BUILDING DEMOLITION		
Furred and nailed	S.F.	0.009
Nailed to framing	"	0.008
Plastered ceiling		
Furred on framing	S.F.	0.020
Suspended system	"	0.027
Roofing removal		
Steel frame		
Corrugated metal roofing	S.F.	0.016
Built-up roof on metal deck	"	0.027
Wood frame		
Built up roof on wood deck	S.F.	0.025
Roof shingles	"	0.013
Roof tiles	"	0.027
Concrete frame	C.F.	0.053
Concrete plank	S.F.	0.040
Built-up roof on concrete	"	0.023
Cut-outs		
Concrete, elevated slabs, mesh reinforcing		
Under 5 cf	C.F.	0.800
Over 5 cf	"	0.667
Bar reinforcing		
Under 5 cf	C.F.	1.333
Over 5 cf	"	1.000
Window removal		
Metal windows, trim included		
2'x3'	EA.	0.800
2'x4'	"	0.889
2'x6'	"	1.000
3'x4'	"	1.000
3'x6'	"	1.143
3'x8'	"	1.333
4'x4'	"	1.333
4'x6'	"	1.600
4'x8'	"	2.000
Wood windows, trim included		
2'x3'	EA.	0.444
2'x4'	"	0.471
2'x6'	"	0.500
3'x4'	"	0.533
3'x6'	"	0.571
3'x8'	"	0.615
6'x4'	"	0.667
6'x6'	"	0.727
6'x8'	"	0.800
Walls, concrete, bar reinforcing		
Small jobs	C.F.	0.533
Large jobs	"	0.444
Brick walls, not including toothing		
4" thick	S.F.	0.040
8" thick	"	0.050
12" thick	"	0.067
16" thick	"	0.100
Concrete block walls, not including toothing		
4" thick	S.F.	0.044

DEMOLITION	UNIT	MAN/ HOURS
02060.10 BUILDING DEMOLITION		
6" thick	S.F.	0.047
8" thick	"	0.050
10" thick	"	0.057
12" thick	"	0.067
Rubbish handling		
Load in dumpster or truck		
Minimum	C.F.	0.018
Maximum	"	0.027
For use of elevators, add		
Minimum	C.F.	0.004
Maximum	"	0.008
Rubbish hauling		
Hand loaded on trucks, 2 mile trip	C.Y.	0.320
Machine loaded on trucks, 2 mile trip	"	0.240

HIGHWAY DEMOLITION	UNIT	MAN/ HOURS
02065.10 PAVEMENT DEMOLITION		
Bituminous pavement, up to 3" thick		
On streets		
Minimum	S.Y.	0.069
Average	"	0.096
Maximum	"	0.160
On pipe trench		
Minimum	S.Y.	0.096
Average	"	0.120
Maximum	"	0.240
Concrete pavement, 6" thick		
No reinforcement		
Minimum	S.Y.	0.120
Average	"	0.160
Maximum	"	0.240
With wire mesh		
Minimum	S.Y.	0.185
Average	"	0.240
Maximum	"	0.300
With rebars		
Minimum	S.Y.	0.240
Average	"	0.300
Maximum	"	0.400
9" thick		
No reinforcement		
Minimum	S.Y.	0.160
Average	"	0.200
Maximum	"	0.240
With wire mesh		
Minimum	S.Y.	0.253

HIGHWAY DEMOLITION

HIGHWAY DEMOLITION	UNIT	MAN/HOURS
02065.10 PAVEMENT DEMOLITION		
Average	S.Y.	0.300
Maximum	"	0.369
With rebars		
Minimum	S.Y.	0.320
Average	"	0.400
Maximum	"	0.533
12" thick		
No reinforcement		
Minimum	S.Y.	0.200
Average	"	0.240
Maximum	"	0.300
With wire mesh		
Minimum	S.Y.	0.282
Average	"	0.343
Maximum	"	0.436
With rebars		
Minimum	S.Y.	0.400
Average	"	0.480
Maximum	"	0.600
Sidewalk, 4" thick, with disposal		
Minimum	S.Y.	0.057
Average	"	0.080
Maximum	"	0.114
Removal of pavement markings by waterblasting		
Minimum	S.F.	0.003
Average	"	0.004
Maximum	"	0.008
02065.15 SAW CUTTING PAVEMENT		
Pavement, bituminous		
2" thick	L.F.	0.016
3" thick	"	0.020
4" thick	"	0.025
5" thick	"	0.027
6" thick	"	0.029
Concrete pavement, with wire mesh		
4" thick	L.F.	0.031
5" thick	"	0.033
6" thick	"	0.036
8" thick	"	0.040
10" thick	"	0.044
Plain concrete, unreinforced		
4" thick	L.F.	0.027
5" thick	"	0.031
6" thick	"	0.033
8" thick	"	0.036
10" thick	"	0.040
02065.80 CURB & GUTTER		
Curb removal		
Concrete, unreinforced		
Minimum	L.F.	0.048
Average	"	0.060

HIGHWAY DEMOLITION	UNIT	MAN/HOURS
02065.80 CURB & GUTTER		
Maximum	L.F.	0.075
Reinforced		
Minimum	L.F.	0.077
Average	"	0.086
Maximum	"	0.096
Combination curb and 2'gutter		
Unreinforced		
Minimum	L.F.	0.063
Average	"	0.083
Maximum	"	0.120
Reinforced		
Minimum	L.F.	0.100
Average	"	0.133
Maximum	"	0.240
Granite curb		
Minimum	L.F.	0.069
Average	"	0.080
Maximum	"	0.092
Asphalt curb		
Minimum	L.F.	0.040
Average	"	0.048
Maximum	"	0.057
02065.85 GUARDRAILS		
Remove standard guardrail		
Steel		
Minimum	L.F.	0.060
Average	"	0.080
Maximum	"	0.120
Wood		
Minimum	L.F.	0.052
Average	"	0.062
Maximum	"	0.100
02075.80 CORE DRILLING		
Concrete		
6" thick		
3" dia.	EA.	0.571
4" dia.	"	0.667
6" dia.	"	0.800
8" dia.	"	1.333
8" thick		
3" dia.	EA.	0.800
4" dia.	"	1.000
6" dia.	"	1.143
8" dia.	"	1.600
10" thick		
3" dia.	EA.	1.000
4" dia.	"	1.143
6" dia.	"	1.333
8" dia.	"	2.000
12" thick		
3" dia.	EA.	1.333

02 SITEWORK

HIGHWAY DEMOLITION

02075.80 CORE DRILLING	UNIT	MAN/HOURS
4" dia.	EA.	1.600
6" dia.	"	2.000
8" dia.	"	2.667

HAZARDOUS WASTE

02080.10 ASBESTOS REMOVAL	UNIT	MAN/HOURS
Enclosure using wood studs & poly, install & remove	S.F.	0.020

02080.12 DUCT INSULATION REMOVAL	UNIT	MAN/HOURS
Remove duct insulation, duct size		
6" x 12"	L.F.	0.044
x 18"	"	0.062
x 24"	"	0.089
8" x 12"	"	0.067
x 18"	"	0.073
x 24"	"	0.100
12" x 12"	"	0.067
x 18"	"	0.089
x 24"	"	0.114

02080.15 PIPE INSULATION REMOVAL	UNIT	MAN/HOURS
Removal, asbestos insulation		
2" thick, pipe		
1" to 3" dia.	L.F.	0.067
4" to 6" dia.	"	0.076
3" thick		
7" to 8" dia.	L.F.	0.080
9" to 10" dia.	"	0.084
11" to 12" dia.	"	0.089
13" to 14" dia.	"	0.094
15" to 18" dia.	"	0.100

SITE DEMOLITION

02105.10 CATCH BASINS/MANHOLES	UNIT	MAN/HOURS
Abandon catch basin or manhole (fill with sand)		
Minimum	EA.	3.000
Average	"	4.800
Maximum	"	8.000
Remove and reset frame and cover		
Minimum	EA.	1.600
Average	"	2.400
Maximum	"	4.000
Remove catch basin, to 10' deep		
Masonry		
Minumum	EA.	4.800
Average	"	6.000
Maximum	"	8.000
Concrete		
Minimum	EA.	6.000
Average	"	8.000
Maximum	"	9.600

02105.20 FENCES	UNIT	MAN/HOURS
Remove fencing		
Chain link, 8' high		
For disposal	L.F.	0.040
For reuse	"	0.100
Wood		
4' high	S.F.	0.027
6' high	"	0.032
8' high	"	0.040
Masonry		
8" thick		
4' high	S.F.	0.080
6' high	"	0.100
8' high	"	0.114
12" thick		
4' high	S.F.	0.133
6' high	"	0.160
8' high	"	0.200
12' high	"	0.267

02105.30 HYDRANTS	UNIT	MAN/HOURS
Remove fire hydrant		
Minimum	EA.	3.000
Average	"	4.000
Maximum	"	6.000
Remove and reset fire hydrant		
Minimum	EA.	8.000
Average	"	12.000
Maximum	"	24.000

02105.42 DRAINAGE PIPING	UNIT	MAN/HOURS
Remove drainage pipe, not including excavation		
12" dia.		
Minimum	L.F.	0.080
Average	"	0.100

SITE DEMOLITION	UNIT	MAN/ HOURS
02105.42 DRAINAGE PIPING		
Maximum	L.F.	0.126
18" dia.		
Minimum	L.F.	0.109
Average	"	0.126
Maximum	"	0.160
24" dia.		
Minimum	L.F.	0.133
Average	"	0.160
Maximum	"	0.200
36" dia.		
Minimum	L.F.	0.160
Average	"	0.200
Maximum	"	0.253
02105.43 GAS PIPING		
Remove welded steel pipe, not including excavation		
4" dia.		
Minimum	L.F.	0.120
Average	"	0.150
Maximum	"	0.200
5" dia.		
Minimum	L.F.	0.200
Average	"	0.240
Maximum	"	0.300
6" dia.		
Minimum	L.F.	0.253
Average	"	0.300
Maximum	"	0.400
8" dia.		
Minimum	L.F.	0.369
Average	"	0.480
Maximum	"	0.632
10" dia.		
Minimum	L.F.	0.480
Average	"	0.600
Maximum	"	0.800
02105.45 SANITARY PIPING		
Remove sewer pipe, not including excavation		
4" dia.		
Minimum	L.F.	0.067
Average	"	0.096
Maximum	"	0.160
6" dia.		
Minimum	L.F.	0.075
Average	"	0.109
Maximum	"	0.200
8" dia.		
Minimum	L.F.	0.080
Average	"	0.120
Maximum	"	0.240
10" dia.		

SITE DEMOLITION	UNIT	MAN/ HOURS
02105.45 SANITARY PIPING		
Minimum	L.F.	0.086
Average	"	0.126
Maximum	"	0.267
12" dia.		
Minimum	L.F.	0.092
Average	"	0.133
Maximum	"	0.300
15" dia.		
Minimum	L.F.	0.100
Average	"	0.141
Maximum	"	0.343
18" dia.		
Minimum	L.F.	0.109
Average	"	0.160
Maximum	"	0.400
24" dia.		
Minimum	L.F.	0.120
Average	"	0.200
Maximum	"	0.480
30" dia.		
Minimum	L.F.	0.133
Average	"	0.240
Maximum	"	0.600
36" dia.		
Minimum	L.F.	0.160
Average	"	0.300
Maximum	"	0.800
02105.48 WATER PIPING		
Remove water pipe, not including excavation		
4" dia.		
Minimum	L.F.	0.096
Average	"	0.109
Maximum	"	0.126
6" dia.		
Minimum	L.F.	0.100
Average	"	0.114
Maximum	"	0.133
8" dia.		
Minimum	L.F.	0.109
Average	"	0.126
Maximum	"	0.150
10" dia.		
Minimum	L.F.	0.114
Average	"	0.133
Maximum	"	0.160
12" dia.		
Minimum	L.F.	0.120
Average	"	0.141
Maximum	"	0.171
14" dia.		
Minimum	L.F.	0.126
Average	"	0.150

02 SITEWORK

SITE DEMOLITION	UNIT	MAN/HOURS
02105.48 WATER PIPING		
Maximum	L.F.	0.185
16" dia.		
Minimum	L.F.	0.133
Average	"	0.160
Maximum	"	0.200
18" dia.		
Minimum	L.F.	0.141
Average	"	0.171
Maximum	"	0.218
20" dia.		
Minimum	L.F.	0.150
Average	"	0.185
Maximum	"	0.240
Remove valves		
6"	EA.	1.200
10"	"	1.333
14"	"	1.500
18"	"	2.000
02105.60 UNDERGROUND TANKS		
Remove underground storage tank, and backfill		
50 to 250 gals	EA.	8.000
600 gals	"	8.000
1000 gals	"	12.000
4000 gals	"	19.200
5000 gals	"	19.200
10,000 gals	"	32.000
12,000 gals	"	40.000
15,000 gals	"	48.000
20,000 gals	"	60.000
02105.66 SEPTIC TANKS		
Remove septic tank		
1000 gals	EA.	2.000
2000 gals	"	2.400
5000 gals	"	3.000
15,000 gals	"	24.000
25,000 gals	"	32.000
40,000 gals	"	48.000
02105.80 WALLS, EXTERIOR		
Concrete wall		
Light reinforcing		
6" thick	S.F.	0.120
8" thick	"	0.126
10" thick	"	0.133
12" thick	"	0.150
Medium reinforcing		
6" thick	S.F.	0.126
8" thick	"	0.133
10" thick	"	0.150
12" thick	"	0.171
Heavy reinforcing		

SITE DEMOLITION	UNIT	MAN/HOURS
02105.80 WALLS, EXTERIOR		
6" thick	S.F.	0.141
8" thick	"	0.150
10" thick	"	0.171
12" thick	"	0.200
Masonry		
No reinforcing		
8" thick	S.F.	0.053
12" thick	"	0.060
16" thick	"	0.069
Horizontal reinforcing		
8" thick	S.F.	0.060
12" thick	"	0.065
16" thick	"	0.077
Vertical reinforcing		
8" thick	S.F.	0.077
12" thick	"	0.089
16" thick	"	0.109
Remove concrete headwall		
15" pipe	EA.	1.714
18" pipe	"	2.000
24" pipe	"	2.182
30" pipe	"	2.400
36" pipe	"	2.667
48" pipe	"	3.429
60" pipe	"	4.800
02110.10 CLEARING AND GRUBBING		
Clear wooded area		
Light density	ACRE	60.000
Medium density	"	80.000
Heavy density	"	96.000
02110.50 TREE CUTTING & CLEARING		
Cut trees and clear out stumps		
9" to 12" dia.	EA.	4.800
To 24" dia.	"	6.000
24" dia. and up	"	8.000
Loading and trucking		
For machine load, per load, round trip		
1 mile	EA.	0.960
3 mile	"	1.091
5 mile	"	1.200
10 mile	"	1.600
20 mile	"	2.400
Hand loaded, round trip		
1 mile	EA.	2.000
3 mile	"	2.286
5 mile	"	2.667
10 mile	"	3.200
20 mile	"	4.000
Tree trimming for pole line construction		
Light cutting	L.F.	0.012
Medium cutting	"	0.016

02 SITEWORK

SITE DEMOLITION

02110.50 TREE CUTTING & CLEARING

Item	UNIT	MAN/HOURS
Heavy cutting	L.F.	0.024

DEWATERING

02144.10 WELLPOINT SYSTEMS

Item	UNIT	MAN/HOURS
Pumping, gas driven, 50' hose		
3" header pipe	DAY	8.000
6" header pipe	"	10.000
Wellpoint system per job		
6" header pipe	L.F.	0.032
8" header pipe	"	0.040
10" header pipe	"	0.053
Jetting wellpoint system		
14' long	EA.	0.533
18' long	"	0.667
Sand filter for wellpoints	L.F.	0.013
Replacement of wellpoint components	EA.	0.160

SHORING AND UNDERPINNING

02162.10 TRENCH SHEETING

Item	UNIT	MAN/HOURS
Closed timber, including pull and salvage, excavation		
8' deep	S.F.	0.064
10' deep	"	0.067
12' deep	"	0.071
14' deep	"	0.075
16' deep	"	0.080
18' deep	"	0.091
20' deep	"	0.098

02170.10 COFFERDAMS

Item	UNIT	MAN/HOURS
Cofferdam, steel, driven from shore		
15' deep	S.F.	0.137
20' deep	"	0.128
25' deep	"	0.120
30' deep	"	0.113
40' deep	"	0.107

SHORING AND UNDERPINNING

02170.10 COFFERDAMS

Item	UNIT	MAN/HOURS
Driven from barge		
20' deep	S.F.	0.148
30' deep	"	0.137
40' deep	"	0.128
50' deep	"	0.120

EARTHWORK

02210.10 HAULING MATERIAL

Item	UNIT	MAN/HOURS
Haul material by 10 cy dump truck, round trip distance		
1 mile	C.Y.	0.044
2 mile	"	0.053
5 mile	"	0.073
10 mile	"	0.080
20 mile	"	0.089
30 mile	"	0.107
Site grading, cut & fill, sandy clay, 200' haul, 75 hp dozer	"	0.032
Spread topsoil by equipment on site	"	0.036
Site grading (cut and fill to 6") less than 1 acre		
75 hp dozer	C.Y.	0.053
1.5 cy backhoe/loader	"	0.080

02210.30 BULK EXCAVATION

Item	UNIT	MAN/HOURS
Excavation, by small dozer		
Large areas	C.Y.	0.016
Small areas	"	0.027
Trim banks	"	0.040
Drag line		
1-1/2 cy bucket		
Sand or gravel	C.Y.	0.040
Light clay	"	0.053
Heavy clay	"	0.060
Unclassified	"	0.064
2 cy bucket		
Sand or gravel	C.Y.	0.037
Light clay	"	0.048
Heavy clay	"	0.053
Unclassified	"	0.056
2-1/2 cy bucket		
Sand or gravel	C.Y.	0.034
Light clay	"	0.044
Heavy clay	"	0.048
Unclassified	"	0.051
3 cy bucket		
Sand or gravel	C.Y.	0.030
Light clay	"	0.040

EARTHWORK

02210.30 BULK EXCAVATION	UNIT	MAN/ HOURS
Heavy clay	C.Y.	0.044
Unclassified	"	0.046
Hydraulic excavator		
1 cy capacity		
Light material	C.Y.	0.040
Medium material	"	0.048
Wet material	"	0.060
Blasted rock	"	0.069
1-1/2 cy capacity		
Light material	C.Y.	0.010
Medium material	"	0.013
Wet material	"	0.016
Blasted rock	"	0.020
2 cy capacity		
Light material	C.Y.	0.009
Medium material	"	0.011
Wet material	"	0.013
Blasted rock	"	0.016
Wheel mounted front-end loader		
7/8 cy capacity		
Light material	C.Y.	0.020
Medium material	"	0.023
Wet material	"	0.027
Blasted rock	"	0.032
1-1/2 cy capacity		
Light material	C.Y.	0.011
Medium material	"	0.012
Wet material	"	0.013
Blasted rock	"	0.015
2-1/2 cy capacity		
Light material	C.Y.	0.009
Medium material	"	0.010
Wet material	"	0.011
Blasted rock	"	0.011
3-1/2 cy capacity		
Light material	C.Y.	0.009
Medium material	"	0.009
Wet material	"	0.010
Blasted rock	"	0.011
6 cy capacity		
Light material	C.Y.	0.005
Medium material	"	0.006
Wet material	"	0.006
Blasted rock	"	0.007
Track mounted front-end loader		
1-1/2 cy capacity		
Light material	C.Y.	0.013
Medium material	"	0.015
Wet material	"	0.016
Blasted rock	"	0.018
2-3/4 cy capacity		
Light material	C.Y.	0.008
Medium material	"	0.009
Wet material	"	0.010

EARTHWORK

02210.30 BULK EXCAVATION	UNIT	MAN/ HOURS
Blasted rock	C.Y.	0.011

02220.10 BORROW	UNIT	MAN/ HOURS
Borrow fill, F.O.B. at pit		
Sand, haul to site, round trip		
10 mile	C.Y.	0.080
20 mile	"	0.133
30 mile	"	0.200
Place borrow fill and compact		
Less than 1 in 4 slope	C.Y.	0.040
Greater than 1 in 4 slope	"	0.053

02220.40 BUILDING EXCAVATION	UNIT	MAN/ HOURS
Structural excavation, unclassified earth		
3/8 cy backhoe	C.Y.	0.107
3/4 cy backhoe	"	0.080
1 cy backhoe	"	0.067
Foundation backfill and compaction by machine	"	0.160

02220.50 UTILITY EXCAVATION	UNIT	MAN/ HOURS
Trencher, sandy clay, 8" wide trench		
18" deep	L.F.	0.018
24" deep	"	0.020
36" deep	"	0.023
Trench backfill, 95% compaction		
Tamp by hand	C.Y.	0.500
Vibratory compaction	"	0.400
Trench backfilling, with borrow sand, place & compact	"	0.400

02220.60 TRENCHING	UNIT	MAN/ HOURS
Trenching and continuous footing excavation		
By gradall		
1 cy capacity		
Light soil	C.Y.	0.023
Medium soil	"	0.025
Heavy/wet soil	"	0.027
Loose rock	"	0.029
Blasted rock	"	0.031
By hydraulic excavator		
1/2 cy capacity		
Light soil	C.Y.	0.027
Medium soil	"	0.029
Heavy/wet soil	"	0.032
Loose rock	"	0.036
Blasted rock	"	0.040
1 cy capacity		
Light soil	C.Y.	0.019
Medium soil	"	0.020
Heavy/wet soil	"	0.021
Loose rock	"	0.023
Blasted rock	"	0.025

02 SITEWORK

EARTHWORK	UNIT	MAN/ HOURS
02220.60 TRENCHING		
1-1/2 cy capacity		
Light soil	C.Y.	0.017
Medium soil	"	0.018
Heavy/wet soil	"	0.019
Loose rock	"	0.020
Blasted rock	"	0.021
2 cy capacity		
Light soil	C.Y.	0.016
Medium soil	"	0.017
Heavy/wet soil	"	0.018
Loose rock	"	0.019
Blasted rock	"	0.020
2-1/2 cy capacity		
Light soil	C.Y.	0.015
Medium soil	"	0.015
Heavy/wet soil	"	0.016
Loose rock	"	0.017
Blasted rock	"	0.018
Trencher, chain, 1' wide to 4' deep		
Light soil	C.Y.	0.020
Medium soil	"	0.023
Heavy soil	"	0.027
Hand excavation		
Bulk, wheeled 100'		
Normal soil	C.Y.	0.889
Sand or gravel	"	0.800
Medium clay	"	1.143
Heavy clay	"	1.600
Loose rock	"	2.000
Trenches, up to 2' deep		
Normal soil	C.Y.	1.000
Sand or gravel	"	0.889
Medium clay	"	1.333
Heavy clay	"	2.000
Loose rock	"	2.667
Trenches, to 6' deep		
Normal soil	C.Y.	1.143
Sand or gravel	"	1.000
Medium clay	"	1.600
Heavy clay	"	2.667
Loose rock	"	4.000
Backfill trenches		
With compaction		
By hand	C.Y.	0.667
By 60 hp tracked dozer	"	0.020
By 200 hp tracked dozer	"	0.009
By small front-end loader	"	0.023
Spread dumped fill or gravel, no compaction		
6" layers	S.Y.	0.013
12" layers	"	0.016
Compaction in 6" layers		
By hand with air tamper	S.Y.	0.016
Backfill trenches, sand bedding, no compaction		
By hand	C.Y.	0.667

EARTHWORK	UNIT	MAN/ HOURS
02220.60 TRENCHING		
By small front-end loader	C.Y.	0.023
02220.70 ROADWAY EXCAVATION		
Roadway excavation		
1/4 mile haul	C.Y.	0.016
2 mile haul	"	0.027
5 mile haul	"	0.040
Excavation of open ditches	"	0.011
Trim banks, swales or ditches	S.Y.	0.013
Bulk swale excavation by dragline		
Small jobs	C.Y.	0.060
Large jobs	"	0.034
Spread base course	"	0.020
Roll and compact	"	0.027
02220.71 BASE COURSE		
Base course, crushed stone		
3" thick	S.Y.	0.004
4" thick	"	0.004
6" thick	"	0.005
8" thick	"	0.005
10" thick	"	0.006
12" thick	"	0.007
Base course, bank run gravel		
4" deep	S.Y.	0.004
6" deep	"	0.005
8" deep	"	0.005
10" deep	"	0.005
12" deep	"	0.006
Prepare and roll sub base		
Minimum	S.Y.	0.004
Average	"	0.005
Maximum	"	0.007
02220.90 HAND EXCAVATION		
Excavation		
To 2' deep		
Normal soil	C.Y.	0.889
Sand and gravel	"	0.800
Medium clay	"	1.000
Heavy clay	"	1.143
Loose rock	"	1.333
To 6' deep		
Normal soil	C.Y.	1.143
Sand and gravel	"	1.000
Medium clay	"	1.333
Heavy clay	"	1.600
Loose rock	"	2.000
Backfilling foundation without compaction, 6" lifts	"	0.500
Compaction of backfill around structures or in trench		

02 SITEWORK

EARTHWORK

EARTHWORK	UNIT	MAN/HOURS
02220.90 HAND EXCAVATION		
By hand with air tamper	C.Y.	0.571
By hand with vibrating plate tamper	"	0.533
1 ton roller	"	0.400
Miscellaneous hand labor		
Trim slopes, sides of excavation	S.F.	0.001
Trim bottom of excavation	"	0.002
Excavation around obstructions and services	C.Y.	2.667
02240.05 SOIL STABILIZATION		
Straw bale secured with rebar	L.F.	0.027
Filter barrier, 18" high filter fabric	"	0.080
Sediment fence, 36" fabric with 6" mesh	"	0.100
Soil stabilization with tar paper, burlap, straw and stakes	S.F.	0.001
02240.30 GEOTEXTILE		
Filter cloth, light reinforcement		
Woven		
12'-6" wide x 50' long	S.F.	0.001
Various lengths	"	0.001
Non-woven		
14'-8" wide x 430' long	S.F.	0.001
Various lengths	"	0.001
02270.10 SLOPE PROTECTION		
Gabions, stone filled		
6" deep	S.Y.	0.200
9" deep	"	0.229
12" deep	"	0.267
18" deep	"	0.320
36" deep	"	0.533
02270.40 RIPRAP		
Riprap		
Crushed stone blanket, max size 2-1/2"	TON	0.533
Stone, quarry run, 300 lb. stones	"	0.492
400 lb. stones	"	0.457
500 lb. stones	"	0.427
750 lb. stones	"	0.400
Dry concrete riprap in bags 3" thick, 80 lb. per bag	BAG	0.027
02280.20 SOIL TREATMENT		
Soil treatment, termite control pretreatment		
Under slabs	S.F.	0.004
By walls	"	0.005
02290.30 WEED CONTROL		
Weed control, bromicil, 15 lb./acre, wettable powder	ACRE	4.000
Vegetation control, by application of plant killer	S.Y.	0.003
Weed killer, lawns and fields	"	0.002

TUNNELING

TUNNELING	UNIT	MAN/HOURS
02300.10 PIPE JACKING		
Pipe casing, horizontal jacking		
18" dia.	L.F.	0.711
21" dia.	"	0.762
24" dia.	"	0.800
27" dia.	"	0.800
30" dia.	"	0.842
36" dia.	"	0.914
42" dia.	"	1.000
48" dia.	"	1.067

PILES AND CAISSONS

PILES AND CAISSONS	UNIT	MAN/HOURS
02360.50 PRESTRESSED PILING		
Prestressed concrete piling, less than 60' long		
10" sq.	L.F.	0.040
12" sq.	"	0.042
14" sq.	"	0.043
16" sq.	"	0.044
18" sq.	"	0.047
20" sq.	"	0.048
24" sq.	"	0.049
More than 60' long		
12" sq.	L.F.	0.034
14" sq.	"	0.035
16" sq.	"	0.036
18" sq.	"	0.036
20" sq.	"	0.037
24" sq.	"	0.038
Straight cylinder, less than 60' long		
12" dia.	L.F.	0.044
14" dia.	"	0.045
16" dia.	"	0.046
18" dia.	"	0.047
20" dia.	"	0.048
24" dia.	"	0.049
More than 60' long		
12" dia.	L.F.	0.035
14" dia.	"	0.036
16" dia.	"	0.036
18" dia.	"	0.037
20" dia.	"	0.038
24" dia.	"	0.038
Concrete sheet piling		
12" thick x 20' long	S.F.	0.096
25' long	"	0.087
30' long	"	0.080

PILES AND CAISSONS	UNIT	MAN/ HOURS
02360.50 PRESTRESSED PILING		
35' long	S.F.	0.074
40' long	"	0.069
16" thick x 40' long	"	0.053
45' long	"	0.051
50' long	"	0.048
55' long	"	0.046
60' long	"	0.044
02360.60 STEEL PILES		
H-section piles		
8x8		
36 lb/ft		
30' long	L.F.	0.080
40' long	"	0.064
50' long	"	0.053
10x10		
42 lb/ft		
30' long	L.F.	0.080
40' long	"	0.064
50' long	"	0.053
57 lb/ft		
30' long	L.F.	0.080
40' long	"	0.064
50' long	"	0.053
12x12		
53 lb/ft		
30' long	L.F.	0.087
40' long	"	0.069
50' long	"	0.053
74 lb/ft		
30' long	L.F.	0.087
40' long	"	0.069
50' long	"	0.053
14x14		
73 lb/ft		
40' long	L.F.	0.087
50' long	"	0.069
60' long	"	0.053
89 lb/ft		
40' long	L.F.	0.087
50' long	"	0.069
60' long	"	0.053
102 lb/ft		
40' long	L.F.	0.087
50' long	"	0.069
60' long	"	0.053
117 lb/ft		
40' long	L.F.	0.091
50' long	"	0.071
60' long	"	0.055
Splice		
8"	EA.	1.333
10"	"	1.600

PILES AND CAISSONS	UNIT	MAN/ HOURS
02360.60 STEEL PILES		
12"	EA.	1.600
14"	"	2.000
Driving cap		
8"	EA.	0.800
10"	"	1.000
12"	"	1.000
14"	"	1.143
Standard point		
8"	EA.	0.800
10"	"	1.000
12"	"	1.143
14"	"	1.333
Heavy duty point		
8"	EA.	0.889
10"	"	1.143
12"	"	1.333
14"	"	1.600
Tapered friction piles, with fluted steel casing, up to 50'		
With 4000 psi concrete no reinforcing		
12" dia.	L.F.	0.048
14" dia.	"	0.049
16" dia.	"	0.051
18" dia.	"	0.056
02360.65 STEEL PIPE PILES		
Concrete filled, 3000# concrete, up to 40'		
8" dia.	L.F.	0.069
10" dia.	"	0.071
12" dia.	"	0.074
14" dia.	"	0.077
16" dia.	"	0.080
18" dia.	"	0.083
Pipe piles, non-filled		
8" dia.	L.F.	0.053
10" dia.	"	0.055
12" dia.	"	0.056
14" dia.	"	0.060
16" dia.	"	0.062
18" dia.	"	0.064
Splice		
8" dia.	EA.	1.600
10" dia.	"	1.600
12" dia.	"	2.000
14" dia.	"	2.000
16" dia.	"	2.667
18" dia.	"	2.667
Standard point		
8" dia.	EA.	1.600
10" dia.	"	1.600
12" dia.	"	2.000
14" dia.	"	2.000
16" dia.	"	2.667
18" dia.	"	2.667

PILES AND CAISSONS

02360.65 STEEL PIPE PILES

PILES AND CAISSONS	UNIT	MAN/HOURS
Heavy duty point		
8" dia.	EA.	2.000
10" dia.	"	2.000
12" dia.	"	2.667
14" dia.	"	2.667
16" dia.	"	3.200
18" dia.	"	3.200

02360.70 STEEL SHEET PILING

PILES AND CAISSONS	UNIT	MAN/HOURS
Steel sheet piling,12" wide		
20' long	S.F.	0.096
35' long	"	0.069
50' long	"	0.048
Over 50' long	"	0.044

02360.80 WOOD AND TIMBER PILES

PILES AND CAISSONS	UNIT	MAN/HOURS
Treated wood piles, 12" butt, 8" tip		
25' long	L.F.	0.096
30' long	"	0.080
35' long	"	0.069
40' long	"	0.060
12" butt, 7" tip		
40' long	L.F.	0.060
45' long	"	0.053
50' long	"	0.048
55' long	"	0.044
60' long	"	0.040

02380.10 CAISSONS

PILES AND CAISSONS	UNIT	MAN/HOURS
Caisson, including 3000# concrete, in stable ground		
18" dia.	L.F.	0.192
24" dia.	"	0.200
30" dia.	"	0.240
36" dia.	"	0.274
48" dia.	"	0.320
60" dia.	"	0.436
72" dia.	"	0.533
84" dia.	"	0.686
Wet ground, casing required but pulled		
18" dia.	L.F.	0.240
24" dia.	"	0.267
30" dia.	"	0.300
36" dia.	"	0.320
48" dia.	"	0.400
60" dia.	"	0.533
72" dia.	"	0.800
84" dia.	"	1.200
Soft rock		
18" dia.	L.F.	0.686
24" dia.	"	1.200
30" dia.	"	1.600
36" dia.	L.F.	2.400
48" dia.	"	3.200
60" dia.	"	4.800
72" dia.	"	5.333
84" dia.	"	6.000

RAILROAD WORK

02450.10 RAILROAD WORK

RAILROAD WORK	UNIT	MAN/HOURS
Rail		
90 lb	L.F.	0.010
100 lb	"	0.010
115 lb	"	0.010
132 lb	"	0.010
Rail relay		
90 lb	L.F.	0.010
100 lb	"	0.010
115 lb	"	0.010
132 lb	"	0.010
New angle bars, per pair		
90 lb	EA.	0.012
100 lb	"	0.012
115 lb	"	0.012
132 lb	"	0.012
Angle bar relay		
90 lb	EA.	0.012
100 lb	"	0.012
115 lb	"	0.012
132 lb	"	0.012
New tie plates		
90 lb	EA.	0.009
100 lb	"	0.009
115 lb	"	0.009
132 lb	"	0.009
Tie plate relay		
90 lb	EA.	0.009
100 lb	"	0.009
115 lb	"	0.009
132 lb	"	0.009
Track accessories		
Wooden cross ties, 8'	EA.	0.060
Concrete cross ties, 8'	"	0.120
Tie plugs, 5"	"	0.006
Track bolts and nuts, 1"	"	0.006
Lockwashers, 1"	"	0.004
Track spikes, 6"	"	0.024

02 SITEWORK

RAILROAD WORK

02450.10 RAILROAD WORK

RAILROAD WORK	UNIT	MAN/HOURS
Wooden switch ties	B.F.	0.006
Rail anchors	EA.	0.022
Ballast	TON	0.120
Gauge rods	EA.	0.096
Compromise splice bars	"	0.160
Turnout		
90 lb	EA.	24.000
100 lb	"	24.000
110 lb	"	24.000
115 lb	"	24.000
132 lb	"	24.000
Turnout relay		
90 lb	EA.	24.000
100 lb	"	24.000
110 lb	"	24.000
115 lb	"	24.000
132 lb	"	24.000
Railroad track in place, complete		
New rail		
90 lb	L.F.	0.240
100 lb	"	0.240
110 lb	"	0.240
115 lb	"	0.240
132 lb	"	0.240
Rail relay		
90 lb	L.F.	0.240
100 lb	"	0.240
110 lb	"	0.240
115 lb	"	0.240
132 lb	"	0.240
No. 8 turnout		
90 lb	EA.	32.000
100 lb	"	32.000
110 lb	"	32.000
115 lb	"	32.000
132 lb	"	32.000
No. 8 turnout relay		
90 lb	EA.	32.000
100 lb	"	32.000
110 lb	"	32.000
115 lb	"	32.000
132 lb	"	32.000
Railroad crossings, asphalt, based on 8" thick x 20'		
Including track and approach		
12' roadway	EA.	6.000
15' roadway	"	6.857
18' roadway	"	8.000
21' roadway	"	9.600
24' roadway	"	12.000
Precast concrete inserts		
12' roadway	EA.	2.400
15' roadway	"	3.000
18' roadway	"	4.000
21' roadway	"	4.800
24' roadway	EA.	5.333
Molded rubber, with headers		
12' roadway	EA.	2.400
15' roadway	"	3.000
18' roadway	"	4.000
21' roadway	"	4.800
24' roadway	"	5.333

PAVING AND SURFACING

02510.20 ASPHALT SURFACES

PAVING AND SURFACING	UNIT	MAN/HOURS
Asphalt wearing surface, for flexible pavement		
1" thick	S.Y.	0.016
1-1/2" thick	"	0.019
2" thick	"	0.024
3" thick	"	0.032
Binder course		
1-1/2" thick	S.Y.	0.018
2" thick	"	0.022
3" thick	"	0.029
4" thick	"	0.032
5" thick	"	0.036
6" thick	"	0.040
Bituminous sidewalk, no base		
2" thick	S.Y.	0.028
3" thick	"	0.030

02520.10 CONCRETE PAVING

PAVING AND SURFACING	UNIT	MAN/HOURS
Concrete paving, reinforced, 5000 psi concrete		
6" thick	S.Y.	0.150
7" thick	"	0.160
8" thick	"	0.171
9" thick	"	0.185
10" thick	"	0.200
11" thick	"	0.218
12" thick	"	0.240
15" thick	"	0.300
Concrete paving, for pipe trench, reinforced		
7" thick	S.Y.	0.240
8" thick	"	0.267
9" thick	"	0.300
10" thick	"	0.343
Fibrous concrete		
5" thick	S.Y.	0.185
8" thick	"	0.200

02 SITEWORK

PAVING AND SURFACING

02520.10 CONCRETE PAVING

	UNIT	MAN/HOURS
Roller compacted concrete, (RCC), place and compact		
8" thick	S.Y.	0.240
12" thick	"	0.300
Steel edge forms up to		
12" deep	L.F.	0.027
15" deep	"	0.032
Paving finishes		
Belt dragged	S.Y.	0.040
Curing	"	0.008

02545.10 ASPHALT REPAIR

	UNIT	MAN/HOURS
Coal tar emulsion seal coat, rubber additive, fuel resistant	S.Y.	0.011
Bituminous surface treatment, single	"	0.008
Double	"	0.001
Bituminous prime coat	"	0.001
Tack coat	"	0.001
Crack sealing, concrete paving	L.F.	0.005
Bituminous paving for pipe trench, 4" thick	S.Y.	0.160
Polypropylene, nonwoven paving fabric	"	0.004
Rubberized asphalt	"	0.073
Asphalt slurry seal	"	0.047

02580.10 PAVEMENT MARKINGS

	UNIT	MAN/HOURS
Pavement line marking, paint		
4" wide	L.F.	0.002
6" wide	"	0.004
8" wide	"	0.007
Reflective paint, 4" wide	"	0.007
Airfield markings, retro-reflective		
White	L.F.	0.007
Yellow	"	0.007
Preformed tape, 4" wide		
Inlaid reflective	L.F.	0.001
Reflective paint	"	0.002
Thermoplastic		
White	L.F.	0.004
Yellow	"	0.004
12" wide, thermoplastic, white	"	0.011
Directional arrows, reflective preformed tape	EA.	0.800
Messages, reflective preformed tape (per letter)	"	0.400
Handicap symbol, preformed tape	"	0.800
Parking stall painting	"	0.160

UTILITIES

02605.30 MANHOLES

	UNIT	MAN/HOURS
Precast sections, 48" dia.		
Base section	EA.	2.000
1'0" riser	"	1.600
1'4" riser	"	1.714
2'8" riser	"	1.846
4'0" riser	"	2.000
2'8" cone top	"	2.400
Precast manholes, 48" dia.		
4' deep	EA.	4.800
6' deep	"	6.000
7' deep	"	6.857
8' deep	"	8.000
10' deep	"	9.600
Cast-in-place, 48" dia., with frame and cover		
5' deep	EA.	12.000
6' deep	"	13.714
8' deep	"	16.000
10' deep	"	19.200
Brick manholes, 48" dia. with cover, 8" thick		
4' deep	EA.	8.000
6' deep	"	8.889
8' deep	"	10.000
10' deep	"	11.429
12' deep	"	13.333
14' deep	"	16.000
Inverts for manholes		
Single channel	EA.	3.200
Triple channel	"	4.000
Frames and covers, 24" diameter		
300 lb	EA.	0.800
400 lb	"	0.889
500 lb	"	1.143
Watertight, 350 lb	"	2.667
For heavy equipment, 1200 lb	"	4.000
Steps for manholes		
7" x 9"	EA.	0.160
8" x 9"	"	0.178
Curb inlet, 4' throat, cast in place		
12"-30" pipe	EA.	12.000
36"-48" pipe	"	13.714
Raise exist frame and cover, when repaving	"	4.800

02610.10 CAST IRON FLANGED PIPE

	UNIT	MAN/HOURS
Cast iron flanged sections		
4" pipe, with one bolt set		
3' section	EA.	0.218
4' section	"	0.240
5' section	"	0.267
6' section	"	0.300
8' section	"	0.343
10' section	"	0.480
12' section	"	0.800
15' section	"	1.200

UTILITIES	UNIT	MAN/ HOURS
02610.10 CAST IRON FLANGED PIPE		
18' section	EA.	1.600
6" pipe, with one bolt set		
3' section	EA.	0.240
4' section	"	0.282
5' section	"	0.320
6' section	"	0.369
8' section	"	0.533
10' section	"	0.600
12' section	"	0.800
15' section	"	1.200
18' section	"	1.714
8" pipe, with one bolt set		
3' section	EA.	0.300
4' section	"	0.343
5' section	"	0.400
6' section	"	0.480
8' section	"	0.686
10' section	"	0.800
12' section	"	1.200
15' section	"	1.600
18' section	"	2.000
10" pipe, with one bolt set		
3' section	EA.	0.308
4' section	"	0.353
5' section	"	0.414
6' section	"	0.500
8' section	"	0.727
10' section	"	0.857
12' section	"	1.333
15' section	"	1.714
18' section	"	2.400
12" pipe, with one bolt set		
3' section	EA.	0.333
4' section	"	0.387
5' section	"	0.462
6' section	"	0.545
8' section	"	0.800
10' section	"	0.923
12' section	"	1.500
15' section	"	2.000
18' section	"	2.667
02610.11 CAST IRON FITTINGS		
Mechanical joint, with 2 bolt kits		
90 deg bend		
4"	EA.	0.533
6"	"	0.615
8"	"	0.800
10"	"	1.143
12"	"	1.600
14"	"	2.000
16"	"	2.667
45 deg bend		

UTILITIES	UNIT	MAN/ HOURS
02610.11 CAST IRON FITTINGS		
4"	EA.	0.533
6"	"	0.615
8"	"	0.800
10"	"	1.143
12"	"	1.600
14"	"	2.000
16"	"	2.667
Tee, with 3 bolt kits		
4" x 4"	EA.	0.800
6" x 6"	"	1.000
8" x 8"	"	1.333
10" x 10"	"	2.000
12" x 12"	"	2.667
Wye, with 3 bolt kits		
6" x 6"	EA.	1.000
8" x 8"	"	1.333
10" x 10"	"	2.000
12" x 12"	"	2.667
Reducer, with 2 bolt kits		
6" x 4"	EA.	1.000
8" x 6"	"	1.333
10" x 8"	"	2.000
12" x 10"	"	2.667
Flanged, 90 deg bend, 125 lb.		
4"	EA.	0.667
6"	"	0.800
8"	"	1.000
10"	"	1.333
12"	"	2.000
14"	"	2.667
16"	"	2.667
Tee		
4"	EA.	1.000
6"	"	1.143
8"	"	1.333
10"	"	1.600
12"	"	2.000
14"	"	2.667
16"	"	4.000
02610.13 GATE VALVES		
Gate valve, (AWWA) mechanical joint, with adjustable box		
4" valve	EA.	0.800
6" valve	"	0.960
8" valve	"	1.200
10" valve	"	1.412
12" valve	"	1.714
14" valve	"	2.000
16" valve	"	2.182
18" valve	"	2.400
Flanged, with box, post indicator (AWWA)		
4" valve	EA.	0.960
6" valve	"	1.091

UTILITIES	UNIT	MAN/HOURS
02610.13 GATE VALVES		
8" valve	EA.	1.333
10" valve	"	1.600
12" valve	"	2.000
14" valve	"	2.400
16" valve	"	3.000
02610.15 WATER METERS		
Water meter, displacement type		
1"	EA.	0.800
1-1/2"	"	0.889
2"	"	1.000
02610.17 CORPORATION STOPS		
Stop for flared copper service pipe		
3/4"	EA.	0.400
1"	"	0.444
1-1/4"	"	0.533
1-1/2"	"	0.667
2"	"	0.800
02610.40 DUCTILE IRON PIPE		
Ductile iron pipe, cement lined, slip-on joints		
4"	L.F.	0.067
6"	"	0.071
8"	"	0.075
10"	"	0.080
12"	"	0.096
14"	"	0.120
16"	"	0.133
18"	"	0.150
20"	"	0.171
Mechanical joint pipe		
4"	L.F.	0.092
6"	"	0.100
8"	"	0.109
10"	"	0.120
12"	"	0.160
14"	"	0.185
16"	"	0.218
18"	"	0.240
20"	"	0.267
Fittings, mechanical joint		
90 degree elbow		
4"	EA.	0.533
6"	"	0.615
8"	"	0.800
10"	"	1.143
12"	"	1.600
14"	"	2.000
16"	"	2.667
18"	"	3.200
20"	"	4.000
45 degree elbow		

UTILITIES	UNIT	MAN/HOURS
02610.40 DUCTILE IRON PIPE		
4"	EA.	0.533
6"	"	0.615
8"	"	0.800
10"	"	1.143
12"	"	1.600
14"	"	2.000
16"	"	2.667
18"	"	4.000
20"	"	4.000
Tee		
4"x3"	EA.	1.000
4"x4"	"	1.000
6"x3"	"	1.143
6"x4"	"	1.143
6"x6"	"	1.143
8"x4"	"	1.333
8"x6"	"	1.333
8"x8"	"	1.333
10"x4"	"	1.600
10"x6"	"	1.600
10"x8"	"	1.600
10"x10"	"	1.600
12"x4"	"	2.000
12"x6"	"	2.000
12"x8"	"	2.000
12"x10"	"	2.000
12"x12"	"	2.133
14"x4"	"	2.286
14"x6"	"	2.286
14"x8"	"	2.286
14"x10"	"	2.286
14"x12"	"	2.462
14"x14"	"	2.462
16"x4"	"	2.667
16"x6"	"	2.667
16"x8"	"	2.667
16"x10"	"	2.667
16"x12"	"	2.667
16"x14"	"	2.667
16"x16"	"	2.667
18"x6"	"	2.909
18"x8"	"	2.909
18"x10"	"	2.909
18"x12"	"	2.909
18"x14"	"	2.909
18"x16"	"	2.909
18"x18"	"	2.909
20"x6"	"	3.200
20"x8"	"	3.200
20"x10"	"	3.200
20"x12"	"	3.200
20"x14"	"	3.200
20"x16"	"	3.200
20"x18"	"	3.200

UTILITIES	UNIT	MAN/HOURS
02610.40 DUCTILE IRON PIPE		
20"x20"	EA.	3.200
Cross		
4"x3"	EA.	1.333
4"x4"	"	1.333
6"x3"	"	1.600
6"x4"	"	1.600
6"x6"	"	1.600
8"x4"	"	1.778
8"x6"	"	1.778
8"x8"	"	1.778
10"x4"	"	2.000
10"x6"	"	2.000
10"x8"	"	2.000
10"x10"	"	2.000
12"x4"	"	2.286
12"x6"	"	2.286
12"x8"	"	2.286
12"x10"	"	2.462
12"x12"	"	2.462
14"x4"	"	2.667
14"x6"	"	2.667
14"x8"	"	2.667
14"x10"	"	2.667
14"x12"	"	2.909
14"x14"	"	2.909
16"x4"	"	3.200
16"x6"	"	3.200
16"x8"	"	3.200
16"x10"	"	3.200
16"x12"	"	3.200
16"x14"	"	3.200
16"x16"	"	3.200
18"x6"	"	3.556
18"x8"	"	3.556
18"x10"	"	3.556
18"x12"	"	3.556
18"x14"	"	3.556
18"x16"	"	3.556
18"x18"	"	3.556
20"x6"	"	3.810
20"x8"	"	3.810
20"x10"	"	3.810
20"x12"	"	3.810
20"x14"	"	3.810
20"x16"	"	3.810
20"x18"	"	4.000
20"x20"	"	4.000
02610.60 PLASTIC PIPE		
PVC, class 150 pipe		
4" dia.	L.F.	0.060
6" dia.	"	0.065
8" dia.	"	0.069

UTILITIES	UNIT	MAN/HOURS
02610.60 PLASTIC PIPE		
10" dia.	L.F.	0.075
12" dia.	"	0.080
Schedule 40 pipe		
1-1/2" dia.	L.F.	0.047
2" dia.	"	0.050
2-1/2" dia.	"	0.053
3" dia.	"	0.057
4" dia.	"	0.067
6" dia.	"	0.080
90 degree elbows		
1"	EA.	0.133
1-1/2"	"	0.133
2"	"	0.145
2-1/2"	"	0.160
3"	"	0.178
4"	"	0.200
6"	"	0.267
45 degree elbows		
1"	EA.	0.133
1-1/2"	"	0.133
2"	"	0.145
2-1/2"	"	0.160
3"	"	0.178
4"	"	0.200
6"	"	0.267
Tees		
1"	EA.	0.160
1-1/2"	"	0.160
2"	"	0.178
2-1/2"	"	0.200
3"	"	0.229
4"	"	0.267
6"	"	0.320
Couplings		
1"	EA.	0.133
1-1/2"	"	0.133
2"	"	0.145
2-1/2"	"	0.160
3"	"	0.178
4"	"	0.200
6"	"	0.267
Drainage pipe		
PVC schedule 80		
1" dia.	L.F.	0.047
1-1/2" dia.	"	0.047
ABS, 2" dia.	"	0.050
2-1/2" dia.	"	0.053
3" dia.	"	0.057
4" dia.	"	0.067
6" dia.	"	0.080
8" dia.	"	0.063
10" dia.	"	0.075
12" dia.	"	0.080
90 degree elbows		

UTILITIES	UNIT	MAN/ HOURS
02610.60 PLASTIC PIPE		
1"	EA.	0.133
1-1/2"	"	0.133
2"	"	0.145
2-1/2"	"	0.160
3"	"	0.178
4"	"	0.200
6"	"	0.267
45 degree elbows		
1"	EA.	0.133
1-1/2"	"	0.133
2"	"	0.145
2-1/2"	"	0.160
3"	"	0.178
4"	"	0.200
6"	"	0.267
Tees		
1"	EA.	0.160
1-1/2"	"	0.160
2"	"	0.178
2-1/2"	"	0.200
3"	"	0.229
4"	"	0.267
6"	"	0.320
Couplings		
1"	EA.	0.133
1-1/2"	"	0.133
2"	"	0.145
2-1/2"	"	0.160
3"	"	0.178
4"	"	0.200
6"	"	0.267
Pressure pipe		
PVC, class 200 pipe		
3/4"	L.F.	0.040
1"	"	0.042
1-1/4"	"	0.044
1-1/2"	"	0.047
2"	"	0.050
2-1/2"	"	0.053
3"	"	0.057
4"	"	0.067
6"	"	0.080
8"	"	0.069
90 degree elbows		
3/4"	EA.	0.133
1"	"	0.133
1-1/4"	"	0.133
1-1/2"	"	0.133
2"	"	0.145
2-1/2"	"	0.160
3"	"	0.178
4"	"	0.200
6"	"	0.267
8"	"	0.400

UTILITIES	UNIT	MAN/ HOURS
02610.60 PLASTIC PIPE		
45 degree elbows		
3/4"	EA.	0.133
1"	"	0.133
1-1/4"	"	0.133
1-1/2"	"	0.133
2"	"	0.145
2-1/2"	"	0.160
3"	"	0.178
4"	"	0.200
6"	"	0.267
8"	"	0.400
Tees		
3/4"	EA.	0.160
1"	"	0.160
1-1/4"	"	0.160
1-1/2"	"	0.160
2"	"	0.178
2-1/2"	"	0.200
3"	"	0.229
4"	"	0.267
6"	"	0.320
8"	"	0.444
Couplings		
3/4"	EA.	0.133
1"	"	0.133
1-1/4"	"	0.133
1-1/2"	"	0.133
2"	"	0.145
2-1/2"	"	0.160
3"	"	0.178
4"	"	0.178
6"	"	0.200
8"	"	0.267
02610.90 VITRIFIED CLAY PIPE		
Vitrified clay pipe, extra strength		
6" dia.	L.F.	0.109
8" dia.	"	0.114
10" dia.	"	0.120
12" dia.	"	0.160
15" dia.	"	0.240
18" dia.	"	0.267
24" dia.	"	0.343
30" dia.	"	0.480
36" dia.	"	0.686
02630.10 TAPPING SADDLES & SLEEVES		
Tapping saddle, tap size to 2"		
4" saddle	EA.	0.400
6" saddle	"	0.500
8" saddle	"	0.667
10" saddle	"	0.800
12" saddle	"	1.143

UTILITIES	UNIT	MAN/HOURS
02630.10 TAPPING SADDLES & SLEEVES		
14" saddle	EA.	1.600
Tapping sleeve		
4x4	EA.	0.533
6x4	"	0.615
6x6	"	0.615
8x4	"	0.800
8x6	"	0.800
10x4	"	0.960
10x6	"	0.960
10x8	"	0.960
10x10	"	1.000
12x4	"	1.000
12x6	"	1.091
12x8	"	1.200
12x10	"	1.333
12x12	"	1.500
Tapping valve, mechanical joint		
4" valve	EA.	3.000
6" valve	"	4.000
8" valve	"	6.000
10" valve	"	8.000
12" valve	"	12.000
Tap hole in pipe		
4" hole	EA.	1.000
6" hole	"	1.600
8" hole	"	2.667
10" hole	"	3.200
12" hole	"	4.000
02640.15 VALVE BOXES		
Valve box, adjustable, for valves up to 20"		
3' deep	EA.	0.267
4' deep	"	0.320
5' deep	"	0.400
02640.19 THRUST BLOCKS		
Thrust block, 3000# concrete		
1/4 c.y.	EA.	1.333
1/2 c.y.	"	1.600
3/4 c.y.	"	2.667
1 c.y.	"	5.333
02645.10 FIRE HYDRANTS		
Standard, 3 way post, 6" mechanical joint		
2' deep	EA.	8.000
4' deep	"	9.600
6' deep	"	12.000
8' deep	"	13.714
02665.10 CHILLED WATER SYSTEMS		
Chilled water pipe, 2" thick insulation, w/casing		
Align and tack weld on sleepers		

UTILITIES	UNIT	MAN/HOURS
02665.10 CHILLED WATER SYSTEMS		
1-1/2" dia.	L.F.	0.022
3" dia.	"	0.034
4" dia.	"	0.048
6" dia.	"	0.060
8" dia.	"	0.069
10" dia.	"	0.080
12" dia.	"	0.096
14" dia.	"	0.104
16" dia.	"	0.120
Align and tack weld on trench bottom		
18" dia.	L.F.	0.133
20" dia.	"	0.150
Preinsulated fittings		
Align and tack weld on sleepers		
Elbows		
1-1/2"	EA.	0.500
3"	"	0.800
4"	"	1.000
6"	"	1.333
8"	"	1.600
Tees		
1-1/2"	EA.	0.533
3"	"	0.889
4"	"	1.143
6"	"	1.600
8"	"	2.000
Reducers		
3"	EA.	0.667
4"	"	0.800
6"	"	1.000
8"	"	1.333
Anchors, not including concrete		
4"	EA.	1.000
6"	"	1.000
Align and tack weld on trench bottom		
Elbows		
10"	EA.	1.500
12"	"	1.714
14"	"	1.846
16"	"	2.000
18"	"	2.182
20"	"	2.400
Tees		
10"	EA.	1.500
12"	"	1.714
14"	"	1.846
16"	"	2.000
18"	"	2.182
20"	"	2.400
Reducers		
10"	EA.	1.000
12"	"	1.091
14"	"	1.200
16"	"	1.333

UTILITIES	UNIT	MAN/ HOURS
02665.10 CHILLED WATER SYSTEMS		
18"	EA.	1.500
20"	"	1.714
Anchors, not including concrete		
10"	EA.	1.000
12"	"	1.091
14"	"	1.200
16"	"	1.333
18"	"	1.500
20"	"	1.714
02670.10 WELLS		
Domestic water, drilled and cased		
4" dia.	L.F.	0.480
6" dia.	"	0.533
8" dia.	"	0.600
02685.10 GAS DISTRIBUTION		
Gas distribution lines		
Polyethylene, 60 psi coils		
1-1/4" dia.	L.F.	0.053
1-1/2" dia.	"	0.057
2" dia.	"	0.067
3" dia.	"	0.080
30' pipe lengths		
3" dia.	L.F.	0.089
4" dia.	"	0.100
6" dia.	"	0.133
8" dia.	"	0.160
Steel, schedule 40, plain end		
1" dia.	L.F.	0.067
2" dia.	"	0.073
3" dia.	"	0.080
4" dia.	"	0.160
5" dia.	"	0.171
6" dia.	"	0.200
8" dia.	"	0.218
Natural gas meters, direct digital reading, threaded		
250 cfh @ 5 lbs	EA.	1.600
425 cfh @ 10 lbs	"	1.600
800 cfh @ 20 lbs	"	2.000
1000 cfh @ 25 lbs	"	2.000
1,400 cfh @ 100 lbs	"	2.667
2,300 cfh @ 100 lbs	"	4.000
5,000 cfh @ 100 lbs	"	8.000
Gas pressure regulators		
Threaded		
3/4"	EA.	1.000
1"	"	1.333
1-1/4"	"	1.333
1-1/2"	"	1.333
2"	"	1.600
Flanged		
3"	EA.	2.000

UTILITIES	UNIT	MAN/ HOURS
02685.10 GAS DISTRIBUTION		
4"	EA.	2.667
02690.10 STORAGE TANKS		
Oil storage tank, underground		
Steel		
500 gals	EA.	3.000
1,000 gals	"	4.000
4,000 gals	"	8.000
5,000 gals	"	12.000
10,000 gals	"	24.000
Fiberglass, double wall		
550 gals	EA.	4.000
1,000 gals	"	4.000
2,000 gals	"	6.000
4,000 gals	"	12.000
6,000 gals	"	16.000
8,000 gals	"	24.000
10,000 gals	"	30.000
12,000 gals	"	40.000
15,000 gals	"	53.333
20,000 gals	"	60.000
Above ground		
Steel		
275 gals	EA.	2.400
500 gals	"	4.000
1,000 gals	"	4.800
1,500 gals	"	6.000
2,000 gals	"	8.000
5,000 gals	"	12.000
Fill cap	"	0.800
Vent cap	"	0.800
Level indicator	"	0.800
02695.40 STEEL PIPE		
Steel pipe, extra heavy, A 53, grade B, seamless		
1/2" dia.	L.F.	0.080
3/4" dia.	"	0.084
1" dia.	"	0.089
1-1/4" dia.	"	0.100
1-1/2" dia.	"	0.114
2" dia.	"	0.133
3" dia.	"	0.120
4" dia.	"	0.133
6" dia.	"	0.150
8" dia.	"	0.171
10" dia.	"	0.200
12" dia.	"	0.240

UTILITIES

02695.80 STEAM METERS

	UNIT	MAN/ HOURS
In-line turbine, direct reading, 300 lb, flanged		
2"	EA.	1.000
3"	"	1.333
4"	"	1.600
Threaded, 2"		
5" line	EA.	8.000
6" line	"	8.000
8" line	"	8.000
10" line	"	8.000
12" line	"	8.000
14" line	"	8.000
16" line	"	8.000

SEWERAGE AND DRAINAGE

02720.10 CATCH BASINS

	UNIT	MAN/ HOURS
Standard concrete catch basin		
Cast in place, 3'8" x 3'8", 6" thick wall		
2' deep	EA.	6.000
3' deep	"	6.000
4' deep	"	8.000
5' deep	"	8.000
6' deep	"	9.600
4'x4', 8" thick wall, cast in place		
2' deep	EA.	6.000
3' deep	"	6.000
4' deep	"	8.000
5' deep	"	8.000
6' deep	"	9.600
Frames and covers, cast iron		
Round		
24" dia.	EA.	2.000
26" dia.	"	2.000
28" dia.	"	2.000
Rectangular		
23"x23"	EA.	2.000
27"x20"	"	2.000
24"x24"	"	2.000
26"x26"	"	2.000
Curb inlet frames and covers		
27"x27"	EA.	2.000
24"x36"	"	2.000
24"x25"	"	2.000
24"x22"	"	2.000
20"x22"	"	2.000
Airfield catch basin frame and grating, galvanized		
2'x4'	EA.	2.000

SEWERAGE AND DRAINAGE

02720.10 CATCH BASINS

	UNIT	MAN/ HOURS
2'x2'	EA.	2.000

02720.40 STORM DRAINAGE

	UNIT	MAN/ HOURS
Headwalls, cast in place, 30 deg wingwall		
12" pipe	EA.	2.000
15" pipe	"	2.000
18" pipe	"	2.286
24" pipe	"	2.286
30" pipe	"	2.667
36" pipe	"	4.000
42" pipe	"	4.000
48" pipe	"	5.333
54" pipe	"	6.667
60" pipe	"	8.000
4" cleanout for storm drain		
4" pipe	EA.	1.000
6" pipe	"	1.000
8" pipe	"	1.000
Connect new drain line		
To existing manhole	EA.	2.667
To new manhole	"	1.600

02720.45 STORM DRAINAGE, CON. PIPE

	UNIT	MAN/ HOURS
Concrete pipe		
Unreinforced, plain, bell and spigot, Class II		
Pipe diameter 6"	L.F.	0.109
8"	"	0.120
10"	"	0.126
12"	"	0.133
15"	"	0.141
18"	"	0.150
21"	"	0.160
24"	"	0.171
Reinforced, bell and spigot, Class III		
Pipe diameter 12"	L.F.	0.133
15"	"	0.141
18"	"	0.150
21"	"	0.160
24"	"	0.171
27"	"	0.185
30"	"	0.200
36"	"	0.218
42"	"	0.240
48"	"	0.267
54"	"	0.300
60"	"	0.343
66"	"	0.400
72"	"	0.480
78"	"	0.533
84"	"	0.600
90"	"	0.649
96"	"	0.686

SEWERAGE AND DRAINAGE

02720.45 STORM DRAINAGE, CON. PIPE

	UNIT	MAN/ HOURS
Class IV		
Pipe diameter 12"	L.F.	0.133
15"	"	0.141
18"	"	0.150
21"	"	0.160
24"	"	0.171
27"	"	0.185
30"	"	0.200
36"	"	0.218
42"	"	0.240
48"	"	0.267
54"	"	0.300
60"	"	0.343
66"	"	0.400
72"	"	0.480
78"	"	0.533
84"	"	0.600
90"	"	0.649
96"	"	0.686
Class V		
Pipe diameter 12"	L.F.	0.133
15"	"	0.141
18"	"	0.150
21"	"	0.160
24"	"	0.171
27"	"	0.185
30"	"	0.200
36"	"	0.218
42"	"	0.240
48"	"	0.267
54"	"	0.300
60"	"	0.343
66"	"	0.400
72"	"	0.480
78"	"	0.533
84"	"	0.600
90"	"	0.649
96"	"	0.686
Eliptical pipe, reinforced		
Class III		
20" x 30"	L.F.	0.171
22" x 34 "	"	0.185
24" x 38"	"	0.200
27" x 42"	"	0.218
29" x 45"	"	0.229
32" x 49"	"	0.240
34" x 54"	"	0.267
38" x 60"	"	0.300
43" x 68"	"	0.343
48" x 76"	"	0.369
53" x 83"	"	0.400
62" x 98"	"	0.480
82" x 128"	"	0.686
Flared end section pipe		

SEWERAGE AND DRAINAGE

02720.45 STORM DRAINAGE, CON. PIPE

	UNIT	MAN/ HOURS
Pipe diameter 12"	L.F.	0.133
15"	"	0.141
18"	"	0.150
24"	"	0.171
30"	"	0.200
36"	"	0.218
42"	"	0.240
48"	"	0.267
54"	"	0.300
60"	"	0.343
Porous concrete pipe, standard strength		
Pipe diameter 4"	L.F.	0.092
6"	"	0.096
8"	"	0.100
10"	"	0.104
12"	"	0.104

02720.50 STORM DRAINAGE, STEEL PIPE

	UNIT	MAN/ HOURS
Steel pipe		
Coated,corrugated metal pipe, paved invert		
16 gauge, pipe diameter 8"	L.F.	0.080
12"	"	0.086
15"	"	0.092
18"	"	0.100
21"	"	0.109
24"	"	0.120
30"	"	0.133
36"	"	0.150
42"	"	0.160
48"	"	0.171
14 gauge, pipe diameter 12"	"	0.086
15"	"	0.092
18"	"	0.100
21"	"	0.109
24"	"	0.120
30"	"	0.133
36"	"	0.150
42"	"	0.160
48"	"	0.171
54"	"	0.185
60"	"	0.200
66"	"	0.218
12 gauge, pipe diameter 18"	"	0.100
21"	"	0.109
24"	"	0.120
30"	"	0.133
36"	"	0.150
42"	"	0.160
48"	"	0.171
54"	"	0.185
60"	"	0.200
66"	"	0.218
72"	"	0.240
78"	"	0.253

SEWERAGE AND DRAINAGE	UNIT	MAN/ HOURS
02720.50 STORM DRAINAGE, STEEL PIPE		
10 gauge, pipe diameter 24"	L.F.	0.120
30"	"	0.133
36"	"	0.150
42"	"	0.160
48"	"	0.171
54"	"	0.185
60"	"	0.200
66"	"	0.218
72"	"	0.240
78"	"	0.253
84"	"	0.267
90"	"	0.282
8 gauge, pipe diameter 48"	"	0.171
54"	"	0.185
60"	"	0.200
66"	"	0.218
72"	"	0.240
78"	"	0.253
84"	"	0.267
90"	"	0.282
96"	"	0.300
Plain,corrugated metal pipe		
16 gauge, pipe diameter 8"	L.F.	0.080
12"	"	0.086
15"	"	0.092
18"	"	0.100
21"	"	0.109
24"	"	0.120
30"	"	0.133
36"	"	0.150
42"	"	0.160
48"	"	0.171
14 gauge, pipe diameter 12"	"	0.086
15"	"	0.092
18"	"	0.100
21"	"	0.109
24"	"	0.120
30"	"	0.133
36"	"	0.150
42"	"	0.160
48"	"	0.171
54"	"	0.185
60"	"	0.200
66"	"	0.218
12 gauge, pipe diameter 18"	"	0.100
21"	"	0.109
24"	"	0.120
30"	"	0.133
36"	"	0.150
42"	"	0.160
48"	"	0.171
54"	"	0.185
60"	"	0.200
66"	"	0.218

SEWERAGE AND DRAINAGE	UNIT	MAN/ HOURS
02720.50 STORM DRAINAGE, STEEL PIPE		
72"	L.F.	0.240
78"	"	0.253
10 gauge, pipe diameter 24"	"	0.120
30"	"	0.133
36"	"	0.150
42"	"	0.160
48"	"	0.171
54"	"	0.185
60"	"	0.200
66"	"	0.218
72"	"	0.240
78"	"	0.253
84"	"	0.267
90"	"	0.282
8 gauge, pipe diameter 48"	"	0.171
54"	"	0.185
60"	"	0.200
66"	"	0.218
72"	"	0.240
78"	"	0.253
84"	"	0.267
90"	"	0.282
96"	"	0.300
Steel arch		
Coated, corrugated		
16 gauge, 17" x 13"	L.F.	0.109
21" x 15"	"	0.120
14 gauge, 29" x 18"	"	0.133
36" x 22"	"	0.171
12 gauge, 43" x 28"	"	0.200
50" x 30"	"	0.218
58" x 36"	"	0.240
66" x 40"	"	0.253
72" x 44"	"	0.267
Plain, corrugated		
16 gauge, 17" x 13"	L.F.	0.109
21" x 15"	"	0.120
14 gauge, 29" x 18"	"	0.133
36" x 22"	"	0.171
12 gauge, 43" x 28"	"	0.200
50" x 30"	"	0.218
58" x 36"	"	0.240
66" x 40"	"	0.253
72" x 44"	"	0.267
Nestable corrugated metal pipe		
16 gauge, pipe diameter 10"	L.F.	0.083
12"	"	0.086
15"	"	0.092
18"	"	0.100
24"	"	0.120
30"	"	0.133
14 gauge, pipe diameter 12"	"	0.086
15"	"	0.092
18"	"	0.100

02 SITEWORK

SEWERAGE AND DRAINAGE	UNIT	MAN/ HOURS
02720.50 STORM DRAINAGE, STEEL PIPE		
24"	L.F.	0.120
30"	"	0.133
36"	"	0.150
02720.70 UNDERDRAIN		
Drain tile, clay		
6" pipe	L.F.	0.053
8" pipe	"	0.056
12" pipe	"	0.060
Porous concrete, standard strength		
6" pipe	L.F.	0.053
8" pipe	"	0.056
12" pipe	"	0.060
15" pipe	"	0.067
18" pipe	"	0.080
Corrugated metal pipe, perforated type		
6" pipe	L.F.	0.060
8" pipe	"	0.063
10" pipe	"	0.067
12" pipe	"	0.071
18" pipe	"	0.075
Perforated clay pipe		
6" pipe	L.F.	0.069
8" pipe	"	0.071
12" pipe	"	0.073
Drain tile, concrete		
6" pipe	L.F.	0.053
8" pipe	"	0.056
12" pipe	"	0.060
Perforated rigid PVC underdrain pipe		
4" pipe	L.F.	0.040
6" pipe	"	0.048
8" pipe	"	0.053
10" pipe	"	0.060
12" pipe	"	0.069
Underslab drainage, crushed stone		
3" thick	S.F.	0.008
4" thick	"	0.009
6" thick	"	0.010
8" thick	"	0.010
Plastic filter fabric for drain lines	"	0.008
Gravel fill in trench, crushed or bank run, 1/2" to 3/4"	C.Y.	0.600
02730.10 SANITARY SEWERS		
Clay		
6" pipe	L.F.	0.080
8" pipe	"	0.086
10" pipe	"	0.092
12" pipe	"	0.100
PVC		
4" pipe	L.F.	0.060
6" pipe	"	0.063

SEWERAGE AND DRAINAGE	UNIT	MAN/ HOURS
02730.10 SANITARY SEWERS		
8" pipe	L.F.	0.067
10" pipe	"	0.071
12" pipe	"	0.075
Cleanout		
4" pipe	EA.	1.000
6" pipe	"	1.000
8" pipe	"	1.000
Connect new sewer line		
To existing manhole	EA.	2.667
To new manhole	"	1.600
02740.10 DRAINAGE FIELDS		
Perforated PVC pipe, for drain field		
4" pipe	L.F.	0.053
6" pipe	"	0.057
02740.50 SEPTIC TANKS		
Septic tank, precast concrete		
1000 gals	EA.	4.000
2000 gals	"	6.000
5000 gals	"	12.000
25,000 gals	"	48.000
40,000 gals	"	80.000
Leaching pit, precast concrete, 72" diameter		
3' deep	EA.	3.000
6' deep	"	3.429
8' deep	"	4.000
02760.10 PIPELINE RESTORATION		
Relining existing water main		
6" dia.	L.F.	0.240
8" dia.	"	0.253
10" dia.	"	0.267
12" dia.	"	0.282
14" dia.	"	0.300
16" dia.	"	0.320
18" dia.	"	0.343
20" dia.	"	0.369
24" dia.	"	0.400
36" dia.	"	0.480
48" dia.	"	0.533
72" dia.	"	0.600
Replacing in line gate valves		
6" valve	EA.	3.200
8" valve	"	4.000
10" valve	"	4.800
12" valve	"	6.000
16" valve	"	6.857
18" valve	"	8.000
20" valve	"	9.600
24" valve	"	12.000
36" valve	"	16.000

POWER & COMMUNICATIONS

02780.20 HIGH VOLTAGE CABLE

	UNIT	MAN/ HOURS
High voltage XLP copper cable, shielded, 5000v		
#6 awg	L.F.	0.013
#4 awg	"	0.016
#2 awg	"	0.019
#1 awg	"	0.021
#1/0 awg	"	0.024
#2/0 awg	"	0.029
#3/0 awg	"	0.034
#4/0 awg	"	0.036
#250 awg	"	0.043
#300 awg	"	0.048
#350 awg	"	0.053
#500 awg	"	0.073
#750 awg	"	0.080
Ungrounded, 15,000v		
#1 awg	L.F.	0.031
#1/0 awg	"	0.034
#2/0 awg	"	0.036
#3/0 awg	"	0.040
#4/0 awg	"	0.046
#250 awg	"	0.048
#300 awg	"	0.053
#350 awg	"	0.062
#500 awg	"	0.080
#750 awg	"	0.098
#1000 awg	"	0.123
Aluminum cable, shielded, 5000v		
#6 awg	L.F.	0.011
#4 awg	"	0.013
#2 awg	"	0.015
#1 awg	"	0.017
#1/0 awg	"	0.019
#2/0 awg	"	0.020
#3/0 awg	"	0.021
#4/0 awg	"	0.024
#250 awg	"	0.026
#300 awg	"	0.031
#350 awg	"	0.034
#500 awg	"	0.036
#750 awg	"	0.044
#1000 awg	"	0.050
Ungrounded, 15,000v		
#1 awg	L.F.	0.021
#1/0 awg	"	0.025
#2/0 awg	"	0.027
#3/0 awg	"	0.028
#4/0 awg	"	0.029
#250 awg	"	0.031
#300 awg	"	0.032
#350 awg	"	0.036
#500 awg	"	0.043
#750 awg	"	0.052
#1000 awg	"	0.064
Indoor terminations, 5000v		

POWER & COMMUNICATIONS

02780.20 HIGH VOLTAGE CABLE

	UNIT	MAN/ HOURS
#6 - #4	EA.	0.157
#2 - #2/0	"	0.157
#3/0 - #250	"	0.157
#300 - #750	"	2.759
#1000	"	3.810
In-line splice, 5000v		
#6 - #4/0	EA.	3.810
#250 - #500	"	10.000
#750 - #1000	"	13.008
T-splice, 5000v		
#2 - #4/0	EA.	11.994
#250 - #500	"	20.000
#750 - #1000	"	25.000
Indoor terminations, 15,000v		
#2 - #2/0	EA.	3.478
#3/0 - #500	"	5.333
#750 - #1000	"	6.154
In-line splice, 15,000v		
#2 - #4/0	EA.	8.999
#250 - #500	"	11.994
#750 - #1000	"	18.018
T-splice, 15,000v		
#4	EA.	18.018
#250 - #500	"	29.963
#750 - #1000	"	44.944
Compression lugs, 15,000v		
#4	EA.	0.400
#2	"	0.533
#1	"	0.533
#1/0	"	0.667
#2/0	"	0.667
#3/0	"	0.851
#4/0	"	0.851
#250	"	0.952
#300	"	0.952
#350	"	1.159
#500	"	1.250
#750	"	1.509
#1000	"	1.905
Compression splices, 15,000v		
#4	EA.	0.667
#2	"	0.727
#1	"	0.899
#1/0	"	1.000
#2/0	"	1.159
#3/0	"	1.250
#4/0	"	1.404
#250	"	1.509
#350	"	1.739
#500	"	2.000
#750	"	2.500

02 SITEWORK

POWER & COMMUNICATIONS	UNIT	MAN/HOURS
02780.40 SUPPORTS & CONNECTORS		
Cable supports for conduit		
1-1/2"	EA.	0.348
2"	"	0.348
2-1/2"	"	0.400
3"	"	0.400
3-1/2"	"	0.500
4"	"	0.500
5"	"	0.667
6"	"	0.727
Split bolt connectors		
#10	EA.	0.200
#8	"	0.200
#6	"	0.200
#4	"	0.400
#3	"	0.400
#2	"	0.400
#1/0	"	0.667
#2/0	"	0.667
#3/0	"	0.667
#4/0	"	0.667
#250	"	1.000
#350	"	1.000
#500	"	1.000
#750	"	1.509
#1000	"	1.509
Single barrel lugs		
#6	EA.	0.250
#1/0	"	0.500
#250	"	0.667
#350	"	0.667
#500	"	0.667
#600	"	0.899
#800	"	0.899
#1000	"	0.899
Double barrel lugs		
#1/0	EA.	0.899
#250	"	1.290
#350	"	1.290
#600	"	1.905
#800	"	1.905
#1000	"	1.905
Three barrel lugs		
#2/0	EA.	1.290
#250	"	1.905
#350	"	1.905
#600	"	2.667
#800	"	2.667
#1000	"	2.667
Four barrel lugs		
#250	EA.	2.759
#350	"	2.759
#600	"	3.478
#800	"	3.478
Compression conductor adapters		

POWER & COMMUNICATIONS	UNIT	MAN/HOURS
02780.40 SUPPORTS & CONNECTORS		
#6	EA.	0.296
#4	"	0.348
#2	"	0.444
#1	"	0.444
#1/0	"	0.533
#250	"	0.800
#350	"	0.851
#500	"	1.096
#750	"	1.143
Terminal blocks, 2 screw		
3 circuit	EA.	0.200
6 circuit	"	0.200
8 circuit	"	0.200
10 circuit	"	0.296
12 circuit	"	0.296
18 circuit	"	0.296
24 circuit	"	0.348
36 circuit	"	0.348
Compression splice		
#8 awg	EA.	0.381
#6 awg	"	0.276
#4 awg	"	0.276
#2 awg	"	0.533
#1 awg	"	0.533
#1/0 awg	"	0.533
#2/0 awg	"	0.851
#3/0 awg	"	0.851
#4/0 awg	"	0.851
#250 awg	"	1.356
#300 awg	"	1.356
#350 awg	"	1.404
#400 awg	"	1.404
#500 awg	"	1.509
#600 awg	"	1.509
#750 awg	"	1.739
#1000 awg	"	1.739

SITE IMPROVEMENTS	UNIT	MAN/HOURS
02830.10 CHAIN LINK FENCE		
Chain link fence, 9 ga., galvanized, with posts 10' o.c.		
4' high	L.F.	0.057
5' high	"	0.073
6' high	"	0.100
7' high	"	0.123
8' high	"	0.160
For barbed wire with hangers, add		

SITE IMPROVEMENTS	UNIT	MAN/ HOURS
02830.10 CHAIN LINK FENCE		
3 strand	L.F.	0.040
6 strand	"	0.067
Corner or gate post, 3" post		
4' high	EA.	0.267
5' high	"	0.296
6' high	"	0.348
7' high	"	0.400
8' high	"	0.444
4" post		
4' high	EA.	0.296
5' high	"	0.348
6' high	"	0.400
7' high	"	0.444
8' high	"	0.500
Gate with gate posts, galvanized, 3' wide		
4' high	EA.	2.000
5' high	"	2.667
6' high	"	2.667
7' high	"	4.000
8' high	"	4.000
Fabric, galvanized chain link, 2" mesh, 9 ga.		
4' high	L.F.	0.027
5' high	"	0.032
6' high	"	0.040
8' high	"	0.053
Line post, no rail fitting, galvanized, 2-1/2" dia.		
4' high	EA.	0.229
5' high	"	0.250
6' high	"	0.267
7' high	"	0.320
8' high	"	0.400
1-7/8" H beam		
4' high	EA.	0.229
5' high	"	0.250
6' high	"	0.267
7' high	"	0.320
8' high	"	0.400
2-1/4" H beam		
4' high	EA.	0.229
5' high	"	0.250
6' high	"	0.267
7' high	"	0.320
8' high	"	0.400
Vinyl coated, 9 ga., with posts 10' o.c.		
4' high	L.F.	0.057
5' high	"	0.073
6' high	"	0.100
7' high	"	0.123
8' high	"	0.160
For barbed wire w/hangers, add		
3 strand	L.F.	0.040
6 Strand	"	0.067
Corner, or gate post, 4' high		
3" dia.	EA.	0.267

SITE IMPROVEMENTS	UNIT	MAN/ HOURS
02830.10 CHAIN LINK FENCE		
4" dia.	EA.	0.267
6" dia.	"	0.320
Gate, with posts, 3' wide		
4' high	EA.	2.000
5' high	"	2.667
6' high	"	2.667
7' high	"	4.000
8' high	"	4.000
Line post, no rail fitting, 2-1/2" dia.		
4' high	EA.	0.229
5' high	"	0.250
6' high	"	0.267
7' high	"	0.320
8' high	"	0.400
Corner post, no top rail fitting, 4" dia.		
4' high	EA.	0.267
5' high	"	0.296
6' high	"	0.348
7' high	"	0.400
8' high	"	0.444
Fabric, vinyl, chain link, 2" mesh, 9 ga.		
4' high	L.F.	0.027
5' high	"	0.032
6' high	"	0.040
8' high	"	0.053
Swing gates, galvanized, 4' high		
Single gate		
3' wide	EA.	2.000
4' wide	"	2.000
Double gate		
10' wide	EA.	3.200
12' wide	"	3.200
14' wide	"	3.200
16' wide	"	3.200
18' wide	"	4.571
20' wide	"	4.571
22' wide	"	4.571
24' wide	"	5.333
26' wide	"	5.333
28' wide	"	6.400
30' wide	"	6.400
5' high		
Single gate		
3' wide	EA.	2.667
4' wide	"	2.667
Double gate		
10' wide	EA.	4.000
12' wide	"	4.000
14' wide	"	4.000
16' wide	"	4.000
18' wide	"	4.571
20' wide	"	4.571
22' wide	"	4.571
24' wide	"	5.333

SITE IMPROVEMENTS	UNIT	MAN/HOURS
02830.10 CHAIN LINK FENCE		
26' wide	EA.	5.333
28' wide	"	6.400
30' wide	"	6.400
6' high		
Single gate		
3' wide	EA.	2.667
4' wide	"	2.667
Double gate		
10' wide	EA.	4.000
12' wide	"	4.000
14' wide	"	4.000
16' wide	"	4.000
18' wide	"	4.571
20' wide	"	4.571
22' wide	"	4.571
24' wide	"	5.333
26' wide	"	5.333
28' wide	"	6.400
30' wide	"	6.400
7' high		
Single gate		
3' wide	EA.	4.000
4' wide	"	4.000
Double gate		
10' wide	EA.	5.333
12' wide	"	5.333
14' wide	"	5.333
16' wide	"	5.333
18' wide	"	6.400
20' wide	"	6.400
22' wide	"	6.400
24' wide	"	8.000
26' wide	"	8.000
28' wide	"	10.000
30' wide	"	10.000
8' high		
Single gate		
3' wide	EA.	4.000
4' wide	"	4.000
Double gate		
10' wide	EA.	5.333
12' wide	"	5.333
14' wide	"	5.333
16' wide	"	5.333
18' wide	"	6.400
20' wide	"	6.400
22' wide	"	6.400
24' wide	"	8.000
26' wide	"	8.000
28' wide	"	10.000
30' wide	"	10.000
Vinyl coated swing gates, 4' high		
Single gate		
3' wide	EA.	2.000

SITE IMPROVEMENTS	UNIT	MAN/HOURS
02830.10 CHAIN LINK FENCE		
4' wide	EA.	2.000
Double gate		
10' wide	EA.	3.200
12' wide	"	3.200
14' wide	"	3.200
16' wide	"	3.200
18' wide	"	4.571
20' wide	"	4.571
22' wide	"	4.571
24' wide	"	5.333
26' wide	"	5.333
28' wide	"	6.400
30' wide	"	6.400
5' high		
Single gate		
3' wide	EA.	2.667
4' wide	"	2.667
Double gate		
10' wide	EA.	4.000
12' wide	"	4.000
14' wide	"	4.000
16' wide	"	4.000
18' wide	"	4.571
20' wide	"	4.571
22' wide	"	4.571
24' wide	"	5.333
26' wide	"	5.333
28' wide	"	6.400
30' wide	"	6.400
6' high		
Single gate		
3' wide	EA.	2.667
4' wide	"	2.667
Double gate		
10' wide	EA.	4.000
12' wide	"	4.000
14' wide	"	4.000
16' wide	"	4.000
18' wide	"	4.571
20' wide	"	4.571
22' wide	"	4.571
24' wide	"	5.333
26' wide	"	5.333
28' wide	"	6.400
30' wide	"	6.400
7' high		
Single gate		
3' wide	EA.	4.000
4' wide	"	4.000
Double gate		
10' wide	EA.	5.333
12' wide	"	5.333
14' wide	"	5.333
16' wide	"	5.333

SITE IMPROVEMENTS	UNIT	MAN/ HOURS
02830.10 CHAIN LINK FENCE		
18' wide	EA.	6.400
20' wide	"	6.400
22' wide	"	6.400
24' wide	"	8.000
26' wide	"	8.000
28' wide	"	10.000
30' wide	"	10.000
8' high		
Single gate		
3' wide	EA.	4.000
4' wide	"	4.000
Double gate		
10' wide	EA.	5.333
12' wide	"	5.333
14' wide	"	5.333
16' wide	"	5.333
18' wide	"	6.400
20' wide	"	6.400
22' wide	"	6.400
24' wide	"	8.000
28' wide	"	8.000
30' wide	"	10.000
Motor operator for gates, no wiring	"	
Drilling fence post holes		
In soil		
By hand	EA.	0.400
By machine auger	"	0.200
In rock		
By jackhammer	EA.	2.667
By rock drill	"	0.800
Aluminum privacy slats, installed vertically	S.F.	0.020
Post hole, dig by hand	EA.	0.533
Set fence post in concrete	"	0.400
02830.70 RECREATIONAL COURTS		
Walls, galvanized steel		
8' high	L.F.	0.160
10' high	"	0.178
12' high	"	0.211
Vinyl coated		
8' high	L.F.	0.160
10' high	"	0.178
12' high	"	0.211
Gates, galvanized steel		
Single, 3' transom		
3'x7'	EA.	4.000
4'x7'	"	4.571
5'x7'	"	5.333
6'x7'	"	6.400
Double, 3' transom		
10'x7'	EA.	16.000
12'x7'	"	17.778
14'x7'	"	20.000

SITE IMPROVEMENTS	UNIT	MAN/ HOURS
02830.70 RECREATIONAL COURTS		
Double, no transom		
10'x10'	EA.	13.333
12'x10'	"	16.000
14'x10'	"	17.778
Vinyl coated		
Single, 3' transom		
3'x7'	EA.	4.000
4'x7'	"	4.571
5'x7'	"	5.333
6'x7'	"	6.400
Double, 3'		
10'x7'	EA.	16.000
12'x7'	"	17.778
14'x7'	"	20.000
Double, no transom		
10'x10'	EA.	13.333
12'x10'	"	16.000
14'x10'	"	17.778
Wire and miscellaneous metal fences		
Chicken wire, post 4' o.c.		
2" mesh		
4' high	L.F.	0.040
6' high	"	0.053
Galvanized steel		
12 gauge, 2" by 4" mesh, posts 5' o.c.		
3' high	L.F.	0.040
5' high	"	0.050
14 gauge, 1" by 2" mesh, posts 5' o.c.		
3' high	L.F.	0.040
5' high	"	0.050
02840.30 GUARDRAILS		
Pipe bollard, steel pipe, concrete filled, painted		
6" dia.	EA.	0.667
8" dia.	"	1.000
12" dia.	"	2.667
Corrugated steel, guardrail, galvanized	L.F.	0.040
End section, wrap around or flared	EA.	0.800
Timber guardrail, 4" x 8"	L.F.	0.030
Guard rail, 3 cables, 3/4" dia.		
Steel posts	L.F.	0.120
Wood posts	"	0.096
Steel box beam		
6" x 6"	L.F.	0.133
6" x 8"	"	0.150
Concrete posts	EA.	0.400
Barrel type impact barrier	"	0.800
Light shield, 6' high	L.F.	0.160
02840.40 PARKING BARRIERS		
Timber, treated, 4' long		
4" x 4"	EA.	0.667
6" x 6"	"	0.800
Precast concrete, 6' long, with dowels		

SITE IMPROVEMENTS	UNIT	MAN/ HOURS
02840.40 PARKING BARRIERS		
12" x 6"	EA.	0.400
12" x 8"	"	0.444
02840.60 SIGNAGE		
Traffic signs		
Reflectorized signs per OSHA standards, including post		
Stop, 24"x24"	EA.	0.533
Yield, 30" triangle	"	0.533
Speed limit, 12"x18"	"	0.533
Directional, 12"x18"	"	0.533
Exit, 12"x18"	"	0.533
Entry, 12"x18"	"	0.533
Warning, 24"x24"	"	0.533
Informational, 12"x18"	"	0.533
Handicap parking, 12"x18"	"	0.533
02860.40 RECREATIONAL FACILITIES		
Bleachers, outdoor, portable, per seat		
10 tiers		
Minimum	EA.	0.150
Maximum	"	0.200
20 tiers		
Minimum	EA.	0.141
Maximum	"	0.185
Grandstands, fixed, wood seat, steel frame, per seat		
15 tiers		
Minimum	EA.	0.240
Maximum	"	0.400
30 tiers		
Minimum	EA.	0.218
Maximum	"	0.343
Seats		
Seat backs only		
Fiberglass	EA.	0.080
Steel and wood seat	"	0.080
Seat restoration, fiberglass on wood		
Seats	EA.	0.160
Plain bench, no backs	"	0.067
Benches		
Park, precast concrete with backs		
4' long	EA.	2.667
8' long	"	4.000
Fiberglass, with backs		
4' long	EA.	2.000
8' long	"	2.667
Wood, with backs and fiberglass supports		
4' long	EA.	2.000
8' long	"	2.667
Steel frame, 6' long		
All steel	EA.	2.000
Hardwood boards	"	2.000
Players bench (no back), steel frame, fir seat, 10' long	"	2.667
Soccer goal posts	PAIR	

SITE IMPROVEMENTS	UNIT	MAN/ HOURS
02860.40 RECREATIONAL FACILITIES		
Running track		
Gravel and cinders over stone base	S.Y.	0.060
Rubber-cork base resilient pavement	"	0.480
For colored surfaces, add	"	0.048
Colored rubberized asphalt	"	0.600
Artificial resilient mat over asphalt	"	1.200
Tennis courts		
Bituminous pavement, 2-1/2" thick	S.Y.	0.150
Colored sealer, acrylic emulsion		
3 coats	S.Y.	0.053
For 2 color seal coating, add	"	0.008
For preparing old courts, add	"	0.005
Net, nylon, 42' long	EA.	1.000
Paint markings on asphalt, 2 coats	"	8.000
Playground equipment		
Basketball backboard		
Minimum	EA.	2.000
Maximum	"	2.286
Bike rack, 10' long	"	1.600
Golf shelter, fiberglass	"	2.000
Ground socket for movable posts		
Minimum	EA.	0.500
Maximum	"	0.500
Horizontal monkey ladder, 14' long	"	1.333
Posts, tether ball	"	0.400
Multiple purpose, 10' long	"	0.800
See-saw, steel		
Minimum	EA.	3.200
Average	"	4.000
Maximum	"	5.333
Slide		
Minimum	EA.	6.400
Maximum	"	7.273
Swings, plain seats		
8' high		
Minimum	EA.	5.333
Maximum	"	6.154
12' high		
Minimum	EA.	6.154
Maximum	"	8.889
02870.10 PREFABRICATED PLANTERS		
Concrete precast, circular		
24" dia., 18" high	EA.	0.800
42" dia., 30" high	"	1.000
Fiberglass, circular		
36" dia., 27" high	EA.	0.400
60" dia., 39" high	"	0.444
Tapered, circular		
24" dia., 36" high	EA.	0.364
40" dia., 36" high	"	0.400
Square		
2' by 2', 17" high	EA.	0.364
4' by 4', 39" high	"	0.444

SITE IMPROVEMENTS

02870.10 PREFABRICATED PLANTERS

	UNIT	MAN/HOURS
Rectangular		
4' by 1', 18" high	EA.	0.400

LANDSCAPING

02910.10 SHRUB & TREE MAINTENANCE

	UNIT	MAN/HOURS
Moving shrubs on site		
12" ball	EA.	1.000
24" ball	"	1.333
3' high	"	0.800
4' high	"	0.889
5' high	"	1.000
18" spread	"	1.143
30" spread	"	1.333
Moving trees on site		
24" ball	EA.	1.200
48" ball	"	1.600
Trees		
3' high	EA.	0.480
6' high	"	0.533
8' high	"	0.600
10' high	"	0.800
Palm trees		
7' high	EA.	0.600
10' high	"	0.800
20' high	"	2.400
40' high	"	4.800
Guying trees		
4" dia.	EA.	0.400
8" dia.	"	0.500

02920.10 TOPSOIL

	UNIT	MAN/HOURS
Spread topsoil, with equipment		
Minimum	C.Y.	0.080
Maximum	"	0.100
By hand		
Minimum	C.Y.	0.800
Maximum	"	1.000
Area preparation for seeding (grade, rake and clean)		
Square yard	S.Y.	0.006
By acre	ACRE	32.000
Remove topsoil and stockpile on site		
4" deep	C.Y.	0.067
6" deep	"	0.062
Spreading topsoil from stock pile		
By loader	C.Y.	0.073
By hand	"	0.800

LANDSCAPING

02920.10 TOPSOIL

	UNIT	MAN/HOURS
Top dress by hand	S.Y.	0.008
Place imported top soil		
By loader		
4" deep	S.Y.	0.008
6" deep	"	0.009
By hand		
4" deep	S.Y.	0.089
6" deep	"	0.100
Plant bed preparation, 18" deep		
With backhoe/loader	S.Y.	0.020
By hand	"	0.133

02930.30 SEEDING

	UNIT	MAN/HOURS
Mechanical seeding, 175 lb/acre		
By square yard	S.Y.	0.002
By acre	ACRE	8.000
450 lb/acre		
By square yard	S.Y.	0.002
By acre	ACRE	10.000
Seeding by hand, 10 lb per 100 s.y.		
By square yard	S.Y.	0.003
By acre	ACRE	13.333
Reseed disturbed areas	S.F.	0.004

02950.10 PLANTS

	UNIT	MAN/HOURS
Euonymus coloratus, 18" (purple wintercreeper)	EA.	0.133
Hedera Helix, 2-1/4" pot (English ivy)	"	0.133
Liriope muscari, 2" clumps	"	0.080
Santolina, 12"	"	0.080
Vinca major or minor, 3" pot	"	0.080
Cortaderia argentia, 2 gallon (pampas grass)	"	0.080
Ophiopogan japonicus, 1 quart (4" pot)	"	0.080
Ajuga reptans, 2-3/4" pot (carpet bugle)	"	0.080
Pachysandra terminalis, 2-3/4" pot (Japanese spurge)	"	0.080

02950.30 SHRUBS

	UNIT	MAN/HOURS
Juniperus conferia litoralis, 18"-24" (Shore Juniper)	EA.	0.320
Horizontalis plumosa, 18"-24" (Andorra Juniper)	"	0.320
Sabina tamar-iscfolia-tamarix juniper, 18"-24"	"	0.320
Chin San Hose, 18"-24" (San Hose Juniper)	"	0.320
Sargenti, 18"-24" (Sargent's Juniper)	"	0.320
Nandina domestica, 18"-24" (Heavenly Bamboo)	"	0.320
Raphiolepis Indica Springtime, 18"-24" Indian Hawthorn	"	0.320
Osmanthus Heterophyllus Gulftide, 18"-24" (Osmanthus)	"	0.320
Ilex Cornuta Burfordi Nana, 18"-24" (Dwarf Burford Holly)	"	0.320
Glabra, 18"-24" (Inkberry Holly)	"	0.320
Azalea, Indica types, 18"-24"	"	0.320
Kurume types, 18"-24"	"	0.320
Berberis Julianae, 18"-24" (Wintergreen Barberry)	"	0.320
Pieris Japonica Japanese, 18"-24" (Japanese Pieris)	"	0.320
Ilex Cornuta Rotunda, 18"-24" (Dwarf Chinese Holly)	"	0.320
Juniperus Horizontalis Plumosa, 24"-30" (Andorra Juniper)	"	0.400

LANDSCAPING	UNIT	MAN/ HOURS
02950.30 SHRUBS		
Rhodopendrow Hybrids, 24"-30"	EA.	0.400
Aucuba Japonica Varigata, 24"-30" (Gold Dust Aucuba)	"	0.400
Ilex Crenata Willow Leaf, 24"-30" (Japanese Holly)	"	0.400
Cleyera Japonica, 30"-36" (Japanese Cleyera)	"	0.500
Pittosporum Tobira, 30"-36"	"	0.500
Prumus Laurocerasus, 30"-36"	"	0.500
Ilex Cornuta Burfordi, 30"-36" (Burford Holly)	"	0.500
Abelia Grandiflora, 24"-36" (Yew Podocarpus)	"	0.400
Podocarpos Macrophylla, 24"-36" (Yew Podocarpus)	"	0.400
Pyracantha Coccinea Lalandi, 3'-4' (Firethorn)	"	0.500
Photinia Frazieri, 3'-4' (Red Photinia)	"	0.500
Forsythia Suspensa, 3'-4' (Weeping Forsythia)	"	0.500
Camellia Japonica, 3'-4' (Common Camellia)	"	0.500
Juniperus Chin Torulosa, 3'-4' (Hollywood Juniper)	"	0.500
Cupressocyparis Leylandi, 3'-4'	"	0.500
Ilex Opaca Fosteri, 5'-6' (Foster's Holly)	"	0.667
Opaca, 5'-6' (American Holly)	"	0.667
Nyrica Cerifera, 4'-5' (Southern Wax Myrtles)	"	0.571
Ligustrum Japonicum, 4'-5' (Japanese Privet)	"	0.571
02950.60 TREES		
Cornus Florida, 5'-6' (White flowering Dogwood)	EA.	0.667
Prunus Serrulata Kwanzan, 6'-8' (Kwanzan Cherry)	"	0.800
Caroliniana, 6'-8' (Carolina Cherry Laurel)	"	0.800
Cercis Canadensis, 6'-8' (Eastern Redbud)	"	0.800
Koelreuteria Paniculata, 8'-10' (Goldenrain tree)	"	1.000
Acer Platanoides, 1-3/4"-2" (11'-13') (Norway Maple)	"	1.333
Rubrum, 1-3/4"-2" (11'-13') (Red Maple)	"	1.333
Saccharum, 1-3/4"-2" (Sugar Maple)	"	1.333
Fraxinus Pennsylvanica, 1-3/4"-2" Laneolata-Green Ash	"	1.333
Celtis Occidentalis, 1-3/4"-2" (American Hackberry)	"	1.333
Glenditsia Triacantos Inermis, 2"	"	1.333
Prunus Cerasifera 'Thundercloud', 6'-8'	"	0.800
Yeodensis, 6'-8' (Yoshino Cherry)	"	0.800
Lagerstroemia Indica, 8'-10' (Crapemyrtle)	"	1.000
Crataegus Phaenopyrum, 8'-10' Washington Hawthorn	"	1.000
Quercus Borealis, 1-3/4"-2" (Northern Red Oak)	"	1.333
Quercus Acutissima, 1-3/4"-2" (8'-10') (Sawtooth Oak)	"	1.333
Saliz Babylonica, 1-3/4"-2" (Weeping Willow)	"	1.333
Tilia Cordata Greenspire, 1-3/4"-2" (10'-12')	"	1.333
Malus, 2"-2-1/2" (8'-10') (Flowering Crabapple)	"	1.333
Platanus Occidentalis, (12'-14')	"	1.600
Pyrus Calleryana Bradford, 2"-2-1/2" (Bradford Pear)	"	1.333
Quercus Palustris, 2"-2-1/2" (12'-14') (Pin Oak)	"	1.333
Phellos, 2-1/2"-3" (Willow Oak)	"	1.600
Nigra, 2"-2-1/2" (Water Oak)	"	1.333
Magnolia Soulangeana, 4'-5' (Saucer Magnolia)	"	0.667
Grandiflora, 6'-8' (Southern Magnolia)	"	0.800
Cedrus Deodara, 10'-12' (Deodare Cedar)	"	1.333
Ginkgo Biloba, 10'-12' (2"-2-1/2") (Maidenhair Tree)	"	1.333
Pinus Thunbergi, 5'-6' (Japanese Black Pine)	"	0.667
Strobus, 6'-8' (White Pine)	"	0.800
Taeda, 6'-8' (Loblolly Pine)	"	0.800

LANDSCAPING	UNIT	MAN/ HOURS
02950.60 TREES		
Quercus Virginiana, 2"-2-1/2" (live oak)	EA.	1.600
02970.10 FERTILIZING		
Fertilizing (23#/1000 sf)		
By square yard	S.Y.	0.002
By acre	ACRE	10.000
Liming (70#/1000 sf)		
By square yard	S.Y.	0.003
By acre	ACRE	13.333
02980.10 LANDSCAPE ACCESSORIES		
Steel edging, 3/16" x 4"	L.F.	0.010
Landscaping stepping stones, 15"x15", white	EA.	0.040
Wood chip mulch	C.Y.	0.533
2" thick	S.Y.	0.016
4" thick	"	0.023
6" thick	"	0.029
Gravel mulch, 3/4" stone	C.Y.	0.800
White marble chips, 1" deep	S.F.	0.008
Peat moss		
2" thick	S.Y.	0.018
4" thick	"	0.027
6" thick	"	0.033
Landscaping timbers, treated lumber		
4" x 4"	L.F.	0.027
6" x 6"	"	0.029
8" x 8"	"	0.033

FORMWORK	UNIT	MAN/ HOURS
03110.05 BEAM FORMWORK		
Beam forms, job built		
Beam bottoms		
1 use	S.F.	0.080
2 uses	"	0.127
3 uses	"	0.123
4 uses	"	0.118
5 uses	"	0.114
Beam sides		
1 use	S.F.	0.089
2 uses	"	0.084
3 uses	"	0.080
4 uses	"	0.076
5 uses	"	0.073
03110.10 BOX CULVERT FORMWORK		
Box culverts, job built		
6' x 6'		
1 use	S.F.	0.080
2 uses	"	0.076
3 uses	"	0.073
4 uses	"	0.070
5 uses	"	0.067
8' x 12'		
1 use	S.F.	0.067
2 uses	"	0.064
3 uses	"	0.062
4 uses	"	0.059
5 uses	"	0.057
03110.15 COLUMN FORMWORK		
Column, square forms, job built		
8" x 8" columns		
1 use	S.F.	0.160
2 uses	"	0.154
3 uses	"	0.148
4 uses	"	0.143
5 uses	"	0.138
12" x 12" columns		
1 use	S.F.	0.145
2 uses	"	0.140
3 uses	"	0.136
4 uses	"	0.131
5 uses	"	0.127
16" x 16" columns		
1 use	S.F.	0.133
2 uses	"	0.129
3 uses	"	0.125
4 uses	"	0.121
5 uses	"	0.118
24" x 24" columns		
1 use	S.F.	0.123
2 uses	"	0.119

FORMWORK	UNIT	MAN/ HOURS
03110.15 COLUMN FORMWORK		
3 uses	S.F.	0.116
4 uses	"	0.113
5 uses	"	0.110
36" x 36" columns		
1 use	S.F.	0.114
2 uses	"	0.111
3 uses	"	0.108
4 uses	"	0.105
5 uses	"	0.103
Round fiber forms, 1 use		
10" dia.	L.F.	0.160
12" dia.	"	0.163
14" dia.	"	0.170
16" dia.	"	0.178
18" dia.	"	0.190
24" dia.	"	0.205
30" dia.	"	0.222
36" dia.	"	0.242
42" dia.	"	0.267
03110.18 CURB FORMWORK		
Curb forms		
Straight, 6" high		
1 use	L.F.	0.080
2 uses	"	0.076
3 uses	"	0.073
4 uses	"	0.070
5 uses	"	0.067
Curved, 6" high		
1 use	L.F.	0.100
2 uses	"	0.094
3 uses	"	0.089
4 uses	"	0.085
5 uses	"	0.082
03110.20 ELEVATED SLAB FORMWORK		
Elevated slab formwork		
Slab, with drop panels		
1 use	S.F.	0.064
2 uses	"	0.062
3 uses	"	0.059
4 uses	"	0.057
5 uses	"	0.055
Floor slab, hung from steel beams		
1 use	S.F.	0.062
2 uses	"	0.059
3 uses	"	0.057
4 uses	"	0.055
5 uses	"	0.053
Floor slab, with pans or domes		
1 use	S.F.	0.073
2 uses	"	0.070

03 CONCRETE

FORMWORK	UNIT	MAN/ HOURS
03110.20 ELEVATED SLAB FORMWORK		
3 uses	S.F.	0.067
4 uses	"	0.064
5 uses	"	0.062
Equipment curbs, 12" high		
1 use	L.F.	0.080
2 uses	"	0.076
3 uses	"	0.073
4 uses	"	0.070
5 uses	"	0.067
03110.25 EQUIPMENT PAD FORMWORK		
Equipment pad, job built		
1 use	S.F.	0.100
2 uses	"	0.094
3 uses	"	0.089
4 uses	"	0.084
5 uses	"	0.080
03110.35 FOOTING FORMWORK		
Wall footings, job built, continuous		
1 use	S.F.	0.080
2 uses	"	0.076
3 uses	"	0.073
4 uses	"	0.070
5 uses	"	0.067
Column footings, spread		
1 use	S.F.	0.100
2 uses	"	0.094
3 uses	"	0.089
4 uses	"	0.084
5 uses	"	0.080
03110.50 GRADE BEAM FORMWORK		
Grade beams, job built		
1 use	S.F.	0.080
2 uses	"	0.076
3 uses	"	0.073
4 uses	"	0.070
5 uses	"	0.067
03110.53 PILE CAP FORMWORK		
Pile cap forms, job built		
Square		
1 use	S.F.	0.100
2 uses	"	0.094
3 uses	"	0.089
4 uses	"	0.084
5 uses	"	0.080
Triangular		
1 use	S.F.	0.114
2 uses	"	0.107
3 uses	"	0.100
4 uses	"	0.094

FORMWORK	UNIT	MAN/ HOURS
03110.53 PILE CAP FORMWORK		
5 uses	S.F.	0.089
03110.55 SLAB/MAT FORMWORK		
Mat foundations, job built		
1 use	S.F.	0.100
2 uses	"	0.094
3 uses	"	0.089
4 uses	"	0.084
5 uses	"	0.080
Edge forms		
6" high		
1 use	L.F.	0.073
2 uses	"	0.070
3 uses	"	0.067
4 uses	"	0.064
5 uses	"	0.062
12" high		
1 use	L.F.	0.080
2 uses	"	0.076
3 uses	"	0.073
4 uses	"	0.070
5 uses	"	0.067
Formwork for openings		
1 use	S.F.	0.160
2 uses	"	0.145
3 uses	"	0.133
4 uses	"	0.123
5 uses	"	0.114
03110.60 STAIR FORMWORK		
Stairway forms, job built		
1 use	S.F.	0.160
2 uses	"	0.145
3 uses	"	0.133
4 uses	"	0.123
5 uses	"	0.114
Stairs, elevated		
1 use	S.F.	0.160
2 uses	"	0.133
3 uses	"	0.114
4 uses	"	0.107
5 uses	"	0.100
03110.65 WALL FORMWORK		
Wall forms, exterior, job built		
Up to 8' high wall		
1 use	S.F.	0.080
2 uses	"	0.076
3 uses	"	0.073
4 uses	"	0.070
5 uses	"	0.067

03 CONCRETE

FORMWORK	UNIT	MAN/ HOURS
03110.65 WALL FORMWORK		
Over 8' high wall		
1 use	S.F.	0.100
2 uses	"	0.094
3 uses	"	0.089
4 uses	"	0.084
5 uses	"	0.080
Over 16' high wall		
1 use	S.F.	0.114
2 uses	"	0.107
3 uses	"	0.100
4 uses	"	0.094
5 uses	"	0.089
Radial wall forms		
1 use	S.F.	0.123
2 uses	"	0.114
3 uses	"	0.107
4 uses	"	0.100
5 uses	"	0.094
Retaining wall forms		
1 use	S.F.	0.089
2 uses	"	0.084
3 uses	"	0.080
4 uses	"	0.076
5 uses	"	0.073
Radial retaining wall forms		
1 use	S.F.	0.133
2 uses	"	0.123
3 uses	"	0.114
4 uses	"	0.107
5 uses	"	0.100
Column pier and pilaster		
1 use	S.F.	0.160
2 uses	"	0.145
3 uses	"	0.133
4 uses	"	0.123
5 uses	"	0.114
Interior wall forms		
Up to 8' high		
1 use	S.F.	0.073
2 uses	"	0.070
3 uses	"	0.067
4 uses	"	0.064
5 uses	"	0.062
Over 8' high		
1 use	S.F.	0.089
2 uses	"	0.084
3 uses	"	0.080
4 uses	"	0.076
5 uses	"	0.073
Over 16' high		
1 use	S.F.	0.100
2 uses	"	0.094
3 uses	"	0.089
4 uses	"	0.084

FORMWORK	UNIT	MAN/ HOURS
03110.65 WALL FORMWORK		
5 uses	S.F.	0.080
Radial wall forms		
1 use	S.F.	0.107
2 uses	"	0.100
3 uses	"	0.094
4 uses	"	0.089
5 uses	"	0.084
Curved wall forms, 24" sections		
1 use	S.F.	0.160
2 uses	"	0.145
3 uses	"	0.133
4 uses	"	0.123
5 uses	"	0.114
PVC form liner, per side, smooth finish		
1 use	S.F.	0.067
2 uses	"	0.064
3 uses	"	0.062
4 uses	"	0.057
5 uses	"	0.053
03110.90 MISCELLANEOUS FORMWORK		
Keyway forms (5 uses)		
2 x 4	L.F.	0.040
2 x 6	"	0.044
Bulkheads		
Walls, with keyways		
2 piece	L.F.	0.073
3 piece	"	0.080
Elevated slab, with keyway		
2 piece	L.F.	0.067
3 piece	"	0.073
Ground slab, with keyway		
2 piece	L.F.	0.057
3 piece	"	0.062
Chamfer strips		
Wood		
1/2" wide	L.F.	0.018
3/4" wide	"	0.018
1" wide	"	0.018
PVC		
1/2" wide	L.F.	0.018
3/4" wide	"	0.018
1" wide	"	0.018
Radius		
1"	L.F.	0.019
1-1/2"	"	0.019
Reglets		
Galvanized steel, 24 ga.	L.F.	0.032
Metal formwork		
Straight edge forms		
4" high	L.F.	0.050
6" high	"	0.053
8" high	"	0.057

FORMWORK

03110.90 MISCELLANEOUS FORMWORK

	UNIT	MAN/ HOURS
12" high	L.F.	0.062
16" high	"	0.067
Curb form, S-shape		
12" x		
1'-6"	L.F.	0.114
2'	"	0.107
2'-6"	"	0.100
3'	"	0.089

REINFORCEMENT

03210.05 BEAM REINFORCING

	UNIT	MAN/ HOURS
Beam-girders		
#3 - #4	TON	20.000
#5 - #6	"	16.000
#7 - #8	"	13.333
#9 - #10	"	11.429
#11 - #12	"	10.667
#13 - #14	"	10.000
Galvanized		
#3 - #4	TON	20.000
#5 - #6	"	16.000
#7 - #8	"	13.333
#9 - #10	"	11.429
#11 - #12	"	10.667
#13 - #14	"	10.000
Epoxy coated		
#3 - #4	TON	22.857
#5 - #6	"	17.778
#7 - #8	"	14.545
#9 - #10	"	12.308
#11 - #12	"	11.429
#13 - #14	"	10.667
Bond Beams		
#3 - #4	TON	26.667
#5 - #6	"	20.000
#7 - #8	"	17.778
Galvanized		
#3 - #4	TON	26.667
#5 - #6	"	20.000
#7 - #8	"	17.778
Epoxy coated		
#3 - #4	TON	32.000
#5 - #6	"	22.857
#7 - #8	"	20.000

REINFORCEMENT

03210.10 BOX CULVERT REINFORCING

	UNIT	MAN/ HOURS
Box culverts		
#3 - #4	TON	10.000
#5 - #6	"	8.889
#7 - #8	"	8.000
#9 - #10	"	7.273
#11 - #12	"	6.667
Galvanized		
#3 - #4	TON	10.000
#5 - #6	"	8.889
#7 - #8	"	8.000
#9 - #10	"	7.273
#11 - #12	"	6.667
Epoxy coated		
#3 - #4	TON	10.667
#5 - #6	"	9.412
#7 - #8	"	8.421
#9 - #10	"	7.619
#11 - #12	"	6.957

03210.15 COLUMN REINFORCING

	UNIT	MAN/ HOURS
Columns		
#3 - #4	TON	22.857
#5 - #6	"	17.778
#7 - #8	"	16.000
#9 - #10	"	14.545
#11 - #12	"	13.333
#13 - #14	"	12.308
#15 - #16	"	11.429
Galvanized		
#3 - #4	TON	22.857
#5 - #6	"	17.778
#7 - #8	"	16.000
#9 - #10	"	14.545
#11 - #12	"	13.333
#13 - #14	"	12.308
#15 - #16	"	11.429
Epoxy coated		
#3 - #4	TON	26.667
#5 - #6	"	20.000
#7 - #8	"	17.778
#9 - #10	"	16.000
#11 - #12	"	14.545
#13 - #14	"	13.333
#15 - #16	"	12.308
Spirals		
8" to 24" dia.	TON	20.000
24" to 48" dia.	"	17.778
48" to 84" dia.	"	16.000

03210.20 ELEVATED SLAB REINFORCING

	UNIT	MAN/ HOURS
Elevated slab		
#3 - #4	TON	10.000
#5 - #6	"	8.889
#7 - #8	"	8.000

03 CONCRETE

REINFORCEMENT	UNIT	MAN/HOURS
03210.20 ELEVATED SLAB REINFORCING		
#9 - #10	TON	7.273
#11 - #12	"	6.667
Galvanized		
#3 - #4	TON	10.000
#5 - #6	"	8.889
#7 - #8	"	8.000
#9 - #10	"	7.273
#11 - #12	"	6.667
Epoxy coated		
#3 - #4	TON	10.667
#5 - #6	"	9.412
#7 - #8	"	8.421
#9 - #10	"	7.619
#11 - #12	"	6.957
03210.25 EQUIP. PAD REINFORCING		
Equipment pad		
#3 - #4	TON	16.000
#5 - #6	"	14.545
#7 - #8	"	13.333
#9 - #10	"	12.308
#11 - #12	"	11.429
03210.35 FOOTING REINFORCING		
Footings		
Grade 50		
#3 - #4	TON	13.333
#5 - #6	"	11.429
#7 - #8	"	10.000
#9 - #10	"	8.889
Grade 60		
#3 - #4	TON	13.333
#5 - #6	"	11.429
#7 - #8	"	10.000
#9 - #10	"	8.889
Grade 70		
#3 - #4	TON	13.333
#5 - #6	"	11.429
#7 - #8	"	10.000
#9 - #10	"	8.889
#11- #12	"	8.000
Straight dowels, 24" long		
1" dia. (#8)	EA.	0.080
3/4" dia. (#6)	"	0.080
5/8" dia. (#5)	"	0.067
1/2" dia. (#4)	"	0.057
03210.45 FOUNDATION REINFORCING		
Foundations		
#3 - #4	TON	13.333
#5 - #6	"	11.429
#7 - #8	"	10.000
#9 - #10	"	8.889
#11 - #12	"	8.000

REINFORCEMENT	UNIT	MAN/HOURS
03210.45 FOUNDATION REINFORCING		
Galvanized		
#3 - #4	TON	13.333
#5 - #6	"	11.429
#7 - #8	"	10.000
#9 - #10	"	8.889
#11 - #12	"	8.000
Epoxy Coated		
#3 - #4	TON	14.545
#5 - #6	"	12.308
#7 - #8	"	10.667
#9 - #10	"	9.412
#11 - #12	"	8.421
03210.50 GRADE BEAM REINFORCING		
Grade beams		
#3 - #4	TON	12.308
#5 - #6	"	10.667
#7 - #8	"	9.412
#9 - #10	"	8.421
#11 - #12	"	7.619
Galvanized		
#3 - #4	TON	12.308
#5 - #6	"	10.667
#7 - #8	"	9.412
#9 - #10	"	8.421
#11 - #12	"	7.619
Epoxy coated		
#3 - #4	TON	13.333
#5 - #6	"	11.429
#7 - #8	"	10.000
#9 - #10	"	8.889
#11 - #12	"	8.000
03210.53 PILE CAP REINFORCING		
Pile caps		
#3 - #4	TON	20.000
#5 - #6	"	17.778
#7 - #8	"	16.000
#9 - #10	"	14.545
#11 - #12	"	13.333
Galvanized		
#3 - #4	TON	20.000
#5 - #6	"	17.778
#7 - #8	"	16.000
#9 - #10	"	14.545
#11 - #12	"	13.333
Epoxy coated		
#3 - #4	TON	22.857
#5 - #6	"	20.000
#7 - #8	"	17.778
#9 - #10	"	16.000
#11 - #12	"	14.545

03 CONCRETE

REINFORCEMENT

03210.55 SLAB/MAT REINFORCING	UNIT	MAN/HOURS
Bars, slabs		
#3 - #4	TON	13.333
#5 - #6	"	11.429
#7 - #8	"	10.000
#9 - #10	"	8.889
#11 - #12	"	8.000
Galvanized		
#3 - #4	TON	13.333
#5 - #6	"	11.429
#7 - #8	"	10.000
#9 - #10	"	8.889
#11 - #12	"	8.000
Epoxy coated		
#3 - #4	TON	14.545
#5 - #6	"	12.308
#7 - #8	"	10.667
#9 - #10	"	9.412
#11 - #12	"	8.421
Wire mesh, slabs		
Galvanized		
4x4		
W1.4xW1.4	S.F.	0.005
W2.0xW2.0	"	0.006
W2.9xW2.9	"	0.006
W4.0xW4.0	"	0.007
6x6		
W1.4xW1.4	S.F.	0.004
W2.0xW2.0	"	0.004
W2.9xW2.9	"	0.005
W4.0xW4.0	"	0.005
Standard		
2x2		
W.9xW.9	S.F.	0.005
4x4		
W1.4xW1.4	S.F.	0.005
W2.0xW2.0	"	0.006
W2.9xW2.9	"	0.006
W4.0xW4.0	"	0.007
6x6		
W1.4xW1.4	S.F.	0.004
W2.0xW2.0	"	0.004
W2.9xW2.9	"	0.005
W4.0xW4.0	"	0.005

03210.60 STAIR REINFORCING	UNIT	MAN/HOURS
Stairs		
#3 - #4	TON	16.000
#5 - #6	"	13.333
#7 - #8	"	11.429
#9 - #10	"	10.000
Galvanized		
#3 - #4	TON	16.000
#5 - #6	"	13.333

REINFORCEMENT

03210.60 STAIR REINFORCING	UNIT	MAN/HOURS
#7 - #8	TON	11.429
#9 - #10	"	10.000
Epoxy coated		
#3 - #4	TON	17.778
#5 - #6	"	14.545
#7 - #8	"	12.308
#9 - #10	"	10.667

03210.65 WALL REINFORCING	UNIT	MAN/HOURS
Walls		
#3 - #4	TON	11.429
#5 - #6	"	10.000
#7 - #8	"	8.889
#9 - #10	"	8.000
Galvanized		
#3 - #4	TON	11.429
#5 - #6	"	10.000
#7 - #8	"	8.889
#9 - #10	"	8.000
Epoxy coated		
#3 - #4	TON	12.308
#5 - #6	"	10.667
#7 - #8	"	9.412
#9 - #10	"	8.421
Masonry wall (horizontal)		
#3 - #4	TON	32.000
#5 - #6	"	26.667
Galvanized		
#3 - #4	TON	32.000
#5 - #6	"	26.667
Masonry wall (vertical)		
#3 - #4	TON	40.000
#5 - #6	"	32.000
Galvanized		
#3 - #4	TON	40.000
#5 - #6	"	32.000

ACCESSORIES

03250.40 CONCRETE ACCESSORIES	UNIT	MAN/HOURS
Expansion joint, poured		
Asphalt		
1/2" x 1"	L.F.	0.016
1" x 2"	"	0.017
Liquid neoprene, cold applied		

03 CONCRETE

ACCESSORIES	UNIT	MAN/ HOURS
03250.40 CONCRETE ACCESSORIES		
1/2" x 1"	L.F.	0.016
1" x 2"	"	0.018
Polyurethane, 2 parts		
1/2" x 1"	L.F.	0.027
1" x 2"	"	0.029
Rubberized asphalt, cold		
1/2" x 1"	L.F.	0.016
1" x 2"	"	0.017
Hot, fuel resistant		
1/2" x 1"	L.F.	0.016
1" x 2"	"	0.017
Expansion joint, premolded, in slabs		
Asphalt		
1/2" x 6"	L.F.	0.020
1" x 12"	"	0.027
Cork		
1/2" x 6"	L.F.	0.020
1" x 12"	"	0.027
Neoprene sponge		
1/2" x 6"	L.F.	0.020
1" x 12"	"	0.027
Polyethylene foam		
1/2" x 6"	L.F.	0.020
1" x 12"	"	0.027
Polyurethane foam		
1/2" x 6"	L.F.	0.020
1" x 12"	"	0.027
Polyvinyl chloride foam		
1/2" x 6"	L.F.	0.020
1" x 12"	"	0.027
Rubber, gray sponge		
1/2" x 6"	L.F.	0.020
1" x 12"	"	0.027
Asphalt felt control joints or bond breaker, screed joints		
4" slab	L.F.	0.016
6" slab	"	0.018
8" slab	"	0.020
10" slab	"	0.023
Keyed cold expansion and control joints, 24 ga.		
4" slab	L.F.	0.050
5" slab	"	0.050
6" slab	"	0.053
8" slab	"	0.057
10" slab	"	0.062
Waterstops		
Polyvinyl chloride		
Ribbed		
3/16" thick x		
4" wide	L.F.	0.040
6" wide	"	0.044
1/2" thick x		
9" wide	L.F.	0.050
Ribbed with center bulb		
3/16" thick x 9" wide	L.F.	0.050

ACCESSORIES	UNIT	MAN/ HOURS
03250.40 CONCRETE ACCESSORIES		
3/8" thick x 9" wide	L.F.	0.050
Dumbbell type, 3/8" thick x 6" wide	"	0.044
Plain, 3/8" thick x 9" wide	"	0.050
Center bulb, 3/8" thick x 9" wide	"	0.050
Rubber		
Flat dumbbell		
3/8" thick x		
6" wide	L.F.	0.044
9" wide	"	0.050
Center bulb		
3/8" thick x		
6" wide	L.F.	0.044
9" wide	"	0.050
Vapor barrier		
4 mil polyethylene	S.F.	0.003
6 mil polyethylene	"	0.003
Gravel porous fill, under floor slabs, 3/4" stone	C.Y.	1.333
Reinforcing accessories		
Beam bolsters		
1-1/2" high, plain	L.F.	0.008
Galvanized	"	0.008
3" high		
Plain	L.F.	0.010
Galvanized	"	0.010
Slab bolsters		
1" high		
Plain	L.F.	0.004
Galvanized	"	0.004
2" high		
Plain	L.F.	0.004
Galvanized	"	0.004
Chairs, high chairs		
3" high		
Plain	EA.	0.020
Galvanized	"	0.020
5" high		
Plain	EA.	0.021
Galvanized	"	0.021
8" high		
Plain	EA.	0.023
Galvanized	"	0.023
12" high		
Plain	EA.	0.027
Galvanized	"	0.027
Continuous, high chair		
3" high		
Plain	L.F.	0.005
Galvanized	"	0.005
5" high		
Plain	L.F.	0.006
Galvanized	"	0.006
8" high		
Plain	L.F.	0.006
Galvanized	"	0.006

03 CONCRETE

ACCESSORIES

03250.40 CONCRETE ACCESSORIES

	UNIT	MAN/HOURS
12" high		
Plain	L.F.	0.007
Galvanized	"	0.007

CAST-IN-PLACE CONCRETE

03350.10 CONCRETE FINISHES

	UNIT	MAN/HOURS
Floor finishes		
Broom	S.F.	0.011
Screed	"	0.010
Darby	"	0.010
Steel float	"	0.013
Granolithic topping		
1/2" thick	S.F.	0.036
1" thick	"	0.040
2" thick	"	0.044
Wall finishes		
Burlap rub, with cement paste	S.F.	0.013
Float finish	"	0.020
Etch with acid	"	0.013
Sandblast		
Minimum	S.F.	0.016
Maximum	"	0.016
Bush hammer		
Green concrete	S.F.	0.040
Cured concrete	"	0.062
Break ties and patch holes	"	0.016
Carborundum		
Dry rub	S.F.	0.027
Wet rub	"	0.040
Floor hardeners		
Metallic		
Light service	S.F.	0.010
Heavy service	"	0.013
Non-metallic		
Light service	S.F.	0.010
Heavy service	"	0.013
Rusticated concrete finish		
Beveled edge	L.F.	0.044
Square edge	"	0.057
Solid board concrete finish		
Standard	S.F.	0.067
Rustic	"	0.080

CAST-IN-PLACE CONCRETE

03360.10 PNEUMATIC CONCRETE

	UNIT	MAN/HOURS
Pneumatic applied concrete (gunite)		
2" thick	S.F.	0.030
3" thick	"	0.040
4" thick	"	0.048
Finish surface		
Minimum	S.F.	0.040
Maximum	"	0.080

03370.10 CURING CONCRETE

	UNIT	MAN/HOURS
Sprayed membrane		
Slabs	S.F.	0.002
Walls	"	0.002
Curing paper		
Slabs	S.F.	0.002
Walls	"	0.002
Burlap		
7.5 oz.	S.F.	0.003
12 oz.	"	0.003

PLACING CONCRETE

03380.05 BEAM CONCRETE

	UNIT	MAN/HOURS
Beams and girders		
2500# or 3000# concrete		
By crane	C.Y.	0.960
By pump	"	0.873
By hand buggy	"	0.800
3500# or 4000# concrete		
By crane	C.Y.	0.960
By pump	"	0.873
By hand buggy	"	0.800
5000# concrete		
By crane	C.Y.	0.960
By pump	"	0.873
By hand buggy	"	0.800
Bond beam, 3000# concrete		
By pump		
8" high		
4" wide	L.F.	0.019
6" wide	"	0.022
8" wide	"	0.024
10" wide	"	0.027
12" wide	"	0.030
16" high		
8" wide	L.F.	0.030

03 CONCRETE

PLACING CONCRETE

03380.05 BEAM CONCRETE

PLACING CONCRETE	UNIT	MAN/ HOURS
10" wide	L.F.	0.034
12" wide	"	0.040
By crane		
8" high		
4" wide	L.F.	0.021
6" wide	"	0.023
8" wide	"	0.024
10" wide	"	0.027
12" wide	"	0.030
16" high		
8" wide	L.F.	0.030
10" wide	"	0.032
12" wide	"	0.037

03380.15 COLUMN CONCRETE

PLACING CONCRETE	UNIT	MAN/ HOURS
Columns		
2500# or 3000# concrete		
By crane	C.Y.	0.873
By pump	"	0.800
3500# or 4000# concrete		
By crane	C.Y.	0.873
By pump	"	0.800
5000# concrete		
By crane	C.Y.	0.873
By pump	"	0.800

03380.20 ELEVATED SLAB CONCRETE

PLACING CONCRETE	UNIT	MAN/ HOURS
Elevated slab		
2500# or 3000# concrete		
By crane	C.Y.	0.480
By pump	"	0.369
By hand buggy	"	0.800
3500# or 4000# concrete		
By crane	C.Y.	0.480
By pump	"	0.369
By hand buggy	"	0.800
5000# concrete		
By crane	C.Y.	0.480
By pump	"	0.369
By hand buggy	"	0.800
Topping		
2500# or 3000# concrete		
By crane	C.Y.	0.480
By pump	"	0.369
By hand buggy	"	0.800
3500# or 4000# concrete		
By crane	C.Y.	0.480
By pump	"	0.369
By hand buggy	"	0.800
5000# concrete		
By crane	C.Y.	0.480
By pump	"	0.369
By hand buggy	"	0.800

03380.25 EQUIPMENT PAD CONCRETE

PLACING CONCRETE	UNIT	MAN/ HOURS
Equipment pad		
2500# or 3000# concrete		
By chute	C.Y.	0.267
By pump	"	0.686
By crane	"	0.800
3500# or 4000# concrete		
By chute	C.Y.	0.267
By pump	"	0.686
By crane	"	0.800
5000# concrete		
By chute	C.Y.	0.267
By pump	"	0.686
By crane	"	0.800

03380.35 FOOTING CONCRETE

PLACING CONCRETE	UNIT	MAN/ HOURS
Continuous footing		
2500# or 3000# concrete		
By chute	C.Y.	0.267
By pump	"	0.600
By crane	"	0.686
3500# or 4000# concrete		
By chute	C.Y.	0.267
By pump	"	0.600
By crane	"	0.686
5000# concrete		
By chute	C.Y.	0.267
By pump	"	0.600
By crane	"	0.686
Spread footing		
2500# or 3000# concrete		
Under 5 cy		
By chute	C.Y.	0.267
By pump	"	0.640
By crane	"	0.738
Over 5 cy		
By chute	C.Y.	0.200
By pump	"	0.565
By crane	"	0.640
3500# or 4000# concrete		
Under 5 c.y.		
By chute	C.Y.	0.267
By pump	"	0.640
By crane	"	0.738
Over 5 c.y.		
By chute	C.Y.	0.200
By pump	"	0.565
By crane	"	0.640
5000# concrete		
Under 5 c.y.		
By chute	C.Y.	0.267
By pump	"	0.640
By crane	"	0.738
Over 5 c.y.		

03 CONCRETE

PLACING CONCRETE

PLACING CONCRETE	UNIT	MAN/HOURS
03380.35 FOOTING CONCRETE		
By chute	C.Y.	0.200
By pump	"	0.565
By crane	"	0.640
03380.50 GRADE BEAM CONCRETE		
Grade beam		
2500# or 3000# concrete		
By chute	C.Y.	0.267
By crane	"	0.686
By pump	"	0.600
By hand buggy	"	0.800
3500# or 4000# concrete		
By chute	C.Y.	0.267
By crane	"	0.686
By pump	"	0.600
By hand buggy	"	0.800
5000# concrete		
By chute	C.Y.	0.267
By crane	"	0.686
By pump	"	0.600
By hand buggy	"	0.800
03380.53 PILE CAP CONCRETE		
Pile cap		
2500# or 3000 concrete		
By chute	C.Y.	0.267
By crane	"	0.800
By pump	"	0.686
By hand buggy	"	0.800
3500# or 4000# concrete		
By chute	C.Y.	0.267
By crane	"	0.800
By pump	"	0.686
By hand buggy	"	0.800
5000# concrete		
By chute	C.Y.	0.267
By crane	"	0.800
By pump	"	0.686
By hand buggy	"	0.800
03380.55 SLAB/MAT CONCRETE		
Slab on grade		
2500# or 3000# concrete		
By chute	C.Y.	0.200
By crane	"	0.400
By pump	"	0.343
By hand buggy	"	0.533
3500# or 4000# concrete		
By chute	C.Y.	0.200
By crane	"	0.400
By pump	"	0.343
By hand buggy	"	0.533
5000# concrete		

PLACING CONCRETE	UNIT	MAN/HOURS
03380.55 SLAB/MAT CONCRETE		
By chute	C.Y.	0.200
By crane	"	0.400
By pump	"	0.343
By hand buggy	"	0.533
Foundation mat		
2500# or 3000# concrete, over 20 cy		
By chute	C.Y.	0.160
By crane	"	0.343
By pump	"	0.300
By hand buggy	"	0.400
03380.58 SIDEWALKS		
Walks, cast in place with wire mesh, base not incl.		
4" thick	S.F.	0.027
5" thick	"	0.032
6" thick	"	0.040
03380.60 STAIR CONCRETE		
Stairs		
2500# or 3000# concrete		
By chute	C.Y.	0.267
By crane	"	0.800
By pump	"	0.686
By hand buggy	"	0.800
3500# or 4000# concrete		
By chute	C.Y.	0.267
By crane	"	0.800
By pump	"	0.686
By hand buggy	"	0.800
5000# concrete		
By chute	C.Y.	0.267
By crane	"	0.800
By pump	"	0.686
By hand buggy	"	0.800
03380.65 WALL CONCRETE		
Walls		
2500# or 3000# concrete		
To 4'		
By chute	C.Y.	0.229
By crane	"	0.800
By pump	"	0.738
To 8'		
By crane	C.Y.	0.873
By pump	"	0.800
To 16'		
By crane	C.Y.	0.960
By pump	"	0.873
Over 16'		
By crane	C.Y.	1.067
By pump	"	0.960
3500# or 4000# concrete		
To 4'		
By chute	C.Y.	0.229

PLACING CONCRETE

03380.65 WALL CONCRETE	UNIT	MAN/ HOURS
By crane	C.Y.	0.800
By pump	"	0.738
To 8'		
By crane	C.Y.	0.873
By pump	"	0.800
To 16'		
By crane	C.Y.	0.960
By pump	"	0.873
Over 16'		
By crane	C.Y.	1.067
By pump	"	0.960
5000# concrete		
To 4'		
By chute	C.Y.	0.229
By crane	"	0.800
By pump	"	0.738
To 8'		
By crane	C.Y.	0.873
By pump	"	0.800
To 16'		
By crane	C.Y.	0.960
By pump	"	0.873
Filled block (CMU)		
3000# concrete, by pump		
4" wide	S.F.	0.034
6" wide	"	0.040
8" wide	"	0.048
10" wide	"	0.056
12" wide	"	0.069
Pilasters, 3000# concrete	C.F.	0.960
Wall cavity, 2" thick, 3000# concrete	S.F.	0.032

PRECAST CONCRETE

03400.10 PRECAST BEAMS	UNIT	MAN/ HOURS
Prestressed, double tee, 24" deep, 8' wide		
35' span		
115 psf	S.F.	0.008
140 psf	"	0.008
40' span		
80 psf	S.F.	0.009
143 psf	"	0.009
45' span		
50 psf	S.F.	0.007
70 psf	"	0.007
100 psf	"	0.007

PRECAST CONCRETE

03400.10 PRECAST BEAMS	UNIT	MAN/ HOURS
130 psf	S.F.	0.007
50' span		
75 psf	S.F.	0.007
100 psf	"	0.007
Precast beams, girders and joists		
1000 lb/lf live load		
10' span	L.F.	0.160
20' span	"	0.096
30' span	"	0.080
3000 lb/lf live load		
10' span	L.F.	0.160
20' span	"	0.096
30' span	"	0.080
5000 lb/lf live load		
10' span	L.F.	0.160
20' span	"	0.096
30' span	"	0.080

03400.20 PRECAST COLUMNS	UNIT	MAN/ HOURS
Prestressed concrete columns		
10" x 10"		
10' long	EA.	0.960
15' long	"	1.000
20' long	"	1.067
25' long	"	1.143
30' long	"	1.200
12" x 12"		
20' long	EA.	1.200
25' long	"	1.297
30' long	"	1.371
16" x 16"		
20' long	EA.	1.200
25' long	"	1.297
30' long	"	1.371
20" x 20"		
20' long	EA.	1.263
25' long	"	1.333
30' long	"	1.412
24" x 24"		
20' long	EA.	1.333
25' long	"	1.412
30' long	"	1.500
28" x 28"		
20' long	EA.	1.500
25' long	"	1.600
30' long	"	1.714
32" x 32"		
20' long	EA.	1.600
25' long	"	1.714
30' long	"	1.846
36" x 36"		
20' long	EA.	1.714
25' long	"	1.846

03 CONCRETE

PRECAST CONCRETE	UNIT	MAN/HOURS
03400.20 PRECAST COLUMNS		
30' long	EA.	2.000
03400.30 PRECAST SLABS		
Prestressed flat slab		
6" thick, 4' wide		
20' span		
80 psf	S.F.	0.020
110 psf	"	0.020
25' span		
80 psf	S.F.	0.019
Cored slab		
6" thick, 4' wide		
20' span		
80 psf	S.F.	0.020
100 psf	"	0.020
130 psf	"	0.020
8" thick, 4' wide		
25' span		
70 psf	S.F.	0.019
125 psf	"	0.019
170 psf	"	0.019
30' span		
70 psf	S.F.	0.016
90 psf	"	0.016
35' span		
70 psf	S.F.	0.015
10" thick, 4' wide		
30' span		
75 psf	S.F.	0.016
100 psf	"	0.016
130 psf	"	0.016
35' span		
60 psf	S.F.	0.015
80 psf	"	0.015
120 psf	"	0.015
40' span		
65 psf	S.F.	0.012
Slabs, roof and floor members, 4' wide		
6" thick, 25' span	S.F.	0.019
8" thick, 30' span	"	0.015
10" thick, 40' span	"	0.013
Tee members		
Multiple tee, roof and floor		
Minimum	S.F.	0.012
Maximum	"	0.024
Double tee wall member		
Minimum	S.F.	0.014
Maximum	"	0.027
Single tee		
Short span, roof members		
Minimum	S.F.	0.015
Maximum	"	0.030

PRECAST CONCRETE	UNIT	MAN/HOURS
03400.30 PRECAST SLABS		
Long span, roof members		
Minimum	S.F.	0.012
Maximum	"	0.024
03400.40 PRECAST WALLS		
Wall panel, 8' x 20'		
Gray cement		
Liner finish		
4" wall	S.F.	0.014
5" wall	"	0.014
6" wall	"	0.015
8" wall	"	0.015
Sandblast finish		
4" wall	S.F.	0.014
5" wall	"	0.014
6" wall	"	0.015
8" wall	"	0.015
White cement		
Liner finish		
4" wall	S.F.	0.014
5" wall	"	0.014
6" wall	"	0.015
8" wall	"	0.015
Sandblast finish		
4" wall	S.F.	0.014
5" wall	"	0.014
6" wall	"	0.015
8" wall	"	0.015
Double tee wall panel, 24" deep		
Gray cement		
Liner finish	S.F.	0.016
Sandblast finish	"	0.016
White cement		
Form liner finish	S.F.	0.016
Sandblast finish	"	0.016
Partition panels		
4" wall	S.F.	0.016
5" wall	"	0.016
6" wall	"	0.016
8" wall	"	0.016
Cladding panels		
4" wall	S.F.	0.017
5" wall	"	0.017
6" wall	"	0.017
8" wall	"	0.017
Sandwich panel, 2.5" cladding panel, 2" insulation		
5" wall	S.F.	0.017
6" wall	"	0.017
8" wall	"	0.017
Adjustable tilt-up brace	EA.	0.200

PRECAST CONCRETE	UNIT	MAN/HOURS
03400.90 PRECAST SPECIALTIES		
Precast concrete, coping, 4' to 8' long		
12" wide	L.F.	0.060
10" wide	"	0.069
Splash block, 30"x12"x4"	EA.	0.400
Stair unit, per riser	"	0.400
Sun screen and trellis, 8' long, 12" high		
4" thick blades	EA.	0.300
5" thick blades	"	0.300
6" thick blades	"	0.320
8" thick blades	"	0.320
Bearing pads for precast members, 2" wide strips		
1/8" thick	L.F.	0.003
1/4" thick	"	0.003
1/2" thick	"	0.003
3/4" thick	"	0.004
1" thick	"	0.004
1-1/2" thick	"	0.004

CEMENTITOUS TOPPINGS	UNIT	MAN/HOURS
03550.10 CONCRETE TOPPINGS		
Gypsum fill		
2" thick	S.F.	0.005
2-1/2" thick	"	0.005
3" thick	"	0.005
3-1/2" thick	"	0.005
4" thick	"	0.006
Formboard		
Mineral fiber board		
1" thick	S.F.	0.020
1-1/2" thick	"	0.023
Cement fiber board		
1" thick	S.F.	0.027
1-1/2" thick	"	0.031
Glass fiber board		
1" thick	S.F.	0.020
1-1/2" thick	"	0.023
Poured deck		
Vermiculite or perlite		
1 to 4 mix	C.Y.	0.800
1 to 6 mix	"	0.738
Vermiculite or perlite		
2" thick		
1 to 4 mix	S.F.	0.005
1 to 6 mix	"	0.005
3" thick		

CEMENTITOUS TOPPINGS	UNIT	MAN/HOURS
03550.10 CONCRETE TOPPINGS		
1 to 4 mix	S.F.	0.007
1 to 6 mix	"	0.007
Concrete plank, lightweight		
2" thick	S.F.	0.024
2-1/2" thick	"	0.024
3-1/2" thick	"	0.027
4" thick	"	0.027
Channel slab, lightweight, straight		
2-3/4" thick	S.F.	0.024
3-1/2" thick	"	0.024
3-3/4" thick	"	0.024
4-3/4" thick	"	0.027
Gypsum plank		
2" thick	S.F.	0.024
3" thick	"	0.024
Cement fiber, T and G planks		
1" thick	S.F.	0.022
1-1/2" thick	"	0.022
2" thick	"	0.024
2-1/2" thick	"	0.024
3" thick	"	0.024
3-1/2" thick	"	0.027
4" thick	"	0.027

GROUT	UNIT	MAN/HOURS
03600.10 GROUTING		
Grouting for bases		
Nonshrink		
Metallic grout		
1" deep	S.F.	0.160
2" deep	"	0.178
Non-metallic grout		
1" deep	S.F.	0.160
2" deep	"	0.178
Fluid type		
Non-metallic		
1" deep	S.F.	0.160
2" deep	"	0.178
Grouting for joints		
Portland cement grout (1 cement to 3 sand, by volume)		
1/2" joint thickness		
6" wide joints	L.F.	0.027
8" wide joints	"	0.032
1" joint thickness		
4" wide joints	L.F.	0.025

03 CONCRETE

GROUT

03600.10 GROUTING

GROUT	UNIT	MAN/HOURS
6" wide joints	L.F.	0.028
8" wide joints	"	0.033
Nonshrink, nonmetallic grout		
1/2" joint thickness		
4" wide joint	L.F.	0.023
6" wide joint	"	0.027
8" wide joint	"	0.032
1" joint thickness		
4" wide joint	L.F.	0.025
6" wide joint	"	0.028
8" wide joint	"	0.033

CONCRETE RESTORATION

03730.10 CONCRETE REPAIR

CONCRETE RESTORATION	UNIT	MAN/HOURS
Epoxy grout floor patch, 1/4" thick	S.F.	0.080
Epoxy gel grout	"	0.800
Injection valve, 1 way, threaded plastic	EA.	0.160
Grout crack seal, 2 component	C.F.	0.800
Grout, non shrink	"	0.800
Concrete, epoxy modified		
Sand mix	C.F.	0.320
Gravel mix	"	0.296
Concrete repair		
Soffit repair		
16" wide	L.F.	0.160
18" wide	"	0.167
24" wide	"	0.178
30" wide	"	0.190
32" wide	"	0.200
Edge repair		
2" spall	L.F.	0.200
3" spall	"	0.211
4" spall	"	0.216
6" spall	"	0.222
8" spall	"	0.235
9" spall	"	0.267
Crack repair, 1/8" crack	"	0.080
Reinforcing steel repair		
1 bar, 4 ft		
#4 bar	L.F.	0.100
#5 bar	"	0.100
#6 bar	"	0.107
#8 bar	"	0.107
#9 bar	"	0.114
#11 bar	"	0.114

CONCRETE RESTORATION

03730.10 CONCRETE REPAIR

CONCRETE RESTORATION	UNIT	MAN/HOURS
Pile repairs		
Polyethylene wrap		
30 mil thick		
60" wide	S.F.	0.267
72" wide	"	0.320
60 mil thick		
60" wide	S.F.	0.267
80" wide	"	0.364
Pile spall, average repair 3'		
18" x 18"	EA.	0.667
20" x 20"	"	0.800

MORTAR AND GROUT

04100.10 MASONRY GROUT

MORTAR AND GROUT	UNIT	MAN/ HOURS
Grout, non shrink, non-metallic, trowelable	C.F.	0.016
Grout door frame, hollow metal		
Single	EA.	0.600
Double	"	0.632
Grout-filled concrete block (CMU)		
4" wide	S.F.	0.020
6" wide	"	0.022
8" wide	"	0.024
12" wide	"	0.025
Grout-filled individual CMU cells		
4" wide	L.F.	0.012
6" wide	"	0.012
8" wide	"	0.012
10" wide	"	0.014
12" wide	"	0.014
Bond beams or lintels, 8" deep		
6" thick	L.F.	0.022
8" thick	"	0.024
10" thick	"	0.027
12" thick	"	0.030
Cavity walls		
2" thick	S.F.	0.032
3" thick	"	0.032
4" thick	"	0.034
6" thick	"	0.040

04150.10 MASONRY ACCESSORIES

MORTAR AND GROUT	UNIT	MAN/ HOURS
Foundation vents	EA.	0.320
Bar reinforcing		
Horizontal		
#3 - #4	Lb.	0.032
#5 - #6	"	0.027
Vertical		
#3 - #4	Lb.	0.040
#5 - #6	"	0.032
Horizontal joint reinforcing		
Truss type		
4" wide, 6" wall	L.F.	0.003
6" wide, 8" wall	"	0.003
8" wide, 10" wall	"	0.003
10" wide, 12" wall	"	0.004
12" wide, 14" wall	"	0.004
Ladder type		
4" wide, 6" wall	L.F.	0.003
6" wide, 8" wall	"	0.003
8" wide, 10" wall	"	0.003
10" wide, 12" wall	"	0.003
Rectangular wall ties		
3/16" dia., galvanized		
2" x 6"	EA.	0.013
2" x 8"	"	0.013
2" x 10"	"	0.013
2" x 12"	"	0.013

MORTAR AND GROUT

04150.10 MASONRY ACCESSORIES

MORTAR AND GROUT	UNIT	MAN/ HOURS
4" x 6"	EA.	0.016
4" x 8"	"	0.016
4" x 10"	"	0.016
4" x 12"	"	0.016
1/4" dia., galvanized		
2" x 6"	EA.	0.013
2" x 8"	"	0.013
2" x 10"	"	0.013
2" x 12"	"	0.013
4" x 6"	"	0.016
4" x 8"	"	0.016
4" x 10"	"	0.016
4" x 12"	"	0.016
"Z" type wall ties, galvanized		
6" long		
1/8" dia.	EA.	0.013
3/16" dia.	"	0.013
1/4" dia.	"	0.013
8" long		
1/8" dia.	EA.	0.013
3/16" dia.	"	0.013
1/4" dia.	"	0.013
10" long		
1/8" dia.	EA.	0.013
3/16" dia.	"	0.013
1/4" dia.	"	0.013
Dovetail anchor slots		
Galvanized steel, filled		
24 ga.	L.F.	0.020
20 ga.	"	0.020
16 oz. copper, foam filled	"	0.020
Dovetail anchors		
16 ga.		
3-1/2" long	EA.	0.013
5-1/2" long	"	0.013
12 ga.		
3-1/2" long	EA.	0.013
5-1/2" long	"	0.013
Dovetail, triangular galvanized ties, 12 ga.		
3" x 3"	EA.	0.013
5" x 5"	"	0.013
7" x 7"	"	0.013
7" x 9"	"	0.013
Brick anchors		
Corrugated, 3-1/2" long		
16 ga.	EA.	0.013
12 ga.	"	0.013
Non-corrugated, 3-1/2" long		
16 ga.	EA.	0.013
12 ga.	"	0.013

04150.20 MASONRY CONTROL JOINTS

MORTAR AND GROUT	UNIT	MAN/ HOURS
Control joint, cross shaped PVC	L.F.	0.020
Closed cell joint filler		

MORTAR AND GROUT

MORTAR AND GROUT	UNIT	MAN/HOURS
04150.20 MASONRY CONTROL JOINTS		
1/2"	L.F.	0.020
3/4"	"	0.020
Rubber, for		
4" wall	L.F.	0.020
6" wall	"	0.021
8" wall	"	0.022
PVC, for		
4" wall	L.F.	0.020
6" wall	"	0.021
8" wall	"	0.022
04150.50 MASONRY FLASHING		
Through-wall flashing		
5 oz. coated copper	S.F.	0.067
0.030" elastomeric	"	0.053

UNIT MASONRY

UNIT MASONRY	UNIT	MAN/HOURS
04210.10 BRICK MASONRY		
Standard size brick, running bond		
Face brick, red (6.4/sf)		
Veneer	S.F.	0.133
Cavity wall	"	0.114
9" solid wall	"	0.229
Back-up		
4" thick	S.F.	0.100
8" thick	"	0.160
Firewall		
12" thick	S.F.	0.267
16" thick	"	0.364
Glazed brick (7.4/sf)		
Veneer	S.F.	0.145
Buff or gray face brick (6.4/sf)		
Veneer	S.F.	0.133
Cavity wall	"	0.114
Jumbo or oversize brick (3/sf)		
4" veneer	S.F.	0.080
4" back-up	"	0.067
8" back-up	"	0.114
12" firewall	"	0.200
16" firewall	"	0.267
Norman brick, red face, (4.5/sf)		
4" veneer	S.F.	0.100
Cavity wall	"	0.089
Chimney, standard brick, including flue		
16" x 16"	L.F.	0.800

UNIT MASONRY

UNIT MASONRY	UNIT	MAN/HOURS
04210.10 BRICK MASONRY		
16" x 20"	L.F.	0.800
16" x 24"	"	0.800
20" x 20"	"	1.000
20" x 24"	"	1.000
20" x 32"	"	1.143
Window sill, face brick on edge	"	0.200
04210.20 STRUCTURAL TILE		
Structural glazed tile		
6T series, 5-1/2" x 12"		
Glazed on one side		
2" thick	S.F.	0.080
4" thick	"	0.080
6" thick	"	0.089
8" thick	"	0.100
Glazed on two sides		
4" thick	S.F.	0.100
6" thick	"	0.114
04210.60 PAVERS, MASONRY		
Brick walk laid on sand, sand joints		
Laid flat, (4.5 per sf)	S.F.	0.089
Laid on edge, (7.2 per sf)	"	0.133
Precast concrete patio blocks		
2" thick		
Natural	S.F.	0.027
Colors	"	0.027
Exposed aggregates, local aggregate		
Natural	S.F.	0.027
Colors	"	0.027
Granite or limestone aggregate	"	0.027
White tumblestone aggregate	"	0.027
Stone pavers, set in mortar		
Bluestone		
1" thick		
Irregular	S.F.	0.200
Snapped rectangular	"	0.160
1-1/2" thick, random rectangular	"	0.200
2" thick, random rectangular	"	0.229
Slate		
Natural cleft		
Irregular, 3/4" thick	S.F.	0.229
Random rectangular		
1-1/4" thick	S.F.	0.200
1-1/2" thick	"	0.222
Granite blocks		
3" thick, 3" to 6" wide		
4" to 12" long	S.F.	0.267
6" to 15" long	"	0.229
Crushed stone, white marble, 3" thick	"	0.016
04220.10 CONCRETE MASONRY UNITS		
Hollow, load bearing		
4"	S.F.	0.059

UNIT MASONRY	UNIT	MAN/ HOURS
04220.10 CONCRETE MASONRY UNITS		
6"	S.F.	0.062
8"	"	0.067
10"	"	0.073
12"	"	0.080
Solid, load bearing		
4"	S.F.	0.059
6"	"	0.062
8"	"	0.067
10"	"	0.073
12"	"	0.080
Back-up block, 8" x 16"		
2"	S.F.	0.046
4"	"	0.047
6"	"	0.050
8"	"	0.053
10"	"	0.057
12"	"	0.062
Foundation wall, 8" x 16"		
6"	S.F.	0.057
8"	"	0.062
10"	"	0.067
12"	"	0.073
Solid		
6"	S.F.	0.062
8"	"	0.067
10"	"	0.073
12"	"	0.080
Exterior, styrofoam inserts, standard weight, 8" x 16"		
6"	S.F.	0.062
8"	"	0.067
10"	"	0.073
12"	"	0.080
Acoustical slotted block		
4"	S.F.	0.073
6"	"	0.073
8"	"	0.080
Filled cavities		
4"	S.F.	0.089
6"	"	0.094
8"	"	0.100
Hollow, split face		
4"	S.F.	0.059
6"	"	0.062
8"	"	0.067
10"	"	0.073
12"	"	0.080
Split rib profile		
4"	S.F.	0.073
6"	"	0.073
8"	"	0.080
10"	"	0.080
12"	"	0.080
High strength block, 3500 psi		
2"	S.F.	0.059

UNIT MASONRY	UNIT	MAN/ HOURS
04220.10 CONCRETE MASONRY UNITS		
4"	S.F.	0.062
6"	"	0.062
8"	"	0.067
10"	"	0.073
12"	"	0.080
Solar screen concrete block		
4" thick		
6" x 6"	S.F.	0.178
8" x 8"	"	0.160
12" x 12"	"	0.123
8" thick		
8" x 16"	S.F.	0.114
Glazed block		
Cove base, glazed 1 side, 2"	L.F.	0.089
4"	"	0.089
6"	"	0.100
8"	"	0.100
Single face		
2"	S.F.	0.067
4"	"	0.067
6"	"	0.073
8"	"	0.080
10"	"	0.089
12"	"	0.094
Double face		
4"	S.F.	0.084
6"	"	0.089
8"	"	0.100
Corner or bullnose		
2"	EA.	0.100
4"	"	0.114
6"	"	0.114
8"	"	0.133
10"	"	0.145
12"	"	0.160
04220.90 BOND BEAMS & LINTELS		
Bond beam, no grout or reinforcement		
8" x 16" x		
4" thick	L.F.	0.062
6" thick	"	0.064
8" thick	"	0.067
10" thick	"	0.070
12" thick	"	0.073
Beam lintel, no grout or reinforcement		
8" x 16" x		
10" thick	L.F.	0.080
12" thick	"	0.089
Precast masonry lintel		
6 lf, 8" high x		
4" thick	L.F.	0.133
6" thick	"	0.133
8" thick	"	0.145

04 MASONRY

UNIT MASONRY

04220.90 BOND BEAMS & LINTELS	UNIT	MAN/HOURS
10" thick	L.F.	0.145
10 lf, 8" high x		
4" thick	L.F.	0.080
6" thick	"	0.080
8" thick	"	0.089
10" thick	"	0.089
Steel angles and plates		
Minimum	Lb.	0.011
Maximum	"	0.020
Various size angle lintels		
1/4" stock		
3" x 3"	L.F.	0.050
3" x 3-1/2"	"	0.050
3/8" stock		
3" x 4"	L.F.	0.050
3-1/2" x 4"	"	0.050
4" x 4"	"	0.050
5" x 3-1/2"	"	0.050
6" x 3-1/2"	"	0.050
1/2" stock		
6" x 4"	L.F.	0.050

04270.10 GLASS BLOCK	UNIT	MAN/HOURS
Glass block, 4" thick		
6" x 6"	S.F.	0.267
8" x 8"	"	0.200
12" x 12"	"	0.160
Replacement glass blocks, 4" x 8" x 8"		
Minimum	S.F.	0.800
Maximum	"	1.600

04295.10 PARGING/MASONRY PLASTER	UNIT	MAN/HOURS
Parging		
1/2" thick	S.F.	0.053
3/4" thick	"	0.067
1" thick	"	0.080

STONE

04400.10 STONE	UNIT	MAN/HOURS
Rubble stone		
Walls set in mortar		
8" thick	S.F.	0.200
12" thick	"	0.320
18" thick	"	0.400
24" thick	S.F.	0.533
Dry set wall		
8" thick	S.F.	0.133
12" thick	"	0.200
18" thick	"	0.267
24" thick	"	0.320
Cut stone		
Facing panels		
3/4" thick	S.F.	0.320
1-1/2" thick	"	0.364

MASONRY RESTORATION

04520.10 RESTORATION AND CLEANING	UNIT	MAN/HOURS
Masonry cleaning		
Washing brick		
Smooth surface	S.F.	0.013
Rough surface	"	0.018
Steam clean masonry		
Smooth face		
Minimum	S.F.	0.010
Maximum	"	0.015
Rough face		
Minimum	S.F.	0.013
Maximum	"	0.020
Sandblast masonry		
Minimum	S.F.	0.016
Maximum	"	0.027
Pointing masonry		
Brick	S.F.	0.032
Concrete block	"	0.023
Cut and repoint		
Brick		
Minimum	S.F.	0.040
Maximum	"	0.080
Stone work	L.F.	0.062
Cut and recaulk		
Oil base caulks	L.F.	0.053
Butyl caulks	"	0.053
Polysulfides and acrylics	"	0.053
Silicones	"	0.053
Cement and sand grout on walls, to 1/8" thick		
Minimum	S.F.	0.032
Maximum	"	0.040
Brick removal and replacement		
Minimum	EA.	0.100
Average	"	0.133

04 MASONRY

MASONRY RESTORATION	UNIT	MAN/ HOURS
04520.10 RESTORATION AND CLEANING		
Maximum	EA.	0.400

05 METALS

METAL FASTENING	UNIT	MAN/ HOURS
05050.10 STRUCTURAL WELDING		
Welding		
Single pass		
1/8"	L.F.	0.040
3/16"	"	0.053
1/4"	"	0.067
Miscellaneous steel shapes		
Plain	Lb.	0.002
Galvanized	"	0.003
Plates		
Plain	Lb.	0.002
Galvanized	"	0.003
05050.95 METAL LINTELS		
Lintels, steel		
Plain	Lb.	0.020
Galvanized	"	0.020
05120.10 STRUCTURAL STEEL		
Beams and girders, A-36		
Welded	TON	4.800
Bolted	"	4.364
Columns		
Pipe		
6" dia.	Lb.	0.005
12" dia.	"	0.004
Purlins and girts		
Welded	TON	8.000
Bolted	"	6.857
Column base plates		
Up to 150 lb each	Lb.	0.005
Over 150 lb each	"	0.007
Structural pipe		
3" to 5" o.d.	TON	9.600
6" to 12" o.d.	"	6.857
Structural tube		
6" square		
Light sections	TON	9.600
Heavy sections	"	6.857
6" wide rectangular		
Light sections	TON	8.000
Heavy sections	"	6.000
Greater than 6" wide rectangular		
Light sections	TON	8.000
Heavy sections	"	6.000
Miscellaneous structural shapes		
Steel angle	TON	12.000
Steel plate	"	8.000
Trusses, field welded		
60 lb/lf	TON	6.000
100 lb/lf	"	4.800
150 lb/lf	"	4.000
Bolted		

METAL FASTENING	UNIT	MAN/ HOURS
05120.10 STRUCTURAL STEEL		
60 lb/lf	TON	5.333
100 lb/lf	"	4.364
150 lb/lf	"	3.692

COLD FORMED FRAMING	UNIT	MAN/ HOURS
05410.10 METAL FRAMING		
Furring channel, galvanized		
Beams and columns, 3/4"		
12" o.c.	S.F.	0.080
16" o.c.	"	0.073
Walls, 3/4"		
12" o.c.	S.F.	0.040
16" o.c.	"	0.033
24" o.c.	"	0.027
1-1/2"		
12" o.c.	S.F.	0.040
16" o.c.	"	0.033
24" o.c.	"	0.027
Stud, load bearing		
16" o.c.		
16 ga.		
2-1/2"	S.F.	0.036
3-5/8"	"	0.036
4"	"	0.036
6"	"	0.040
24" o.c.		
16 ga.		
2-1/2"	S.F.	0.031
3-5/8"	"	0.031
4"	"	0.031
6"	"	0.033
8"	"	0.033

05 METALS

METAL FABRICATIONS

METAL FABRICATIONS	UNIT	MAN/ HOURS
05510.10 STAIRS		
Stock unit, steel, complete, per riser		
Tread		
3'-6" wide	EA.	1.000
4' wide	"	1.143
5' wide	"	1.333
Metal pan stair, cement filled, per riser		
3'-6" wide	EA.	0.800
4' wide	"	0.889
5' wide	"	1.000
Landing, steel pan	S.F.	0.200
Cast iron tread, steel stringers, stock units, per riser		
Tread		
3'-6" wide	EA.	1.000
4' wide	"	1.143
5' wide	"	1.333
Stair treads, abrasive, 12" x 3'-6"		
Cast iron		
3/8"	EA.	0.400
1/2"	"	0.400
Cast aluminum		
5/16"	EA.	0.400
3/8"	"	0.400
1/2"	"	0.400
05515.10 LADDERS		
Ladder, 18" wide		
With cage	L.F.	0.533
Without cage	"	0.400
05520.10 RAILINGS		
Railing, pipe		
1-1/4" diameter, welded steel		
2-rail		
Primed	L.F.	0.160
Galvanized	"	0.160
3-rail		
Primed	L.F.	0.200
Galvanized	"	0.200
Wall mounted, single rail, welded steel		
Primed	L.F.	0.123
Galvanized	"	0.123
1-1/2" diameter, welded steel		
2-rail		
Primed	L.F.	0.160
Galvanized	"	0.160
3-rail		
Primed	L.F.	0.200
Galvanized	"	0.200
Wall mounted, single rail, welded steel		
Primed	L.F.	0.123
Galvanized	"	0.123
2" diameter, welded steel		

METAL FABRICATIONS	UNIT	MAN/ HOURS
05520.10 RAILINGS		
2-rail		
Primed	L.F.	0.178
Galvanized	"	0.178
3-rail		
Primed	L.F.	0.229
Galvanized	"	0.229
Wall mounted, single rail, welded steel		
Primed	L.F.	0.133
Galvanized	"	0.133
05530.10 METAL GRATING		
Floor plate, checkered, steel		
1/4"		
Primed	S.F.	0.011
Galvanized	"	0.011
3/8"		
Primed	S.F.	0.012
Galvanized	"	0.012
Aluminum grating, pressure-locked bearing bars		
3/4" x 1/8"	S.F.	0.020
1" x 1/8"	"	0.020
1-1/4" x 1/8"	"	0.020
1-1/4" x 3/16"	"	0.020
1-1/2" x 1/8"	"	0.020
1-3/4" x 3/16"	"	0.020
Miscellaneous expenses		
Cutting		
Minimum	L.F.	0.053
Maximum	"	0.080
Banding		
Minimum	L.F.	0.133
Maximum	"	0.160
Toe plates		
Minimum	L.F.	0.160
Maximum	"	0.200
Steel grating, primed		
3/4" x 1/8"	S.F.	0.027
1" x 1/8"	"	0.027
1-1/4" x 1/8"	"	0.027
1-1/4" x 3/16"	"	0.027
1-1/2" x 1/8"	"	0.027
1-3/4" x 3/16"	"	0.027
Galvanized		
3/4" x 1/8"	S.F.	0.027
1" x 1/8"	"	0.027
1-1/4" x 1/8"	"	0.027
1-1/4" x 3/16"	"	0.027
1-1/2" x 1/8"	"	0.027
1-3/4" x 3/16"	"	0.027
Miscellaneous expenses		
Cutting		
Minimum	L.F.	0.057
Maximum	"	0.089

05 METALS

METAL FABRICATIONS

05530.10 METAL GRATING	UNIT	MAN/ HOURS
Banding		
Minimum	L.F.	0.145
Maximum	"	0.178
Toe plates		
Minimum	L.F.	0.178
Maximum	"	0.229

05540.10 CASTINGS	UNIT	MAN/ HOURS
Miscellaneous castings		
Light sections	Lb.	0.016
Heavy sections	"	0.011
Manhole covers and frames		
Regular, city type		
18" dia.		
100 lb	EA.	1.600
24" dia.		
200 lb	EA.	1.600
300 lb	"	1.778
400 lb	"	1.778
26" dia., 475 lb	"	2.000
30" dia., 600 lb	"	2.286
8" square, 75 lb	"	0.320
24" square		
126 lb	EA.	1.600
500 lb	"	2.000
Watertight type		
20" dia., 200 lb	EA.	2.000
24" dia., 350 lb	"	2.667
Steps, cast iron		
7" x 9"	EA.	0.160
8" x 9"	"	0.178
Manhole covers and frames, aluminum		
12" x 12"	EA.	0.320
18" x 18"	"	0.320
24" x 24"	"	0.400
Corner protection		
Steel angle guard with anchors		
2" x 2" x 3/16"	L.F.	0.114
2" x 3" x 1/4"	"	0.114
3" x 3" x 5/16"	"	0.114
3" x 4" x 5/16"	"	0.123
4" x 4" x 5/16"	"	0.123

MISC. FABRICATIONS

05580.10 METAL SPECIALTIES	UNIT	MAN/ HOURS
Kick plate		
4" high x 1/4" thick		
Primed	L.F.	0.160
Galvanized	"	0.160
6" high x 1/4" thick		
Primed	L.F.	0.178
Galvanized	"	0.178

05700.10 ORNAMENTAL METAL	UNIT	MAN/ HOURS
Railings, vertical square bars, 6" o.c., with shaped top rails		
Steel	L.F.	0.400
Aluminum	"	0.400
Bronze	"	0.533
Stainless steel	"	0.533
Laminated metal or wood handrails with metal supports		
2-1/2" round or oval shape	L.F.	0.400

05800.10 EXPANSION CONTROL	UNIT	MAN/ HOURS
Expansion joints with covers, floor assembly type		
With 1" space		
Aluminum	L.F.	0.133
Bronze	"	0.133
Stainless steel	"	0.133
With 2" space		
Aluminum	L.F.	0.133
Bronze	"	0.133
Stainless steel	"	0.133
Ceiling and wall assembly type		
With 1" space		
Aluminum	L.F.	0.160
Bronze	"	0.160
Stainless steel	"	0.160
With 2" space		
Aluminum	L.F.	0.160
Bronze	"	0.160
Stainless steel	"	0.160
Exterior roof and wall, aluminum		
Roof to roof		
With 1" space	L.F.	0.133
With 2" space	"	0.133
Roof to wall		
With 1" space	L.F.	0.145
With 2" space	"	0.145

06 WOOD AND PLASTICS

FASTENERS AND ADHESIVES

06050.10 ACCESSORIES

	UNIT	MAN/ HOURS
Column/post base, cast aluminum		
4" x 4"	EA.	0.200
6" x 6"	"	0.200
Bridging, metal, per pair		
12" o.c.	EA.	0.080
16" o.c.	"	0.073
Anchors		
Bolts, threaded two ends, with nuts and washers		
1/2" dia.		
4" long	EA.	0.050
7-1/2" long	"	0.050
15" long	"	0.050
Bolts, carriage		
1/4 x 4	EA.	0.080
5/16 x 6	"	0.084
Joist and beam hangers		
18 ga.		
2 x 4	EA.	0.080
2 x 6	"	0.080
2 x 8	"	0.080
2 x 10	"	0.089
2 x 12	"	0.100
16 ga.		
3 x 6	EA.	0.089
3 x 8	"	0.089
3 x 10	"	0.094
3 x 12	"	0.107
3 x 14	"	0.114
4 x 6	"	0.089
4 x 8	"	0.089
4 x 10	"	0.094
4 x 12	"	0.107
4 x 14	"	0.114
Rafter anchors, 18 ga., 1-1/2" wide		
5-1/4" long	EA.	0.067
10-3/4" long	"	0.067
Sill anchors		
Embedded in concrete	EA.	0.080
Strap ties, 14 ga., 1-3/8" wide		
12" long	EA.	0.067

ROUGH CARPENTRY

06110.10 BLOCKING

	UNIT	MAN/ HOURS
Steel construction		
Walls		
2x4	L.F.	0.053
2x6	L.F.	0.062
2x8	"	0.067
2x10	"	0.073
2x12	"	0.080
Ceilings		
2x4	L.F.	0.062
2x6	"	0.073
2x8	"	0.080
2x10	"	0.089
2x12	"	0.100
Wood construction		
Walls		
2x4	L.F.	0.044
2x6	"	0.050
2x8	"	0.053
2x10	"	0.057
2x12	"	0.062
Ceilings		
2x4	L.F.	0.050
2x6	"	0.057
2x8	"	0.062
2x10	"	0.067
2x12	"	0.073

06110.20 CEILING FRAMING

	UNIT	MAN/ HOURS
Ceiling joists		
16" o.c.		
2x4	S.F.	0.015
2x6	"	0.016
2x8	"	0.017
2x10	"	0.017
2x12	"	0.018
Headers and nailers		
2x4	L.F.	0.026
2x6	"	0.027
2x8	"	0.029
2x10	"	0.031
2x12	"	0.033
Sister joists for ceilings		
2x4	L.F.	0.057
2x6	"	0.067
2x8	"	0.080
2x10	"	0.100
2x12	"	0.133

06110.30 FLOOR FRAMING

	UNIT	MAN/ HOURS
Floor joists		
16" o.c.		
2x6	S.F.	0.013
2x8	"	0.014
2x10	"	0.014
2x12	"	0.014

06 WOOD AND PLASTICS

ROUGH CARPENTRY	UNIT	MAN/HOURS
06110.30 FLOOR FRAMING		
2x14	S.F.	0.015
3x6	"	0.014
3x8	"	0.014
3x10	"	0.015
3x12	"	0.015
3x14	"	0.016
4x6	"	0.014
4x8	"	0.014
4x10	"	0.015
4x12	"	0.015
4x14	"	0.016
Sister joists for floors		
2x4	L.F.	0.050
2x6	"	0.057
2x8	"	0.067
2x10	"	0.080
2x12	"	0.100
06110.40 FURRING		
Furring, wood strips		
Walls		
On masonry or concrete walls		
1x2 furring		
12" o.c.	S.F.	0.025
16" o.c.	"	0.023
24" o.c.	"	0.021
1x3 furring		
12" o.c.	S.F.	0.025
16" o.c.	"	0.023
24" o.c.	"	0.021
On wood walls		
1x2 furring		
12" o.c.	S.F.	0.018
16" o.c.	"	0.016
24" o.c.	"	0.015
1x3 furring		
12" o.c.	S.F.	0.018
16" o.c.	"	0.016
24" o.c.	"	0.015
Ceilings		
On masonry or concrete ceilings		
1x2 furring		
12" o.c.	S.F.	0.044
16" o.c.	"	0.040
24" o.c.	"	0.036
1x3 furring		
12" o.c.	S.F.	0.044
16" o.c.	"	0.040
24" o.c.	"	0.036
On wood ceilings		
1x2 furring		
12" o.c.	S.F.	0.030
16" o.c.	"	0.027

ROUGH CARPENTRY	UNIT	MAN/HOURS
06110.40 FURRING		
24" o.c.	S.F.	0.024
1x3		
12" o.c.	S.F.	0.030
16" o.c.	"	0.027
24" o.c.	"	0.024
06110.50 ROOF FRAMING		
Roof framing		
Rafters, gable end		
4-6 pitch (4-in-12 to 6-in-12)		
16" o.c.		
2x6	S.F.	0.015
2x8	"	0.015
2x10	"	0.016
2x12	"	0.017
24" o.c.		
2x6	S.F.	0.013
2x8	"	0.013
2x10	"	0.014
2x12	"	0.015
Ridge boards		
2x6	L.F.	0.040
2x8	"	0.044
2x10	"	0.050
2x12	"	0.057
Hip rafters		
2x6	L.F.	0.029
2x8	"	0.030
2x10	"	0.031
2x12	"	0.032
Jack rafters		
4-6 pitch (4-in-12 to 6-in-12)		
16" o.c.		
2x6	S.F.	0.024
2x8	"	0.024
2x10	"	0.026
2x12	"	0.027
24" o.c.		
2x6	S.F.	0.018
2x8	"	0.019
2x10	"	0.020
2x12	"	0.020
Sister rafters		
2x4	L.F.	0.057
2x6	"	0.067
2x8	"	0.080
2x10	"	0.100
2x12	"	0.133
Fascia boards		
2x4	L.F.	0.040
2x6	"	0.040
2x8	"	0.044
2x10	"	0.044

06 WOOD AND PLASTICS

ROUGH CARPENTRY	UNIT	MAN/ HOURS
06110.50 ROOF FRAMING		
2x12	L.F.	0.050
Cant strips		
Fiber		
3x3	L.F.	0.023
4x4	"	0.024
Wood		
3x3	L.F.	0.024
06110.60 SLEEPERS		
Sleepers, over concrete		
12" o.c.		
1x2	S.F.	0.018
1x3	"	0.019
2x4	"	0.022
2x6	"	0.024
16" o.c.		
1x2	S.F.	0.016
1x3	"	0.016
2x4	"	0.019
2x6	"	0.020
06110.65 SOFFITS		
Soffit framing		
2x3	L.F.	0.057
2x4	"	0.062
2x6	"	0.067
2x8	"	0.073
06110.70 WALL FRAMING		
Framing wall, studs		
12" o.c.		
2x3	S.F.	0.015
2x4	"	0.015
2x6	"	0.016
2x8	"	0.017
16" o.c.		
2x3	S.F.	0.013
2x4	"	0.013
2x6	"	0.013
2x8	"	0.014
24" o.c.		
2x3	S.F.	0.011
2x4	"	0.011
2x6	"	0.011
2x8	"	0.012
Plates, top or bottom		
2x3	L.F.	0.024
2x4	"	0.025
2x6	"	0.027
2x8	"	0.029
Headers, door or window		

ROUGH CARPENTRY	UNIT	MAN/ HOURS
06110.70 WALL FRAMING		
2x8		
Single		
4' long	EA.	0.500
8' long	"	0.615
Double		
4' long	EA.	0.571
8' long	"	0.727
2x12		
Single		
6' long	EA.	0.615
12' long	"	0.800
Double		
6' long	EA.	0.727
12' long	"	0.889
06115.10 FLOOR SHEATHING		
Sub-flooring, plywood, CDX		
1/2" thick	S.F.	0.010
5/8" thick	"	0.011
3/4" thick	"	0.013
Structural plywood		
1/2" thick	S.F.	0.010
5/8" thick	"	0.011
3/4" thick	"	0.012
Underlayment		
Hardboard, 1/4" tempered	S.F.	0.010
Plywood, CDX		
3/8" thick	S.F.	0.010
1/2" thick	"	0.011
5/8" thick	"	0.011
3/4" thick	"	0.012
06115.20 ROOF SHEATHING		
Sheathing		
Plywood, CDX		
3/8" thick	S.F.	0.010
1/2" thick	"	0.011
5/8" thick	"	0.011
3/4" thick	"	0.012
06115.30 WALL SHEATHING		
Sheathing		
Plywood, CDX		
3/8" thick	S.F.	0.012
1/2" thick	"	0.012
5/8" thick	"	0.013
3/4" thick	"	0.015
06125.10 WOOD DECKING		
Decking, T&G solid		
Fir		
3" thick	S.F.	0.020
4" thick	"	0.021

06 WOOD AND PLASTICS

ROUGH CARPENTRY

06125.10 WOOD DECKING

ROUGH CARPENTRY	UNIT	MAN/HOURS
Southern yellow pine		
3" thick	S.F.	0.023
4" thick	"	0.025
White pine		
3" thick	S.F.	0.020
4" thick	"	0.021

06130.10 HEAVY TIMBER

ROUGH CARPENTRY	UNIT	MAN/HOURS
Mill framed structures		
Beams to 20' long		
Douglas fir		
6x8	L.F.	0.080
6x10	"	0.083
Southern yellow pine		
6x8	L.F.	0.080
6x10	"	0.083
Columns to 12' high		
6x6	L.F.	0.120
10x10	"	0.133

06190.20 WOOD TRUSSES

ROUGH CARPENTRY	UNIT	MAN/HOURS
Truss, fink, 2x4 members		
3-in-12 slope		
24' span	EA.	0.686
30' span	"	0.727

FINISH CARPENTRY

06200.10 FINISH CARPENTRY

FINISH CARPENTRY	UNIT	MAN/HOURS
Casing		
11/16 x 2-1/2	L.F.	0.036
11/16 x 3-1/2	"	0.038
Half round		
1/2	L.F.	0.032
5/8	"	0.032
Railings, balusters		
1-1/8 x 1-1/8	L.F.	0.080
1-1/2 x 1-1/2	"	0.073
Stop		
5/8 x 1-5/8		
Colonial	L.F.	0.050
Ranch	"	0.050
Exterior trim, casing, select pine, 1x3	"	0.040
Douglas fir		
1x3	L.F.	0.040
1x4	"	0.040

FINISH CARPENTRY

06200.10 FINISH CARPENTRY

FINISH CARPENTRY	UNIT	MAN/HOURS
1x6	L.F.	0.044
Cornices, white pine, #2 or better		
1x2	L.F.	0.040
1x4	"	0.040
1x8	"	0.047
Shelving, pine		
1x8	L.F.	0.062
1x10	"	0.064
1x12	"	0.067

06220.10 MILLWORK

FINISH CARPENTRY	UNIT	MAN/HOURS
Countertop, laminated plastic		
25" x 7/8" thick		
Minimum	L.F.	0.200
Average	"	0.267
Maximum	"	0.320
Base cabinets, 34-1/2" high, 24" deep, hardwood, no tops		
Minimum	L.F.	0.320
Average	"	0.400
Maximum	"	0.533
Wall cabinets		
Minimum	L.F.	0.267
Average	"	0.320
Maximum	"	0.400
Oil borne		
Water borne		

ARCHITECTURAL WOODWORK

06420.10 PANEL WORK

ARCHITECTURAL WOODWORK	UNIT	MAN/HOURS
Plywood unfinished, 1/4" thick		
Birch		
Natural	S.F.	0.027
Select	"	0.027
Knotty pine	"	0.027
Plywood, prefinished, 1/4" thick, premium grade		
Birch veneer	S.F.	0.032
Cherry veneer	"	0.032

06430.10 STAIRWORK

ARCHITECTURAL WOODWORK	UNIT	MAN/HOURS
Risers, 1x8, 42" wide		
White oak	EA.	0.400
Pine	"	0.400
Treads, 1-1/16" x 9-1/2" x 42"		
White oak	EA.	0.500

ARCHITECTURAL WOODWORK	UNIT	MAN/ HOURS
06440.10 COLUMNS		
Column, hollow, round wood		
12" diameter		
10' high	EA.	0.800
12' high	"	0.857
24" diameter		
16' high	EA.	1.200
18' high	"	1.263

MOISTURE PROTECTION	UNIT	MAN/ HOURS
07100.10 WATERPROOFING		
Membrane waterproofing, elastomeric		
Butyl		
1/32" thick	S.F.	0.032
1/16" thick	"	0.033
Butyl with nylon		
1/32" thick	S.F.	0.032
1/16" thick	"	0.033
Neoprene		
1/32" thick	S.F.	0.032
1/16" thick	"	0.033
Neoprene with nylon		
1/32" thick	S.F.	0.032
1/16" thick	"	0.033
Plastic vapor barrier (polyethylene)		
4 mil	S.F.	0.003
6 mil	"	0.003
10 mil	"	0.004
Bituminous membrane waterproofing, asphalt felt, 15 lb.		
One ply	S.F.	0.020
Two ply	"	0.024
Three ply	"	0.029
Four ply	"	0.033
Five ply	"	0.042
Modified asphalt membrane waterproofing, fibrous asphalt		
One ply	S.F.	0.033
Two ply	"	0.040
Three ply	"	0.044
Four ply	"	0.053
Five ply	"	0.064
Asphalt coated protective board		
1/8" thick	S.F.	0.020
1/4" thick	"	0.020
3/8" thick	"	0.020
1/2" thick	"	0.021
Cement protective board		
3/8" thick	S.F.	0.027
1/2" thick	"	0.027
Fluid applied, neoprene		
50 mil	S.F.	0.027
90 mil	"	0.027
Tab extended polyurethane		
.050" thick	S.F.	0.020
Fluid applied rubber based polyurethane		
6 mil	S.F.	0.025
15 mil	"	0.020
Bentonite waterproofing, panels		
3/16" thick	S.F.	0.020
1/4" thick	"	0.020
5/8" thick	"	0.021
Granular admixtures, trowel on, 3/8" thick	"	0.020
Metallic oxide waterproofing, iron compound, troweled		
5/8" thick	S.F.	0.020
3/4" thick	"	0.023

MOISTURE PROTECTION	UNIT	MAN/ HOURS
07150.10 DAMPPROOFING		
Silicone dampproofing, sprayed on		
Concrete surface		
1 coat	S.F.	0.004
2 coats	"	0.006
Concrete block		
1 coat	S.F.	0.005
2 coats	"	0.007
Brick		
1 coat	S.F.	0.006
2 coats	"	0.008
07160.10 BITUMINOUS DAMPPROOFING		
Building paper, asphalt felt		
15 lb	S.F.	0.032
30 lb	"	0.033
Asphalt dampproofing, troweled, cold, primer plus		
1 coat	S.F.	0.027
2 coats	"	0.040
3 coats	"	0.050
Fibrous asphalt dampproofing, hot troweled, primer plus		
1 coat	S.F.	0.032
2 coats	"	0.044
3 coats	"	0.057
Asphaltic paint dampproofing, per coat		
Brush on	S.F.	0.011
Spray on	"	0.009
07190.10 VAPOR BARRIERS		
Vapor barrier, polyethylene		
2 mil	S.F.	0.004
6 mil	"	0.004
8 mil	"	0.004
10 mil	"	0.004

INSULATION	UNIT	MAN/ HOURS
07210.10 BATT INSULATION		
Ceiling, fiberglass, unfaced		
3-1/2" thick, R11	S.F.	0.009
6" thick, R19	"	0.011
9" thick, R30	"	0.012
Suspended ceiling, unfaced		
3-1/2" thick, R11	S.F.	0.009
6" thick, R19	"	0.010
9" thick, R30	"	0.011
Wall, fiberglass		

07 THERMAL AND MOISTURE

INSULATION	UNIT	MAN/ HOURS
07210.10 BATT INSULATION		
Paper backed		
2" thick, R7	S.F.	0.008
3" thick, R8	"	0.009
4" thick, R11	"	0.009
6" thick, R19	"	0.010
Foil backed, 1 side		
2" thick, R7	S.F.	0.008
3" thick, R11	"	0.009
4" thick, R14	"	0.009
Foil backed, 2 sides		
Unfaced		
2" thick, R7	S.F.	0.008
3" thick, R9	"	0.009
4" thick, R11	"	0.009
6" thick, R19	"	0.010
07210.20 BOARD INSULATION		
Insulation, rigid		
0.75" thick, R2.78	S.F.	0.007
1.06" thick, R4.17	"	0.008
1.31" thick, R5.26	"	0.008
1.63" thick, R6.67	"	0.008
2.25" thick, R8.33	"	0.009
Perlite board, roof		
1.00" thick, R2.78	S.F.	0.007
1.50" thick, R4.17	"	0.007
2.00" thick, R5.92	"	0.007
2.50" thick, R6.67	"	0.008
Rigid urethane		
Roof		
1" thick, R6.67	S.F.	0.007
1.20" thick, R8.33	"	0.007
1.50" thick, R11.11	"	0.007
2" thick, R14.29	"	0.007
2.25" thick, R16.67	"	0.008
Polystyrene		
Roof		
1.0" thick, R4.17	S.F.	0.007
1.5" thick, R6.26	"	0.007
2.0" thick, R8.33	"	0.007
Wall		
1.0" thick, R4.17	S.F.	0.008
1.5" thick, R6.26	"	0.009
2.0" thick, R8.33	"	0.009
07210.60 LOOSE FILL INSULATION		
Blown-in type		
Fiberglass		
5" thick, R11	S.F.	0.007
6" thick, R13	"	0.008
9" thick, R19	"	0.011
Poured type		

INSULATION	UNIT	MAN/ HOURS
07210.60 LOOSE FILL INSULATION		
Fiberglass		
1" thick, R4	S.F.	0.005
2" thick, R8	"	0.006
3" thick, R12	"	0.007
4" thick, R16	"	0.008
Vermiculite or perlite		
2" thick, R4.8	S.F.	0.006
3" thick, R7.2	"	0.007
4" thick, R9.6	"	0.008
Masonry, poured vermiculite or perlite		
4" block	S.F.	0.004
6" block	"	0.005
8" block	"	0.006
10" block	"	0.006
12" block	"	0.007
07210.70 SPRAYED INSULATION		
Foam, sprayed on		
Polystyrene		
1" thick, R4	S.F.	0.008
2" thick, R8	"	0.011
Urethane		
1" thick, R4	S.F.	0.008
2" thick, R8	"	0.011
07250.10 FIREPROOFING		
Sprayed on		
1" thick		
On beams	S.F.	0.018
On columns	"	0.016
On decks		
Flat surface	S.F.	0.008
Fluted surface	"	0.010
1-1/2" thick		
On beams	S.F.	0.023
On columns	"	0.020
On decks		
Flat surface	S.F.	0.010
Fluted surface	"	0.013

07 THERMAL AND MOISTURE

SHINGLES AND TILES

SHINGLES AND TILES	UNIT	MAN/ HOURS
07310.10 ASPHALT SHINGLES		
Standard asphalt shingles, strip shingles		
210 lb/square	SQ.	0.800
240 lb/square	"	1.000
Roll roofing, mineral surface		
90 lb	SQ.	0.571
140 lb	"	0.800
07310.30 METAL SHINGLES		
Aluminum, .020" thick		
Plain	SQ.	1.600
Steel, galvanized		
Plain	SQ.	1.600
07310.60 SLATE SHINGLES		
Slate shingles		
Ribbon	SQ.	4.000
Clear	"	4.000
Replacement shingles		
Small jobs	EA.	0.267
Large jobs	S.F.	0.133
07310.70 WOOD SHINGLES		
Wood shingles, on roofs		
White cedar, #1 shingles		
4" exposure	SQ.	2.667
5" exposure	"	2.000
#2 shingles		
4" exposure	SQ.	2.667
5" exposure	"	2.000
Resquared and rebutted		
4" exposure	SQ.	2.667
5" exposure	"	2.000
07310.80 WOOD SHAKES		
Shakes, hand split, 24" red cedar, on roofs		
5" exposure	SQ.	4.000
7" exposure	"	3.200
9" exposure	"	2.667

ROOFING AND SIDING

ROOFING AND SIDING	UNIT	MAN/ HOURS
07410.10 MANUFACTURED ROOFS		
Aluminum roof panels, for structural steel framing		
Corrugated		
Unpainted finish		
.024"	S.F.	0.020
.030"	"	0.020
Painted finish		
.024"	S.F.	0.020
.030"	"	0.020
Steel roof panels, for structural steel framing		
Corrugated, painted		
18 ga.	S.F.	0.020
20 ga.	"	0.020
07460.10 METAL SIDING PANELS		
Aluminum siding panels		
Corrugated		
Plain finish		
.024"	S.F.	0.032
.032"	"	0.032
Painted finish		
.024"	S.F.	0.032
.032"	"	0.032
Steel siding panels		
Corrugated		
22 ga.	S.F.	0.053
24 ga.	"	0.053
07460.70 STEEL SIDING		
Ribbed, sheets, galvanized		
22 ga.	S.F.	0.032
24 ga.	"	0.032
Primed		
24 ga.	S.F.	0.032
26 ga.	"	0.032

MEMBRANE ROOFING

MEMBRANE ROOFING	UNIT	MAN/ HOURS
07510.10 BUILT-UP ASPHALT ROOFING		
Built-up roofing, asphalt felt, including gravel		
2 ply	SQ.	2.000
3 ply	"	2.667
4 ply	"	3.200
Walkway, for built-up roofs		
3' x 3' x		
1/2" thick	S.F.	0.027
3/4" thick	"	0.027
1" thick	"	0.027
Cant strip, 4" x 4"		
Treated wood	L.F.	0.023
Foamglass	"	0.020
Mineral fiber	"	0.020

07 THERMAL AND MOISTURE

MEMBRANE ROOFING

07510.10 BUILT-UP ASPHALT ROOFING	UNIT	MAN/ HOURS
New gravel for built-up roofing, 400 lb/sq	SQ.	1.600
Roof gravel (ballast)	C.Y.	4.000
Aluminum coating, top surfacing, for built-up roofing	SQ.	1.333
Remove 4-ply built-up roof (includes gravel)	"	4.000
Remove & replace gravel, includes flood coat	"	2.667

07530.10 SINGLE-PLY ROOFING	UNIT	MAN/ HOURS
Elastic sheet roofing		
Neoprene, 1/16" thick	S.F.	0.010
EPDM rubber		
45 mil	S.F.	0.010
PVC		
45 mil	S.F.	0.010
Flashing		
Pipe flashing, 90 mil thick		
1" pipe	EA.	0.200
Neoprene flashing, 60 mil thick strip		
6" wide	L.F.	0.067
12" wide	"	0.100
18" wide	"	0.133
24" wide	"	0.200
Adhesives		
Mastic sealer, applied at joints only		
1/4" bead	L.F.	0.004
Fluid applied roofing		
Urethane, 2 components, elastomeric top membrane		
1" thick	S.F.	0.013
Vinyl liquid roofing, 2 coats, 2 mils per coat	"	0.011
Silicone roofing, 2 coats sprayed, 16 mil per coat	"	0.013
Inverted roof system		
Insulated membrane with coarse gravel ballast		
3 ply with 2" polystyrene	S.F.	0.013
Ballast, 3/4" through 1-1/2" dia. river gravel, 100lb/sf	"	0.800
Walkway for membrane roofs, 1/2" thick	"	0.027

FLASHING AND SHEET METAL

07610.10 METAL ROOFING	UNIT	MAN/ HOURS
Sheet metal roofing, copper, 16 oz, batten seam	SQ.	5.333
Standing seam	"	5.000
Aluminum roofing, natural finish		
Corrugated, on steel frame		
.0175" thick	SQ.	2.286
.0215" thick	"	2.286
.024" thick	"	2.286
.032" thick	SQ.	2.286
Ridge cap		
.019" thick	L.F.	0.027
Corrugated galvanized steel roofing, on steel frame		
28 ga.	SQ.	2.286
26 ga.	"	2.286
24 ga.	"	2.286
22 ga.	"	2.286

07620.10 FLASHING AND TRIM	UNIT	MAN/ HOURS
Counter flashing		
Aluminum, .032"	S.F.	0.080
Stainless steel, .015"	"	0.080
Copper		
16 oz.	S.F.	0.080
20 oz.	"	0.080
24 oz.	"	0.080
32 oz.	"	0.080
Valley flashing		
Aluminum, .032"	S.F.	0.050
Stainless steel, .015	"	0.050
Copper		
16 oz.	S.F.	0.050
20 oz.	"	0.067
24 oz.	"	0.050
32 oz.	"	0.050
Base flashing		
Aluminum, .040"	S.F.	0.067
Stainless steel, .018"	"	0.067
Copper		
16 oz.	S.F.	0.067
20 oz.	"	0.050
24 oz.	"	0.067
32 oz.	"	0.067
Flashing and trim, aluminum		
.019" thick	S.F.	0.057
.032" thick	"	0.057
.040" thick	"	0.062
Reglets, copper 10 oz.	L.F.	0.053
Stainless steel, .020"	"	0.053
Gravel stop		
Aluminum, .032"		
4"	L.F.	0.027
10"	"	0.031
Copper, 16 oz.		
4"	L.F.	0.027
10"	"	0.031

07700.10 MANUFACTURED SPECIALTIES	UNIT	MAN/ HOURS
Smoke vent, 48" x 48"		
Aluminum	EA.	2.000
Galvanized steel	"	2.000
Heat/smoke vent, 48" x 96"		

FLASHING AND SHEET METAL	UNIT	MAN/ HOURS
07700.10 MANUFACTURED SPECIALTIES		
Aluminum	EA.	2.667
Galvanized steel	"	2.667
Ridge vent strips		
Mill finish	L.F.	0.053
Soffit vents		
Mill finish		
2-1/2" wide	L.F.	0.032
Roof hatches		
Steel, plain, primed		
2'6" x 3'0"	EA.	2.000
Galvanized steel		
2'6" x 3'0"	EA.	2.000
Aluminum		
2'6" x 3'0"	EA.	2.000
Gravity ventilators, with curb, base, damper and screen		
Wind driven spinner		
6" dia.	EA.	0.533
12" dia.	"	0.533

SKYLIGHTS	UNIT	MAN/ HOURS
07810.10 PLASTIC SKYLIGHTS		
Single thickness, not including mounting curb		
2' x 4'	EA.	1.000
4' x 4'	"	1.333
Double thickness, not including mounting curb		
2' x 4'	EA.	1.000
4' x 4'	"	1.333

08 DOORS AND WINDOWS

METAL	UNIT	MAN/HOURS
08110.10 METAL DOORS		
Flush hollow metal, standard duty, 20 ga., 1-3/8" thick		
2-6 x 6-8	EA.	0.889
2-8 x 6-8	"	0.889
3-0 x 6-8	"	0.889
1-3/4" thick		
2-6 x 6-8	EA.	0.889
2-8 x 6-8	"	0.889
3-0 x 6-8	"	0.889
Heavy duty, 20 ga., unrated, 1-3/4"		
2-8 x 6-8	EA.	0.889
3-0 x 6-8	"	0.889
08110.40 METAL DOOR FRAMES		
Hollow metal, stock, 18 ga., 4-3/4" x 1-3/4"		
2-0 x 7-0	EA.	1.000
2-4 x 7-0	"	1.000
2-6 x 7-0	"	1.000
3-0 x 7-0	"	1.000
08120.10 ALUMINUM DOORS		
Aluminum doors, commercial		
Narrow stile		
2-6 x 7-0	EA.	4.000
3-0 x 7-0	"	4.000
3-6 x 7-0	"	4.000
Wide stile		
2-6 x 7-0	EA.	4.000
3-0 x 7-0	"	4.000
3-6 x 7-0	"	4.000
08300.10 SPECIAL DOORS		
Overhead door, coiling insulated		
Chain gear, no frame, 12' x 12'	EA.	10.000
Sliding metal fire doors, motorized, fusible link, 3 hr.		
3-0 x 6-8	EA.	16.000
3-8 x 6-8	"	16.000
4-0 x 8-0	"	16.000
5-0 x 8-0	"	16.000
Counter doors, (roll-up shutters), standard, manual		
Opening, 4' high		
4' wide	EA.	6.667
6' wide	"	6.667
8' wide	"	7.273
10' wide	"	10.000
14' wide	"	10.000
6' high		
4' wide	EA.	6.667
6' wide	"	7.273
8' wide	"	8.000
10' wide	"	10.000
14' wide	"	11.429

METAL	UNIT	MAN/HOURS
08300.10 SPECIAL DOORS		
Service doors, (roll up shutters), standard, manual		
Opening		
8' high x 8' wide	EA.	4.444
10' high x 10' wide	"	6.667
12' high x 12' wide	"	10.000
14' high x 14' wide	"	13.333
16' high x 14' wide	"	13.333
20' high x 14' wide	"	20.000
24' high x 16' wide	"	17.778
Roll-up doors		
13-0 high x 14-0 wide	EA.	11.429
12-0 high x 14-0 wide	"	11.429
Top coiling grilles, manually operated, steel or aluminum		
Opening, 4' high x		
4' wide	EA.	3.200
6' wide	"	3.200
8' wide	"	4.444
12' wide	"	4.444
16' wide	"	6.667
6' high x		
4' wide	EA.	6.667
6' wide	"	7.273
8' wide	"	8.000
12' wide	"	8.889
16' wide	"	11.429
Side coiling grilles, manually operated, aluminum		
Opening, 8' high x		
18' wide	EA.	60.000
24' wide	"	68.571
12' high x		
12' wide	EA.	60.000
18' wide	"	68.571
24' wide	"	80.000

STOREFRONTS	UNIT	MAN/HOURS
08410.10 STOREFRONTS		
Storefront, aluminum and glass		
Minimum	S.F.	0.100
Average	"	0.114
Maximum	"	0.133

08 DOORS AND WINDOWS

METAL WINDOWS

08510.10 STEEL WINDOWS

	UNIT	MAN/HOURS
Steel windows, primed		
Casements		
Operable		
Minimum	S.F.	0.047
Maximum	"	0.053
Fixed sash	"	0.040
Double hung	"	0.044
Industrial windows		
Horizontally pivoted sash	S.F.	0.053
Fixed sash	"	0.044
Security sash		
Operable	S.F.	0.053
Fixed	"	0.044
Picture window	"	0.044
Projecting sash		
Minimum	S.F.	0.050
Maximum	"	0.050
Mullions	L.F.	0.040

08520.10 ALUMINUM WINDOWS

	UNIT	MAN/HOURS
Fixed window		
6 sf to 8 sf	S.F.	0.114
12 sf to 16 sf	"	0.089
Projecting window		
6 sf to 8 sf	S.F.	0.200
12 sf to 16 sf	"	0.133
Horizontal sliding		
6 sf to 8 sf	S.F.	0.100
12 sf to 16 sf	"	0.080
Double hung		
6 sf to 8 sf	S.F.	0.160
10 sf to 12 sf	"	0.133

HARDWARE

08710.20 LOCKSETS

	UNIT	MAN/HOURS
Latchset, heavy duty		
Cylindrical	EA.	0.500
Mortise	"	0.800
Lockset, heavy duty		
Cylindrical	EA.	0.500
Mortise	"	0.800

08710.30 CLOSERS

	UNIT	MAN/HOURS
Door closers		
Surface mounted, traditional type, parallel arm		
Standard	EA.	1.000

HARDWARE

08710.30 CLOSERS

	UNIT	MAN/HOURS
Heavy duty	EA.	1.000

08710.40 DOOR TRIM

	UNIT	MAN/HOURS
Panic device		
Mortise	EA.	2.000
Vertical rod	"	2.000
Labelled, rim type	"	2.000
Mortise	"	2.000
Vertical rod	"	2.000
Door plates		
Kick plate, aluminum, 3 beveled edges		
10" x 28"	EA.	0.400
10" x 38"	"	0.400
Push plate, 4" x 16"		
Aluminum	EA.	0.160
Bronze	"	0.160
Stainless steel	"	0.160

08710.60 WEATHERSTRIPPING

	UNIT	MAN/HOURS
Weatherstrip, head and jamb, metal strip, neoprene bulb		
Standard duty	L.F.	0.044
Heavy duty	"	0.050
Thresholds		
Bronze	L.F.	0.200
Aluminum		
Plain	L.F.	0.200
Vinyl insert	"	0.200
Aluminum with grit	"	0.200
Steel		
Plain	L.F.	0.200
Interlocking	"	0.667

GLAZING

08810.10 GLAZING

	UNIT	MAN/HOURS
Sheet glass, 1/8" thick	S.F.	0.044
Plate glass, bronze or grey, 1/4" thick	"	0.073
Clear	"	0.073
Polished	"	0.073
Plexiglass		
1/8" thick	S.F.	0.073
1/4" thick	"	0.044
Float glass, clear		
1/4" thick	S.F.	0.073
1/2" thick	"	0.133

GLAZING	UNIT	MAN/ HOURS
08810.10 GLAZING		
3/4" thick	S.F.	0.200
1" thick	"	0.267
Tinted glass, polished plate, twin ground		
1/4" thick	S.F.	0.073
1/2" thick	"	0.133
Total, full vision, all glass window system		
To 10' high		
Minimum	S.F.	0.200
Average	"	0.200
Maximum	"	0.200
10' to 20' high		
Minimum	S.F.	0.200
Average	"	0.200
Maximum	"	0.200
Insulated glass, bronze or gray		
1/2" thick	S.F.	0.133
1" thick	"	0.200
Spandrel glass, polished bronze/grey, 1 side, 1/4" thick	"	0.073
Tempered glass (safety)		
Clear sheet glass		
1/8" thick	S.F.	0.044
3/16" thick	"	0.062
Clear float glass		
1/4" thick	S.F.	0.067
1/2" thick	"	0.133
3/4" thick	"	0.267
Tinted float glass		
3/16" thick	S.F.	0.062
1/4" thick	"	0.067
3/8" thick	"	0.100
1/2" thick	"	0.133
Laminated glass		
Float safety glass with polyvinyl plastic interlayer		
1/4", sheet or float		
Two lites, 1/8" thick, clear glass	S.F.	0.067
1/2" thick, float glass		
Two lites, 1/4" thick, clear glass	S.F.	0.133
Tinted glass	"	0.133
Insulating glass, two lites, clear float glass		
1/2" thick	S.F.	0.133
3/4" thick	"	0.200
1" thick	"	0.267
Glass seal edge		
3/8" thick	S.F.	0.133
Tinted glass		
1/2" thick	S.F.	0.133
1" thick	"	0.267
Tempered, clear		
1" thick	S.F.	0.267
Wire reinforced	"	0.267
Plate mirror glass		
1/4" thick		
15 sf	S.F.	0.080
Over 15 sf	"	0.073

GLAZING	UNIT	MAN/ HOURS
08810.10 GLAZING		
Door type, 1/4" thick	S.F.	0.080
Transparent, one way vision, 1/4" thick	"	0.080
Sheet mirror glass		
3/16" thick	S.F.	0.080
1/4" thick	"	0.067
Wall tiles, 12" x 12"		
Clear glass	S.F.	0.044
Veined glass	"	0.044
Wire glass, 1/4" thick		
Clear	S.F.	0.267
Hammered	"	0.267
Obscure	"	0.267
Glazing accessories		
Neoprene glazing gaskets		
1/4" glass	L.F.	0.032
1/2" glass	"	0.035
3/4" glass	"	0.036
1" glass	"	0.040
Mullion section		
1/4" glass	L.F.	0.016
3/8" glass	"	0.020
1/2" glass	"	0.023
3/4" glass	"	0.027
1" glass	"	0.032
Molded corners	EA.	0.533

GLAZED CURTAIN WALLS	UNIT	MAN/ HOURS
08910.10 GLAZED CURTAIN WALLS		
Curtain wall, aluminum system, framing sections		
2" x 3"		
Jamb	L.F.	0.067
Horizontal	"	0.067
Mullion	"	0.067
2" x 4"		
Jamb	L.F.	0.100
Horizontal	"	0.100
Mullion	"	0.100
3" x 5-1/2"		
Jamb	L.F.	0.100
Horizontal	"	0.100
Mullion	"	0.100
4" corner mullion	"	0.133
Coping sections		
1/8" x 8"	L.F.	0.133
1/8" x 9"	"	0.133

GLAZED CURTAIN WALLS	UNIT	MAN/ HOURS
08910.10 GLAZED CURTAIN WALLS		
1/8" x 12-1/2"	L.F.	0.160
Sill section		
1/8" x 6"	L.F.	0.080
1/8" x 7"	"	0.080
1/8" x 8-1/2"	"	0.080
Column covers, aluminum		
1/8" x 26"	L.F.	0.200
1/8" x 34"	"	0.211
1/8" x 38"	"	0.211
Doors		
Aluminum framed, standard hardware		
Narrow stile		
2-6 x 7-0	EA.	4.000
3-0 x 7-0	"	4.000
3-6 x 7-0	"	4.000
Wide stile		
2-6 x 7-0	EA.	4.000
3-0 x 7-0	"	4.000
3-6 x 7-0	"	4.000
Window wall system, complete		
Minimum	S.F.	0.080
Average	"	0.089
Maximum	"	0.114

09 FINISHES

SUPPORT SYSTEMS

09110.10 METAL STUDS

SUPPORT SYSTEMS	UNIT	MAN/HOURS
Studs, non load bearing, galvanized		
2-1/2", 20 ga.		
12" o.c.	S.F.	0.017
16" o.c.	"	0.013
25 ga.		
12" o.c.	S.F.	0.017
16" o.c.	"	0.013
24" o.c.	"	0.011
3-5/8", 20 ga.		
12" o.c.	S.F.	0.020
16" o.c.	"	0.016
24" o.c.	"	0.013
25 ga.		
12" o.c.	S.F.	0.020
16" o.c.	"	0.016
24" o.c.	"	0.013
4", 20 ga.		
12" o.c.	S.F.	0.020
16" o.c.	"	0.016
24" o.c.	"	0.013
25 ga.		
12" o.c.	S.F.	0.020
16" o.c.	"	0.016
24" o.c.	"	0.013
6", 20 ga.		
12" o.c.	S.F.	0.025
16" o.c.	"	0.020
24" o.c.	"	0.017
25 ga.		
12" o.c.	S.F.	0.025
16" o.c.	"	0.020
24" o.c.	"	0.017
Load bearing studs, galvanized		
3-5/8", 16 ga.		
12" o.c.	S.F.	0.020
16" o.c.	"	0.016
18 ga.		
12" o.c.	S.F.	0.013
16" o.c.	"	0.016
4", 16 ga.		
12" o.c.	S.F.	0.020
16" o.c.	"	0.016
6", 16 ga.		
12" o.c.	S.F.	0.025
16" o.c.	"	0.020
Furring		
On beams and columns		
7/8" channel	L.F.	0.053
1-1/2" channel	"	0.062
On ceilings		
3/4" furring channels		
12" o.c.	S.F.	0.033
16" o.c.	"	0.032
24" o.c.	"	0.029
1-1/2" furring channels		
12" o.c.	S.F.	0.036
16" o.c.	"	0.033
24" o.c.	"	0.031
On walls		
3/4" furring channels		
12" o.c.	S.F.	0.027
16" o.c.	"	0.025
24" o.c.	"	0.024
1-1/2" furring channels		
12" o.c.	S.F.	0.029
16" o.c.	"	0.027

LATH AND PLASTER

09205.10 GYPSUM LATH

LATH AND PLASTER	UNIT	MAN/HOURS
Gypsum lath, 1/2" thick		
Clipped	S.Y.	0.044
Nailed	"	0.050

09205.20 METAL LATH

LATH AND PLASTER	UNIT	MAN/HOURS
Stucco lath		
1.8 lb.	S.Y.	0.100
3.6 lb.	"	0.100
Paper backed		
Minimum	S.Y.	0.080
Maximum	"	0.114

09210.10 PLASTER

LATH AND PLASTER	UNIT	MAN/HOURS
Gypsum plaster, trowel finish, 2 coats		
Ceilings	S.Y.	0.250
Walls	"	0.235
3 coats		
Ceilings	S.Y.	0.348
Walls	"	0.308
Patch holes, average size holes		
1 sf to 5 sf		
Minimum	S.F.	0.133
Average	"	0.160
Maximum	"	0.200
Over 5 sf		
Minimum	S.F.	0.080
Average	"	0.114
Maximum	"	0.133
Patch cracks		
Minimum	S.F.	0.027

09 FINISHES

LATH AND PLASTER	UNIT	MAN/ HOURS
09210.10 PLASTER		
average	S.F.	0.040
Maximum	"	0.080
09220.10 PORTLAND CEMENT PLASTER		
Stucco, portland, gray, 3 coat, 1" thick		
Sand finish	S.Y.	0.348
Trowel finish	"	0.364
White cement		
Sand finish	S.Y.	0.364
Trowel finish	"	0.400
Scratch coat		
For ceramic tile	S.Y.	0.080
For quarry tile	"	0.080
Portland cement plaster		
2 coats, 1/2"	S.Y.	0.160
3 coats, 7/8"	"	0.200
09250.10 GYPSUM BOARD		
Drywall, plasterboard, 3/8" clipped to		
Metal furred ceiling	S.F.	0.009
Columns and beams	"	0.020
Walls	"	0.008
Nailed or screwed to		
Wood framed ceiling	S.F.	0.008
Columns and beams	"	0.018
Walls	"	0.007
1/2", clipped to		
Metal furred ceiling	S.F.	0.009
Columns and beams	"	0.020
Walls	"	0.008
Nailed or screwed to		
Wood framed ceiling	S.F.	0.008
Columns and beams	"	0.018
Walls	"	0.007
5/8", clipped to		
Metal furred ceiling	S.F.	0.010
Columns and beams	"	0.022
Walls	"	0.009
Nailed or screwed to		
Wood framed ceiling	S.F.	0.010
Columns and beams	"	0.022
Walls	"	0.009
Taping and finishing joints		
Minimum	S.F.	0.005
Average	"	0.007
Maximum	"	0.008
Casing bead		
Minimum	L.F.	0.023
Average	"	0.027
Maximum	"	0.040
Corner bead		
Minimum	L.F.	0.023
Average	"	0.027
Maximum	"	0.040

TILE	UNIT	MAN/ HOURS
09310.10 CERAMIC TILE		
Glazed wall tile, 4-1/4" x 4-1/4"		
Minimum	S.F.	0.057
Average	"	0.067
Maximum	"	0.080
Unglazed floor tile		
Portland cement bed, cushion edge, face mounted		
1" x 1"	S.F.	0.073
1" x 2"	"	0.070
2" x 2"	"	0.067
Adhesive bed, with white grout		
1" x 1"	S.F.	0.073
1" x 2"	"	0.070
2" x 2"	"	0.067
09330.10 QUARRY TILE		
Floor		
4 x 4 x 1/2"	S.F.	0.107
6 x 6 x 1/2"	"	0.100
6 x 6 x 3/4"	"	0.100
Wall, applied to 3/4" portland cement bed		
4 x 4 x 1/2"	S.F.	0.160
6 x 6 x 3/4"	"	0.133
Cove base		
5 x 6 x 1/2" straight top	L.F.	0.133
6 x 6 x 3/4" round top	"	0.133
Stair treads 6 x 6 x 3/4"	"	0.200
Window sill 6 x 8 x 3/4"	"	0.160
For abrasive surface, add to material, 25%		
09410.10 TERRAZZO		
Floors on concrete, 1-3/4" thick, 5/8" topping		
Gray cement	S.F.	0.114
White cement	"	0.114
Sand cushion, 3" thick, 5/8" top, 1/4"		
Gray cement	S.F.	0.133
White cement	"	0.133
Monolithic terrazzo, 3-1/2" base slab, 5/8" topping	"	0.100
Terrazzo wainscot, cast-in-place, 1/2" thick	"	0.200
Base, cast in place, terrazzo cove type, 6" high	L.F.	0.114
Curb, cast in place, 6" wide x 6" high, polished top	"	0.400
Stairs, cast-in-place, topping on concrete or metal		
1-1/2" thick treads, 12" wide	L.F.	0.400
Combined tread and riser	"	1.000
Precast terrazzo, thin set		
Terrazzo tiles, non-slip surface		
9" x 9" x 1" thick	S.F.	0.114
12" x 12"		
1" thick	S.F.	0.107
1-1/2" thick	"	0.114
18" x 18" x 1-1/2" thick	"	0.114
24" x 24" x 1-1/2" thick	"	0.094
Terrazzo wainscot		

09 FINISHES

TILE

09410.10 TERRAZZO	UNIT	MAN/ HOURS
12" x 12" x 1" thick	S.F.	0.200
18" x 18" x 1-1/2" thick	"	0.229
Base		
6" high		
Straight	L.F.	0.062
Coved	"	0.062
8" high		
Straight	L.F.	0.067
Coved	"	0.067
Terrazzo curbs		
8" wide x 8" high	L.F.	0.320
6" wide x 6" high	"	0.267
Precast terrazzo stair treads, 12" wide		
1-1/2" thick		
Diamond pattern	L.F.	0.145
Non-slip surface	"	0.145
2" thick		
Diamond pattern	L.F.	0.145
Non-slip surface	"	0.160
Stair risers, 1" thick to 6" high		
Straight sections	L.F.	0.080
Cove sections	"	0.080
Combined tread and riser		
Straight sections		
1-1/2" tread, 3/4" riser	L.F.	0.229
3" tread, 1" riser	"	0.229
Curved sections		
2" tread, 1" riser	L.F.	0.267
3" tread, 1" riser	"	0.267
Stair stringers, notched for treads and risers		
1" thick	L.F.	0.200
2" thick	"	0.267
Landings, structural, nonslip		
1-1/2" thick	S.F.	0.133
3" thick	"	0.160
Conductive terrazzo, spark proof industrial floor		
Epoxy terrazzo		
Floor	S.F.	0.050
Base	"	0.067
Polyacrylate		
Floor	S.F.	0.050
Base	"	0.067
Polyester		
Floor	S.F.	0.032
Base	"	0.040
Synthetic latex mastic		
Floor	S.F.	0.050
Base	"	0.067

ACOUSTICAL TREATMENT

09510.10 CEILINGS AND WALLS	UNIT	MAN/ HOURS
Acoustical panels, suspension system not included		
Fiberglass panels		
5/8" thick		
2' x 2'	S.F.	0.011
2' x 4'	"	0.009
3/4" thick		
2' x 2'	S.F.	0.011
2' x 4'	"	0.009
Mineral fiber panels		
5/8" thick		
2' x 2'	S.F.	0.011
2' x 4'	"	0.009
3/4" thick		
2' x 2'	S.F.	0.011
2' x 4'	"	0.009
Ceiling suspension systems		
T bar system		
2' x 4'	S.F.	0.008
2' x 2'	"	0.009

FLOORING

09550.10 WOOD FLOORING	UNIT	MAN/ HOURS
Wood block industrial flooring		
Creosoted		
2" thick	S.F.	0.021
2-1/2" thick	"	0.025
3" thick	"	0.027
Gym floor, 2 ply felt, 25/32" maple, finished, in mastic	"	0.044
Over wood sleepers	"	0.050
Finishing, sand, fill, finish, and wax	"	0.020
Refinish sand, seal, and 2 coats of polyurethane	"	0.027
Clean and wax floors	"	0.004

09630.10 UNIT MASONRY FLOORING	UNIT	MAN/ HOURS
Clay brick		
9 x 4-1/2 x 3" thick		
Glazed	S.F.	0.067
Unglazed	"	0.067
8 x 4 x 3/4" thick		
Glazed	S.F.	0.070
Unglazed	"	0.070

09 FINISHES

FLOORING

	UNIT	MAN/ HOURS
09660.10 RESILIENT TILE FLOORING		
Solid vinyl tile, 1/8" thick, 12" x 12"		
Marble patterns	S.F.	0.020
Solid colors	"	0.020
Travertine patterns	"	0.020
09665.10 RESILIENT SHEET FLOORING		
Vinyl sheet flooring		
Minimum	S.F.	0.008
Average	"	0.010
Maximum	"	0.013
Cove, to 6"	L.F.	0.016
Fluid applied resilient flooring		
Polyurethane, poured in place, 3/8" thick	S.F.	0.067
Wall base, vinyl		
4" high	L.F.	0.027
6" high	"	0.027
Stair accessories		
Treads, 1/4" x 12", rubber diamond surface		
Marbled	L.F.	0.067
Plain	"	0.067
Grit strip safety tread, 12" wide, colors		
3/16" thick	L.F.	0.067
5/16" thick	"	0.067
Risers, 7" high, 1/8" thick, colors		
Flat	L.F.	0.040
Coved	"	0.040

CARPET

	UNIT	MAN/ HOURS
09680.10 FLOOR LEVELING		
Repair and level floors to receive new flooring		
Minimum	S.Y.	0.027
Average	"	0.067
Maximum	"	0.080
09682.10 CARPET PADDING		
Carpet padding		
Jute padding		
Minimum	S.Y.	0.036
Average	"	0.040
Maximum	"	0.044
Sponge rubber cushion		
Minimum	S.Y.	0.036
Average	"	0.040
Maximum	"	0.044
Urethane cushion, 3/8" thick		
Minimum	S.Y.	0.036

CARPET

	UNIT	MAN/ HOURS
09682.10 CARPET PADDING		
Average	S.Y.	0.040
Maximum	"	0.044
09685.10 CARPET		
Carpet, acrylic		
24 oz., light traffic	S.Y.	0.044
28 oz., medium traffic	"	0.044
Commercial		
Nylon		
28 oz., medium traffic	S.Y.	0.044
35 oz., heavy traffic	"	0.044
Wool		
30 oz., medium traffic	S.Y.	0.044
36 oz., medium traffic	"	0.044
42 oz., heavy traffic	"	0.044
Carpet tile		
Foam backed		
Clean and vacuum carpet		
Minimum	S.Y.	0.004
Average	"	0.005
Maximum	"	0.008
09700.10 SPECIAL FLOORING		
Epoxy flooring, marble chips		
Epoxy with colored quartz chips in 1/4" base	S.F.	0.044
Heavy duty epoxy topping, 3/16" thick	"	0.044
Epoxy terrazzo		
1/4" thick chemical resistant	S.F.	0.050

PAINTING

	UNIT	MAN/ HOURS
09910.10 EXTERIOR PAINTING		
Exterior painting		
Wood surfaces, 1 coat primer, two coats paint		
Door and frame	EA.	1.333
Windows	S.F.	0.020
Wood trim	"	0.020
Wood siding	"	0.010
Hardboard surfaces		
One coat primer, two coats paint	S.F.	0.010
Asbestos cement surfaces		
One coat primer, two coats paint	S.F.	0.010
Galvanized surfaces, galvanized primer		
One coat primer, two coats paint	S.F.	0.009
Stucco surfaces, acrylic primer, acrylic latex paint		

09 FINISHES

PAINTING

PAINTING	UNIT	MAN/ HOURS
09910.10 EXTERIOR PAINTING		
One coat primer, two coats paint	S.F.	0.013
Concrete masonry unit surfaces, brush work		
One coat filler, one coat paint	S.F.	0.010
Two coats epoxy	"	0.013
Texture coating	"	0.008
Concrete surfaces		
One coat filler, one coat paint	S.F.	0.010
Two coats paint	"	0.013
Structural steel		
One field coat paint, brush work		
Light framing	S.F.	0.007
Heavy framing	"	0.004
One field coat paint, spray work		
Light framing	S.F.	0.002
Heavy framing	"	0.002
Miscellaneous steel items, spray work, one coat		
Exposed decking	S.F.	0.007
Joist	"	0.010
Columns	"	0.010
Pipes, one coat primer, one coat paint		
4" dia.	L.F.	0.010
8" dia.	"	0.013
12" dia.	"	0.020
Paint letters on pipe with brush	"	0.020
Paint pipe insulation cloth cover	S.F.	0.010
Sprinkler system piping	L.F.	0.007
Miscellaneous surfaces		
Stair pipe rails		
Two rails	L.F.	0.027
One rail	"	0.016
Stair to 4' wide, including rails, per riser	EA.	0.114
Gratings and frames	S.F.	0.027
Ladders	L.F.	0.023
Miscellaneous exposed metal	S.F.	0.008
09920.10 INTERIOR PAINTING		
Walls, concrete and masonry, brush, primer, acrylic		
One coat primer, one coat paint	S.F.	0.010
Two coats paint	"	0.013
Plywood, paint	"	0.004
Natural finish	"	0.005
Wood, paint	"	0.005
Natural finish	"	0.006
Metal		
One coat filler	S.F.	0.005
One coat primer, one coat paint	"	0.010
Two coats paint	"	0.013
Plaster or gypsum board, paint	"	0.004
Epoxy	"	0.005
Ceilings, one coat paint, wood	"	0.006
Concrete	"	0.005
Plaster	"	0.004
Miscellaneous metal, brushwork	"	0.010

PAINTING	UNIT	MAN/ HOURS
09920.30 DOORS AND MILLWORK		
Painting, doors		
Minimum	S.F.	0.027
Average	"	0.040
Maximum	"	0.053
Cabinets, shelves, and millwork		
Minimum	S.F.	0.013
Average	"	0.023
Maximum	"	0.040
09920.60 WINDOWS		
Painting, windows		
Minimum	S.F.	0.016
Average	"	0.020
Maximum	"	0.032
09955.10 WALL COVERING		
Vinyl wall covering		
Medium duty	S.F.	0.011
Heavy duty	"	0.013
Over pipes and irregular shapes		
Flexible gypsum coated wall fabric, fire resistant	S.F.	0.008
Vinyl corner guards		
3/4" x 3/4" x 8'	EA.	0.100
2-3/4" x 2-3/4" x 4'	"	0.100
09980.10 PAINTING PREPARATION		
Cleaning, light		
Wood	S.F.	0.002
Plaster or gypsum wallboard	"	0.002
Prepare sprinkler piping for painting	L.F.	0.002
Prepare insulated pipe for painting	S.F.	0.002
Normal painting prep, masonry and concrete		
Unpainted	S.F.	0.001
Painted	"	0.002
Plaster or gypsum		
Unpainted	S.F.	0.001
Painted	"	0.002
Wood		
Unpainted	S.F.	0.001
Painted	"	0.002
Painted steel, light rusting	"	0.003
Sandblasting		
Brush off blast	S.F.	0.005
Commercial blast	"	0.013
Near white metal blast	"	0.023
White metal blast	"	0.027

10 SPECIALTIES

SPECIALTIES	UNIT	MAN/HOURS
10110.10 CHALKBOARDS		
Chalkboard, metal frame, 1/4" thick		
48"x60"	EA.	0.800
48"x96"	"	0.889
48"x144"	"	1.000
48"x192"	"	1.143
Liquid chalkboard		
48"x60"	EA.	0.800
48"x96"	"	0.889
48"x144"	"	1.000
48"x192"	"	1.143
Map rail, deluxe	L.F.	0.040
Average	PCT.	
10165.10 TOILET PARTITIONS		
Toilet partition, plastic laminate		
Ceiling mounted	EA.	2.667
Floor mounted	"	2.000
Metal		
Ceiling mounted	EA.	2.667
Floor mounted	"	2.000
Wheel chair partition, plastic laminate		
Ceiling mounted	EA.	2.667
Floor mounted	"	2.000
Painted metal		
Ceiling mounted	EA.	2.667
Floor mounted	"	2.000
Urinal screen, plastic laminate		
Wall hung	EA.	1.000
Floor mounted	"	1.000
Porcelain enameled steel, floor mounted	"	1.000
Painted metal, floor mounted	"	1.000
Stainless steel, floor mounted	"	1.000
Metal toilet partitions		
Toilet partitions, front door and side divider, floor mounted		
Porcelain enameled steel	EA.	2.000
Painted steel	"	2.000
Stainless steel	"	2.000
10185.10 SHOWER STALLS		
Shower receptors		
Precast, terrazzo		
32" x 32"	EA.	0.667
32" x 48"	"	0.800
Concrete		
32" x 32"	EA.	0.667
48" x 48"	"	0.889
Shower door, trim and hardware		
Porcelain enameled steel, flush	EA.	0.800
Baked enameled steel, flush	"	0.800
Aluminum frame, tempered glass, 48" wide, sliding	"	1.000
Folding	"	1.000
Shower compartment, precast concrete receptor		
Single entry type		

SPECIALTIES	UNIT	MAN/HOURS
10185.10 SHOWER STALLS		
Porcelain enameled steel	EA.	8.000
Baked enameled steel	"	8.000
Stainless steel	"	8.000
Double entry type		
Porcelain enameled steel	EA.	10.000
Baked enameled steel	"	10.000
Stainless steel	"	10.000
10210.10 VENTS AND WALL LOUVERS		
Vents w/screen, 4" deep, 8" wide, 5" high		
Modular	EA.	0.250
Aluminum gable louvers	S.F.	0.133
Vent screen aluminum, 4" wide, continuous	L.F.	0.027
Wall louver, aluminum mill finish		
Under, 2 sf	S.F.	0.100
2 to 4 sf	"	0.089
5 to 10 sf	"	0.089
Galvanized steel		
Under 2 sf	S.F.	0.100
2 to 4 sf	"	0.089
5 to 10 sf	"	0.089
10225.10 DOOR LOUVERS		
Fixed, 1" thick, enameled steel		
8"x8"	EA.	0.100
12"x12"	"	0.114
20"x20"	"	0.320
24"x24"	"	0.364
10290.10 PEST CONTROL		
Termite control		
Under slab spraying		
Minimum	S.F.	0.002
Average	"	0.004
Maximum	"	0.008
10350.10 FLAGPOLES		
Installed in concrete base		
Fiberglass		
25' high	EA.	5.333
50' high	"	13.333
Aluminum		
25' high	EA.	5.333
50' high	"	13.333
Bonderized steel		
25' high	EA.	6.154
50' high	"	16.000
Freestanding tapered, fiberglass		
30' high	EA.	5.714
40' high	"	7.273
50' high	"	8.000

SPECIALTIES	UNIT	MAN/ HOURS
10350.10 FLAGPOLES		
60' high	EA.	9.412
Wall mounted, with collar, brushed aluminum finish		
15' long	EA.	4.000
18' long	"	4.000
20' long	"	4.211
24' long	"	4.706
Outrigger, wall, including base		
10' long	EA.	5.333
20' long	"	6.667
10400.10 IDENTIFYING DEVICES		
Directory and bulletin boards		
Open face boards		
Chrome plated steel frame	S.F.	0.400
Aluminum framed	"	0.400
Bronze framed	"	0.400
Stainless steel framed	"	0.400
Tack board, aluminum framed	"	0.400
Visual aid board, aluminum framed	"	0.400
Glass encased boards, hinged and keyed		
Aluminum framed	S.F.	1.000
Bronze framed	"	1.000
Stainless steel framed	"	1.000
Chrome plated steel framed	"	1.000
Metal plaque		
Cast bronze	S.F.	0.667
Aluminum	"	0.667
Metal engraved plaque		
Porcelain steel	S.F.	0.667
Stainless steel	"	0.667
Brass	"	0.667
Aluminum	"	0.667
Metal built-up plaque		
Bronze	S.F.	0.800
Copper and bronze	"	0.800
Copper and aluminum	"	0.800
Metal nameplate plaques		
Cast bronze	S.F.	0.500
Cast aluminum	"	0.500
Engraved, 1-1/2" x 6"		
Bronze	EA.	0.500
Aluminum	"	0.500
Letters, on masonry or concrete, aluminum, satin finish		
1/2" thick		
2" high	EA.	0.320
4" high	"	0.400
6" high	"	0.444
3/4" thick		
8" high	EA.	0.500
10" high	"	0.571
1" thick		
12" high	EA.	0.667
14" high	"	0.800

SPECIALTIES	UNIT	MAN/ HOURS
10400.10 IDENTIFYING DEVICES		
16" high	EA.	1.000
3/8" thick		
2" high	EA.	0.320
4" high	"	0.400
1/2" thick, 6" high	"	0.444
5/8" thick, 8" high	"	0.500
1" thick		
10" high	EA.	0.571
12" high	"	0.667
14" high	"	0.800
16" high	"	1.000
Interior door signs, adhesive, flexible		
2" x 8"	EA.	0.200
4" x 4"	"	0.200
6" x 7"	"	0.200
6" x 9"	"	0.200
10" x 9"	"	0.200
10" x 12"	"	0.200
Hard plastic type, no frame		
3" x 8"	EA.	0.200
4" x 4"	"	0.200
4" x 12"	"	0.200
Hard plastic type, with frame		
3" x 8"	EA.	0.200
4" x 4"	"	0.200
4" x 12"	"	0.200
10450.10 CONTROL		
Access control, 7' high, indoor or outdoor impenetrability		
Remote or card control, type B	EA.	10.667
Free passage, type B	"	10.667
Remote or card control, type AA	"	10.667
Free passage, type AA	"	10.667
10500.10 LOCKERS		
Locker bench, floor mounted, laminated maple		
4'	EA.	0.667
6'	"	0.667
Wardrobe locker, 12" x 60" x 15", baked on enamel		
1-tier	EA.	0.400
2-tier	"	0.400
3-tier	"	0.421
4-tier	"	0.421
12" x 72" x 15", baked on enamel		
1-tier	EA.	0.400
2-tier	"	0.400
4-tier	"	0.421
5-tier	"	0.421
15" x 60" x 15", baked on enamel		
1-tier	EA.	0.400
4-tier	"	0.421
Wardrobe locker, single tier type		
12" x 15" x 72"	EA.	0.800

10 SPECIALTIES

SPECIALTIES	UNIT	MAN/ HOURS
10500.10 LOCKERS		
18" x 15" x 72"	EA.	0.842
12" x 18" x 72"	"	0.889
18" x 18" x 72"	"	0.941
Double tier type		
12" x 15" x 36"	EA.	0.400
18" x 15" x 36"	"	0.400
12" x 18" x 36"	"	0.400
18" x 18" x 36"	"	0.400
Two person unit		
18" x 15" x 72"	EA.	1.333
18" x 18" x 72"	"	1.600
Duplex unit		
15" x 15" x 72"	EA.	0.800
15" x 21" x 72"	"	0.800
Basket lockers, basket sets with baskets		
24 basket set	SET	4.000
30 basket set	"	5.000
36 basket set	"	6.667
42 basket set	"	8.000
10520.10 FIRE PROTECTION		
Portable fire extinguishers		
Water pump tank type		
2.5 gal.		
Red enameled galvanized	EA.	0.533
Red enameled copper	"	0.533
Polished copper	"	0.533
Carbon dioxide type, red enamel steel		
Squeeze grip with hose and horn		
2.5 lb	EA.	0.533
5 lb	"	0.615
10 lb	"	0.800
15 lb	"	1.000
20 lb	"	1.000
Wheeled type		
125 lb	EA.	1.600
250 lb	"	1.600
500 lb	"	1.600
Dry chemical, pressurized type		
Red enameled steel		
2.5 lb	EA.	0.533
5 lb	"	0.615
10 lb	"	0.800
20 lb	"	1.000
30 lb	"	1.000
Chrome plated steel, 2.5 lb	"	0.533
Other type extinguishers		
2.5 gal, stainless steel, pressurized water tanks	EA.	0.533
Soda and acid type	"	0.533
Cartridge operated, water type	"	0.533
Loaded stream, water type	"	0.533
Foam type	"	0.533
40 gal, wheeled foam type	"	1.600

SPECIALTIES	UNIT	MAN/ HOURS
10520.10 FIRE PROTECTION		
Fire extinguisher cabinets		
Enameled steel		
8" x 12" x 27"	EA.	1.600
8" x 16" x 38"	"	1.600
Aluminum		
8" x 12" x 27"	EA.	1.600
8" x 16" x 38"	"	1.600
8" x 12" x 27"	"	1.600
Stainless steel		
8" x 16" x 38"	EA.	1.600
10550.10 POSTAL SPECIALTIES		
Mail chutes		
Single mail chute		
Finished aluminum	L.F.	2.000
Bronze	"	2.000
Single mail chute receiving box		
Finished aluminum	EA.	4.000
Bronze	"	4.000
Twin mail chute, double parallel		
Finished aluminum	FLR	4.000
Bronze	"	4.000
Receiving box, 36" x 20" x 12"		
Finished aluminum	EA.	6.667
Bronze	"	6.667
Locked receiving mail box		
Finished aluminum	EA.	4.000
Bronze	"	4.000
Commercial postal accessories for mail chutes		
Letter slot, brass	EA.	1.333
Bulk mail slot, brass	"	1.333
Mail boxes		
Residential postal accessories		
Letter slot	EA.	0.400
Rural letter box	"	1.000
10670.10 SHELVING		
Shelving, enamel, closed side and back, 12" x 36"		
5 shelves	EA.	1.333
8 shelves	"	1.778
Open		
5 shelves	EA.	1.333
8 shelves	"	1.778
Metal storage shelving, baked enamel		
7 shelf unit, 72" or 84" high		
12" shelf	L.F.	0.842
24" shelf	"	1.000
36" shelf	"	1.143
4 shelf unit, 40" high		
12" shelf	L.F.	0.727
24" shelf	"	0.889
3 shelf unit, 32" high		
12" shelf	L.F.	0.421

SPECIALTIES	UNIT	MAN/ HOURS
10670.10 SHELVING		
24" shelf	L.F.	0.500
Single shelf unit, attached to masonry		
12" shelf	L.F.	0.145
24" shelf	"	0.174
10800.10 BATH ACCESSORIES		
Ash receiver, wall mounted, aluminum	EA.	0.400
Grab bar, 1-1/2" dia., stainless steel, wall mounted		
24" long	EA.	0.400
36" long	"	0.421
48" long	"	0.471
1" dia., stainless steel		
12" long	EA.	0.348
24" long	"	0.400
36" long	"	0.444
48" long	"	0.471
Hand dryer, surface mounted, 110 volt	"	1.000
Medicine cabinet, 16 x 22, baked enamel, steel, lighted	"	0.320
With mirror, lighted	"	0.533
Mirror, 1/4" plate glass, up to 10 sf	S.F.	0.080
Mirror, stainless steel frame		
18"x24"	EA.	0.267
24"x30"	"	0.400
24"x48"	"	0.667
30"x30"	"	0.800
48"x72"	"	1.333
With shelf, 18"x24"	"	0.320
Sanitary napkin dispenser, stainless steel, wall mounted	"	0.533
Shower rod, 1" diameter		
Chrome finish over brass	EA.	0.400
Stainless steel	"	0.400
Soap dish, stainless steel, wall mounted	"	0.533
Toilet tissue dispenser, stainless, wall mounted		
Single roll	EA.	0.200
Double roll	"	0.229
Towel dispenser, stainless steel		
Flush mounted	EA.	0.444
Surface mounted	"	0.400
Combination towel dispenser and waste receptacle	"	0.533
Towel bar, stainless steel		
18" long	EA.	0.320
24" long	"	0.364
30" long	"	0.400
36" long	"	0.444
Waste receptacle, stainless steel, wall mounted	"	0.667

11 EQUIPMENT

ARCHITECTURAL EQUIPMENT	UNIT	MAN/ HOURS
11020.10 SECURITY EQUIPMENT		
Office safes, 30" x 20" x 20", 1 hr rating	EA.	2.000
30" x 16" x 15", 2 hr rating	"	1.600
30" x 28" x 20", H&G rating	"	1.000
Surveillance system		
Minimum	EA.	16.000
Maximum	"	80.000
Insulated file room door		
1 hr rating		
32" wide	EA.	8.000
40" wide	"	8.889
11161.10 LOADING DOCK EQUIPMENT		
Dock leveler, 10 ton capacity		
6' x 8'	EA.	8.000
7' x 8'	"	8.000
Bumpers, laminated rubber		
4-1/2" thick		
6" x 14"	EA.	0.160
10" x 14"	"	0.200
10" x 36"	"	0.267
12" x 14"	"	0.211
12" x 36"	"	0.296
6" thick		
10" x 14"	EA.	0.229
10" x 24"	"	0.276
10" x 36"	"	0.400
Extruded rubber bumpers		
T-section, 22" x 22" x 3"	EA.	0.160
Molded rubber bumpers		
24" x 12" x 3" thick	EA.	0.400
Door seal, 12" x 12", vinyl covered	L.F.	0.200
Dock boards, heavy duty, 5' x 5'		
5000 lb		
Minimum	EA.	6.667
Maximum	"	6.667
9000 lb		
Minimum	EA.	6.667
Maximum	"	7.273
15,000 lb	"	7.273
Truck shelters		
Minimum	EA.	6.154
Maximum	"	11.429
11170.10 WASTE HANDLING		
Incinerator, electric		
100 lb/hr		
Minimum	EA.	8.000
Maximum	"	8.000
400 lb/hr		
Minimum	EA.	16.000
Maximum	"	16.000
1000 lb/hr		
Minimum	EA.	24.242
Maximum	"	24.242

ARCHITECTURAL EQUIPMENT	UNIT	MAN/ HOURS
11480.10 ATHLETIC EQUIPMENT		
Basketball backboard		
Fixed	EA.	10.000
Swing-up	"	16.000
Portable, hydraulic	"	4.000
Suspended type, standard	"	16.000
Bleacher, telescoping, manual		
15 tier, minimum	SEAT	0.160
Maximum	"	0.160
20 tier, minimum	"	0.178
Maximum	"	0.178
30 tier, minimum	"	0.267
Maximum	"	0.267
Boxing ring elevated, complete, 22' x 22'	EA.	80.000
Gym divider curtain		
Minimum	S.F.	0.011
Maximum	"	0.011
Scoreboards, single face		
Minimum	EA.	8.000
Maximum	"	40.000
Parallel bars		
Minimum	EA.	8.000
Maximum	"	13.333
11500.10 INDUSTRIAL EQUIPMENT		
Vehicular paint spray booth, solid back, 14'4" x 9'6"		
24' deep	EA.	8.000
26'6" deep	"	8.000
28'6" deep	"	8.000
Drive through, 14'9" x 9'6"		
24' deep	EA.	8.000
26'6" deep	"	8.000
28'6" deep	"	8.000
Water wash, paint spray booth		
5' x 11'2" x 10'8"	EA.	8.000
6' x 11'2" x 10'8"	"	8.000
8' x 11'2" x 10'8"	"	8.000
10' x 11'2" x 11'2"	"	8.000
12' x 12'2" x 11'2"	"	8.000
14' x 12'2" x 11'2"	"	8.000
16' x 12'2" x 11'2"	"	8.000
20' x 12'2" x 11'2"	"	8.000
Dry type spray booth, with paint arrestors		
5'4" x 7'2" x 6'8"	EA.	8.000
6'4" x 7'2" x 6'8"	"	8.000
8'4" x 7'2" x 9'2"	"	8.000
10'4" x 7'2" x 9'2"	"	8.000
12'4" x 7'6" x 9'2"	"	8.000
14'4" x 7'6" x 9'8"	"	8.000
16'4" x 7'7" x 9'8"	"	8.000
20'4" x 7'7" x 10'8"	"	8.000
Air compressor, electric		
1 hp		
115 volt	EA.	5.333
5 hp		

ARCHITECTURAL EQUIPMENT	UNIT	MAN/ HOURS
11500.10 INDUSTRIAL EQUIPMENT		
115 volt	EA.	8.000
230 volt	"	8.000
Hydraulic lifts		
8,000 lb capacity	EA.	20.000
11,000 lb capacity	"	32.000
24,000 lb capacity	"	53.333
Power tools		
Band saws		
10"	EA.	0.667
14"	"	0.800
Motorized shaper	"	0.615
Motorized lathe	"	0.667
Bench saws		
9" saw	EA.	0.533
10" saw	"	0.571
12" saw	"	0.667
Electric grinders		
1/3 hp	EA.	0.320
1/2 hp	"	0.348
3/4 hp	"	0.348

INTERIOR		UNIT	MAN/ HOURS
12690.40	FLOOR MATS		
Recessed entrance mat, 3/8" thick, aluminum link		S.F.	0.400
Steel, flexible		"	0.400

CONSTRUCTION	UNIT	MAN/ HOURS
13121.10 PRE-ENGINEERED BUILDINGS		
Pre-engineered metal building, 40'x100'		
14' eave height	S.F.	0.032
16' eave height	"	0.037
20' eave height	"	0.048
60'x100'		
14' eave height	S.F.	0.032
16' eave height	"	0.037
20' eave height	"	0.048
80'x100'		
14' eave height	S.F.	0.032
16' eave height	"	0.037
20' eave height	"	0.048
100'x100'		
14' eave height	S.F.	0.032
16' eave height	"	0.037
20' eave height	"	0.048
100'x150'		
14' eave height	S.F.	0.032
16' eave height	"	0.037
20' eave height	"	0.048
120'x150'		
14' eave height	S.F.	0.032
16' eave height	"	0.037
20' eave height	"	0.048
140'x150'		
14' eave height	S.F.	0.032
16' eave height	"	0.037
20' eave height	"	0.048
160'x200'		
14' eave height	S.F.	0.032
16' eave height	"	0.037
20' eave height	"	0.048
200'x200'		
14' eave height	S.F.	0.032
16' eave height	"	0.037
20' eave height	"	0.048
Liner panel, 26 ga, painted steel	"	0.020
Wall panel insulated, 26 ga. steel, foam core	"	0.020
Roof panel, 26 ga. painted steel	"	0.011
Plastic (sky light)	"	0.011
Insulation, 3-1/2" thick blanket, R11	"	0.005

14 CONVEYING

ELEVATORS

14210.10 ELEVATORS	UNIT	MAN/ HOURS
Passenger elevators, electric, geared		
Based on a shaft of 6 stops and 6 openings		
50 fpm, 2000 lb	EA.	24.000
100 fpm, 2000 lb	"	26.667
150 fpm		
2000 lb	EA.	30.000
3000 lb	"	34.286
4000 lb	"	40.000

LIFTS

14410.10 PERSONNEL LIFTS	UNIT	MAN/ HOURS
Residential stair climber, per story	EA.	6.667

14450.10 VEHICLE LIFTS	UNIT	MAN/ HOURS
Automotive hoist, one post, semi-hydraulic, 8,000 lb	EA.	24.000
Full hydraulic, 8,000 lb	"	24.000
2 post, semi-hydraulic, 10,000 lb	"	34.286
Full hydraulic		
10,000 lb	EA.	34.286
13,000 lb	"	60.000
18,500 lb	"	60.000
24,000 lb	"	60.000
26,000 lb	"	60.000
Pneumatic hoist, fully hydraulic		
11,000 lb	EA.	80.000
24,000 lb	"	80.000

HOISTS AND CRANES

14600.10 INDUSTRIAL HOISTS	UNIT	MAN/ HOURS
Industrial hoists, electric, light to medium duty		
500 lb	EA.	4.000
1000 lb	"	4.211
2000 lb	"	4.444
5000 lb	"	5.333
10,000 lb	"	5.926
20,000 lb	EA.	6.667
30,000 lb	"	8.000
Heavy duty		
500 lb	EA.	4.000
1000 lb	"	4.211
2000 lb	"	4.444
5000 lb	"	5.333
10,000 lb	"	5.926
20,000 lb	"	6.667
30,000 lb	"	8.000
Air powered hoists		
500 lb	EA.	4.000
1000 lb	"	4.000
2000 lb	"	4.211
4000 lb	"	4.706
6000 lb	"	6.154
Overhead traveling bridge crane		
Single girder, 20' span		
3 ton	EA.	12.000
5 ton	"	12.000
7.5 ton	"	12.000
10 ton	"	15.000
15 ton	"	15.000
30' span		
3 ton	EA.	12.000
5 ton	"	12.000
10 ton	"	15.000
15 ton	"	15.000
Double girder, 40' span		
3 ton	EA.	26.667
5 ton	"	26.667
7.5 ton	"	26.667
10 ton	"	34.286
15 ton	"	34.286
25 ton	"	34.286
50' span		
3 ton	EA.	26.667
5 ton	"	26.667
7.5 ton	"	26.667
10 ton	"	34.286
15 ton	"	34.286
25 ton	"	34.286
Rail for bridge crane, including splice bars	Lb.	

14650.10 JIB CRANES	UNIT	MAN/ HOURS
Self supporting, swinging 8' boom, 200 deg rotation		
2000 lb	EA.	6.667
4000 lb	"	13.333
10,000 lb	"	13.333
Wall mounted, 180 deg rotation		
2000 lb	EA.	6.667
4000 lb	"	13.333
10,000 lb	"	13.333

BASIC MATERIALS	UNIT	MAN/ HOURS
15100.10 SPECIALTIES		
Wall penetration		
Concrete wall, 6" thick		
2" dia.	EA.	0.267
4" dia.	"	0.400
12" thick		
2" dia.	EA.	0.364
4" dia.	"	0.571
15120.10 BACKFLOW PREVENTERS		
Backflow preventer, flanged, cast iron, with valves		
3" pipe	EA.	4.000
4" pipe	"	4.444
Threaded		
3/4" pipe	EA.	0.500
2" pipe	"	0.800
15140.11 PIPE HANGERS, LIGHT		
A band, black iron		
1/2"	EA.	0.057
1"	"	0.059
1-1/4"	"	0.062
1-1/2"	"	0.067
2"	"	0.073
2-1/2"	"	0.080
3"	"	0.089
4"	"	0.100
Copper		
1/2"	EA.	0.057
3/4"	"	0.059
1"	"	0.059
1-1/4"	"	0.062
1-1/2"	"	0.067
2"	"	0.073
2-1/2"	"	0.080
3"	"	0.089
4"	"	0.100
2 hole clips, galvanized		
3/4"	EA.	0.053
1"	"	0.055
1-1/4"	"	0.057
1-1/2"	"	0.059
2"	"	0.062
2-1/2"	"	0.064
3"	"	0.067
4"	"	0.073
Perforated strap		
3/4"		
Galvanized, 20 ga.	L.F.	0.040
Copper, 22 ga.	"	0.040
J-Hooks		
1/2"	EA.	0.036
3/4"	"	0.036
1"	"	0.038

BASIC MATERIALS	UNIT	MAN/ HOURS
15140.11 PIPE HANGERS, LIGHT		
1-1/4"	EA.	0.039
1-1/2"	"	0.040
2"	"	0.040
3"	"	0.042
4"	"	0.042
PVC coated hangers, galvanized, 28 ga.		
1-1/2" x 12"	EA.	0.053
2" x 12"	"	0.057
3" x 12"	"	0.062
4" x 12"	"	0.067
Copper, 30 ga.		
1-1/2" x 12"	EA.	0.053
2" x 12"	"	0.057
3" x 12"	"	0.062
4" x 12"	"	0.067
Wire hook hangers		
Black wire, 1/2" x		
4"	EA.	0.040
6"	"	0.042
Copper wire hooks		
1/2" x		
4"	EA.	0.040
6"	"	0.042
8"	"	0.044
10"	"	0.047
12"	"	0.050
15240.10 VIBRATION CONTROL		
Vibration isolator, in-line, stainless connector, screwed		
3/4"	EA.	0.471
1"	"	0.500
2"	"	0.615
3"	"	0.727
4"	"	0.800

INSULATION	UNIT	MAN/ HOURS
15260.10 FIBERGLASS PIPE INSULATION		
Fiberglass insulation on 1/2" pipe		
1" thick	L.F.	0.027
1-1/2" thick	"	0.033
3/4" pipe		
1" thick	L.F.	0.027
1-1/2" thick	"	0.033
1" pipe		
1" thick	L.F.	0.027
1-1/2" thick	"	0.033

INSULATION	UNIT	MAN/ HOURS
15260.10 FIBERGLASS PIPE INSULATION		
2" thick	L.F.	0.040
1-1/4" dia. pipe		
1" thick	L.F.	0.033
1-1/2" thick	"	0.036
1-1/2" pipe		
1" thick	L.F.	0.033
1-1/2" thick	"	0.036
2" pipe		
1" thick	L.F.	0.033
1-1/2" thick	"	0.036
2-1/2" pipe		
1" thick	L.F.	0.033
1-1/2" thick	"	0.036
3" pipe		
1" thick	L.F.	0.038
1-1/2" thick	"	0.040
4" pipe		
1" thick	L.F.	0.038
1-1/2" thick	"	0.040
6" pipe		
1" thick	L.F.	0.042
2" thick	"	0.044
10" pipe		
2" thick	L.F.	0.042
3" thick	"	0.044
15260.20 CALCIUM SILICATE		
Calcium silicate insulation, 6" pipe		
2" thick	L.F.	0.057
3" thick	"	0.067
6" thick	"	0.080
12" pipe		
2" thick	L.F.	0.062
3" thick	"	0.073
6" thick	"	0.089
15260.60 EXTERIOR PIPE INSULATION		
Fiberglass insulation, aluminum jacket		
1/2" pipe		
1" thick	L.F.	0.062
1-1/2" thick	"	0.067
1" pipe		
1" thick	L.F.	0.062
1-1/2" thick	"	0.067
2" pipe		
1" thick	L.F.	0.073
1-1/2" thick	"	0.076
3" pipe		
1" thick	L.F.	0.080
1-1/2" thick	"	0.084
4" pipe		
1" thick	L.F.	0.080
1-1/2" thick	"	0.084

INSULATION	UNIT	MAN/ HOURS
15260.60 EXTERIOR PIPE INSULATION		
6" pipe		
1" thick	L.F.	0.089
2" thick	"	0.094
10" pipe		
2" thick	L.F.	0.089
3" thick	"	0.094
15260.90 PIPE INSULATION FITTINGS		
Insulation protection saddle		
1" thick covering		
1/2" pipe	EA.	0.320
3/4" pipe	"	0.320
1" pipe	"	0.320
2" pipe	"	0.320
3" pipe	"	0.364
6" pipe	"	0.500
1-1/2" thick covering		
3/4" pipe	EA.	0.320
1" pipe	"	0.320
2" pipe	"	0.320
3" pipe	"	0.320
6" pipe	"	0.500
10" pipe	"	0.667
15280.10 EQUIPMENT INSULATION		
Equipment insulation, 2" thick, cellular glass	S.F.	0.050
Urethane, rigid, field applied jacket, plastered finish	"	0.100
Fiberglass, rigid, with vapor barrier	"	0.044
15290.10 DUCTWORK INSULATION		
Fiberglass duct insulation, plain blanket		
1-1/2" thick	S.F.	0.010
2" thick	"	0.013
With vapor barrier		
1-1/2" thick	S.F.	0.010
2" thick	"	0.013
Rigid with vapor barrier		
2" thick	S.F.	0.027

FIRE PROTECTION	UNIT	MAN/ HOURS
15330.10 WET SPRINKLER SYSTEM		
Sprinkler head, 212 deg, brass, exposed piping	EA.	0.320
Chrome, concealed piping	"	0.444
Water motor alarm	"	1.333

FIRE PROTECTION	UNIT	MAN/ HOURS
15330.10 WET SPRINKLER SYSTEM		
Fire department inlet connection	EA.	1.600
Wall plate for fire dept connection	"	0.667
Swing check valve flanged iron body, 4"	"	2.667
Check valve, 6"	"	4.000
Wet pipe valve, flange to groove, 4"	"	0.889
Flange to flange		
6"	EA.	1.333
8"	"	2.667
Alarm valve, flange to flange, (wet valve)		
4"	EA.	0.889
8"	"	6.667
Inspector's test connection	"	0.667
Wall hydrant, polished brass, 2-1/2" x 2-1/2", single	"	0.571
2-way	"	0.571
3-way	"	0.571
Wet valve trim, includes retard chamber & gauges, 4"-6"	"	0.667
Retard pressure switch for wet systems	"	1.600
Air maintenance device	"	0.667
Wall hydrant non-freeze, 8" thick wall, vacuum breaker	"	0.400
12" thick wall	"	0.400

PLUMBING	UNIT	MAN/ HOURS
15410.05 C.I. PIPE, ABOVE GROUND		
No hub pipe		
1-1/2" pipe	L.F.	0.057
2" pipe	"	0.067
3" pipe	"	0.080
4" pipe	"	0.133
No hub fittings, 1-1/2" pipe		
1/4 bend	EA.	0.267
1/8 bend	"	0.267
Sanitary tee	"	0.400
Sanitary cross	"	0.400
Wye	"	0.400
2" pipe		
1/4 bend	EA.	0.320
1/8 bend	"	0.320
Sanitary tee	"	0.533
Wye	"	0.667
3" pipe		
1/4 bend	EA.	0.400
1/8 bend	"	0.400
Sanitary tee	"	0.500
Wye	"	0.667
4" pipe		
1/4 bend	EA.	0.400

PLUMBING	UNIT	MAN/ HOURS
15410.05 C.I. PIPE, ABOVE GROUND		
1/8 bend	EA.	0.400
Sanitary tee	"	0.667
Wye	"	0.667
15410.06 C.I. PIPE, BELOW GROUND		
No hub pipe		
1-1/2" pipe	L.F.	0.040
2" pipe	"	0.044
3" pipe	"	0.050
4" pipe	"	0.067
Fittings, 1-1/2"		
1/4 bend	EA.	0.229
1/8 bend	"	0.229
Wye	"	0.320
Wye & 1/8 bend	"	0.229
P-trap	"	0.229
2"		
1/4 bend	EA.	0.267
1/8 bend	"	0.267
Double wye	"	0.500
Wye & 1/8 bend	"	0.400
Double wye & 1/8 bend	"	0.500
P-trap	"	0.267
3"		
1/4 bend	EA.	0.320
1/8 bend	"	0.320
Wye	"	0.500
3x2" wye	"	0.500
Wye & 1/8 bend	"	0.500
Double wye & 1/8 bend	"	0.500
3x2" double wye & 1/8 bend	"	0.500
3x2" reducer	"	0.320
P-trap	"	0.320
4"		
1/4 bend	EA.	0.320
1/8 bend	"	0.320
Wye	"	0.500
15410.09 SERVICE WEIGHT PIPE		
Service weight pipe, single hub		
3" x 5'	EA.	0.170
4" x 5'	"	0.178
6" x 5'	"	0.200
1/8 bend		
3"	EA.	0.320
4"	"	0.364
6"	"	0.400
1/4 bend		
3"	EA.	0.320
4"	"	0.364
6"	"	0.400
Sweep		
3"	EA.	0.320

PLUMBING	UNIT	MAN/ HOURS
15410.09 SERVICE WEIGHT PIPE		
4"	EA.	0.364
6"	"	0.400
Sanitary T		
3"	EA.	0.571
4"	"	0.667
6"	"	0.727
Wye		
3"	EA.	0.444
4"	"	0.471
6"	"	0.571
15410.10 COPPER PIPE		
Type "K" copper		
1/2"	L.F.	0.025
3/4"	"	0.027
1"	"	0.029
DWV, copper		
1-1/4"	L.F.	0.033
1-1/2"	"	0.036
2"	"	0.040
3"	"	0.044
4"	"	0.050
6"	"	0.057
Type "L" copper		
1/4"	L.F.	0.024
3/8"	"	0.024
1/2"	"	0.025
3/4"	"	0.027
1"	"	0.029
Type "M" copper		
1/2"	L.F.	0.025
3/4"	"	0.027
1"	"	0.029
15410.11 COPPER FITTINGS		
DWV fittings, coupling with stop		
1-1/4"	EA.	0.471
1-1/2"	"	0.500
1-1/2" x 1-1/4"	"	0.500
2"	"	0.533
2" x 1-1/4"	"	0.533
2" x 1-1/2"	"	0.533
3"	"	0.667
3" x 1-1/2"	"	0.667
3" x 2"	"	0.667
4"	"	0.800
Slip coupling		
1-1/2"	EA.	0.500
2"	"	0.533
3"	"	0.667
90 ells		
1-1/2"	EA.	0.500
1-1/2" x 1-1/4"	"	0.500

PLUMBING	UNIT	MAN/ HOURS
15410.11 COPPER FITTINGS		
2"	EA.	0.533
2" x 1-1/2"	"	0.533
3"	"	0.667
4"	"	0.800
Street, 90 elbows		
1-1/2"	EA.	0.500
2"	"	0.533
3"	"	0.667
4"	"	0.800
45 ells		
1-1/4"	EA.	0.471
1-1/2"	"	0.500
2"	"	0.533
3"	"	0.667
4"	"	0.800
Street, 45 ell		
1-1/2"	EA.	0.500
2"	"	0.533
3"	"	0.667
Wye		
1-1/4"	EA.	0.471
1-1/2"	"	0.500
2"	"	0.533
3"	"	0.667
4"	"	0.800
Sanitary tee		
1-1/4"	EA.	0.471
1-1/2"	"	0.500
2"	"	0.533
3"	"	0.667
4"	"	0.800
No-hub adapters		
1-1/2" x 2"	EA.	0.500
2"	"	0.533
2" x 3"	"	0.533
3"	"	0.667
3" x 4"	"	0.667
4"	"	0.800
Fitting reducers		
1-1/2" x 1-1/4"	EA.	0.500
2" x 1-1/2"	"	0.533
3" x 1-1/2"	"	0.667
3" x 2"	"	0.667
Copper caps		
1-1/2"	EA.	0.500
2"	"	0.533
Copper pipe fittings		
1/2"		
90 deg ell	EA.	0.178
45 deg ell	"	0.178
Tee	"	0.229
Cap	"	0.089
Coupling	"	0.178
Union	"	0.200

PLUMBING	UNIT	MAN/HOURS
15410.11 COPPER FITTINGS		
3/4"		
90 deg ell	EA.	0.200
45 deg ell	"	0.200
Tee	"	0.267
Cap	"	0.094
Coupling	"	0.200
Union	"	0.229
1"		
90 deg ell	EA.	0.267
45 deg ell	"	0.267
Tee	"	0.320
Cap	"	0.133
Coupling	"	0.267
Union	"	0.267
1-1/4"		
90 deg ell	EA.	0.229
45 deg ell	"	0.229
Tee	"	0.400
Cap	"	0.133
Union	"	0.286
1-1/2"		
90 deg ell	EA.	0.286
45 deg ell	"	0.286
Tee	"	0.444
Cap	"	0.133
Coupling	"	0.267
Union	"	0.364
2"		
90 deg ell	EA.	0.320
45 deg ell	"	0.500
Tee	"	0.500
Cap	"	0.160
Coupling	"	0.320
Union	"	0.400
2-1/2"		
90 deg ell	EA.	0.400
45 deg ell	"	0.400
Tee	"	0.571
Cap	"	0.200
Coupling	"	0.400
Union	"	0.444
15410.15 BRASS FITTINGS		
Compression fittings, union		
3/8"	EA.	0.133
1/2"	"	0.133
5/8"	"	0.133
Union elbow		
3/8"	EA.	0.133
1/2"	"	0.133
5/8"	"	0.133
Union tee		
3/8"	EA.	0.133

PLUMBING	UNIT	MAN/HOURS
15410.15 BRASS FITTINGS		
1/2"	EA.	0.133
5/8"	"	0.133
Male connector		
3/8"	EA.	0.133
1/2"	"	0.133
5/8"	"	0.133
Female connector		
3/8"	EA.	0.133
1/2"	"	0.133
5/8"	"	0.133
15410.30 PVC/CPVC PIPE		
PVC schedule 40		
1/2" pipe	L.F.	0.033
3/4" pipe	"	0.036
1" pipe	"	0.040
1-1/4" pipe	"	0.044
1-1/2" pipe	"	0.050
2" pipe	"	0.057
2-1/2" pipe	"	0.067
3" pipe	"	0.080
4" pipe	"	0.100
6" pipe	"	0.200
8" pipe	"	0.267
Fittings, 1/2"		
90 deg ell	EA.	0.100
45 deg ell	"	0.100
Tee	"	0.114
Polypropylene, acid resistant, DWV pipe		
Schedule 40		
1-1/2" pipe	L.F.	0.057
2" pipe	"	0.067
3" pipe	"	0.080
4" pipe	"	0.100
6" pipe	"	0.200
Polyethylene pipe and fittings		
SDR-21		
3" pipe	L.F.	0.100
4" pipe	"	0.133
6" pipe	"	0.200
8" pipe	"	0.229
10" pipe	"	0.267
12" pipe	"	0.320
14" pipe	"	0.400
16" pipe	"	0.500
18" pipe	"	0.615
20" pipe	"	0.800
22" pipe	"	0.889
24" pipe	"	1.000
Fittings, 3"		
90 deg elbow	EA.	0.400
45 deg elbow	"	0.400
Tee	"	0.667
45 deg wye	"	0.667

PLUMBING	UNIT	MAN/HOURS
15410.30 PVC/CPVC PIPE		
Reducer	EA.	0.500
Flange assembly	"	0.400
4"		
90 deg elbow	EA.	0.500
45 deg elbow	"	0.500
Tee	"	0.800
45 deg wye	"	0.800
Reducer	"	0.667
Flange assembly	"	0.500
8"		
90 deg elbow	EA.	1.000
45 deg elbow	"	1.000
Tee	"	1.600
45 deg wye	"	1.600
Reducer	"	1.333
Flange assembly	"	1.000
10"		
90 deg elbow	EA.	1.333
45 deg elbow	"	1.333
Tee	"	2.000
45 deg wye	"	2.000
Reducer	"	1.600
Flange assembly	"	1.333
12"		
90 deg elbow	EA.	1.600
45 deg elbow	"	1.600
Tee	"	2.667
45 deg wye	"	2.667
Reducer	"	2.000
Flange assembly	"	1.600
14"		
90 deg elbow	EA.	2.000
45 deg elbow	"	2.000
Tee	"	3.200
45 deg wye	"	3.200
Reducer	"	2.667
Flange assembly	"	2.000
16"		
90 deg elbow	EA.	2.000
45 deg elbow	"	2.000
Tee	"	3.200
45 deg wye	"	3.200
Reducer	"	2.667
Flange assembly	"	2.000
18"		
90 deg elbow	EA.	2.667
45 deg elbow	"	2.667
Tee	"	4.000
45 deg wye	"	4.000
Reducer	"	2.667
Flange assembly	"	2.667
20"		
90 deg elbow	EA.	2.667
45 deg elbow	"	2.667

PLUMBING	UNIT	MAN/HOURS
15410.33 ABS DWV PIPE		
Schedule 40 ABS		
1-1/2" pipe	L.F.	0.040
2" pipe	"	0.044
3" pipe	"	0.057
4" pipe	"	0.080
6" pipe	"	0.100
15410.35 PLASTIC PIPE		
Fiberglass reinforced pipe		
2" pipe	L.F.	0.062
3" pipe	"	0.067
4" pipe	"	0.073
6" pipe	"	0.080
8" pipe	"	0.133
10" pipe	"	0.160
12" pipe	"	0.200
Fittings		
90 deg elbow, flanged		
2"	EA.	0.800
3"	"	0.889
4"	"	1.000
6"	"	1.333
8"	"	1.600
10"	"	2.000
12"	"	2.667
45 deg elbow, flanged		
2"	EA.	0.667
3"	"	0.800
4"	"	1.000
6"	"	1.333
8"	"	1.600
10"	"	2.000
12"	"	2.667
Tee, flanged		
2"	EA.	1.000
3"	"	1.143
4"	"	1.333
6"	"	1.600
8"	"	2.000
10"	"	2.667
12"	"	4.000
Wye, flanged		
2"	EA.	1.000
3"	"	1.143
4"	"	1.333
6"	"	1.600
8"	"	2.000
10"	"	2.667
12"	"	4.000
Concentric reducer, flanged		
2"	EA.	0.667
4"	"	0.800
6"	"	1.143
8"	"	1.600

PLUMBING	UNIT	MAN/ HOURS
15410.35 PLASTIC PIPE		
10"	EA.	2.000
12"	"	2.667
Adapter, bell x male or female		
2"	EA.	0.667
3"	"	0.727
4"	"	0.800
6"	"	1.143
8"	"	1.600
10"	"	2.000
12"	"	2.667
Nipples		
2" x 6"	EA.	0.080
2" x 12"	"	0.100
3" x 8"	"	0.123
3" x 12"	"	0.133
4" x 8"	"	0.133
4" x 12"	"	0.160
6" x 12"	"	0.200
8" x 18"	"	0.200
8" x 24"	"	0.229
10" x 18"	"	0.267
10" x 24"	"	0.320
12" x 18"	"	0.364
12" x 24"	"	0.400
Sleeve coupling		
2"	EA.	0.667
3"	"	0.800
4"	"	1.143
6"	"	1.600
8"	"	2.000
10"	"	2.667
Flanges		
2"	EA.	0.667
3"	"	0.800
4"	"	1.143
6"	"	1.600
8"	"	2.000
10"	"	2.667
12"	"	2.667
15410.70 STAINLESS STEEL PIPE		
Stainless steel, schedule 40, threaded		
1/2" pipe	L.F.	0.114
1" pipe	"	0.123
1-1/2" pipe	"	0.133
2" pipe	"	0.145
2-1/2" pipe	"	0.160
3" pipe	"	0.178
4" pipe	"	0.200
15410.80 STEEL PIPE		
Black steel, extra heavy pipe, threaded		
1/2" pipe	L.F.	0.032

PLUMBING	UNIT	MAN/ HOURS
15410.80 STEEL PIPE		
3/4" pipe	L.F.	0.032
1" pipe	"	0.040
1-1/2" pipe	"	0.044
2-1/2" pipe	"	0.100
3" pipe	"	0.133
4" pipe	"	0.160
5" pipe	"	0.200
6" pipe	"	0.200
8" pipe	"	0.267
10" pipe	"	0.320
12" pipe	"	0.400
Fittings, malleable iron, threaded, 1/2" pipe		
90 deg ell	EA.	0.267
45 deg ell	"	0.267
Tee	"	0.400
3/4" pipe		
90 deg ell	EA.	0.267
45 deg ell	"	0.400
Tee	"	0.400
1-1/2" pipe		
90 deg ell	EA.	0.400
45 deg ell	"	0.400
Tee	"	0.571
2-1/2" pipe		
90 deg ell	EA.	1.000
45 deg ell	"	1.000
Tee	"	1.333
3" pipe		
90 deg ell	EA.	1.333
45 deg ell	"	1.333
Tee	"	2.000
4" pipe		
90 deg ell	EA.	1.600
45 deg ell	"	1.600
Tee	"	2.667
6" pipe		
90 deg ell	EA.	1.600
45 deg ell	"	1.600
Tee	"	2.667
8" pipe		
90 deg ell	EA.	3.200
45 deg ell	"	3.200
Tee	"	5.000
10" pipe		
90 deg ell	EA.	4.000
45 deg ell	"	4.000
Tee	"	5.000
12" pipe		
90 deg ell	EA.	5.000
45 deg ell	"	5.000
Tee	"	6.667
Butt welded, 1/2" pipe		
90 deg ell	EA.	0.267
45 deg ell	"	0.267

15 MECHANICAL

PLUMBING	UNIT	MAN/HOURS
15410.80 STEEL PIPE		
Tee	EA.	0.400
3/4" pipe		
90 deg ell	EA.	0.267
45 deg. ell	"	0.267
Tee	"	0.400
1" pipe		
90 deg ell	EA.	0.320
45 deg ell	"	0.320
Tee	"	0.444
1-1/2" pipe		
90 deg ell	EA.	0.400
45 deg. ell	"	0.400
Tee	"	0.571
Reducing tee	"	0.571
Cap	"	0.320
2-1/2" pipe		
90 deg. ell	EA.	0.800
45 deg. ell	"	0.800
Tee	"	1.143
Reducing tee	"	1.143
Cap	"	0.400
3" pipe		
90 deg ell	EA.	1.000
45 deg. ell	"	1.000
Tee	"	1.333
Reducing tee	"	1.333
Cap	"	0.667
4" pipe		
90 deg ell	EA.	1.333
45 deg. ell	"	1.333
Tee	"	2.000
Reducing tee	"	2.000
Cap	"	0.667
6" pipe		
90 deg. ell	EA.	1.600
45 deg. ell	"	1.600
Tee	"	2.667
Reducing tee	"	2.667
Cap	"	0.800
8" pipe		
90 deg. ell	EA.	2.667
45 deg. ell	"	2.667
Tee	"	4.000
Reducing tee	"	4.000
Cap	"	1.600
10" pipe		
90 deg ell	EA.	2.667
45 deg. ell	"	2.667
Tee	"	4.000
Reducing tee	"	4.000
Cap	"	2.000
12" pipe		
90 deg. ell	EA.	3.200
45 deg. ell	"	3.200

PLUMBING	UNIT	MAN/HOURS
15410.80 STEEL PIPE		
Tee	EA.	5.714
Reducing tee	"	5.714
Cap	"	2.000
Cast iron fittings		
1/2" pipe		
90 deg. ell	EA.	0.267
45 deg. ell	"	0.267
Tee	"	0.400
Reducing tee	"	0.400
3/4" pipe		
90 deg. ell	EA.	0.267
45 deg. ell	"	0.267
Tee	"	0.400
Reducing tee	"	0.400
1" pipe		
90 deg. ell	EA.	0.320
45 deg. ell	"	0.320
Tee	"	0.444
Reducing tee	"	0.444
1-1/2" pipe		
90 deg. ell	EA.	0.400
45 deg. ell	"	0.400
Tee	"	0.571
Reducing tee	"	0.571
2-1/2" pipe		
90 deg. ell	EA.	0.800
45 deg. ell	"	0.800
Tee	"	1.143
Reducing tee	"	1.143
3" pipe		
90 deg. ell	EA.	1.000
45 deg. ell	"	1.000
Tee	"	1.600
Reducing tee	"	1.600
4" pipe		
90 deg. ell	EA.	1.333
45 deg. ell	"	1.333
Tee	"	2.000
Reducing tee	"	2.000
6" pipe		
90 deg. ell	EA.	1.333
45 deg. ell	"	1.333
Tee	"	2.000
Reducing tee	"	2.000
8" pipe		
90 deg. ell	EA.	2.667
45 deg. ell	"	2.667
Tee	"	4.000
Reducing tee	"	4.000
15410.82 GALVANIZED STEEL PIPE		
Galvanized pipe		
1/2" pipe	L.F.	0.080

PLUMBING	UNIT	MAN/HOURS
15410.82 GALVANIZED STEEL PIPE		
3/4" pipe	L.F.	0.100
1" pipe	"	0.114
1-1/4" pipe	"	0.133
1-1/2" pipe	"	0.160
2" pipe	"	0.200
2-1/2" pipe	"	0.267
3" pipe	"	0.286
4" pipe	"	0.333
6" pipe	"	0.667
15430.23 CLEANOUTS		
Cleanout, wall		
2"	EA.	0.533
3"	"	0.533
4"	"	0.667
6"	"	0.800
8"	"	1.000
Floor		
2"	EA.	0.667
3"	"	0.667
4"	"	0.800
6"	"	1.000
8"	"	1.143
15430.24 GREASE TRAPS		
Grease traps, cast iron, 3" pipe		
35 gpm, 70 lb capacity	EA.	8.000
50 gpm, 100 lb capacity	"	10.000
15430.25 HOSE BIBBS		
Hose bibb		
1/2"	EA.	0.267
3/4"	"	0.267
15430.60 VALVES		
Gate valve, 125 lb, bronze, soldered		
1/2"	EA.	0.200
3/4"	"	0.200
1"	"	0.267
1-1/2"	"	0.320
2"	"	0.400
2-1/2"	"	0.500
Threaded		
1/4", 125 lb	EA.	0.320
1/2"		
125 lb	EA.	0.320
150 lb	"	0.320
300 lb	"	0.320
3/4"		
125 lb	EA.	0.320
150 lb	"	0.320

PLUMBING	UNIT	MAN/HOURS
15430.60 VALVES		
300 lb	EA.	0.320
1"		
125 lb	EA.	0.320
150 lb	"	0.320
300 lb	"	0.400
1-1/2"		
125 lb	EA.	0.400
150 lb	"	0.400
300 lb	"	0.444
2"		
125 lb	EA.	0.571
150 lb	"	0.571
300 lb	"	0.667
Cast iron, flanged		
2", 150 lb	EA.	0.667
2-1/2"		
125 lb	EA.	0.667
150 lb	"	0.667
250 lb	"	0.667
3"		
125 lb	EA.	0.800
150 lb	"	0.800
250 lb	"	0.800
4"		
125 lb	EA.	1.143
150 lb	"	1.143
250 lb	"	1.143
6"		
125 lb	EA.	1.600
250 lb	"	1.600
8"		
125 lb	EA.	2.000
250 lb	"	2.000
OS&Y, flanged		
2"		
125 lb	EA.	0.667
250 lb	"	0.667
2-1/2"		
125 lb	EA.	0.667
250 lb	"	0.800
3"		
125 lb	EA.	0.800
250 lb	"	0.800
4"		
125 lb	EA.	1.333
250 lb	"	1.333
6"		
125 lb	EA.	1.600
250 lb	"	1.600
Check valve, bronze, soldered, 125 lb		
1/2"	EA.	0.200
3/4"	"	0.200
1"	"	0.267
1-1/4"	"	0.320

PLUMBING	UNIT	MAN/ HOURS
15430.60 VALVES		
1-1/2"	EA.	0.320
2"	"	0.400
Threaded		
1/2"		
125 lb	EA.	0.267
150 lb	"	0.267
200 lb	"	0.267
3/4"		
125 lb	EA.	0.320
150 lb	"	0.320
200 lb	"	0.320
1"		
125 lb	EA.	0.400
150 lb	"	0.400
200 lb	"	0.400
Flow check valve, cast iron, threaded		
1"	EA.	0.320
1-1/4"	"	0.400
1-1/2"		
125 lb	EA.	0.400
150 lb	"	0.400
200 lb	"	0.444
2"		
125 lb	EA.	0.444
150 lb	"	0.444
200 lb	"	0.500
2-1/2"		
125 lb	EA.	0.667
250 lb	"	0.800
3"		
125 lb	EA.	0.800
250 lb	"	1.000
4"		
125 lb	EA.	1.143
250 lb	"	1.333
6"		
125 lb	EA.	1.600
250 lb	"	1.600
Vertical check valve, bronze, 125 lb, threaded		
1/2"	EA.	0.320
3/4"	"	0.364
1"	"	0.400
1-1/4"	"	0.444
1-1/2"	"	0.500
2"	"	0.571
Cast iron, flanged		
2-1/2"	EA.	0.800
3"	"	1.000
4"	"	1.333
6	"	1.600
8"	"	2.000
10"	"	2.667
12"	"	3.200
Globe valve, bronze, soldered, 125 lb		

PLUMBING	UNIT	MAN/ HOURS
15430.60 VALVES		
1/2"	EA.	0.229
3/4"	"	0.250
1"	"	0.267
1-1/4"	"	0.286
1-1/2"	"	0.333
2"	"	0.400
Threaded		
1/2"		
125 lb	EA.	0.267
150 lb	"	0.267
300 lb	"	0.267
3/4"		
125 lb	EA.	0.320
150 lb	"	0.320
300 lb	"	0.320
1"		
125 lb	EA.	0.400
150 lb	"	0.400
300 lb	"	0.400
1-1/4"		
125 lb	EA.	0.400
150 lb	"	0.400
300 lb	"	0.400
1-1/2"		
125 lb	EA.	0.444
150 lb	"	0.444
300 lb	"	0.444
2"		
125 lb	EA.	0.533
150 lb	"	0.533
300 lb	"	0.533
Cast iron flanged		
2-1/2"		
125 lb	EA.	0.800
250 lb	"	0.800
3"		
125 lb	EA.	1.000
250 lb	"	1.000
4"		
125 lb	EA.	1.333
250 lb	"	1.333
6"		
125 lb	EA.	1.600
250 lb	"	1.600
8"		
125 lb	EA.	2.000
250 lb	"	2.000
Butterfly valve, cast iron, wafer type		
2"		
150 lb	EA.	0.571
200 lb	"	0.667
2-1/2"		
150 lb	EA.	0.667
200 lb	"	0.727

PLUMBING	UNIT	MAN/ HOURS
15430.60 VALVES		
3"		
150 lb	EA.	0.800
200 lb	"	0.889
4"		
150 lb	EA.	1.143
200 lb	"	1.333
6"		
150 lb	EA.	1.600
200 lb	"	1.600
8"		
150 lb	EA.	1.778
200 lb	"	2.000
10"		
150 lb	EA.	2.000
200 lb	"	2.667
Ball valve, bronze, 250 lb, threaded		
1/2"	EA.	0.320
3/4"	"	0.320
1"	"	0.400
1-1/4"	"	0.444
1-1/2"	"	0.500
2"	"	0.571
Angle valve, bronze, 150 lb, threaded		
1/2"	EA.	0.286
3/4"	"	0.320
1"	"	0.320
1-1/4"	"	0.400
1-1/2"	"	0.444
Balancing valve, with meter connections, circuit setter		
1/2"	EA.	0.320
3/4"	"	0.364
1"	"	0.400
1-1/4"	"	0.444
1-1/2"	"	0.533
2"	"	0.667
2-1/2"	"	0.800
3"	"	1.000
4"	"	1.333
Balancing valve, straight type		
1/2"	EA.	0.320
3/4"	"	0.320
Angle type		
1/2"	EA.	0.320
3/4"	"	0.320
Square head cock, 125 lb, bronze body		
1/2"	EA.	0.267
3/4"	"	0.320
1"	"	0.364
1-1/4"	"	0.400
Pressure regulating valve, bronze, class 300		
1"	EA.	0.500
1-1/2"	"	0.615
2"	"	0.800
3"	"	1.143

PLUMBING	UNIT	MAN/ HOURS
15430.60 VALVES		
4"	EA.	1.600
5"	"	2.000
6"	"	2.667
15430.68 STRAINERS		
Strainer, Y pattern, 125 psi, cast iron body, threaded		
3/4"	EA.	0.286
1"	"	0.320
1-1/4"	"	0.400
1-1/2"	"	0.400
2"	"	0.500
250 psi, brass body, threaded		
3/4"	EA.	0.320
1"	"	0.320
1-1/4"	"	0.400
1-1/2"	"	0.400
2"	"	0.500
Cast iron body, threaded		
3/4"	EA.	0.320
1"	"	0.320
1-1/4"	"	0.400
1-1/2"	"	0.400
2"	"	0.500
15430.70 DRAINS, ROOF & FLOOR		
Floor drain, cast iron, with cast iron top		
2"	EA.	0.667
3"	"	0.667
4"	"	0.667
6"	"	0.800
Roof drain, cast iron		
2"	EA.	0.667
3"	"	0.667
4"	"	0.667
5"	"	0.800
6"	"	0.800

PLUMBING FIXTURES	UNIT	MAN/ HOURS
15440.15 FAUCETS		
Washroom		
Minimum	EA.	1.333
Average	"	1.600
Maximum	"	2.000
Handicapped		

15 MECHANICAL

PLUMBING FIXTURES	UNIT	MAN/HOURS
15440.15 FAUCETS		
Minimum	EA.	1.600
Average	"	2.000
Maximum	"	2.667
For trim and rough-in		
Minimum	EA.	1.600
Average	"	2.000
Maximum	"	4.000
15440.18 HYDRANTS		
Wall hydrant		
8" thick	EA.	1.333
12" thick	"	1.600
18" thick	"	1.778
24" thick	"	2.000
Ground hydrant		
2' deep	EA.	1.000
4' deep	"	1.143
6' deep	"	1.333
8' deep	"	2.000
15440.20 LAVATORIES		
Lavatory, counter top, porcelain enamel on cast iron		
Minimum	EA.	1.600
Average	"	2.000
Maximum	"	2.667
Wall hung, china		
Minimum	EA.	1.600
Average	"	2.000
Maximum	"	2.667
Handicapped		
Minimum	EA.	2.000
Average	"	2.667
Maximum	"	4.000
For trim and rough-in		
Minimum	EA.	2.000
Average	"	2.667
Maximum	"	4.000
15440.30 SHOWERS		
Shower, fiberglass, 36"x34"x84"		
Minimum	EA.	5.714
Average	"	8.000
Maximum	"	8.000
Steel, 1 piece, 36"x36"		
Minimum	EA.	5.714
Average	"	8.000
Maximum	"	8.000
Receptor, molded stone, 36"x36"		
Minimum	EA.	2.667
Average	"	4.000
Maximum	"	6.667
For trim and rough-in		
Minimum	EA.	3.636

PLUMBING FIXTURES	UNIT	MAN/HOURS
15440.30 SHOWERS		
Average	EA.	4.444
Maximum	"	8.000
15440.40 SINKS		
Service sink, 24"x29"		
Minimum	EA.	2.000
Average	"	2.667
Maximum	"	4.000
Mop sink, 24"x36"x10"		
Minimum	EA.	1.600
Average	"	2.000
Maximum	"	2.667
For trim and rough-in		
Minimum	EA.	2.667
Average	"	4.000
Maximum	"	5.333
15440.50 URINALS		
Urinal, flush valve, floor mounted		
Minimum	EA.	2.000
Average	"	2.667
Maximum	"	4.000
Wall mounted		
Minimum	EA.	2.000
Average	"	2.667
Maximum	"	4.000
For trim and rough-in		
Minimum	EA.	2.000
Average	"	4.000
Maximum	"	5.333
15440.60 WATER CLOSETS		
Water closet flush tank, floor mounted		
Minimum	EA.	2.000
Average	"	2.667
Maximum	"	4.000
Handicapped		
Minimum	EA.	2.667
Average	"	4.000
Maximum	"	8.000
Bowl, with flush valve, floor mounted		
Minimum	EA.	2.000
Average	"	2.667
Maximum	"	4.000
Wall mounted		
Minimum	EA.	2.000
Average	"	2.667
Maximum	"	4.000
For trim and rough-in		
Minimum	EA.	2.000
Average	"	2.667
Maximum	"	4.000

PLUMBING FIXTURES	UNIT	MAN/ HOURS
15440.70 WATER HEATERS		
Water heater, electric		
6 gal	EA.	1.333
10 gal	"	1.333
20 gal	"	1.600
40 gal	"	1.600
80 gal	"	2.000
100 gal	"	2.667
120 gal	"	2.667
15440.90 MISCELLANEOUS FIXTURES		
Electric water cooler		
Floor mounted	EA.	2.667
Wall mounted	"	2.667
Wash fountain		
Wall mounted	EA.	4.000
Circular, floor supported	"	8.000
Deluge shower and eye wash	"	4.000
15440.95 FIXTURE CARRIERS		
Water fountain, wall carrier		
Minimum	EA.	0.800
Average	"	1.000
Maximum	"	1.333
Lavatory, wall carrier		
Minimum	EA.	0.800
Average	"	1.000
Maximum	"	1.333
Sink, industrial, wall carrier		
Minimum	EA.	0.800
Average	"	1.000
Maximum	"	1.333
Toilets, water closets, wall carrier		
Minimum	EA.	0.800
Average	"	1.000
Maximum	"	1.333
Floor support		
Minimum	EA.	0.667
Average	"	0.800
Maximum	"	1.000
Urinals, wall carrier		
Minimum	EA.	0.800
Average	"	1.000
Maximum	"	1.333
Floor support		
Minimum	EA.	0.667
Average	"	0.800
Maximum	"	1.000
15450.30 PUMPS		
In-line pump, bronze, centrifugal		
5 gpm, 20' head	EA.	0.500
20 gpm, 40' head	"	0.500
50 gpm		

PLUMBING FIXTURES	UNIT	MAN/ HOURS
15450.30 PUMPS		
50' head	EA.	1.000
100' head	"	1.000
70 gpm, 100' head	"	1.333
100 gpm, 80' head	"	1.333
250 gpm, 150' head	"	2.000
Cast iron, centrifugal		
50 gpm, 200' head	EA.	1.000
100 gpm		
100' head	EA.	1.333
200' head	"	1.333
200 gpm		
100' head	EA.	2.000
200' head	"	2.000
Centrifugal, close coupled, c.i., single stage		
50 gpm, 100' head	EA.	1.000
100 gpm, 100' head	"	1.333
Base mounted		
50 gpm, 100' head	EA.	1.000
100 gpm, 50' head	"	1.333
200 gpm, 100' head	"	2.000
300 gpm, 175' head	"	2.000
Suction diffuser, flanged, strainer		
3" inlet, 2-1/2" outlet	EA.	1.000
3" outlet	"	1.000
4" inlet		
3" outlet	EA.	1.333
4" outlet	"	1.333
6" inlet		
4" outlet	EA.	1.600
5" outlet	"	1.600
6" Outlet	"	1.600
8" inlet		
6" outlet	EA.	2.000
8" outlet	"	2.000
10" inlet		
8" outlet	EA.	2.667
Vertical turbine		
Single stage, C.I., 3550 rpm, 200 gpm, 50'head	EA.	2.667
Multi stage, 3550 rpm		
50 gpm, 100' head	EA.	2.000
100 gpm		
100' head	EA.	2.000
200 gpm		
50' head	EA.	2.667
100' head	"	2.667
Bronze		
Single stage, 3550 rpm, 100 gpm, 50' head	EA.	2.000
Multi stage, 3550 rpm, 50 gpm, 100' head	"	2.000
100 gpm		
100' head	EA.	2.000
200 gpm		
50' head	EA.	2.667
100' head	"	2.667
Sump pump, bronze, 1750 rpm, 25 gpm		

15 MECHANICAL

PLUMBING FIXTURES

15450.30 PUMPS	UNIT	MAN/ HOURS
20' head	EA.	10.000
150' head	"	13.333
50 gpm		
100' head	EA.	10.000
100 gpm		
50' head	EA.	10.000

15480.10 SPECIAL SYSTEMS	UNIT	MAN/ HOURS
Air compressor, air cooled, two stage		
5.0 cfm, 175 psi	EA.	16.000
10 cfm, 175 psi	"	17.778
20 cfm, 175 psi	"	19.048
50 cfm, 125 psi	"	21.053
80 cfm, 125 psi	"	22.857
Single stage, 125 psi		
1.0 cfm	EA.	11.429
1.5 cfm	"	11.429
2.0 cfm	"	11.429
Automotive compressor, hose reel, air and water, 50' hose	"	6.667
Lube equipment, 3 reel, with pumps	"	32.000
Tire changer		
Truck	EA.	11.429
Passenger car	"	6.154
Air hose reel, includes, 50' hose	"	6.154
Hose reel, 5 reel, motor oil, gear oil, lube, air & water	"	32.000
Water hose reel, 50' hose	"	6.154
Pump, air operated, for motor or gear oil, fits 55 gal drum	"	0.800
For chassis lube	"	0.800
Fuel dispensing pump, lighted dial, one product		
One hose	EA.	6.667
Two hose	"	6.667
Two products, two hose	"	6.667

HEATING & VENTILATING

15610.10 FURNACES	UNIT	MAN/ HOURS
Electric, hot air		
40 mbh	EA.	4.000
80 mbh	"	4.444
100 mbh	"	4.706
160 mbh	"	5.000
200 mbh	"	5.161
400 mbh	"	5.333
Gas fired hot air		
40 mbh	EA.	4.000
80 mbh	"	4.444
100 mbh	"	4.706

HEATING & VENTILATING

15610.10 FURNACES	UNIT	MAN/ HOURS
160 mbh	EA.	5.000
200 mbh	"	5.161
400 mbh	"	5.333
Oil fired hot air		
40 mbh	EA.	4.000
80 mbh	"	4.444
100 mbh	"	4.706
160 mbh	"	5.000
200 mbh	"	5.161
400 mbh	"	5.333

15780.20 ROOFTOP UNITS	UNIT	MAN/ HOURS
Packaged, single zone rooftop unit, with roof curb		
2 ton	EA.	8.000
3 ton	"	8.000
4 ton	"	10.000
5 ton	"	13.333
7.5 ton	"	16.000

15830.70 UNIT HEATERS	UNIT	MAN/ HOURS
Steam unit heater, horizontal		
12,500 btuh, 200 cfm	EA.	1.333
17,000 btuh, 300 cfm	"	1.333
40,000 btuh, 500 cfm	"	1.333
60,000 btuh, 700 cfm	"	1.333
70,000 btuh, 1000 cfm	"	2.000
Vertical		
12,500 btuh, 200 cfm	EA.	1.333
17,000 btuh, 300 cfm	"	1.333
40,000 btuh, 500 cfm	"	1.333
60,000 btuh, 700 cfm	"	1.333
70,000 btuh, 1000 cfm	"	1.333
Gas unit heater, horizontal		
27,400 btuh	EA.	3.200
38,000 btuh	"	3.200
56,000 btuh	"	3.200
82,200 btuh	"	3.200
103,900 btuh	"	5.000
125,700 btuh	"	5.000
133,200 btuh	"	5.000
149,000 btuh	"	5.000
172,000 btuh	"	5.000
190,000 btuh	"	5.000
225,000 btuh	"	5.000
Hot water unit heater, horizontal		
12,500 btuh, 200 cfm	EA.	1.333
17,000 btuh, 300 cfm	"	1.333
25,000 btuh, 500 cfm	"	1.333
30,000 btuh, 700 cfm	"	1.333
50,000 btuh, 1000 cfm	"	2.000
60,000 btuh, 1300 cfm	"	2.000
Vertical		
12,500 btuh, 200 cfm	EA.	1.333
17,000 btuh, 300 cfm	"	1.333

15 MECHANICAL

HEATING & VENTILATING

15830.70 UNIT HEATERS	UNIT	MAN/HOURS
25,000 btuh, 500 cfm	EA.	1.333
30,000 btuh, 700 cfm	"	1.333
50,000 btuh, 1000 cfm	"	1.333
60,000 btuh, 1300 cfm	"	1.333
Cabinet unit heaters, ceiling, exposed, hot water		
200 cfm	EA.	2.667
300 cfm	"	3.200
400 cfm	"	3.810
600 cfm	"	4.211
800 cfm	"	5.000
1000 cfm	"	5.714
1200 cfm	"	6.667
2000 cfm	"	8.889

AIR HANDLING

15855.10 AIR HANDLING UNITS	UNIT	MAN/HOURS
Air handling unit, medium pressure, single zone		
1500 cfm	EA.	5.000
3000 cfm	"	8.889
4000 cfm	"	10.000
5000 cfm	"	10.667
6000 cfm	"	11.429
7000 cfm	"	12.308
8500 cfm	"	13.333
Rooftop air handling units		
4950 cfm	EA.	8.889
7370 cfm	"	11.429

15870.20 EXHAUST FANS	UNIT	MAN/HOURS
Belt drive roof exhaust fans		
640 cfm, 2618 fpm	EA.	1.000
940 cfm, 2604 fpm	"	1.000
1050 cfm, 3325 fpm	"	1.000
1170 cfm, 2373 fpm	"	1.000
2440 cfm, 4501 fpm	"	1.000

AIR DISTRIBUTION

15890.10 METAL DUCTWORK	UNIT	MAN/HOURS
Rectangular duct		
Galvanized steel		
Minimum	Lb.	0.073
Average	"	0.089
Maximum	"	0.133
Aluminum		
Minimum	Lb.	0.160
Average	"	0.200
Maximum	"	0.267
Fittings		
Minimum	EA.	0.267
Average	"	0.400
Maximum	"	0.800

15890.30 FLEXIBLE DUCTWORK	UNIT	MAN/HOURS
Flexible duct, 1.25" fiberglass		
6" dia.	L.F.	0.044
8" dia.	"	0.050
12" dia.	"	0.062
16" dia.	"	0.073
Flexible duct connector, 3" wide fabric	"	0.133

15910.10 DAMPERS	UNIT	MAN/HOURS
Horizontal parallel aluminum backdraft damper		
12" x 12"	EA.	0.200
24" x 24"	"	0.400
36" x 36"	"	0.571

15940.10 DIFFUSERS	UNIT	MAN/HOURS
Ceiling diffusers, round, baked enamel finish		
6" dia.	EA.	0.267
8" dia.	"	0.333
12" dia.	"	0.333
16" dia.	"	0.364
20" dia.	"	0.400
Rectangular		
6x6"	EA.	0.267
12x12"	"	0.400
18x18"	"	0.400
24x24"	"	0.500

15940.40 REGISTERS AND GRILLES	UNIT	MAN/HOURS
Lay in flush mounted, perforated face, return		
6x6/24x24	EA.	0.320
8x8/24x24	"	0.320
9x9/24x24	"	0.320
10x10/24x24	"	0.320
12x12/24x24	"	0.320
Rectangular, ceiling return, single deflection		
10x10	EA.	0.400
12x12	"	0.400
16x16	"	0.400

AIR DISTRIBUTION	UNIT	MAN/ HOURS
15940.40 REGISTERS AND GRILLES		
20x20	EA.	0.400
24x18	"	0.400
36x24	"	0.444
36x30	"	0.444
Wall, return air register		
12x12	EA.	0.200
16x16	"	0.200
18x18	"	0.200
20x20	"	0.200
24x24	"	0.200

BASIC MATERIALS	UNIT	MAN/ HOURS
16050.30 BUS DUCT		
Bus duct, 100a, plug-in		
10', 600v	EA.	2.759
With ground	"	4.211
Circuit breakers, with enclosure		
1 pole		
15a-60a	EA.	1.000
70a-100a	"	1.250
2 pole		
15a-60a	EA.	1.100
70a-100a	"	1.301
Circuit breaker, adapter cubicle		
225a	EA.	1.509
200a	"	1.600
Fusible switches, 240v, 3 phase		
30a	EA.	1.000
60a	"	1.250
100a	"	1.509
200a	"	2.105
16110.12 CABLE TRAY		
Cable tray, 6"	L.F.	0.059
Ventilated cover	"	0.030
Solid cover	"	0.030
16110.20 CONDUIT SPECIALTIES		
Rod beam clamp, 1/2"	EA.	0.050
Hanger rod		
3/8"	L.F.	0.040
1/2"	"	0.050
Hanger channel, 1-1/2"		
No holes	EA.	0.030
Holes	"	0.030
Channel strap		
1/2"	EA.	0.050
1"	"	0.050
2"	"	0.080
3"	"	0.123
4"	"	0.145
5"	"	0.145
6"	"	0.145
Conduit penetrations, roof and wall, 8" thick		
1/2"	EA.	0.615
1"	"	0.800
2"	"	1.600
3"	"	1.600
4"	"	2.000
Fireproofing, for conduit penetrations		
1/2"	EA.	0.500
1"	"	0.500
2"	"	0.727
3"	"	0.899
4"	"	1.509

BASIC MATERIALS	UNIT	MAN/ HOURS
16110.21 ALUMINUM CONDUIT		
Aluminum conduit		
1/2"	L.F.	0.030
3/4"	"	0.040
1"	"	0.050
1-1/4"	"	0.059
1-1/2"	"	0.080
2"	"	0.089
2-1/2"	"	0.100
3"	"	0.107
3-1/2"	"	0.123
4"	"	0.145
5"	"	0.182
6"	"	0.200
16110.22 EMT CONDUIT		
EMT conduit		
1/2"	L.F.	0.030
3/4"	"	0.040
1"	"	0.050
1-1/4"	"	0.059
1-1/2"	"	0.080
2"	"	0.089
2-1/2"	"	0.100
3"	"	0.123
3-1/2"	"	0.145
4"	"	0.182
16110.23 FLEXIBLE CONDUIT		
Flexible conduit, steel		
3/8"	L.F.	0.030
1/2	"	0.030
3/4"	"	0.040
1"	"	0.040
1-1/4"	"	0.050
1-1/2"	"	0.059
2"	"	0.080
2-1/2"	"	0.089
3"	"	0.107
16110.24 GALVANIZED CONDUIT		
Galvanized rigid steel conduit		
1/2"	L.F.	0.040
3/4"	"	0.050
1"	"	0.059
1-1/4"	"	0.080
1-1/2"	"	0.089
2"	"	0.100
2-1/2"	"	0.145
3"	"	0.182
3-1/2"	"	0.190
4"	"	0.211
5"	"	0.286

16 ELECTRICAL

BASIC MATERIALS	UNIT	MAN/ HOURS
16110.24 GALVANIZED CONDUIT		
6"	L.F.	0.381
16110.25 PLASTIC CONDUIT		
PVC conduit, schedule 40		
1/2"	L.F.	0.030
3/4"	"	0.030
1"	"	0.040
1-1/4"	"	0.040
1-1/2"	"	0.050
2"	"	0.050
2-1/2"	"	0.059
3"	"	0.059
3-1/2"	"	0.080
4"	"	0.080
5"	"	0.089
6"	"	0.100
16110.27 PLASTIC COATED CONDUIT		
Rigid steel conduit, plastic coated		
1/2"	L.F.	0.050
3/4"	"	0.059
1"	"	0.080
1-1/4"	"	0.100
1-1/2"	"	0.123
2"	"	0.145
2-1/2"	"	0.190
3"	"	0.222
3-1/2"	"	0.250
4"	"	0.308
5"	"	0.381
90 degree elbows		
1/2"	EA.	0.308
3/4"	"	0.381
1"	"	0.444
1-1/4"	"	0.500
1-1/2"	"	0.615
2"	"	0.800
2-1/2"	"	1.143
3"	"	1.333
3-1/2"	"	1.633
4"	"	2.000
5"	"	2.500
Couplings		
1/2"	EA.	0.059
3/4"	"	0.080
1"	"	0.089
1-1/4"	"	0.107
1-1/2"	"	0.123
2"	"	0.145
2-1/2"	"	0.182
3"	"	0.190

BASIC MATERIALS	UNIT	MAN/ HOURS
16110.27 PLASTIC COATED CONDUIT		
3-1/2"	EA.	0.200
4"	"	0.222
5"	"	0.250
1 hole conduit straps		
3/4"	EA.	0.050
1"	"	0.050
1-1/4"	"	0.059
1-1/2"	"	0.059
2"	"	0.059
3"	"	0.080
3-1/2"	"	0.080
4"	"	0.100
16110.28 STEEL CONDUIT		
Intermediate metal conduit (IMC)		
1/2"	L.F.	0.030
3/4"	"	0.040
1"	"	0.050
1-1/4"	"	0.059
1-1/2"	"	0.080
2"	"	0.089
2-1/2"	"	0.119
3"	"	0.145
3-1/2"	"	0.182
4"	"	0.190
90 degree ell		
1/2"	EA.	0.250
3/4"	"	0.308
1"	"	0.381
1-1/4"	"	0.444
1-1/2"	"	0.500
2"	"	0.571
2-1/2"	"	0.667
3"	"	0.889
3-1/2"	"	1.143
4"	"	1.333
Couplings		
1/2"	EA.	0.050
3/4"	"	0.059
1"	"	0.080
1-1/4"	"	0.089
1-1/2"	"	0.100
2"	"	0.107
2-1/2"	"	0.123
3"	"	0.145
3-1/2"	"	0.145
4"	"	0.160
16110.35 WIREMOLD		
Wiremold raceway with fittings, surface mounted		
#200	L.F.	0.030
#500	"	0.030
#700	"	0.040

BASIC MATERIALS	UNIT	MAN/ HOURS
16110.35 WIREMOLD		
#800	L.F.	0.040
Fittings, #200, 90 degree flat elbow	EA.	0.050
Internal elbow	"	0.050
Extension adapter	"	0.059
#200, #500, #700		
Single pole switch and box	EA.	0.400
Duplex receptacle with box	"	0.308
#500, #700		
90 deg. flat elbow	EA.	0.080
Internal elbow	"	0.080
Junction box	"	0.133
Fixture box	"	0.133
Shallow switch and receptacle box	"	0.080
#800		
90 deg. flat elbow	EA.	0.080
Internal elbow	"	0.080
Junction box	"	0.123
16110.60 TRENCH DUCT		
Trench duct, with cover		
9"	L.F.	0.170
12"	"	0.200
18"	"	0.267
Tees		
9"	EA.	1.739
12"	"	2.000
18"	"	2.222
Vertical elbows		
9"	EA.	0.800
12"	"	1.096
18"	"	1.356
Cabinet connectors		
9"	EA.	2.000
12"	"	2.105
18"	"	2.424
End closers		
9"	EA.	0.615
12"	"	0.667
18"	"	0.800
Horizontal elbows		
9"	EA.	1.509
12"	"	1.739
18"	"	2.105
Crosses		
9"	EA.	2.000
12"	"	2.222
18"	"	2.500
16110.80 WIREWAYS		
Wireway, hinge cover type		
2-1/2" x 2-1/2"		
1' section	EA.	0.154
2'	"	0.190

BASIC MATERIALS	UNIT	MAN/ HOURS
16110.80 WIREWAYS		
3'	EA.	0.250
16120.41 ALUMINUM CONDUCTORS		
Type XHHW, stranded aluminum, 600v		
#8	L.F.	0.005
#6	"	0.006
#4	"	0.008
#2	"	0.009
1/0	"	0.011
2/0	"	0.012
3/0	"	0.014
4/0	"	0.015
THW, stranded		
#8	L.F.	0.005
#6	"	0.006
#4	"	0.008
#3	"	0.009
#1	"	0.010
1/0	"	0.011
2/0	"	0.012
3/0	"	0.012
4/0	"	0.015
16120.43 COPPER CONDUCTORS		
Copper conductors, type THW, solid		
#14	L.F.	0.004
#12	"	0.005
#10	"	0.006
Stranded		
#14	L.F.	0.004
#12	"	0.005
#10	"	0.006
#8	"	0.008
#6	"	0.009
#4	"	0.010
#3	"	0.010
#2	"	0.012
#1	"	0.014
1/0	"	0.016
2/0	"	0.020
3/0	"	0.025
4/0	"	0.028
THHN-THWN, solid		
#14	L.F.	0.004
#12	"	0.005
#10	"	0.006
Stranded		
#14	L.F.	0.004
#12	"	0.005
#10	"	0.006
#8	"	0.008

16 ELECTRICAL

BASIC MATERIALS

16120.43 COPPER CONDUCTORS

	UNIT	MAN/ HOURS
#6	L.F.	0.009
#4	"	0.010
#2	"	0.012
#1	"	0.014
1/0	"	0.016
2/0	"	0.020
3/0	"	0.025
4/0	"	0.028
XLP, 600v		
#12	L.F.	0.005
#10	"	0.006
#8	"	0.008
#6	"	0.009
#4	"	0.010
#3	"	0.011
#2	"	0.012
#1	"	0.014
1/0	"	0.016
2/0	"	0.020
3/0	"	0.026
4/0	"	0.028
Bare solid wire		
#14	L.F.	0.004
#12	"	0.005
#10	"	0.006
#8	"	0.008
#6	"	0.009
#4	"	0.010
#2	"	0.012
Bare stranded wire		
#8	L.F.	0.008
#6	"	0.010
#4	"	0.010
#2	"	0.011
#1	"	0.014
1/0	"	0.018
2/0	"	0.020
3/0	"	0.025
4/0	"	0.028
Type "BX" solid armored cable		
#14/2	L.F.	0.025
#14/3	"	0.028
#14/4	"	0.031
#12/2	"	0.028
#12/3	"	0.031
#12/4	"	0.035
#10/2	"	0.031
#10/3	"	0.035
#10/4	"	0.040
#8/2	"	0.035
#8/3	"	0.040
Steel type, metal clad cable, solid, with ground		
#14/2	L.F.	0.018
#14/3	"	0.020

BASIC MATERIALS

16120.43 COPPER CONDUCTORS

	UNIT	MAN/ HOURS
#14/4	L.F.	0.023
#12/2	"	0.020
#12/3	"	0.025
#12/4	"	0.030
#10/2	"	0.023
#10/3	"	0.028
#10/4	"	0.033
Metal clad cable, stranded, with ground		
#8/2	L.F.	0.028
#8/3	"	0.035
#8/4	"	0.042
#6/2	"	0.030
#6/3	"	0.038
#6/4	"	0.044
#4/2	"	0.040
#4/3	"	0.044
#4/4	"	0.055
#3/3	"	0.050
#3/4	"	0.059
#2/3	"	0.057
#2/4	"	0.067
#1/3	"	0.076
#1/4	"	0.084

16120.47 SHEATHED CABLE

	UNIT	MAN/ HOURS
Non-metallic sheathed cable		
Type NM cable with ground		
#14/2	L.F.	0.015
#12/2	"	0.016
#10/2	"	0.018
#8/2	"	0.020
#6/2	"	0.025
#14/3	"	0.026
#12/3	"	0.027
#10/3	"	0.027
#8/3	"	0.028
#6/3	"	0.028
#4/3	"	0.032
#2/3	"	0.035
Type U.F. cable with ground		
#14/2	L.F.	0.016
#12/2	"	0.019
#10/2	"	0.020
#8/2	"	0.023
#6/2	"	0.027
#14/3	"	0.020
#12/3	"	0.022
#10/3	"	0.025
#8/3	"	0.028
#6/3	"	0.032
Type S.F.U. cable, 3 conductor		
#8	L.F.	0.028
#6	"	0.031

BASIC MATERIALS

16120.47 SHEATHED CABLE

BASIC MATERIALS	UNIT	MAN/HOURS
Type SER cable, 4 conductor		
#6	L.F.	0.036
#4	"	0.039
Flexible cord, type STO cord		
#18/2	L.F.	0.004
#18/3	"	0.005
#18/4	"	0.006
#16/2	"	0.004
#16/3	"	0.004
#16/4	"	0.005
#14/2	"	0.005
#14/3	"	0.006
#14/4	"	0.007
#12/2	"	0.006
#12/3	"	0.007
#12/4	"	0.008
#10/2	"	0.007
#10/3	"	0.008
#10/4	"	0.009
#8/2	"	0.008
#8/3	"	0.009
#8/4	"	0.010

16130.40 BOXES

BASIC MATERIALS	UNIT	MAN/HOURS
Round cast box, type SEH		
1/2"	EA.	0.348
3/4"	"	0.421
SEHC		
1/2"	EA.	0.348
3/4"	"	0.421
SEHL		
1/2"	EA.	0.348
3/4"	"	0.444
SEHT		
1/2"	EA.	0.421
3/4"	"	0.500
SEHX		
1/2"	EA.	0.500
3/4"	"	0.615
Blank cover	"	0.145
1/2", hub cover	"	0.145
Cover with gasket	"	0.178
Rectangle, type FS boxes		
1/2"	EA.	0.348
3/4"	"	0.400
1"	"	0.500
FSA		
1/2"	EA.	0.348
3/4"	"	0.400
FSC		
1/2"	EA.	0.348
3/4"	"	0.421
1"	"	0.500
FSL		

BASIC MATERIALS

16130.40 BOXES

BASIC MATERIALS	UNIT	MAN/HOURS
1/2"	EA.	0.348
3/4"	"	0.400
FSR		
1/2"	EA.	0.348
3/4"	"	0.400
FSS		
1/2"	EA.	0.348
3/4"	"	0.400
FSLA		
1/2"	EA.	0.348
3/4"	"	0.400
FSCA		
1/2"	EA.	0.348
3/4"	"	0.400
FSCC		
1/2"	EA.	0.400
3/4"	"	0.500
FSCT		
1/2"	EA.	0.400
3/4"	"	0.500
1"	"	0.571
FST		
1/2"	EA.	0.500
3/4"	"	0.571
FSX		
1/2"	EA.	0.615
3/4"	"	0.727
FSCD boxes		
1/2"	EA.	0.615
3/4"	"	0.727
Rectangle, type FS, 2 gang boxes		
1/2"	EA.	0.348
3/4"	"	0.400
1"	"	0.500
Weatherproof cast aluminum boxes, 1 gang, 3 outlets		
1/2"	EA.	0.400
3/4"	"	0.500
2 gang, 3 outlets		
1/2"	EA.	0.500
3/4"	"	0.533
1 gang, 4 outlets		
1/2"	EA.	0.615
3/4"	"	0.727
2 gang, 4 outlets		
1/2"	EA.	0.615
3/4"	"	0.727
Weatherproof and type FS box covers, blank, 1 gang	"	0.145
Tumbler switch, 1 gang	"	0.145
1 gang, single recept	"	0.145
Duplex recept	"	0.145
Despard	"	0.145
Red pilot light	"	0.145
SW and		
Single recept	EA.	0.200

16 ELECTRICAL

BASIC MATERIALS	UNIT	MAN/ HOURS
16130.40 BOXES		
Duplex recept	EA.	0.200
2 gang		
Blank	EA.	0.182
Tumbler switch	"	0.182
Single recept	"	0.182
Duplex recept	"	0.182
Box covers		
Surface	EA.	0.200
Sealing	"	0.200
Dome	"	0.200
1/2" nipple	"	0.200
3/4" nipple	"	0.200
16130.60 PULL AND JUNCTION BOXES		
4"		
Octagon box	EA.	0.114
Box extension	"	0.059
Plaster ring	"	0.059
Cover blank	"	0.059
Square box	"	0.114
Box extension	"	0.059
Plaster ring	"	0.059
Cover blank	"	0.059
4-11/16"		
Square box	EA.	0.114
Box extension	"	0.059
Plaster ring	"	0.059
Cover blank	"	0.059
Switch and device boxes		
2 gang	EA.	0.114
3 gang	"	0.114
4 gang	"	0.160
Device covers		
2 gang	EA.	0.059
3 gang	"	0.059
4 gang	"	0.059
Handy box	"	0.114
Extension	"	0.059
Switch cover	"	0.059
Switch box with knockout	"	0.145
Weatherproof cover, spring type	"	0.080
Cover plate, dryer receptacle 1 gang plastic	"	0.100
For 4" receptacle, 2 gang	"	0.100
Duplex receptacle cover plate, plastic	"	0.059
4", vertical bracket box, 1-1/2" with		
RMX clamps	EA.	0.145
BX clamps	"	0.145
4", octagon device cover		
1 switch	EA.	0.059
1 duplex recept	"	0.059
4" octagon adjustable bar hangers		
18-1/2"	EA.	0.050
26-1/2"	"	0.050
With clip		

BASIC MATERIALS	UNIT	MAN/ HOURS
16130.60 PULL AND JUNCTION BOXES		
18-1/2"	EA.	0.050
26-1/2"	"	0.050
4" square to round plaster rings	"	0.059
2,gang device plaster rings	"	0.059
Surface covers		
1 gang switch	EA.	0.059
2 gang switch	"	0.059
1 single recept	"	0.059
1 20a twist lock recept	"	0.059
1 30a twist lock recept	"	0.059
1 duplex recept	"	0.059
2 duplex recept	"	0.059
Switch and duplex recept	"	0.059
4" plastic round boxes, ground straps		
Box only	EA.	0.145
Box w/clamps	"	0.200
Box w/16" bar	"	0.229
Box w/24" bar	"	0.250
4" plastic round box covers		
Blank cover	EA.	0.059
Plaster ring	"	0.059
4" plastic square boxes		
Box only	EA.	0.145
Box w/clamps	"	0.200
Box w/hanger	"	0.250
Box w/nails and clamp	"	0.250
4" plastic square box covers		
Blank cover	EA.	0.059
1 gang ring	"	0.059
2 gang ring	"	0.059
Round ring	"	0.059
16130.65 PULL BOXES AND CABINETS		
Galvanized pull boxes, screw cover		
4x4x4	EA.	0.190
4x6x4	"	0.190
16130.80 RECEPTACLES		
Contractor grade duplex receptacles, 15 a 120v		
Duplex	EA.	0.200
125 volt, 20a, duplex, grounding type, standard grade	"	0.200
Ground fault interrupter type	"	0.296
250 volt, 20a, 2 pole, single receptacle, ground type	"	0.200
120/208v, 4 pole, single receptacle, twist lock		
20a	EA.	0.348
50a	"	0.348
125/250v, 3 pole, flush receptacle		
30a	EA.	0.296
50a	"	0.296
60a	"	0.348
Clock receptacle, 2 pole, grounding type	"	0.200
125/250v, 3 pole, 3 wire surface recepts		

BASIC MATERIALS	UNIT	MAN/ HOURS
16130.80 RECEPTACLES		
30a	EA.	0.296
50a	"	0.296
60a	"	0.348
Cord set, 3 wire, 6' cord		
30a	EA.	0.296
50a	"	0.296
125/250v, 3 pole, 3 wire cap		
30a	EA.	0.400
50a	"	0.400
60a	"	0.444
16198.10 ELECTRIC MANHOLES		
Precast, handhole, 4' deep		
2'x2'	EA.	3.478
3'x3'	"	5.556
4'x4'	"	10.256
Power manhole, complete, precast, 8' deep		
4'x4'	EA.	14.035
6'x6'	"	20.000
8'x8'	"	21.053
6' deep, 9' x 12'	"	25.000
Cast in place, power manhole, 8' deep		
4'x4'	EA.	14.035
6'x6'	"	20.000
8'x8'	"	21.053
16199.10 UTILITY POLES & FITTINGS		
Wood pole, creosoted		
25'	EA.	2.353
30'	"	2.963
35'	"	3.478
40'	"	3.791
45'	"	6.957
50'	"	7.207
55'	"	7.547
Treated, wood preservative, 6"x6"		
8'	EA.	0.500
10'	"	0.800
12'	"	0.889
14'	"	1.333
16'	"	1.600
18'	"	2.000
20'	"	2.000
Aluminum, brushed, no base		
8'	EA.	2.000
10'	"	2.667
15'	"	2.759
20'	"	3.200
25'	"	3.810
30'	"	4.396
35'	"	5.000
40'	"	6.250
Steel, no base		

BASIC MATERIALS	UNIT	MAN/ HOURS
16199.10 UTILITY POLES & FITTINGS		
10'	EA.	2.500
15'	"	2.963
20'	"	3.810
25'	"	4.520
30'	"	5.096
35'	"	6.250
Concrete, no base		
13'	EA.	5.517
16'	"	7.273
18'	"	8.791
25'	"	10.000
30'	"	12.121
35'	"	14.035
40'	"	16.000
45'	"	17.021
50'	"	18.182
55'	"	19.048
60'	"	20.000
Pole line hardware		
Wood crossarm		
4'	EA.	1.333
8'	"	1.667
10'	"	2.051
Angle steel brace		
1 piece	EA.	0.250
2 piece	"	0.348
Eye nut, 5/8"	"	0.050
Bolt (14-16"), 5/8"	"	0.200
Transformer, ground connection	"	0.250
Stirrup	"	0.308
Secondary lead support	"	0.400
Spool insulator	"	0.200
Guy grip, preformed		
7/16"	EA.	0.145
1/2"	"	0.145
Hook	"	0.250
Strain insulator	"	0.364
Wire		
5/16"	L.F.	0.005
7/16"	"	0.006
1/2"	"	0.008
Soft drawn ground, copper, #8	"	0.008
Ground clamp	EA.	0.308
Perforated strapping for conduit, 1-1/2"	L.F.	0.145
Hot line clamp	EA.	0.800
Lightning arrester		
3kv	EA.	1.000
10kv	"	1.600
30kv	"	2.000
36kv	"	2.500
Fittings		
Plastic molding	L.F.	0.145
Molding staples	EA.	0.050
Ground wires staples	"	0.030

16 ELECTRICAL

BASIC MATERIALS

16199.10 UTILITY POLES & FITTINGS

	UNIT	MAN/HOURS
Copper butt plate	EA.	0.296
Anchor bond clamp	"	0.145
Guy wire		
1/4"	L.F.	0.030
3/8"	"	0.050
Guy grip		
1/4"	EA.	0.050
3/8"	"	0.050

POWER GENERATION

16210.10 GENERATORS

	UNIT	MAN/HOURS
Diesel generator, with auto transfer switch		
50kw	EA.	30.769
125kw	"	50.000
300kw	"	100
750kw	"	200

16320.10 TRANSFORMERS

	UNIT	MAN/HOURS
Floor mounted, single phase, int. dry, 480v-120/240v		
3 kva	EA.	1.818
5 kva	"	3.077
7.5 kva	"	3.478
10 kva	"	3.810
15 kva	"	4.301
100 kva	"	11.594
Three phase, 480v-120/208v		
15 kva	EA.	6.015
30 kva	"	9.412
45 kva	"	10.811
225 kva	"	15.385

16350.10 CIRCUIT BREAKERS

	UNIT	MAN/HOURS
Molded case, 240v, 15-60a, bolt-on		
1 pole	EA.	0.250
2 pole	"	0.348
70-100a, 2 pole	"	0.533
15-60a, 3 pole	"	0.400
70-100a, 3 pole	"	0.615
480v, 2 pole		
15-60a	EA.	0.296
70-100a	"	0.400
3 pole		
15-60a	EA.	0.400
70-100a	"	0.444
70-225a	"	0.615

POWER GENERATION

16350.10 CIRCUIT BREAKERS

	UNIT	MAN/HOURS
Load center circuit breakers, 240v		
1 pole, 10-60a	EA.	0.250
2 pole		
10-60a	EA.	0.400
70-100a	"	0.667
110-150a	"	0.727
3 pole		
10-60a	EA.	0.500
70-100a	"	0.727
Load center, G.F.I. breakers, 240v		
1 pole, 15-30a	EA.	0.296
2 pole, 15-30a	"	0.400
Key operated breakers, 240v, 1 pole, 10-30a	"	0.296
Tandem breakers, 240v		
1 pole, 15-30a	EA.	0.400
2 pole, 15-30a	"	0.533
Bolt-on, G.F.I. breakers, 240v, 1 pole, 15-30a	"	0.348

16360.10 SAFETY SWITCHES

	UNIT	MAN/HOURS
Fused, 3 phase, 30 amp, 600v, heavy duty		
NEMA 1	EA.	1.143
NEMA 3r	"	1.143
NEMA 4	"	1.600
NEMA 12	"	1.739
60a		
NEMA 1	EA.	1.143
NEMA 3r	"	1.143
NEMA 4	"	1.600
NEMA 12	"	1.739
100a		
NEMA 1	EA.	1.739
NEMA 3r	"	1.739
NEMA 4	"	2.000
NEMA 12	"	2.500
200a		
NEMA 1	EA.	2.500
NEMA 3r	"	2.500
NEMA 4	"	2.759
NEMA 12	"	3.478
Non-fused, 240-600v, heavy duty, 3 phase, 30 amp		
NEMA 1	EA.	1.143
NEMA 3r	"	1.143
NEMA 4	"	1.739
NEMA 12	"	1.739
60a		
NEMA1	EA.	1.143
NEMA 3r	"	1.143
NEMA 4	"	1.739
NEMA 12	"	1.739
100a		
NEMA 1	EA.	1.739
NEMA 3r	"	1.739
NEMA 4	"	2.500

16 ELECTRICAL

POWER GENERATION

16360.10 SAFETY SWITCHES	UNIT	MAN/ HOURS
NEMA 12	EA.	2.500
200a, NEMA 1	"	2.500
600a, NEMA 12	"	12.308

16365.10 FUSES	UNIT	MAN/ HOURS
Fuse, one-time, 250v		
30a	EA.	0.050
60a	"	0.050
100a	"	0.050
200a	"	0.050
400a	"	0.050
600a	"	0.050
600v		
30a	EA.	0.050
60a	"	0.050
100a	"	0.050
200a	"	0.050
400a	"	0.050

16395.10 GROUNDING	UNIT	MAN/ HOURS
Ground rods, copper clad, 1/2" x		
6'	EA.	0.667
8'	"	0.727
10'	"	1.000
5/8" x		
5'	EA.	0.615
6'	"	0.727
8'	"	1.000
10'	"	1.250
3/4" x		
8'	EA.	0.727
10'	"	0.800
Ground rod clamp		
5/8"	EA.	0.123
3/4"	"	0.123
Ground rod couplings		
1/2"	EA.	0.100
5/8"	"	0.100
Ground rod, driving stud		
1/2"	EA.	0.100
5/8"	"	0.100
3/4"	"	0.100
Ground rod clamps, #8-2 to		
1" pipe	EA.	0.200
2" pipe	"	0.250
3" pipe	"	0.296
5" pipe	"	0.348
6" pipe	"	0.444

SERVICE AND DISTRIBUTION

16425.10 SWITCHBOARDS	UNIT	MAN/ HOURS
Switchboard, 90" high, no main disconnect, 208/120v		
400a	EA.	7.921
600a	"	8.000
1000a	"	8.000
1200a	"	10.000
1600a	"	11.940
2000a	"	14.035
2500a	"	16.000
277/480v		
600a	EA.	8.163
800a	"	8.163
1600a	"	11.940
2000a	"	14.035
2500a	"	16.000
3000a	"	27.586
4000a	"	29.630

16430.20 METERING	UNIT	MAN/ HOURS
Outdoor wp meter sockets, 1 gang, 240v, 1 phase		
Includes sealing ring, 100a	EA.	1.509
150a	"	1.778
200a	"	2.000
Die cast hubs, 1-1/4"	"	0.320
1-1/2"	"	0.320
2"	"	0.320

16470.10 PANELBOARDS	UNIT	MAN/ HOURS
Indoor load center, 1 phase 240v main lug only		
30a - 2 spaces	EA.	2.000
100a - 8 spaces	"	2.424
150a - 16 spaces	"	2.963
200a - 24 spaces	"	3.478
200a - 42 spaces	"	4.000
Main circuit breaker		
100a - 8 spaces	EA.	2.424
100a - 16 spaces	"	2.759
150a - 16 spaces	"	2.963
150a - 24 spaces	"	3.200
200a - 24 spaces	"	3.478
200a - 42 spaces	"	3.636
3 phase, 480/277v, main lugs only, 120a, 30 circuits	"	3.478
277/480v, 4 wire, flush surface		
225a, 30 circuits	EA.	4.000
400a, 30 circuits	"	5.000
600a, 42 circuits	"	6.015
208/120v, main circuit breaker, 3 phase, 4 wire		
100a		
12 circuits	EA.	5.096
20 circuits	"	6.299
30 circuits	"	7.018
400a		
30 circuits	EA.	14.815
42 circuits	"	16.000

SERVICE AND DISTRIBUTION	UNIT	MAN/ HOURS
16470.10 PANELBOARDS		
600a, 42 circuits	EA.	18.182
120/208v, flush, 3 ph., 4 wire, main only		
100a		
12 circuits	EA.	5.096
20 circuits	"	6.299
30 circuits	"	7.018
225a		
30 circuits	EA.	7.767
42 circuits	"	9.524
400a		
30 circuits	EA.	14.815
42 circuits	"	16.000
600a, 42 circuits	"	18.182
16480.10 MOTOR CONTROLS		
Motor generator set, 3 phase, 480/277v, w/controls		
10kw	EA.	27.586
15kw	"	30.769
20kw	"	32.000
40kw	"	38.095
100kw	"	61.538
200kw	"	72.727
300kw	"	80.000
2 pole, 230 volt starter, w/NEMA-1		
1 hp, 9 amp, size 00	EA.	1.000
2 hp, 18amp, size 0	"	1.000
3 hp, 27amp, size 1	"	1.000
5 hp, 45amp, size 1p	"	1.000
7-1/2 hp, 45a, size 2	"	1.000
15 hp, 90a, size 3	"	1.000
16490.10 SWITCHES		
Fused interrupter load, 35kv		
20A		
1 pole	EA.	16.000
2 pole	"	17.021
3 way	"	17.021
4 way	"	18.182
30a, 1 pole	"	16.000
3 way	"	17.021
4 way	"	18.182
Weatherproof switch, including box & cover, 20a		
1 pole	EA.	16.000
2 pole	"	17.021
3 way	"	18.182
4 way	"	18.182
Photo electric switches		
1000 watt		
105-135v	EA.	0.727
Dimmer switch and switch plate		
600 w	EA.	0.308
Contractor grade wall switch 15a, 120v		

SERVICE AND DISTRIBUTION	UNIT	MAN/ HOURS
16490.10 SWITCHES		
Single pole	EA.	0.160
Three way	"	0.200
Four way	"	0.267
Specification grade toggle switches, 20a, 120-277v		
Single pole	EA.	0.200
Double pole	"	0.296
3 way	"	0.250
4 way	"	0.296
Switch plates, plastic ivory		
1 gang	EA.	0.080
2 gang	"	0.100
3 gang	"	0.119
4 gang	"	0.145
5 gang	"	0.160
6 gang	"	0.182
Stainless steel		
1 gang	EA.	0.080
2 gang	"	0.100
3 gang	"	0.123
4 gang	"	0.145
5 gang	"	0.160
6 gang	"	0.182
Brass		
1 gang	EA.	0.080
2 gang	"	0.100
3 gang	"	0.123
4 gang	"	0.145
5 gang	"	0.160
6 gang	"	0.182
16490.20 TRANSFER SWITCHES		
Automatic transfer switch 600v, 3 pole		
30a	EA.	3.478
100a	"	4.762
400a	"	10.000
800a	"	18.182
1200a	"	22.857
2600a	"	42.105
16490.80 SAFETY SWITCHES		
Safety switch, 600v, 3 pole, heavy duty, NEMA-1		
30a	EA.	1.000
60a	"	1.143
100a	"	1.600
200a	"	2.500
400a	"	5.517
600a	"	8.000
800a	"	10.526
1200a	"	14.286

LIGHTING	UNIT	MAN/HOURS
16510.05 INTERIOR LIGHTING		
Recessed fluorescent fixtures, 2'x2'		
2 lamp	EA.	0.727
4 lamp	"	0.727
2 lamp w/flange	"	1.000
4 lamp w/flange	"	1.000
1'x4'		
2 lamp	EA.	0.667
3 lamp	"	0.667
2 lamp w/flange	"	0.727
3 lamp w/flange	"	0.727
2'x4'		
2 lamp	EA.	0.727
3 lamp	"	0.727
4 lamp	"	0.727
2 lamp w/flange	"	1.000
3 lamp w/flange	"	1.000
4 lamp w/flange	"	1.000
4'x4'		
4 lamp	EA.	1.000
6 lamp	"	1.000
8 lamp	"	1.000
4 lamp w/flange	"	1.509
6 lamp w/flange	"	1.509
8 lamp, w/flange	"	1.509
Surface mounted incandescent fixtures		
40w	EA.	0.667
75w	"	0.667
100w	"	0.667
150w	"	0.667
Pendant		
40w	EA.	0.800
75w	"	0.800
100w	"	0.800
150w	"	0.800
Recessed incandescent fixtures		
40w	EA.	1.509
75w	"	1.509
100w	"	1.509
150w	"	1.509
Exit lights, 120v		
Recessed	EA.	1.250
Back mount	"	0.727
Universal mount	"	0.727
Emergency battery units, 6v-120v, 50 unit	"	1.509
With 1 head	"	1.509
With 2 heads	"	1.509
Mounting bucket	"	0.727
Light track single circuit		
2'	EA.	0.500
4'	"	0.500
8'	"	1.000
12'	"	1.509
Fixtures, square		
R-20	EA.	0.145

LIGHTING	UNIT	MAN/HOURS
16510.05 INTERIOR LIGHTING		
R-30	EA.	0.145
Mini spot	"	0.145
16510.10 LIGHTING INDUSTRIAL		
Surface mounted fluorescent, wrap around lens		
1 lamp	EA.	0.800
2 lamps	"	0.889
4 lamps	"	1.000
Wall mounted fluorescent		
2-20w lamps	EA.	0.500
2-30w lamps	"	0.500
2-40w lamps	"	0.667
Indirect, with wood shielding, 2049w lamps		
4'	EA.	1.000
8'	"	1.600
Industrial fluorescent, 2 lamp		
4'	EA.	0.727
8'	"	1.333
Strip fluorescent		
4'		
1 lamp	EA.	0.667
2 lamps	"	0.667
8'		
1 lamp	EA.	0.727
2 lamps	"	0.889
Wire guard for strip fixture, 4' long	"	0.348
Strip fluorescent, 8' long, two 4' lamps	"	1.333
With four 4' lamps	"	1.600
Wet location fluorescent, plastic housing		
4' long		
1 lamp	EA.	1.000
2 lamps	"	1.333
8' long		
2 lamps	EA.	1.600
4 lamps	"	1.739
Parabolic troffer, 2'x2'		
With 2 "U" lamps	EA.	1.000
With 3 "U" lamps	"	1.143
2'x4'		
With 2 40w lamps	EA.	1.143
With 3 40w lamps	"	1.333
With 4 40w lamps	"	1.333
1'x4'		
With 1 T-12 lamp, 9 cell	EA.	0.727
With 2 T-12 lamps	"	0.889
With 1 T-12 lamp, 20 cell	"	0.727
With 2 T-12 lamps	"	0.889
Steel sided surface fluorescent, 2'x4'		
3 lamps	EA.	1.333
4 lamps	"	1.333
Outdoor sign fluor., 1 lamp, remote ballast		
4' long	EA.	6.015
6' long	"	8.000
Recess mounted, commercial, 2'x2', 13" high		

LIGHTING	UNIT	MAN/ HOURS
16510.10 LIGHTING INDUSTRIAL		
100w	EA.	4.000
250w	"	4.494
High pressure sodium, hi-bay open		
400w	EA.	1.739
1000w	"	2.424
Enclosed		
400w	EA.	2.424
1000w	"	2.963
Metal halide hi-bay, open		
400w	EA.	1.739
1000w	"	2.424
Enclosed		
400w	EA.	2.424
1000w	"	2.963
High pressure sodium, low bay, surface mounted		
100w	EA.	1.000
150w	"	1.143
250w	"	1.333
400w	"	1.600
Metal halide, low bay, pendant mounted		
175w	EA.	1.333
250w	"	1.600
400w	"	2.222
Indirect luminare, square, metal halide, freestanding		
175w	EA.	1.000
250w	"	1.000
400w	"	1.000
High pressure sodium		
150w	EA.	1.000
250w	"	1.000
400w	"	1.000
Round, metal halide		
175w	EA.	1.000
250w	"	1.000
400w	"	1.000
High pressure sodium		
150w	EA.	1.000
250w	"	1.000
400w	"	1.000
Wall mounted, metal halide		
175w	EA.	2.500
250w	"	2.500
400w	"	3.200
High pressure sodium		
150w	EA.	2.500
250w	"	2.500
400w	"	3.200
Wall pack lithonia, high pressure sodium		
35w	EA.	0.889
55w	"	1.000
150w	"	1.600
250w	"	1.739
Low pressure sodium		
35w	EA.	1.739

LIGHTING	UNIT	MAN/ HOURS
16510.10 LIGHTING INDUSTRIAL		
55w	EA.	2.000
Wall pack hubbell, high pressure sodium		
35w	EA.	0.889
150w	"	1.600
250w	"	1.739
Compact fluorescent		
2-7w	EA.	1.000
2-13w	"	1.333
1-18w	"	1.333
Handball & racquet ball court, 2'x2', metal halide		
250w	EA.	2.500
400w	"	2.759
High pressure sodium		
250w	EA.	2.500
400w	"	2.759
Bollard light, 42" w/found., high pressure sodium		
70w	EA.	2.581
100w	"	2.581
150w	"	2.581
Light fixture lamps		
Lamp		
20w med. bipin base, cool white, 24"	EA.	0.145
30w cool white, rapid start, 36"	"	0.145
40w cool white "U", 3"	"	0.145
40w cool white, rapid start, 48"	"	0.145
70w high pressure sodium, mogul base	"	0.200
75w slimline, 96"	"	0.200
100w		
Incandescent, 100a, inside frost	EA.	0.100
Mercury vapor, clear, mogul base	"	0.200
High pressure sodium, mogul base	"	0.200
150w		
Par 38 flood or spot, incandescent	EA.	0.100
High pressure sodium, 1/2 mogul base	"	0.200
175w		
Mercury vapor, clear, mogul base	EA.	0.200
Metal halide, clear, mogul base	"	0.200
250w		
High pressure sodium, mogul base	EA.	0.200
Mercury vapor, clear, mogul base	"	0.200
Metal halide, clear, mogul base	"	0.200
High pressure sodium, mogul base	"	0.200
400w		
Mercury vapor, clear, mogul base	EA.	0.200
Metal halide, clear, mogul base	"	0.200
High pressure sodium, mogul base	"	0.200
1000w		
Mercury vapor, clear, mogul base	EA.	0.250
High pressure sodium, mogul base	"	0.250
16510.30 EXTERIOR LIGHTING		
Exterior light fixtures		
Rectangle, high pressure sodium		

16 ELECTRICAL

LIGHTING

16510.30 EXTERIOR LIGHTING	UNIT	MAN/ HOURS
70w	EA.	2.500
100w	"	2.581
150w	"	2.581
250w	"	2.759
400w	"	3.478
Flood, rectangular, high pressure sodium		
70w	EA.	2.500
100w	"	2.581
150w	"	2.581
400w	"	3.478
1000w	"	4.494
Round		
400w	EA.	3.478
1000w	"	4.494
Round, metal halide		
400w	EA.	3.478
1000w	"	4.494
Light fixture arms, cobra head, 6', high press. sodium		
100w	EA.	2.000
150w	"	2.500
250w	"	2.500
400w	"	2.963
Flood, metal halide		
400w	EA.	3.478
1000w	"	4.494
1500w	"	6.015
Mercury vapor		
250w	EA.	2.759
400w	"	3.478
Incandescent		
300w	EA.	1.739
500w	"	2.000
1000w	"	3.200

16510.90 POWER LINE FILTERS	UNIT	MAN/ HOURS
Heavy duty power line filter, 240v		
100a	EA.	10.000
300a	"	16.000
600a	"	24.242

16610.30 UNINTERRUPTIBLE POWER	UNIT	MAN/ HOURS
Uninterruptible power systems, (U.P.S.), 3kva	EA.	8.000
5 kva	"	11.004
7.5 kva	"	16.000
10 kva	"	21.978
15 kva	"	22.857
20 kva	"	24.024
25 kva	"	25.000
30 kva	"	25.974
35 kva	"	27.027
40 kva	"	27.972
45 kva	"	28.986
50 kva	"	29.963

LIGHTING

16610.30 UNINTERRUPTIBLE POWER	UNIT	MAN/ HOURS
62.5 kva	EA.	32.000
75 kva	"	34.934
100 kva	"	36.036
150 kva	"	50.000
200 kva	"	55.172
300 kva	"	74.766
400 kva	"	89.888
500 kva	"	110

16670.10 LIGHTNING PROTECTION	UNIT	MAN/ HOURS
Lightning protection		
Copper point, nickel plated, 12'		
1/2" dia.	EA.	1.000
5/8" dia.	"	1.000

COMMUNICATIONS

16720.10 FIRE ALARM SYSTEMS	UNIT	MAN/ HOURS
Master fire alarm box, pedestal mounted	EA.	16.000
Master fire alarm box	"	6.015
Box light	"	0.500
Ground assembly for box	"	0.667
Bracket for pole type box	"	0.727
Pull station		
Waterproof	EA.	0.500
Manual	"	0.400
Horn, waterproof	"	1.000
Interior alarm	"	0.727
Coded transmitter, automatic	"	2.000
Control panel, 8 zone	"	8.000
Battery charger and cabinet	"	2.000
Batteries, nickel cadmium or lead calcium	"	5.000
CO2 pressure switch connection	"	0.727
Annunciator panels		
Fire detection annunciator, remote type, 8 zone	EA.	1.818
12 zone	"	2.000
16 zone	"	2.500
Fire alarm systems		
Bell	EA.	0.615
Weatherproof bell	"	0.667
Horn	"	0.727
Siren	"	2.000
Chime	"	0.615
Audio/visual	"	0.727
Strobe light	"	0.727
Smoke detector	"	0.667
Heat detection	"	0.500

16 ELECTRICAL

COMMUNICATIONS

16720.10 FIRE ALARM SYSTEMS

COMMUNICATIONS	UNIT	MAN/ HOURS
Thermal detector	EA.	0.500
Ionization detector	"	0.533
Duct detector	"	2.759
Test switch	"	0.500
Remote indicator	"	0.571
Door holder	"	0.727
Telephone jack	"	0.296
Fireman phone	"	1.000
Speaker	"	0.800
Remote fire alarm annunciator panel		
24 zone	EA.	6.667
48 zone	"	13.008
Control panel		
12 zone	EA.	2.963
16 zone	"	4.444
24 zone	"	6.667
48 zone	"	16.000
Power supply	"	1.509
Status command	"	5.000
Printer	"	1.509
Transponder	"	0.899
Transformer	"	0.667
Transceiver	"	0.727
Relays	"	0.500
Flow switch	"	2.000
Tamper switch	"	2.963
End of line resistor	"	0.348
Printed ckt. card	"	0.500
Central processing unit	"	6.154
UPS backup to c.p.u.	"	8.999
Smoke detector, fixed temp. & rate of rise comb.	"	1.600

16720.50 SECURITY SYSTEMS

COMMUNICATIONS	UNIT	MAN/ HOURS
Sensors	EA.	
Balanced magnetic door switch, surface mounted	"	0.500
With remote test	"	1.000
Flush mounted	"	1.860
Mounted bracket	"	0.348
Mounted bracket spacer	"	0.348
Photoelectric sensor, for fence	"	
6 beam	"	2.759
9 beam	"	4.255
Photoelectric sensor, 12 volt dc	"	
500' range	"	1.600
800' range	"	2.000
Monitor cabinet, wall mounted		
1 zone	EA.	1.000
5 zone	"	1.600
10 zone	"	1.739
20 zone	"	2.000

COMMUNICATIONS

16730.20 CLOCK SYSTEMS

COMMUNICATIONS	UNIT	MAN/ HOURS
Clock systems		
Single face	EA.	0.800
Double face	"	0.800
Skeleton	"	2.759
Master	"	5.000
Signal generator	"	4.000
Elapsed time indicator	"	0.800
Controller	"	0.533
Clock and speaker	"	1.096
Bell		
Standard	EA.	0.533
Weatherproof	"	0.800
Horn		
Standard	EA.	0.727
Weatherproof	"	0.952
Chime	"	0.533
Buzzer	"	0.533
Flasher	"	0.615
Control Board	"	3.478
Program unit	"	5.000
Block back box	"	0.500
Double clock back box	"	0.667
Wire guard	"	0.200

16740.10 TELEPHONE SYSTEMS

COMMUNICATIONS	UNIT	MAN/ HOURS
Communication cable		
25 pair	L.F.	0.026
100 pair	"	0.029
150 pair	"	0.033
200 pair	"	0.040
300 pair	"	0.042
400 pair	"	0.044
Cable tap in manhole or junction box		
25 pair cable	EA.	3.810
50 pair cable	"	7.547
75 pair cable	"	11.268
100 pair cable	"	15.094
150 pair cable	"	22.222
200 pair cable	"	29.630
300 pair cable	"	44.444
400 pair cable	"	61.538
Cable terminations, manhole or junction box		
25 pair cable	EA.	3.756
50 pair cable	"	7.477
100 pair cable	"	15.094
150 pair cable	"	22.222
200 pair cable	"	29.630
300 pair cable	"	44.444
400 pair cable	"	61.538
Telephones, standard		
1 button	EA.	2.963
2 button	"	3.478
6 button	"	5.333

COMMUNICATIONS	UNIT	MAN/HOURS
16740.10 TELEPHONE SYSTEMS		
12 button	EA.	7.619
18 button	"	8.889
Hazardous area		
Desk	EA.	7.273
Wall	"	5.000
Accessories		
Standard ground	EA.	1.600
Push button	"	1.600
Buzzer	"	1.600
Interface device	"	0.800
Long cord	"	0.800
Interior jack	"	0.400
Exterior jack	"	0.615
Hazardous area		
Selector switch	EA.	3.200
Bell	"	3.200
Horn	"	4.211
Horn relay	"	3.077
16770.30 SOUND SYSTEMS		
Power amplifiers	EA.	3.478
Pre-amplifiers	"	2.759
Tuner	"	1.455
Horn		
Equilizer	EA.	1.600
Mixer	"	2.222
Tape recorder	"	1.860
Microphone	"	1.000
Cassette Player	"	2.162
Record player	"	1.905
Equipment rack	"	1.290
Speaker		
Wall	EA.	4.000
Paging	"	0.800
Column	"	0.533
Single	"	0.615
Double	"	4.444
Volume control	"	0.533
Plug-in	"	0.800
Desk	"	0.400
Outlet	"	0.400
Stand	"	0.296
Console	"	8.000
Power supply	"	1.290
16780.10 ANTENNAS AND TOWERS		
Guy cable, alumaweld		
1x3, 7/32"	L.F.	0.050
1x3, 1/4"	"	0.050
1x3, 25/64"	"	0.059
1x19, 1/2"	"	0.070
1x7, 35/64"	"	0.080
1x19, 13/16"	"	0.100
Preformed alumaweld end grip		

COMMUNICATIONS	UNIT	MAN/HOURS
16780.10 ANTENNAS AND TOWERS		
1/4" cable	EA.	0.100
3/8" cable	"	0.100
1/2" cable	"	0.145
9/16" cable	"	0.200
5/8" cable	"	0.250
Fiberglass guy rod, white epoxy coated		
1/4" dia.	L.F.	0.145
3/8" dia	"	0.145
1/2" dia	"	0.200
5/8" dia	"	0.250
Preformed glass grip end grip, guy rod		
1/4" dia.	EA.	0.145
3/8" dia.	"	0.200
1/2" dia.	"	0.250
5/8" dia.	"	0.250
Spelter socket end grip, 1/4" dia. guy rod		
Standard strength	EA.	0.500
High performance	"	0.500
3/8" dia. guy rod		
Standard strength	EA.	0.348
High performance	"	0.500
Timber pole, Douglas Fir		
80-85 ft	EA.	19.512
90-95 ft	"	22.222
Southern yellow pine		
35-45 ft	EA.	10.959
50-55 ft	"	14.035
16780.50 TELEVISION SYSTEMS		
TV outlet, self terminating, w/cover plate	EA.	0.308
Thru splitter	"	1.600
End of line	"	1.333
In line splitter multitap		
4 way	EA.	1.818
2 way	"	1.702
Equipment cabinet	"	1.600
Antenna		
Broad band uhf	EA.	3.478
Lightning arrester	"	0.727
TV cable	L.F.	0.005
Coaxial cable rg	"	0.005
Cable drill, with replacement tip	EA.	0.500
Cable blocks for in-line taps	"	0.727
In-line taps ptu-series 36 tv system	"	1.143
Control receptacles	"	0.449
Coupler	"	2.424
Head end equipment	"	6.667
TV camera	"	1.667
TV power bracket	"	0.800
TV monitor	"	1.455
Video recorder	"	2.105
Console	"	8.502
Selector switch	"	1.379

COMMUNICATIONS

16780.50 TELEVISION SYSTEMS

	UNIT	MAN/HOURS
TV controller	EA.	1.404

RESISTANCE HEATING

16850.10 ELECTRIC HEATING

	UNIT	MAN/HOURS
Baseboard heater		
2', 375w	EA.	1.000
3', 500w	"	1.000
4', 750w	"	1.143
5', 935w	"	1.333
6', 1125w	"	1.600
7', 1310w	"	1.818
8', 1500w	"	2.000
9', 1680w	"	2.222
10', 1875w	"	2.286
Unit heater, wall mounted		
750w	EA.	1.600
1500w	"	1.667
2000w	"	1.739
2500w	"	1.818
3000w	"	2.000
4000w	"	2.286
Thermostat		
Integral	EA.	0.500
Line voltage	"	0.500
Electric heater connection	"	0.250
Fittings		
Inside corner	EA.	0.400
Outside corner	"	0.400
Receptacle section	"	0.400
Blank section	"	0.400
Infrared heaters		
600w	EA.	1.000
2000w	"	1.194
3000w	"	2.000
4000w	"	2.500
Controller	"	0.667
Wall bracket	"	0.727
Radiant ceiling heater panels		
500w	EA.	1.000
750w	"	1.000
Unit heaters, suspended, single phase		
3.0 kw	EA.	2.759
5.0 kw	"	2.759
7.5 kw	"	3.200

RESISTANCE HEATING

16850.10 ELECTRIC HEATING

	UNIT	MAN/HOURS
10.0 kw	EA.	3.810
Three phase		
5 kw	EA.	2.759
7.5 kw	"	3.200
10 kw	"	3.810
15 kw	"	4.211
20 kw	"	5.333
25 kw	"	6.400
30 kw	"	8.000
35 kw	"	8.000
Unit heater thermostat	"	0.533
Mounting bracket	"	0.727
Relay	"	0.615
Duct heaters, three phase		
10 kw	EA.	3.810
15 kw	"	3.810
17.5 kw	"	4.000
20 kw	"	6.154

CONTROLS

16910.40 CONTROL CABLE

	UNIT	MAN/HOURS
Control cable, 600v, #14 THWN, PVC jacket		
2 wire	L.F.	0.008
4 wire	"	0.010
6 wire	"	0.131
8 wire	"	0.145
10 wire	"	0.160
12 wire	"	0.182
14 wire	"	0.211
16 wire	"	0.222
18 wire	"	0.242
20 wire	"	0.250
22 wire	"	0.286
Audio cables, shielded, #24 gauge		
3 conductor	L.F.	0.004
4 conductor	"	0.006
5 conductor	"	0.007
6 conductor	"	0.009
7 conductor	"	0.011
8 conductor	"	0.012
9 conductor	"	0.014
10 conductor	"	0.015
15 conductor	"	0.018
20 conductor	"	0.023
25 conductor	"	0.027
30 conductor	"	0.030

CONTROLS	UNIT	MAN/ HOURS
16910.40 CONTROL CABLE		
40 conductor	L.F.	0.036
50 conductor	"	0.042
#22 gauge		
3 conductor	L.F.	0.004
4 conductor	"	0.006
#20 gauge		
3 conductor	L.F.	0.004
10 conductor	"	0.015
15 conductor	"	0.018
#18 gauge		
3 conductor	L.F.	0.004
4 conductor	"	0.006
Microphone cables, #24 gauge		
2 conductor	L.F.	0.004
3 conductor	"	0.005
#20 gauge		
1 conductor	L.F.	0.004
2 conductor	"	0.004
2 conductor	"	0.004
3 conductor	"	0.006
4 conductor	"	0.007
5 conductor	"	0.009
7 conductor	"	0.011
8 conductor	"	0.012
Computer cables shielded, #24 gauge		
1 pair	L.F.	0.004
2 pair	"	0.004
3 pair	"	0.006
4 pair	"	0.007
5 pair	"	0.009
6 pair	"	0.011
7 pair	"	0.012
8 pair	"	0.014
50 pair	"	0.039
Fire alarm cables, #22 gauge		
6 conductor	L.F.	0.010
9 conductor	"	0.015
12 conductor	"	0.016
#18 gauge		
2 conductor	L.F.	0.005
4 conductor	"	0.007
#16 gauge		
2 conductor	L.F.	0.007
4 conductor	"	0.008
#14 gauge		
2 conductor	L.F.	0.008
#12 gauge		
2 conductor	L.F.	0.010
Plastic jacketed thermostat cable		
2 conductor	L.F.	0.004
3 conductor	"	0.005
4 conductor	"	0.006
5 conductor	"	0.008
6 conductor	"	0.009

CONTROLS	UNIT	MAN/ HOURS
16910.40 CONTROL CABLE		
7 conductor	L.F.	0.012
8 conductor	"	0.013

BNi® Building News

Supporting Construction Reference Data

This section contains information, text, charts and tables on various aspects of construction. The intent is to provide the user with a better understanding of unfamiliar areas in order to be able to estimate better. This information includes actual takeoff data for some areas and also selected explanations of common construction materials, methods and common practices.

TYPICAL BUILDING COST BROKEN DOWN BY CSI FORMAT

(Commercial Construction)

Division		New Construction	Remodeling Construction
1.	General Requirements	6 to 8%	Up to 30%
2.	Sitework	4 to 6%	
3.	Concrete	15 to 20%	
4.	Masonry	8 to 12%	
5.	Metals	5 to 7%	
6.	Wood And Plastics	1 to 5%	
7.	Thermal And Moisture Protection	4 to 6%	
8.	Doors And Windows	5 to 7%	Up to 30% (Divisions 8–9)
9.	Finishes	8 to 12%	
10.	Specialties	6 to 10% (Divisions 10–14)	
11.	Architectural Equipment		
12.	Furnishings		
13.	Special Construction		
14.	Conveying Systems		
15.	Mechanical	15 to 25%	Up to 40% (Divisions 15–16)
16.	Electrical	8 to 12%	
	Total Cost	100%	

CONVERSION FACTORS

Change	To	Multiply By
Atmospheres	Pounds per square inch	14.696
Atmospheres	Inches of mercury	29.92
Atmospheres	Feet of water	34
Barrels, oil	Gallons, of oil	42
Barrels, cement	Pounds of cement	376
Bags or sacks, cement	Pounds of cement	94
Btu/min.	Foot-pounds/sec.	12.96
Btu/min.	Horse-power	0.02356
Btu/min.	Kilowatts	0.01757
Btu/min.	Watts	17.57
Centimeters	Inches	0.3937
Centimeters of mercury	Atmospheres	0.01316
Centimeters of mercury	Feet of water	0.4461
Cubic inches	Cubic feet	0.00058
Cubic feet	Cubic inches	1728
cubic feet	Cubic yards	0.03703
Cubic yards	Cubic feet	27
Cubic inches	Gallons	0.00433
Cubic feet	Gallons	7.48
Feet	Inches	12
Feet	Yards	0.3333
Yards	Feet	3
Feet of water	Atmospheres	0.02950
Feet of water	Inches of mercury	0.8826
Gallons	Cubic Inches	231
Gallons	Cubic feet	0.1337
Gallons	Pounds of water	8.33
Gallons	Quarts	4
Gallons per min.	Cubic feet sec.	0.002228
Gallons per min.	Cubic feet hour	8.0208
Gallons water per min.	Tons water/24 hours	6.0086
Horse-power	Foot-lbs./sec.	550
Inches	Centimeters	2.540
Inches	Feet	0.0833
Inches	Millimeters	25.4
Inches of water	Pounds per Sq. inch	0.0361
Inches of water	Inches of mercury	0.0735
Inches of water	Ounces per square inch	0.578
Inches of water	Ounces per square foot	5.2
Inches of mercury	Inches of water	13.6
Inches of mercury	Feet of water	1.1333
Inches of mercury	Pounds per square inch	0.4914
Kilometers	Miles	0.6214
Meters	Inches	39.37
Miles	Feet	5280
Millimeters	Centimeters	0.1
Millimeters	Inches	0.03937
Ounces (fluid)	Cubic inches	1.805
Ounces	Pounds	0.0625
Pounds	Ounces	16
Pounds per square inch	Inches of water	27.72
Pounds per square inch	Feet of water	2.310
Pounds per square inch	Inches of mercury	2.04
Pounds per square inch	Atmospheres	0.0681
Quarts	Cubic Inches	67.20
Square Inches	Square feet	0.00694
Square Feet	Square inches	144
Square Feet	Square yards	0.11111
Square yards	Square feet	9
Square miles	Acres	640
Short tons	Pounds	2000
Short tons	Long tons	0.89285
Tons of water/24 hours	Gallons per minute	0.16643
Yards	Feet	3
Yards	Centimeters	91.44
Yards	Inches	36

CONVERSION CALCULATIONS

Commercial Measure

16 drams	=1 ounce
16 ounces	=1 pound
2,000 pounds	=1 ton

Long Measure

12 inches	=1 foot
3 feet	=1 yard
16½ feet	=1 rod
40 rods	=1 furlong
8 furlongs (5,280 ft) =	=1 mile
3 miles	=1 league

Square Measure

144 square inches	=1 square foot
9 square feet	=1 square yard
30¼ square yards	=1 square rod
160 square rods	=1 acre
4840 square yards	=1 acre
640 acres	=1 square mile
36 square miles	=1 township

Surveyors Measure

7.92 inches	=1 link
25 links	=1 rod
4 rods (66 ft.)	=1 chain
10 chains	=1 furlong
8 furlongs	=1 mile
1 square mile	=1 section

Cubic Measure

1728 cubic inches	=1 cubic foot
27 cubic feet	=1 cubic yard
128 cubic feet	=1 cord (wood/stone)
231 cubic inches	=1 U.S. gallon
7.48 U.S. Gallons	=1 cubic foot
2150.4 cubic inches	=1 U.S. bushel

Liquid Measure

4 fluid ounces	=1 gill
4 gills	=1 pint
2 pints	=1 quart
4 quarts	=1 gallon
9 gallons	=1 firkin
31½ gallons	=1 barrel
2 barrels	=1 hogshead

Dry Measure

2 pints	=1 quart
8 quarts	=1 peck
4 pecks	=1 bushel
2150.42 cubic inches	=1 bushel

SQUARE

$A = a^2$

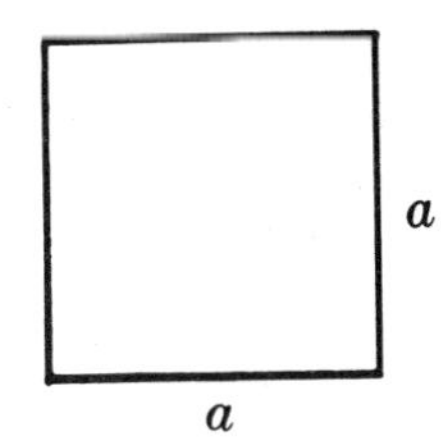

RECTANGLE

$A = bh$

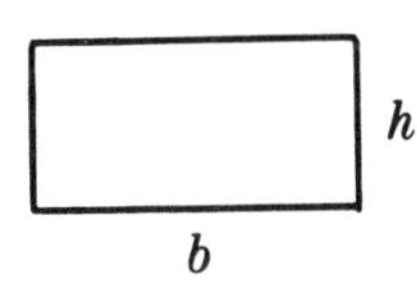

TRIANGLE

$A = \frac{1}{2} bh$

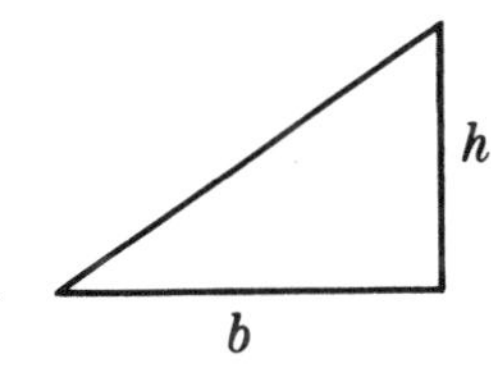

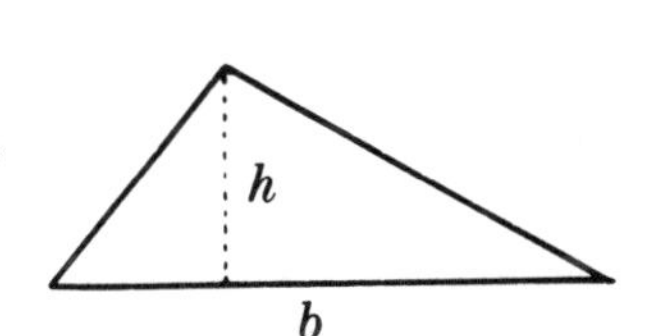

PARALLELOGRAM

$A = bh = ab \ Sin\phi$

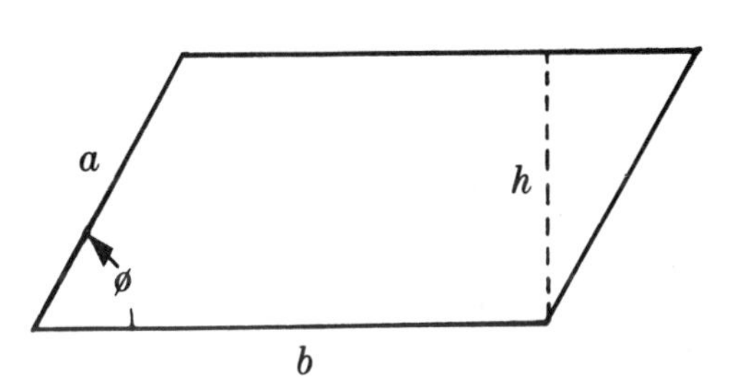

TRAPEZOID

$A = \left(\frac{a+b}{2}\right)h$

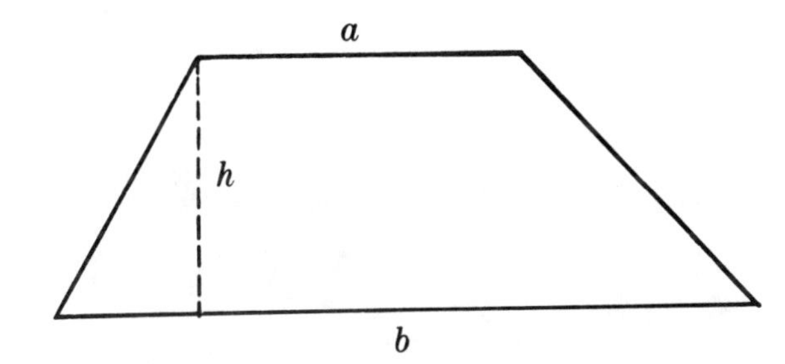

CIRCLE

$A = \pi r^2 = \frac{\pi d^2}{4}$

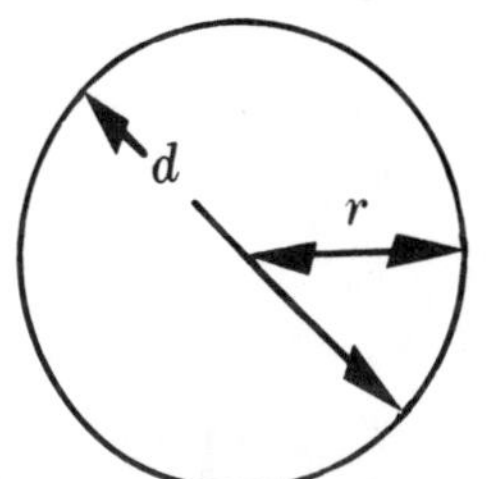

$Circumference = C = 2\pi r = \pi d$

ELLIPSE

$A = 0.7854\ ab$

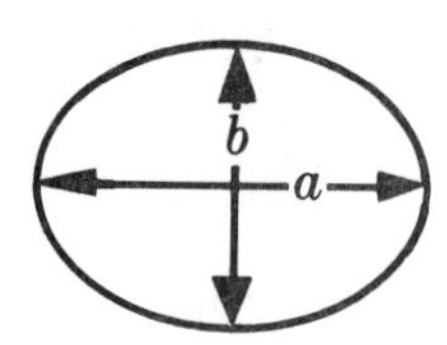

PARABOLA

$A = \frac{2}{3} bh$

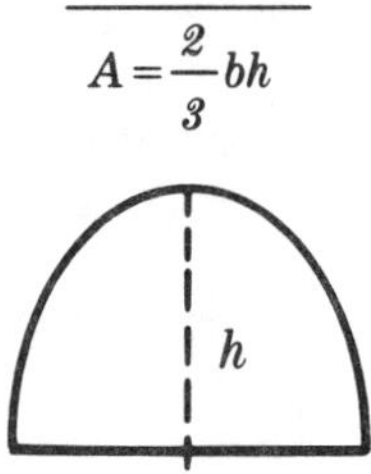

VOLUMES

CUBE

$V = a^3$

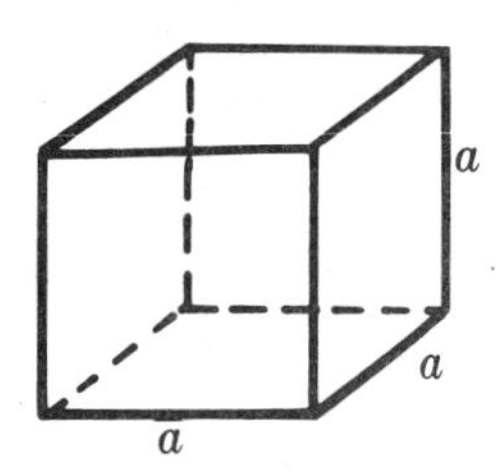

CYLINDER

$V = \pi r^3 h$

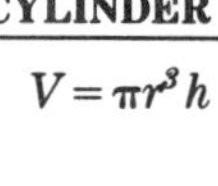

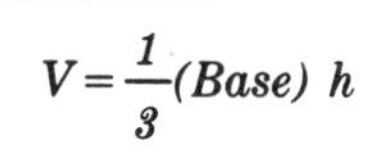

PYRAMID

$V = \frac{1}{3} (Base)\ h$

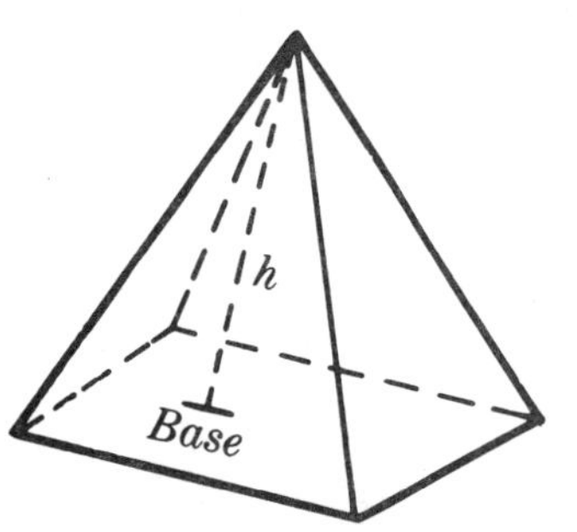

CONE

$V = \frac{1}{3} \pi r^2 h$

SPHERE

$V = \frac{4}{3} \pi r^3 = \frac{1}{6} \pi d^3$

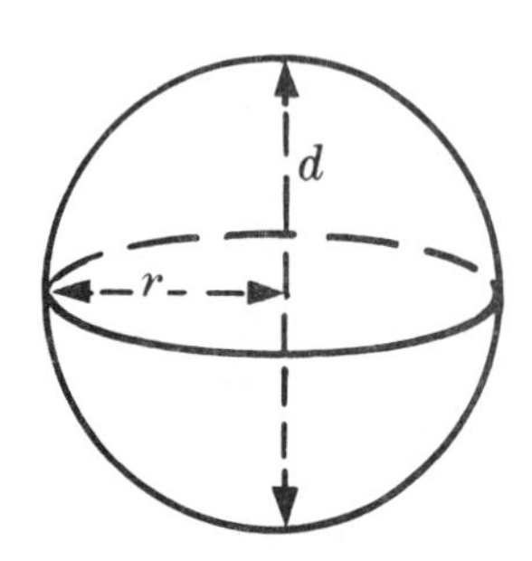

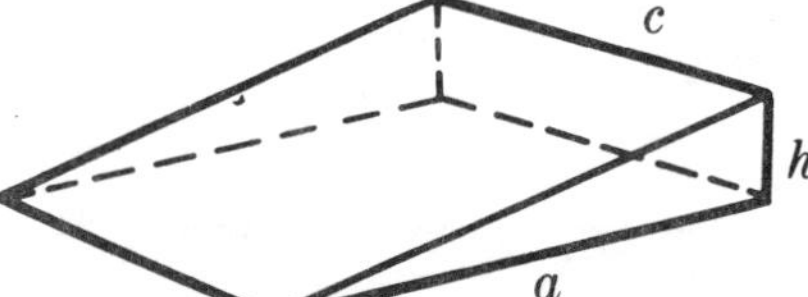

WEDGE

$V = \frac{1}{2} abc$

CONVERSION FACTORS

ENGLISH TO SI (SYSTEM INTERNATIONAL)

To Convert from	To	Multiply by
LENGTH		
Inches	Millimetres	25.4[a]
Feet	Metres	0.3048[a]
Yards	Metres	0.9144[a]
Miles (statute)	Kilometres	1.609
AREA		
Square inches	Square millimetres	645.2
Square feet	Square metres	0.0929
Square yards	Square metres	0.8361
VOLUME		
Cubic inches	Cubic millimetres	16.387
Cubic feet	Cubic metres	0.02832
Cubic yards	Cubic metres	0.7646
Gallons (U.S. liquid)[b]	Cubic metres[c]	0.003785
Gallons (Canadian liquid)[b]	Cubic metres[c]	0.004546
Ounces (U.S. liquid)[b]	Millilitres[c, d]	29.57
Quarts (U.S. liquid)[b]	Litres[c, d]	0.9464
Gallons (U.S. liquid)[b]	Litres[c]	3.785
FORCE		
Kilograms force	Newtons	9.807
Pounds force	Newtons	4.448
Pounds force	Kilograms force[d]	0.4536
Kips	Newtons	4448
Kips	Kilograms force[d]	453.6
PRESSURE, STRESS, STRENGTH (FORCE PER UNIT AREA)		
Kilograms force per sq. centimetre	Megapascals	0.09807
Pounds force per square inch (psi)	Megapascals	6895
Kips per square inch	Megapascals	6.895
Pounds force per square inch (psi)	Kilograms force per square centimetre[d]	0.07031
Pounds force per square foot	Pascals	47.88
Pounds force per square foot	Kilograms force per square metre[d]	4.882
BENDING MOMENT OR TORQUE		
Inch-pounds force	Metre-kilog. force[d]	0.01152
Inch-pounds force	Newton-metres	0.1130
Foot-pounds force	Metre-kilog. force[d]	0.1383
Foot-pounds force	Newton-metres	1.356
Metre-kilograms force	Newton-metres	9.807
MASS		
Ounce (avoirdupois)	Grams	28.35
Pounds (avoirdupois)	Kilograms	0.4536
Tons (metric)	Kilograms	1000[a]
Tons, short (2000 pounds)	Kilograms	907.2
Tons, short (2000 pounds)	Megagrams[e]	0.9072
MASS PER UNIT VOLUME		
Pounds mass per cubic foot	Kilog. per cubic metre	16.02
Pounds mass per cubic yard	Kilog. per cubic metre	0.5933
Pds. mass per gallon (U.S. liquid)[b]	Kilog. per cubic metre	119.8
Pds. mass p/gal. (Canadian liquid)[b]	Kilog. per cubic metre	99.78
TEMPERATURE		
Degrees Fahrenheit	Degrees Celsius	tK = (1F − 32)/1.8
Degrees Fahrenheit	Degrees Kelvin	tK = (1F + 459.67)/1.8
Degree Celsius	Degree Kelvin	tK = 1C + 273.15

[a] The factor given is exact.
[b] One U.S. gallon equals 0.8327 Canadian gallon.
[c] 1 litre = 1000 millilitres = 10,000 cubic centimetres = 1 cubic decimetre = 0.001 cubic metre.
[d] Metric but not SI unit.
[e] Called "tonne" in England. Called "metric ton" in other metric systems.

TRENCH BRACING

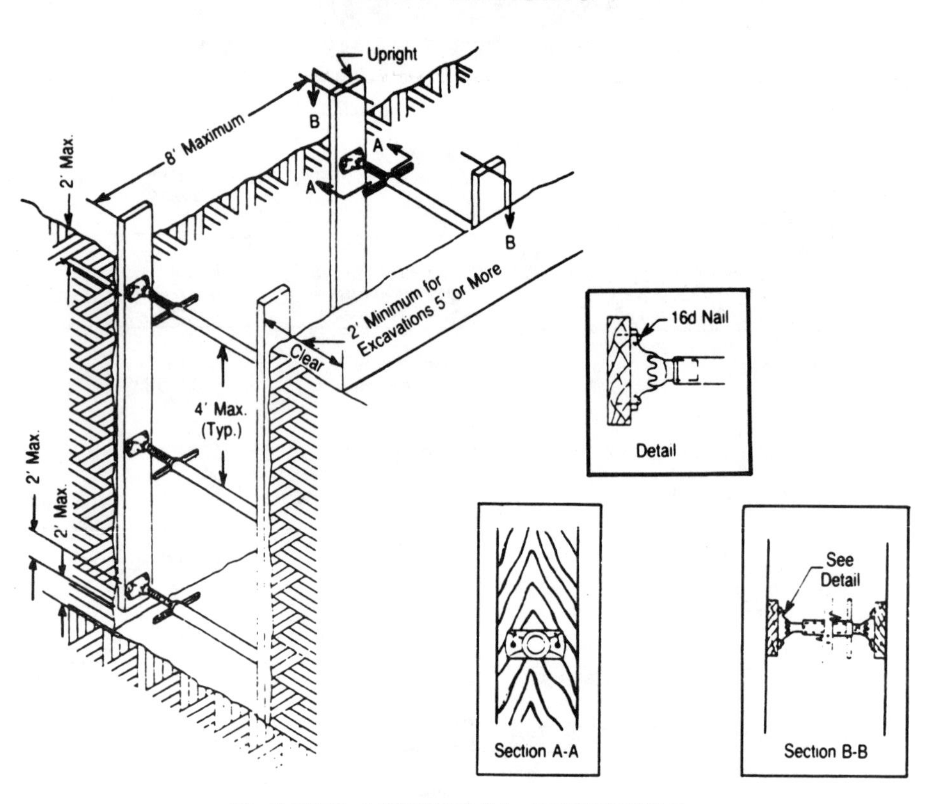

CLOSED VERTICAL SHEETING

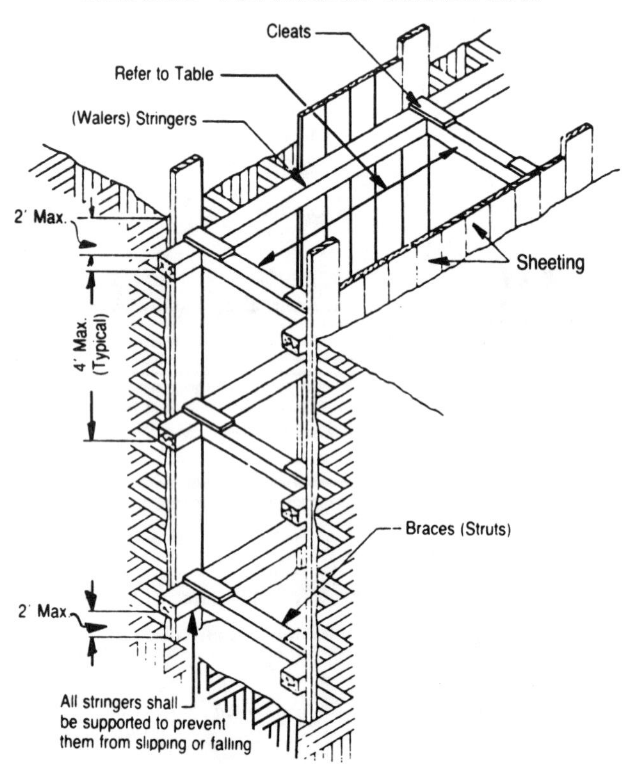

Soil Identification. For most purposes, soils can usually be identified visually and by texture, as described in the chart that follows. For design purposes, however, soils must be formally identified and their performance characteristics determined in a laboratory by skilled soil mechanics.

Classification	Identifying Characteristics
Gravel	Rounded or water-worn pebbles or bulk rock grains. No cohesion or plasticity. Gritty, granular and crunchy underfoot.
Sand	Granular, gritty, loose grains, passing a No. 4 sieve and between .002 and .079 inches in diameter. Individual grains readily seen and felt. No plasticity or cohesion. When dry, it cannot be molded but will crumble when touched. The coarse grains are rounded; the fine grains are visible and angular.
Silt	Fine, barely visible grains passing a No. 200 sieve and between .0002 and .002 inches in diameter. Little or no plasticity and no cohesion. A dried cast is easily crushed. Is permeable and movement of water through the voids occurs easily and is visible. Feels gritty when bitten and will not form a thread.
Clay	Invisible particles under .0002 inches in diameter. Cohesive and highly plastic when moist. Will form a long, thin, flexible thread when rolled between the hands. Does not feel gritty when bitten. Will form hard lumps or clods when dry which resist crushing. Impermeable, with no apparent movement of water through voids.
Muck and Organic Silt	Thoroughly decomposed organic material often mixed with other soils of mineral origin. Usually black with fibrous remains. Odorous when dried and burnt. Found as deposits in swamps, peat bogs and muskeg flats.
Peat	Partly decayed plant material. Mostly organic. Highly fibrous with visible plant remains. Spongy and easily identified.

Classification by Particle Size. Soils can be classified in general terms by the nature of their predominant particle size or the grading of the particle sizes. These particle sizes are usually grouped into gravel, coarse sand, medium sand, fine sand, silt, clay and colloids.

The major divisions of soils are:

Coarse-Grained (Granular)		Fine-Grained		Organic	
Gravel	Sand	Silt	Clay	Muck	Peat

Soils comprised primarily of sand particles are referred to as "granular soils," while fine-grained soils are commonly called "heavy soils." It is accepted practice in the field to refer to a particular soil as a coarse sand, or silt, or by any of the particle size groupings which describe the soil generally from a visual examination.

Classification of Soil Mixtures.

Class	% Sand	% Silt	% Clay
Sand	80-100	0- 20	0- 20
Sandy clay loam	50- 80	0- 30	20- 30
Sandy loam	50- 80	0- 50	0- 20
Loam	30- 50	30- 50	0- 20
Silty loam	0- 50	50- 80	0- 20
Silt	0- 20	80-100	0- 20
Silty clay loam	0- 30	50- 80	20- 30
Silty clay	0- 20	50- 70	30- 50
Clay loam	20- 50	20- 50	20- 30
Sandy clay	50- 70	0- 20	30- 50
Clay	0- 50	0- 50	30-100

Material	Approx. In-Bank Weight (lbs. per cu. yd.)	Percent Swell
Clay, dry	2300	40
Clay, wet	3000	40
Granite, decomposed	4500	65
Gravel, dry	3250	10-15
Gravel, wet	3600	10-15
Loam, dry	2800	15-35
Loam, wet	3370	25
Rock, well blasted	4200	65
Sand, dry	3250	10-15
Sand, wet	3600	10-15
Shale and soft rock	3000	65
Slate	4700	65

PILES AND PILE DRIVING

General. A pile is a column driven or jetted into the ground which derives its supporting capabilities from end-bearing on the underlying strata, skin friction between the pile surface and the soil, or from a combination of end-bearing and skin friction.

Piles can be divided into two major classes: **Sheet piles** and **load-bearing piles.** Sheet piling is used primarily to restrain lateral forces as in trench sheeting and bulkheads, or to resist the flow of water as in cofferdams. It is prefabricated and is available in steel, wood or concrete. Load-bearing piles are used primarily to transmit loads through soil formations of low bearing values to formations that are capable of supporting the designed loads. If the load is supported predominantly by the action of soil friction on the surface of the pile, it is called a **friction pile.** If the load is transmitted to the soil primarily through the lower tip, it is called an **end-bearing pile.**

There are several load-bearing pile types, which can be classified according to the material from which they are fabricated:

Timber (Treated and untreated)
Concrete (Precast and cast in place)
Steel (H-Section and steel pipe)
Composite (A combination of two or more materials)

Some of the additional uses of piling are to: eliminate or control settlement of structures, support bridge piers and abutments and protect them from scour, anchor structures against uplift or overturning, and for numerous marine structures such as docks, wharves, fenders, anchorages, piers, trestles and jetties.

Timber Piles. Timber piles, treated or untreated, are the piles most commonly used thoroughout the world, primarily because they are readily available, economical, easily handled, can be easily cut off to any desired length after driving and can be easily removed if necessary. On the other hand, they have some serious disadvantages which include: difficulty in securing straight piles of long length, problems in driving them into hard formations and difficulty in splicing to increase their length. They are generally not suitable for use as end-bearing piles under heavy load and they are subject to decay and insect attack. Timber piles are resilient and particularly adaptable for use in waterfront structures such as wharves, docks and piers for anchorages since they will bend or give under load or impact where other materials may break. The ease with which they can be worked and their economy makes them popular for trestle construction and for temporary structures such as falsework or centering. Where timber piles can be driven and cut off below the permanent ground-water level, they will last indefinitely; but above this level in the soil, a timber pile will rot or will be attacked by insects and eventually destroyed. In sea water, marine borers and fungus will act to deteriorate timber piles. Treatment of timber piles increases their life but does not protect them indefinitely.

Concrete Piles. Concrete piles are of two general types, precast and cast-in-place. The advantages in the use of concrete piles are that they can be fabricated to meet the most exacting conditions of design, can be cast in any desired shape or length, possess high strength and have excellent resistance to chemical and biological attack. Certain disadvantages are encountered in the use of precast piles, such as:

(a) Their heavy weight and bulk (which introduces problems in handling and driving).

(b) Problems with hair cracks which often develop in the concrete as a result of shrinkage after curing (which may expose the steel reinforcement to deterioration).

(c) Difficulty encountered in cut-off or splicing.

(d) Susceptibility to damage or breakage in handling and driving.

(e) They are more expensive to fabricate, transport and drive.

Precast piles are fabricated in casting yards. Centrifugally spun piles (or piles with square or octagonal cross-sections) are cast in horizontal forms, while round piles are usually cast in vertical forms. With the exception of relatively short lengths, precast piles must be reinforced to provide the designed column strengths and to resist damage or breakage while being transported or driven.

Precast piles can be tapered or have parallel sides. The reinforcement can be of deformed bars or be prestressed or poststressed with high strength steel tendons. Prestressing or prestressing eliminates the problem of open shrinkage cracks in the concrete. Otherwise, the pile must be protected by coating it with a bituminous or plastic material to prevent ultimate deterioration of the reinforcement. Proper curing of the precast concrete in piles is essential.

Cast-in-place pile types are numerous and vary according to the manufacturer of the shell or inventor of the method. In general, they can be classified into two groups: shell-less types and the shell types. The shell-less type is constructed by driving a steel shell into the ground and filling it with concrete as the shell is pulled from the ground. The shell type is constructed by driving a steel shell into the ground and filling it in place with concrete. Some of the advantages of cast-in-place concrete piles are: lightweight shells are handled and driven easily, lengths of the shell may be increased or decreased easily, shells may be transported in short lengths and quickly assembled, the problem of breakage is eliminated and a driven shell may be inspected for shell damage or an uncased hole for "pinching off." Among the disadvantages are problems encountered in the proper centering of the reinforcement cages, in placing and consolidating the concrete without displacement of the reinforcement steel or segregation of the concrete, and shell damage or "pinching-off" of uncased holes.

Shell type piles are fabricated of heavy gage metal or are fluted, corrugated or spirally reinforced with heavy wire to make them strong enough to be driven without a mandrel. Other thin-shell types are driven with a collapsible steel mandrel or core inside the casing. In addition to making the driving of a long thin shell possible, the mandrel prevents or minimizes damage to the shell from tearing, buckling, collapsing or from hard objects encountered in driving.

Some shell type piles are fabricated of heavy gauge metal with enlargement at the lower end to increase the end bearing.

These enlargements are formed by withdrawing the casing two to three feet after placing concrete in the lower end of the shell. This wet concrete is then struck by a blow of the pile hammer on a core in the casing and the enlargement is formed. As the shell is withdrawn, the core is used to consolidate the concrete after each batch is placed in the shell. The procedure results in completely filling the hole left by the withdrawal of the shell.

Steel Piles. A steel pile is any pile fabricated entirely of steel. They are usually formed of rolled steel H sections, but heavy steel pipe or box piles (fabricated from sections of steel sheet piles welded together) are also used. The advantages of steel piles are that they are readily available, have a thin uniform section and high strength, will take hard driving, will develop high load-bearing values, are easily cut off or extended, are easily adapted to the structure they are to support, and breakage is eliminated. Some disadvantages are: they will rust and deteriorate unless protected from the elements; acid, soils or water will result in corrosion of the pile; and greater lengths may be required than for other types of piles to achieve the same bearing value unless bearing on rock strata. Pipe pile can either be driven open-end or closed-end and can be unfilled, sand filled or concrete filled. After open-end pipe piles are driven, the material from inside can be removed by an earth auger, air or water jets, or other means, inspected, and then filled with concrete. Concrete filled pipe piles are subject to corrosion on the outside surface only.

Composite Piles. Any pile that is fabricated of two or more materials is called a composite pile. There are three general classes of composite piles: wood with concrete, steel with concrete, and wood with steel. Composite piles are usually used for a special purpose or for reasons of economy.

Where a permanent ground-water table exists and a composite pile is to be used, it will generally be of concrete and wood. The wood portion is driven to below the water table level and the concrete upper portion eliminates problems of decay and insect infestation above the water table. Composite piles of steel and concrete are used where high bearing loads are desired or where driving in hard or rocky soils is expected. Composite wood and steel piles are relatively uncommon. It is important that the pile design provides for a permanent joint between the two materials used, so constructed that the parts do not separate or shift out of axial alignment during driving operations.

Sheet Piles. Sheet piles are made from the same basic materials as other piling: wood, steel and concrete. They are ordinarily designed so as to interlock along the edges of adjacent piles.

Sheet piles are used where support of a vertical wall of earth is required, such as trench walls, bulkheads, waterfront structures or cofferdams. Wood sheet piling is generally used in temporary installations, but is seldom used where water-tightness is required or hard driving expected. Concrete sheet piling has the capability of resisting much larger lateral loads than wood sheet piling, but considerable difficulty is experienced in securing water-tight joints. The type referred to as "fish-mouth" type is designed to permit jetting out the joint and filling with grout, but a seal is not always effected unless the adjacent piles are wedged tightly together. Concrete sheet piling has the advantage that it is the most permanent of all types of sheet piling.

Steel sheet piling is manufactured with a tension-type interlock along its edges. Several different shapes are available to permit versatility in its use. It has the advantages that it can take hard driving, has reasonably water-tight joints and can be easily cut, patched, lengthened or reinforced. It can also be easily extracted and reused. Its principal disadvantage is its vulnerability to corrosion.

Types of Pile Driving Hammers. A pile-driving hammer is used to drive load-bearing or sheet piles. The commonly used types are: drop, single-acting, double-acting, differential acting and diesel hammers. The most recent development is a type of hammer that utilizes high-frequency sound and a dead load as the principal sources of driving energy.

Drop Hammers. These hammers employ the principle of lifting a heavy weight by a cable and releasing it to fall on top of the pile. This type of hammer is rapidly disappearing from use, primarily because other types of pile driving hammers are more efficient. Its disadvantages are that it has a slow rate of driving (four to eight blows per minute), that there is some risk of damaging the pile from excessive impact, that damage may occur in adjacent structures from heavy vibration and that it cannot be used directly for driving piles under water. Drop hammers have the advantages of simplicity of operation, ability to vary the energy by changing the height of fall and they represent a small investment in equipment.

Single-Acting Hammers. These hammers can be operated either on steam or compressed air. The driving energy is provided by a free-falling weight (called a ram) which is raised after each stroke by the action of steam or air on a piston. They are manufactured as either open or closed types. Single-acting hammers are best suited for jobs where dense or elastic soil materials must be penetrated or where long heavy timber or precast concrete piles must be driven. The closed type can be used for underwater pile driving. Its advantages include: faster driving (50 blows or more per minute), reduction in skin friction as a result of more frequent blows, lower velocity of the ram which transmits a greater proportion of its energy to the pile and minimizes piles damage during driving, and it has underwater driving capability. Some of its disadvantages are: requires higher investment in equipment (i.e. steam boiler, air compressor, etc.), higher maintenance costs, greater set-up and moving time required, and a larger operating crew.

Double-Acting Hammers. These hammers are similar to the single-acting hammers except that steam or compressed air is used both to lift the ram and to impart energy to the falling ram. While the action is approximately twice as fast as the single-acting hammer (100 blows per minute or more), the ram is much lighter and operates at a greater velocity, thereby making it particularly useful in high production driving of light or medium-weight piles of moderate lengths in granular soils. The hammer is nearly always fully encased by a steel housing which also permits direct driving of piles under water.

Some of its advantages are: faster driving rate, less static skin friction develops between blows, has underwater driving capability and piles can be driven more easily without leads.

Among its disadvantages are: it is less suitable for driving heavy piles in high-friction soils and the more complicated mechanism results in higher maintenance costs.

Differential-Acting Hammers. This type of hammer is, in effect, a modified double-acting hammer with the actuating mechanism having two different diameters. A large-diameter piston operates in an upper cylinder to accelerate the ram on the downstroke and a small-diameter piston operates in a lower cylinder to raise the ram. Thc additional energy added to the falling ram is the difference in areas of the two pistons multiplied by the unit pressure of the steam or air used. This hammer is a short-stroke, fast-acting hammer with a cycle rate approximately that of the double-acting hammer. Its advantages are that it has the speed and characteristics of the double-acting hammer with a ram weight comparable to the single-acting type, and it uses from 25 to 35 percent less steam or air. It is also more suitable for driving heavy piles under more difficult driving conditions than is the double-acting hammer. It is available in the open or closed-type cases, the latter permitting direct underwater pile driving. Its principal disadvantage is higher maintenance costs.

Diesel Hammers. This hammer is a self-contained driving unit which does not require an auxilliary steam boiler or air compressor. It consists essentially of a ram operating as a piston in a cylinder. When the ram is lifted and allowed to fall in the cylinder, diesel fuel is injected in the compression space between the ram and an anvil placed on top of the pile. The continued downstroke of the ram compresses the air and fuel to ignition heat and the resultant explosion drives the pile downward and the ram upward to start another cycle. This hammer is capable of driving at a rate of from 80 to 100 blows per minute. Its advantages are that it has a low equipment investment cost, is easily moved, requires a small crew, has a high driving rate, does not require a steam boiler or air compressor and can be used with or without leads for most work. Its disadvantages are that it is not self-starting (the ram must be mechanically lifted to start the action) and it does not deliver a uniform blow. The latter disadvantge arises from the fact that as the reaction of the pile to driving increases, the reaction to the ram increases correspondingly. That is, when the pile encounters considerable resistance, the rebound of the ram is higher and the energy is increased automatically. The operator is required to observe the driving operations closely to identify changing driving conditions and compensate for such changes with his controls to avoid damaging the pile.

Diesel hammers can be used on all types of piles and they are best suited to jobs where mobility or frequent relocation of the pile driving equipment is necessary.

PILE CHART

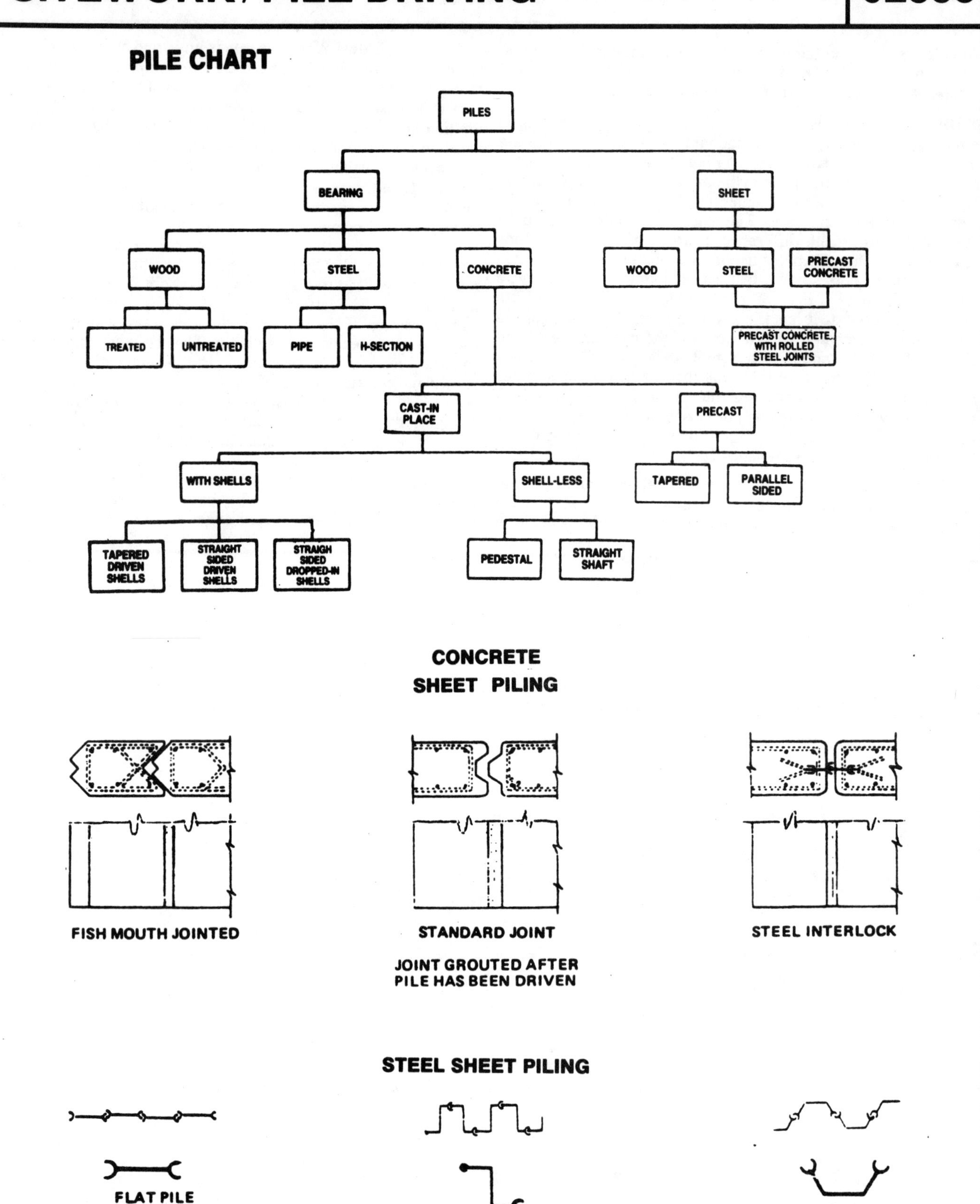
PILES
BEARING
SHEET
WOOD
STEEL
CONCRETE
WOOD
STEEL
PRECAST CONCRETE
TREATED
UNTREATED
PIPE
H-SECTION
PRECAST CONCRETE WITH ROLLED STEEL JOINTS
CAST-IN PLACE
PRECAST
WITH SHELLS
SHELL-LESS
TAPERED
PARALLEL SIDED
TAPERED DRIVEN SHELLS
STRAIGHT SIDED DRIVEN SHELLS
STRAIGH SIDED DROPPED-IN SHELLS
PEDESTAL
STRAIGHT SHAFT
CONCRETE SHEET PILING
FISH MOUTH JOINTED
STANDARD JOINT
JOINT GROUTED AFTER PILE HAS BEEN DRIVEN
STEEL INTERLOCK
STEEL SHEET PILING
FLAT PILE
ZEE PILE
DEEP WEB PILE

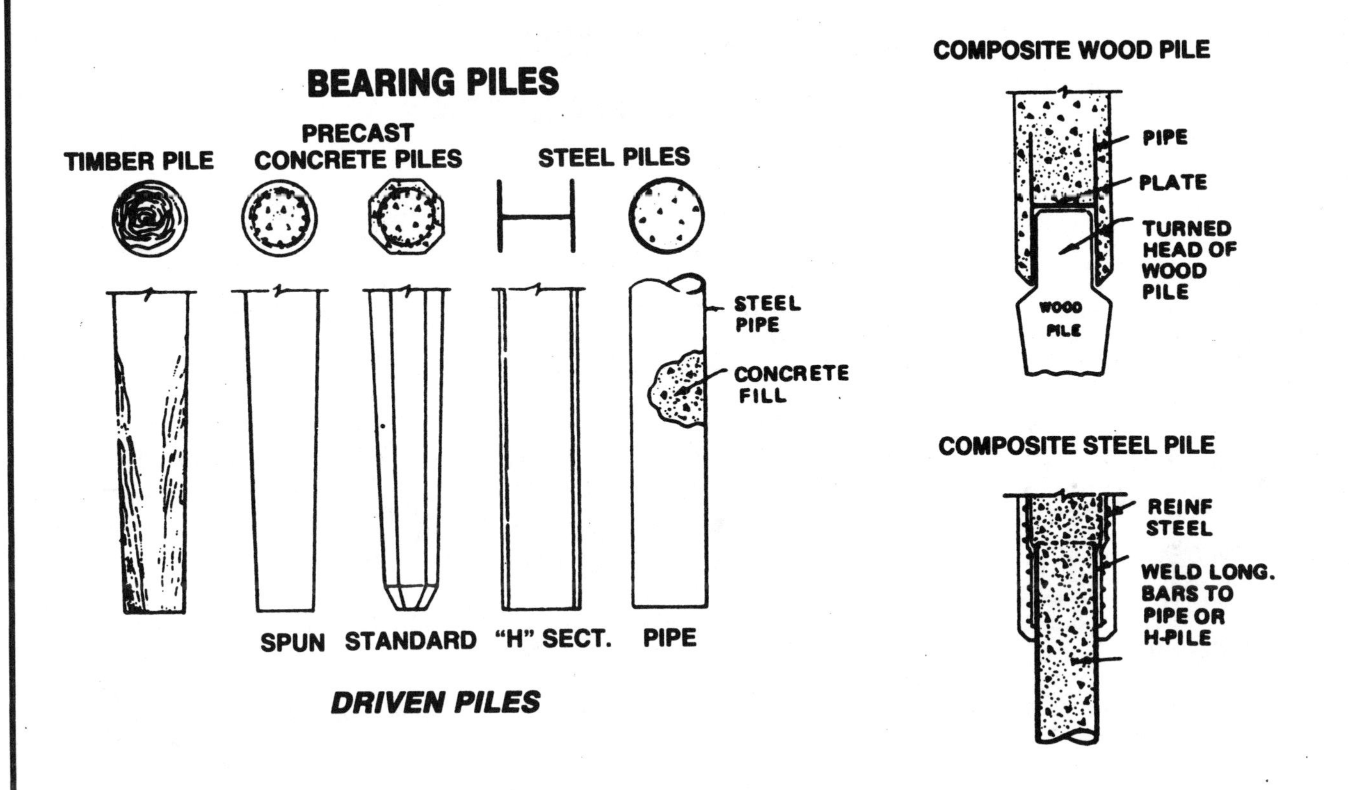

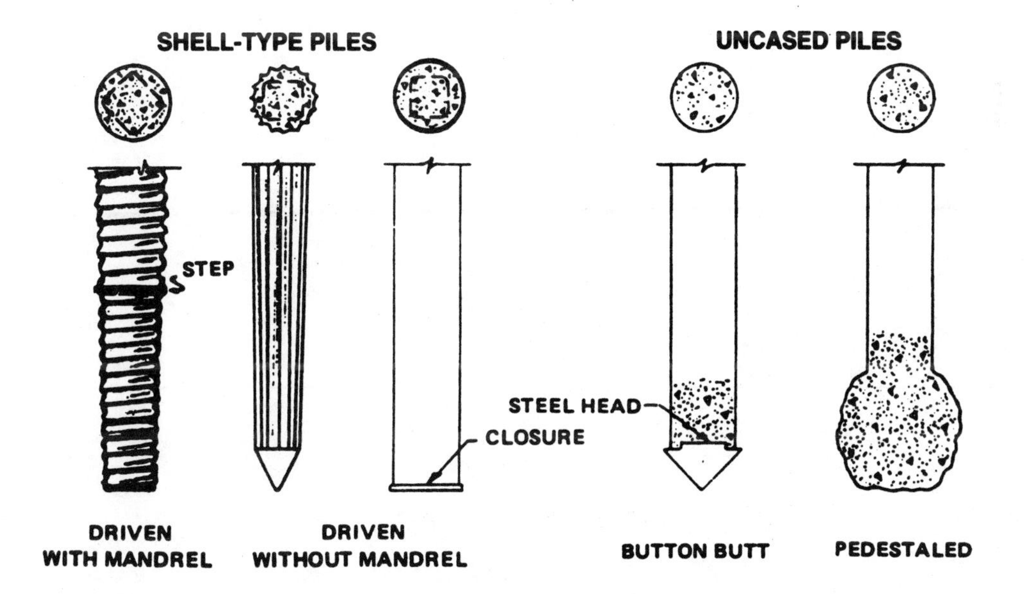

CAST-IN-PLACE PILES

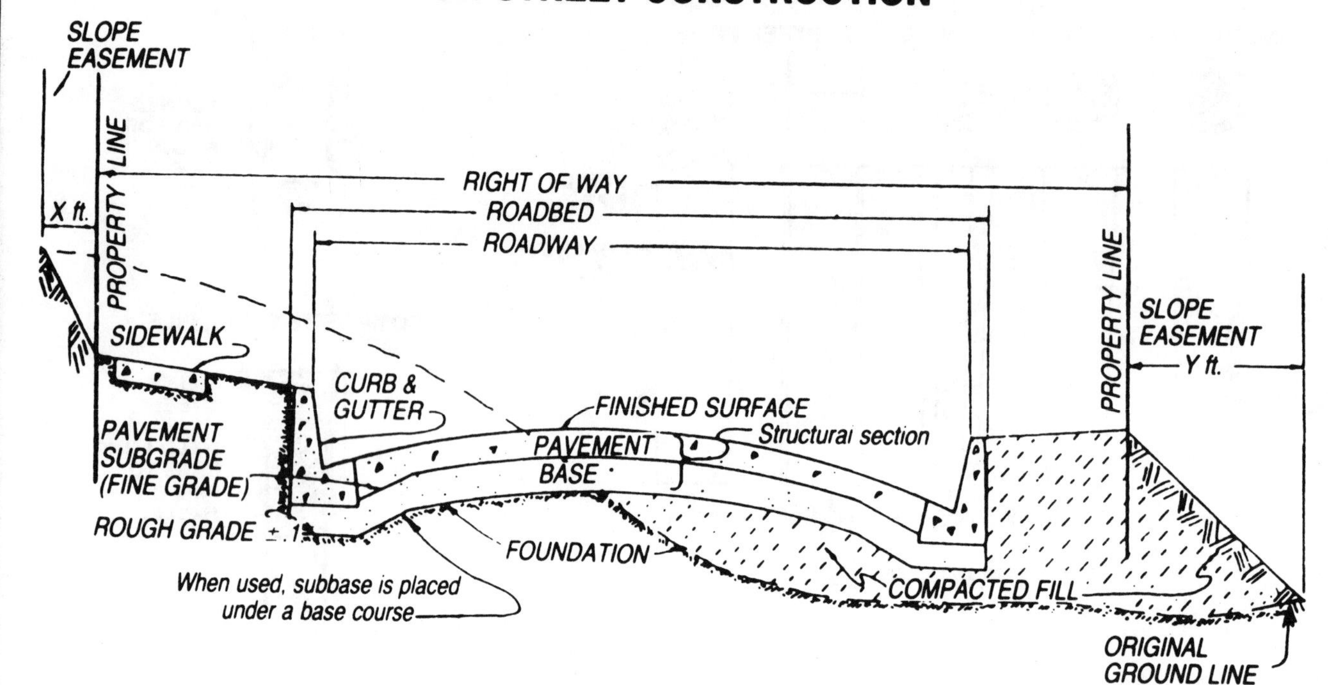
STANDARD NOMENCLATURE
FOR STREET CONSTRUCTION
SLOPE
EASEMENT
X ft.
PROPERTY LINE
RIGHT OF WAY
ROADBED
ROADWAY
PROPERTY LINE
SLOPE
EASEMENT
Y ft.
SIDEWALK
CURB &
GUTTER
FINISHED SURFACE
Structural section
PAVEMENT
BASE
PAVEMENT
SUBGRADE
(FINE GRADE)
ROUGH GRADE ±.1
FOUNDATION
When used, subbase is placed
under a base course
COMPACTED FILL
ORIGINAL
GROUND LINE

CONCRETE MASONRY PAVING UNITS

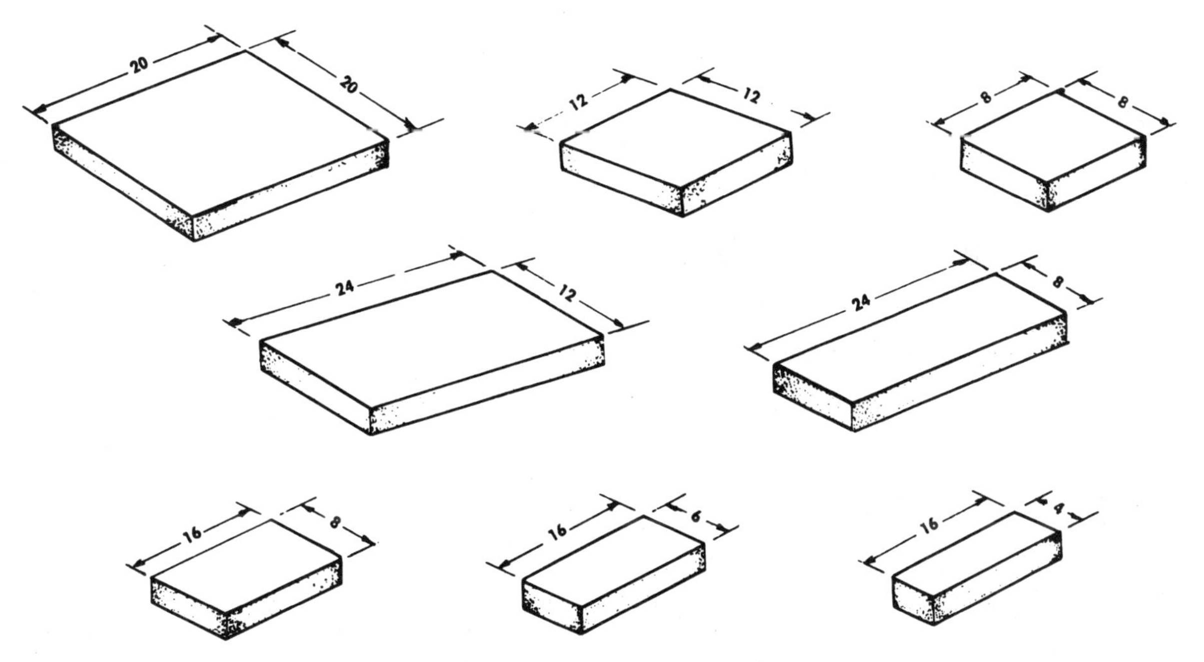

NOTE: Sizes are nominal and will vary by manufacturer.

HEXAGON PAVER UNITS
Various Sizes Available

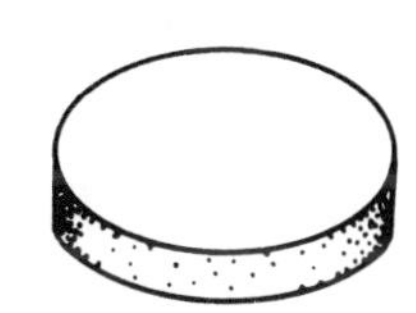

ROUND PAVING UNITS
Various Sizes Available

VEHICULAR PAVING UNITS

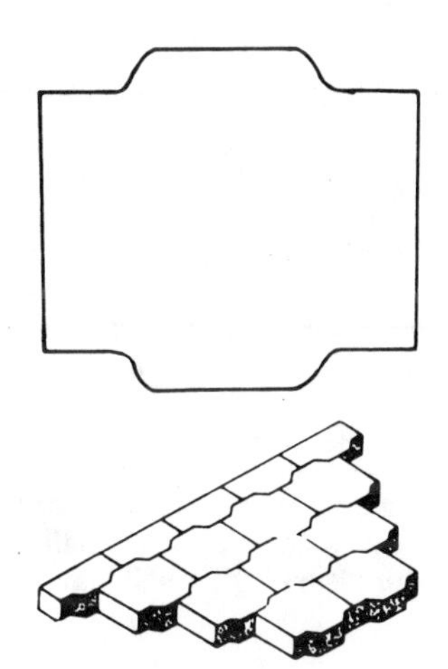

INTERLOCKING PAVER
7¼" x 3" x 8½"

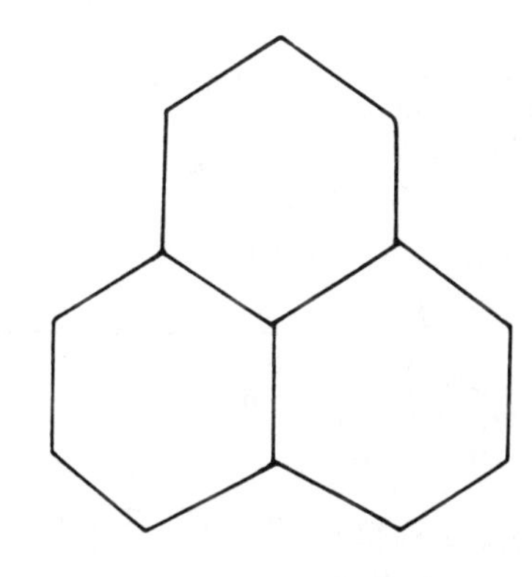

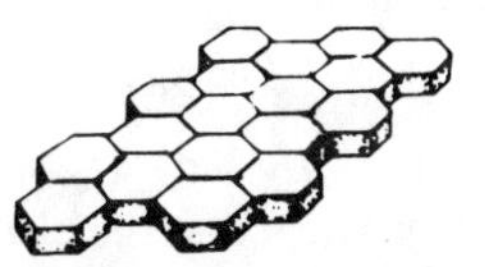

INTERLOCKING PAVER
12" x 3⅝" x 12"

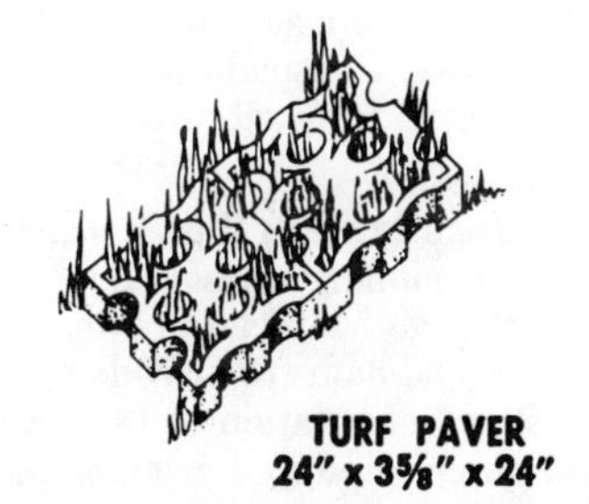

TURF PAVER
24" x 3⅝" x 24"

PIPE

Clay Pipe. Clay pipe is manufactured by blending various clays together, milling, mixing, extruding and firing in a kiln to obtain vitrification. The physical properties of the pipe can be changed by varying the proportions of the several clays used. The pipe is supplied in two basic styles: spigot and socket; and plain end.

Spigot and Socket Pipe has a spigot on one end and a socket on the other, and is commonly referred to as "bell and spigot" pipe. The plans generally specify the type of joint to be used from the several types of jointing methods available. This type of pipe is manufactured with matching polyurethane gaskets molded on the spigot and socket which form a tight seal when the pipe is jointed.

Plain End Pipe is without a socket on either end and is joined with special couplings. The coupling consists of a circular rubber sleeve, two stainless steel compression bands with tightening devices and a corrosion resistant shear ring. Sometimes this joint is supplied with a cardboard form, open at the top, which is filled with portland cement mortar to resist shear and prevent future corrosion of the bands.

Concrete Pipe. Unreinforced and reinforced concrete pipe is manufactured by casting in stationary or revolving metal molds. At the present time, the design practice is to specify reinforced concrete pipe for all purposes.

Unreinforced Concrete Pipe is cast in vertical steel molds, usually in pipe sizes of 21 inches or less, and is of the spigot and socket type. No steel reinforcement is used and the pipe is usually intended for use in irrigation systems and under light loading conditions.

Reinforced Concrete Pipe (RCP) is made in a number of different manufacturing processes and for a wide variety of pressure and non-pressure classes. It is available in standard sizes or it can be made to order to any diameter desired. Some of the larger diameters include diameters of 12 and 14 feet. A large variety of joint details are used with RCP. Tongue and groove joints are used for storm drain pipelines.

Reinforced concrete pipe for wastewater pipeline projects is supplied with gasketed joints and a polyvinyl chloride (PVC) plastic liner cast into the pipe.

(a) **Cast Pipe** is cast vertically in steel forms with the reinforcing cage securely held in place. The reinforcement is generally elliptical in shape to provide the maximum structural strength to resist the loads imposed on the pipe by the backfill and other stresses. Consolidation of the concrete is obtained by the use of external form vibrators.

(b) **Centrifugally Spun Pipe** is manufactured by introducing concrete into a spinning horizontal steel cylinder into which the reinforcement cage has been previously installed and which is equipped with end dams to provide the proper pipe wall thickness. The speed of rotation of the mold is increased and the centrifugal force produces a smooth, dense concrete pipe.

(c) **Pressure Pipe** may be cast or centrifugally spun pipe but it usually has a circular steel reinforcement cage (or cages) designed not only to resist the trench loading, but also the internal pressures exerted on the pipe from the fluid under pressure in the line.

Concrete Cylinder Pipe. This class of pipe is generally used for high pressure water lines and sewer force mains and is available in sizes ranging from 10 inches to 60 inches and larger in special cases.

A sheet steel cylinder is wrapped with the designed steel reinforcement and a concrete lining is centrifugally spun in the interior of the steel cylinder. An exterior coating of concrete is applied generally by the gunite process, while the cylinder is slowly rotated. These coatings vary in thickness from ½ to ¾ of an inch. The joints are commonly of the steel ring and rubber gasket type, but are generally designed for the special purpose for which the pipe line is intended.

Definition of Terms. In general, the terms used to designate types of reinforced concrete pipe refer to the process used in manufacture.

Cast RCP (Cast Reinforced Concrete Pipe). A concrete pipe having one or more cylindrical or elliptical cages of reinforcement steel embedded in it, the concrete for which is cast with the forms in a vertical position.

CSRCP (Centrifugally Spun Reinforced Concrete Pipe). A concrete pipe having one or more cylindrical or elliptical cages of reinforcement steel embedded in it, and cast in a horizontal position while the forms are spinning rapidly. This type of pipe may be designated as Spun RCP or as CCP (Centrifugal Concrete Pipe).

RCP (Reinforced Concrete Pipe). A reinforced concrete pipe manufactured by either the casting or spinning method.

Steel Reinforcement. Steel for reinforcing concrete pipe is generally furnished in large coils which will permit the use of machines to fabricate the "cages." The continuous steel rod is wound spirally at a prescribed pitch on a drum of the proper diameter. Where the rod crosses a longitudinal spacer rod, it is electrically welded to it so that the complete cage is relatively rigid.

Reinforcement cages for pipe designed for external loading are generally elliptical in shape to take full advantage of the steel in tension. Pipe to be used with relatively small external loads or pipe designed for pressure lines will have circular cages.

Reinforcement cages must be rigidly fixed in the forms so that the placement of concrete or the effects of centrifugal spinning will not result in distortion or displacement of the steel. The orientation of an elliptical cage must be marked on the forms to assure that the minor axis can be located after the concrete is placed.

CAPACITIES FOR SEPTIC TANKS SERVING AN INDIVIDUAL DWELLING

No. of bedrooms	Capacity of tank (gals.)
2 or less	750
3	900
4	1,000

CONCRETE / FORMWORK | 03110

FORM NOMENCLAURE

1. SHEATHING
2. STUDS
3. WALES
4. FORM BOLTS
5. NUT WASHER
6. TOP PLATE
7. BOTTOM PLATE
8. KEY-WAY
9. SPREADER
10. STRONGBACK
11. BRACE
12. STRUT
13. CLEATS
14. SCAB
14. POUR STRIP

FALSEWORK NOMENCLATURE

1. SHEATHING
2. JOIST
3. STRINGER
4. CAP
5. CORBEL
6. POST
7. SILL
8. FOOTING
9. SWAY BRACE
10. LONGITUDINAL BRACE
11. SCAB
12. BLOCKING
13. BRIDGING

TYPICAL PAN-JOIST FORM CONSTRUCTION

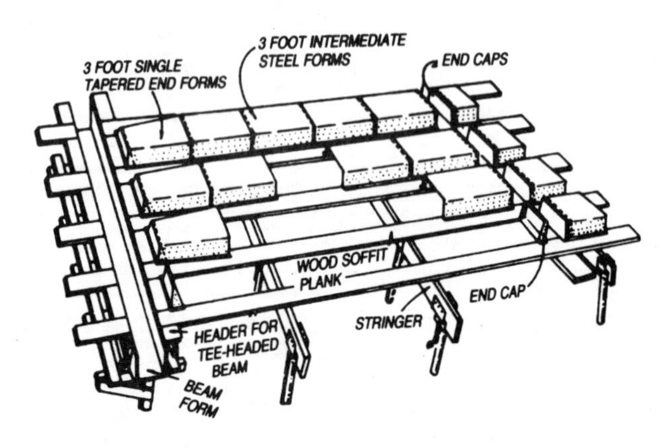

TYPICAL WAFFLE SLAB FORM CONSTRUCTION

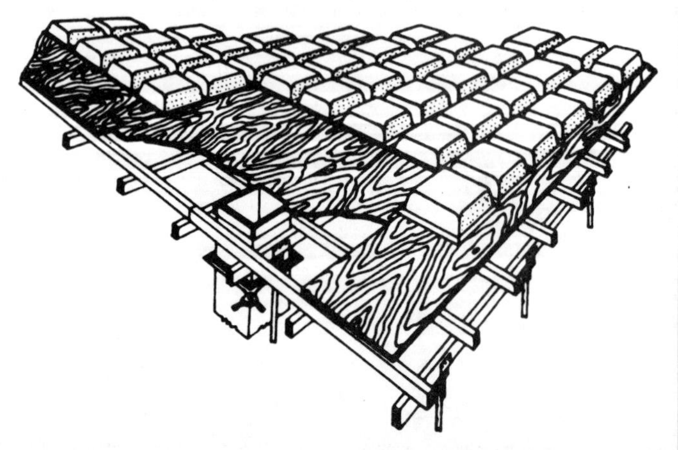

COMMON TYPES OF STEEL REINFORCEMENT BARS

ASTM specifications for billet steel reinforcing bars (A 615) require identification marks to be rolled into the surface of one side of the bar to denote the producer's mill designation, bar size and type of steel. For Grade 60 and Grade 75 bars, grade marks indicating yield strength must be show. Grade 40 bars show only three marks (no grade mark) in the following order:

1st — Producing Mill (usually an initial)
2nd — Bar Size Number (#3 through #18)
3rd — Type (N for New Billet)

NUMBER SYSTEM — GRADE MARKS

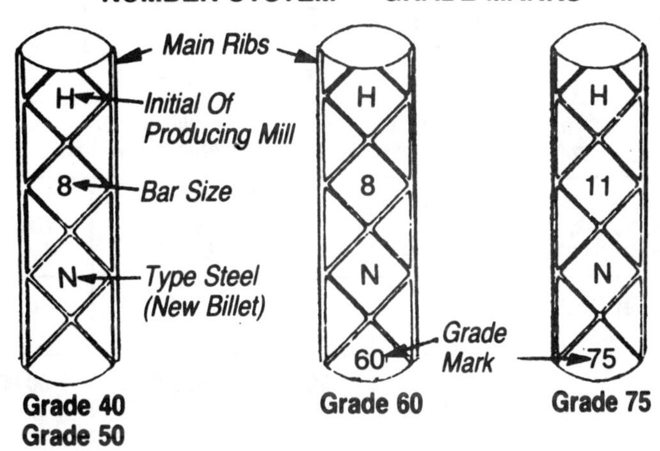

LINE SYSTEM — GRADE MARKS

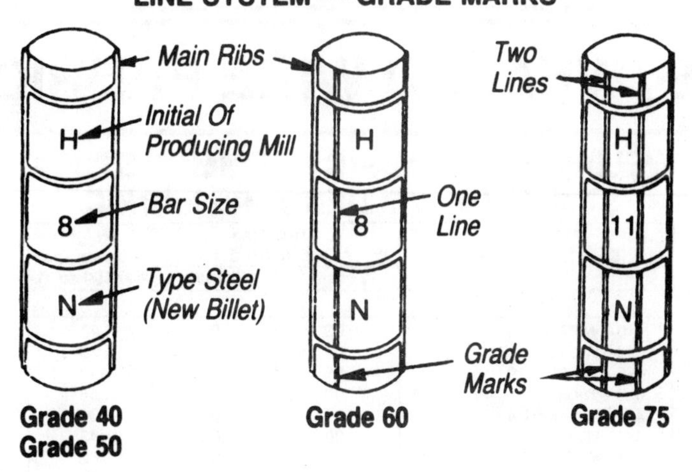

CONCRETE / REINFORCING STEEL 03210

STANDARD SIZES OF STEEL REINFORCEMENT BARS

STANDARD REINFORCEMENT BARS				
		Nominal Dimensions		
Bar Designation Number*	Nominal Weight, lb. per ft.	Diameter, in.	Cross Sectional Area, sq. in.	Perimeter, in.
3	0.376	0.375	0.11	1.178
4	0.668	0.500	0.20	1.571
5	1.043	0.625	0.31	1.963
6	1.502	0.750	0.44	2.356
7	2.044	0.875	0.60	2.749
8	2.670	1.000	0.79	3.142
9	3.400	1.128	1.00	3.544
10	4.303	1.270	1.27	3.990
11	5.313	1.410	1.56	4.430
14	7.65	1.693	2.25	5.32
18	13.60	2.257	4.00	7.09

*The bar numbers are based on the number of ⅛ inches included in the nominal diameter of the bar.

Type of Steel and ASTM Specification No.	Size Nos. Inclusive	Grade	Tensile Strength Min., psi	Yield (a) Min., psi
Billet Steel A 615	3-11	40	70,000	40,000
	3-11 14, 18	60	90,000	60,000
	11, 14, 18	75	100,000	75,000

CONCRETE / WELDED WIRE FABRIC 03220

WELDED WIRE FABRIC – COMMON STOCK STYLES OF WELDED WIRE FABRIC

Style Designation	Steel Area sq. in. per ft.		Weight Approx. lbs. per 100 sq. ft.
	Longit.	Transv.	
Rolls			
6x6—W1.4xW1.4	.03	.03	21
6x6—W2xW2	.04	.04	29
6x6—W2.9xW2.9	.06	.06	42
6x6—W4xW4	.08	.08	58
4x4—W1.4xW1.4	.04	.04	31
4x4—W2xW2	.06	.06	43
4x4—W2.9xW2.9	.09	.09	62
4x4—W4xW4	.12	.12	86
Sheets			
6x6—W2.9xW2.9	.06	.06	42
6x6—W4xW4	.08	.08	58
6x6—W5.5xW5.5	.11	.11	80
4x4—W4xW4	.12	.12	86

Insofar as is possible, the moisture content should be kept uniform to avoid problems in determining the proper amount of water to be added for mixing. Mixing water must be reduced to compensate for moisture in the aggregate in order to control the slump of the concrete and avoid exceeding the specified water-cement ratio.

Handling Concrete by Pumping Methods. Transportation and placement of concrete by pumping is another method gaining increased popularity. Pumps have several advantages, the primary one being that a pump will high-lift concrete without the need for an expensive crane and bucket. Since the concrete is delivered through pipe and hoses, concrete can be conveyed to remote locations in buildings, in tunnels, to locations otherwise inaccessible on steep hillside slopes for anchor walls, pipe bedding or encasement, or for placing concrete for chain link fence post bases. Concrete pumps have been found to be economical and expedient in the placement of concrete, and this has promoted the use and acceptance of this development. The essence of proper concrete pumping is the placement of the concrete in its final location without segregation.

Modern concrete pumps, depending on the mix design and size of line, can pump to a height of 200 feet or a horizontal distance of 1,000 feet. They can handle, economically, structural mixes, standard mixes, low slump mixes, mixes with two-inch maximum size aggregate and light weight concrete. When a special pump mix is required for structural concrete in a major structure, the mix design must be approved by the Engineer and checked and confirmed by the Supervisor of the Materials Control Group. The Inspector should obtain the pump manufacturer's printed information and evaluate its characteristics and ability to handle the concrete mixture specified for the project.

If concrete is being placed for a major reinforced structure, it is important that the placement continue without interruption. The Inspector should be sure that the contractor has ready access to a back-up pump to be used in the event of a breakdown. In order to further insure the success of the concrete placement by the pumping method, the user should be aware of the following points:

(a) A protective grating over the receiving hopper of the pump is necessary to exclude large pieces of aggregate or foreign material.

(b) The pump and lines require lubrication with a grout of cement and water. All of the excess grout is to be wasted prior to pumping the concrete.

(c) All changes in direction must be made by a large radius bend with a maximum bend of 90 degrees. Wye connections induce segregation and shall not be used.

(d) Pump lines should be made of a material capable of resisting abrasion and with a smooth interior surface having a low coefficient of friction. Steel is commonly used for pump lines, ance. Aluminum pipe should not be used for pumping concrete and some of the new plastic or rubber tubing is gaining acceptbecause a chemical reaction occurs between the concrete and the aluminum. Hydrogen is generated which results in a swelling of the concrete, causing a significant reduction in compressive strength. This reaction is aggravated by any of the following: abrasive coarse aggregate, non-uniformly graded sand, low-slump concrete, low sand-aggregate ratio, high-alkali cement or when no air-entraining agent is used.

(e) During temporary interruptions in pumping, the hopper must remain nearly full, with an occasional turning and pumping to avoid developing a hard slug of concrete in the lines.

(f) Excessive line pressures must be avoided. When this occurs, check these points as the probable cause: segregation caused by too low a slump or too high a slump; large particle contamination caused by large pieces of aggregate or frozen lumps not eliminated by the grating; poor gradation of aggregates or particle shape; rich or lean spots caused by improper mixing.

(g) Corrections must be made to correct excessive slump loss as measured at the transit-mixed concrete truck and as measured at the hose outlet. This may be attributable to porous aggregate, high temperature or rapid setting mixes.

(h) Two transit-mix concrete trucks must be used simultaneously to deliver concrete into the pump hopper. These trucks must be discharged alternately to assure a continuous flow of concrete as trucks are replaced.

(i) Samples of concrete for test specimens prepared to determine the acceptance of the concrete quality are to be taken as required for conventional concrete.

Sampling is done before the concrete is deposited in the pump hopper. However, it is suggested that, where possible, the effect of pumping on the compressive strength be checked by taking companion samples, so identified, from the end of the pump line at the same time. The Record of Test must be properly noted as being a special mix used for pumping purposes. This will enable the Materials Control Group to compile a complete history of mix designs and their respective compressive strengths.

The prudent use of pumped concrete can result in economy and improved quality. However, only the control exercised by the operator will assure continued high standards of quality concrete.

Pump lines must be properly fastened to supports to eliminate excessive vibration. Couplings must be easily and securely fastened in a manner that will prevent mortar leakage. It is preferable to use the flexible hose only at the discharge point. This hose must be moved in such a manner as to avoid kinks or sharp bends. The pump line should be protected from excessive heat during hot weather by water sprinkling or shade.

COMPRESSIVE STRENGTH FOR VARIOUS WATER-CEMENT RATIOS

(The strengths listed are based on the use of normal portland cement)

WATER/CEMENT RATIO		PROBABLE 28-DAY STRENGTH	
WEIGHT	GALS./100#	PSI	MEGAPASCALS*
.40	4.8	5000	34
.45	5.4	4500	31
.50	6.0	4000	28
.55	6.6	3500	24
.60	7.2	3000	21
.65	7.8	2500	17
.70	8.4	2000	14

* International system equivalent.

APPROXIMATE CONTENT OF SAND, CEMENT AND WATER PER CUBIC YARD OF CONCRETE

Based on aggregates of average grading and physical characteristics in concrete mixes having a water-cement ratio (W/C) of about .65 by weight (or 7.8 gallons) per sack of cement; 3-in. slump; and a medium natural sand having a fineness modulus of about 2.75.

COURSE AGGREGATE MAX. SIZE	WATER		CEMENT	% SAND
	POUNDS	GALLONS		
3/8	385	46	590	57
1/2	365	44	560	50
3/4	340	41	525	43
1	325	39	500	39
1 1/2	300	36	460	37

It can be noted from the above chart that, for a given slump, the amount of mixing water increases as the size of the course aggregate decreases. The size of the course aggregate controls the sand content in the same way; that is, the amount of sand required in the mix increases as the size of the course aggregate decreases.

Other typical examples are contained in the pamphlet published by the Portland Cement Association entitled "Design and Control of Concrete Mixtures."

Effects of Temperature on Concrete. Concrete mixtures gain strength rapidly in the first few days after placement. While the rate of gain in strength diminishes, concrete continues to become stronger with time over a period of many years, so long as drying of the concrete is prevented. Its strength at 28 days is considered to be the compressive strength upon which the Engineer bases his calculations. The temperature of the atmosphere has a significant effect upon the development of strength in concrete. Lower temperatures retard and higher temperatures accelerate the gain in strength.

Most destructive of the natural forces is freezing and thawing action. While the concrete is still wet or moist, expansion of the water as it is converted into ice results in severe damage to the fresh concrete. In situations where freezing may be encountered, high early strength cement may be used. Also, the mixing water or the aggregate (or both) may be preheated before mixing. Covering the concrete, and using steam or salamanders to heat the concrete under the covering, will help prevent freezing. Air-entraining agents help to diminish the effects of freezing of fresh concrete as well as in subsequent freezing and thawing cycles throughout the life of the concrete.

Hot weather will present problems of a different nature in placing concrete. Concrete will set up faster and tend to shrink and crack at the surface. To minimize this problem, the concrete should be placed without delay after mixing. Avoid the use of accelerators (perhaps even use a retarding agent), dampen all subgrade and forms, protect the freshly placed concrete from hot dry winds, and provide for adequate curing. Crushed ice or chilled water can be used as part of the mixing water to reduce the temperature of the mix in extremely hot areas.

Admixture	Purpose	Effects on Concrete	Advantages	Disadvantages
Accelerator	Hasten setting.	Improves cement dispersion and increases early strength.	Permits earlier finishing, form removal, and use of the structure.	Increases shrinkage, decreases sulfate resistance, tends to clog mixing and handling equipment.
Air-Entraining Agent	Increase workability and reduce mixing water.	Reduces segregation, bleeding and increases freeze-thaw resistance. Increases strength	Increases workability and reduces finishing time.	Excess will reduce strength and increase slump. Bulks concrete volume.
Bonding Agent	Increase bond to old concrete.	Produces a non-dusting, slip-resistant finish.	Permits a thin topping without roughening old concrete, self-curing, ready in one day.	Quick setting and susceptible to damage from fats, oils and solvents.
Densifier	To obtain dense concrete.	Increased workability and strength.	Increases workability and increases waterproofing characteristics, more impermeable.	Care must be used to reduce mixing water in proportion to amount used.
Foaming Agent	Reduce weight.	Increases insulating properties.	Produces a more plastic mix, reduces dead weight loads.	Its use must be very carefully regulated — following instructions explicitly.
Retarder	Retard setting.	Increases control of setting.	Provides more time to work and finish concrete.	Performance varies with cement used — adds to slump. Requires stronger forms.
Water Reducer and Retarder	Increase compressive and flexural strength.	Reduces segregation, bleeding, absorption, shrinkage, and increases cement dispersion.	Easier to place work, provides better control.	Performance varies with cement. Of no use in cold weather.
Water Reducer, Retarder and Air-Entraining Agent	Increases workability.	Improves cohesiveness. Reduces bleeding and segregation.	Easier to place and work.	Care must be taken to avoid excessive air entrainment.

ARCHITECTURAL WALL PATTERNS (BONDS)

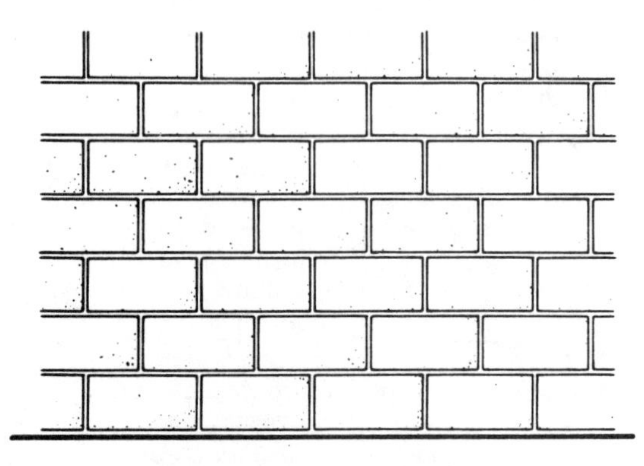

COMMON BOND
8"x 16" UNITS

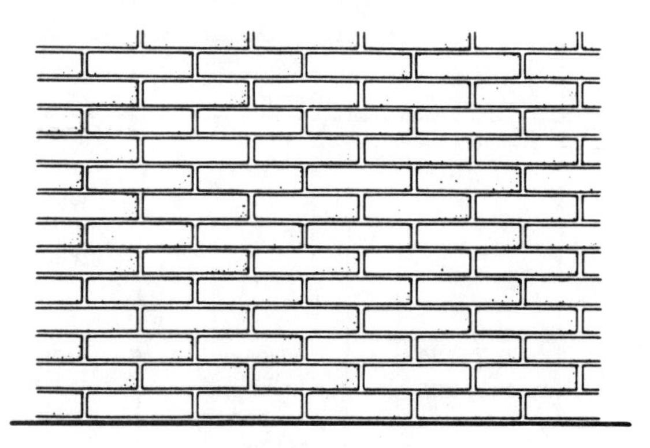

COMMON BOND
4"x 16" UNITS

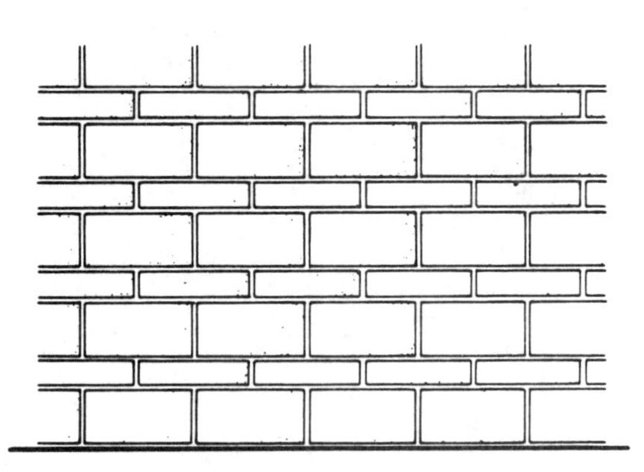

COURSED ASHLAR
8"x 16" & 4"x16" UNITS

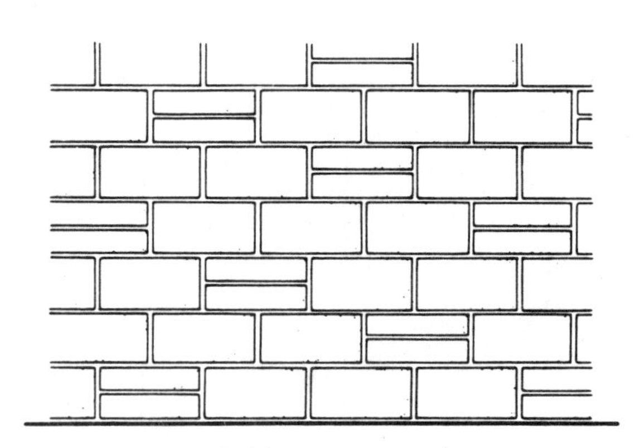

COURSED ASHLAR
8"x 16" & 4"x 16" UNITS

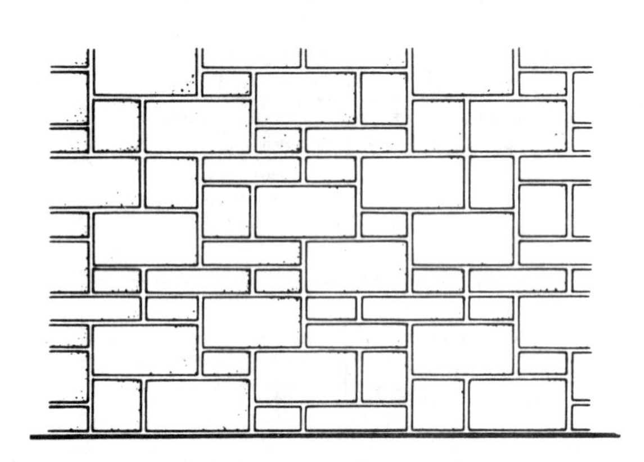

RANDOM ASHLAR
8"x 16", 8"x8", 4"x 16" AND 4"x 8" UNITS

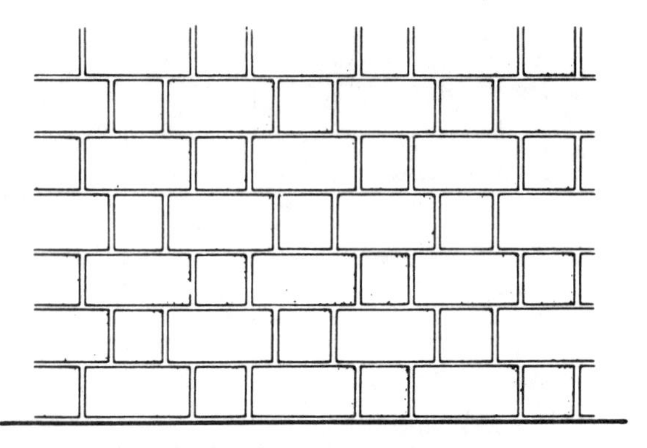

COURSED ASHLAR
8"x 16" & 8"x8" UNITS

ARCHITECTURAL WALL PATTERNS (BONDS) — (Continued)

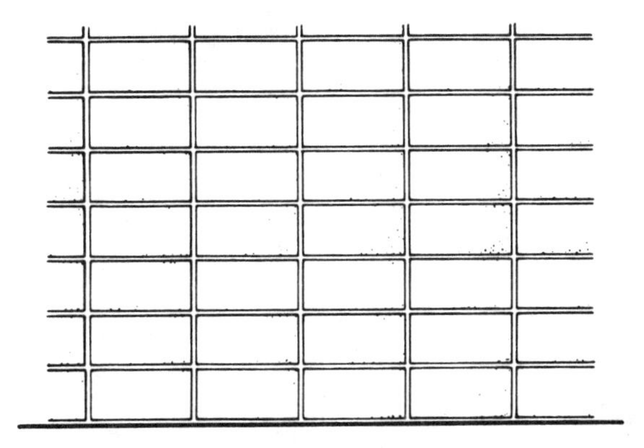

STACKED BOND
8" x 16" UNITS

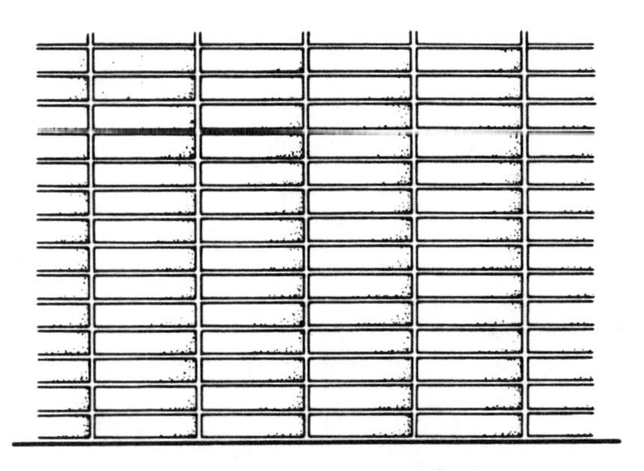

STACKED BOND
4" x 16" UNITS

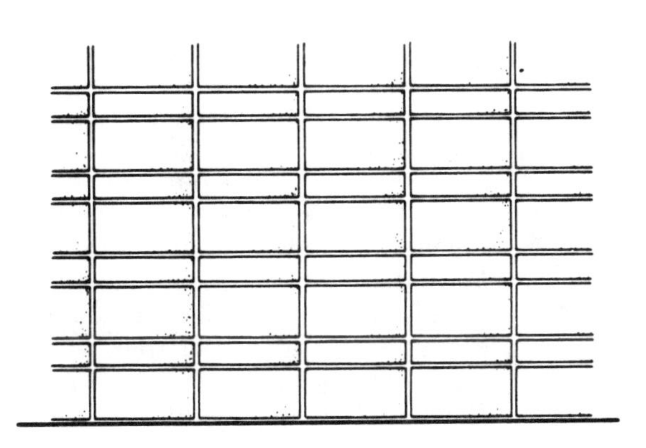

STACKED BOND
8" x 16" & 4" x 16" UNITS

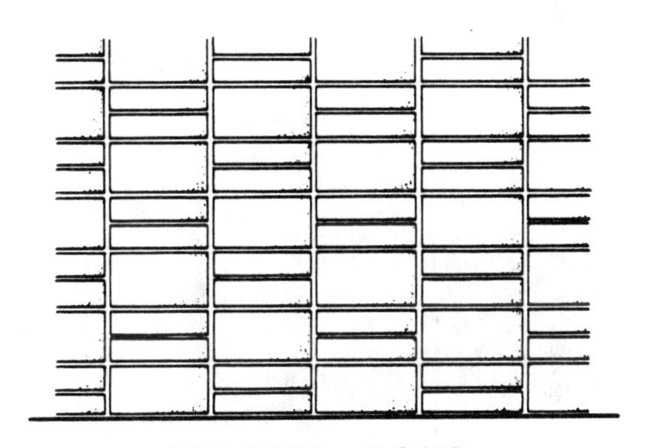

STACKED BOND
8" x 16" & 4" x 16" UNITS

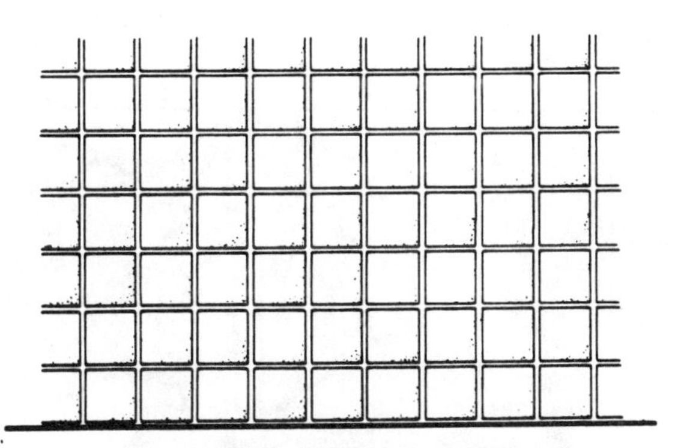

STACKED BOND VERTICAL SCORED UNITS

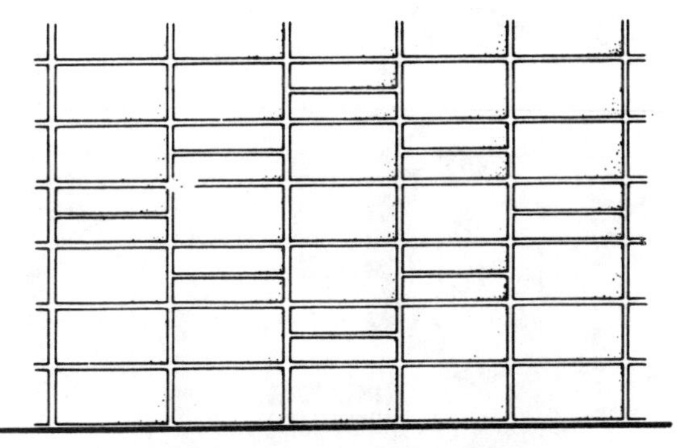

USE OF BLOCK DESIGN IN STACKED BOND

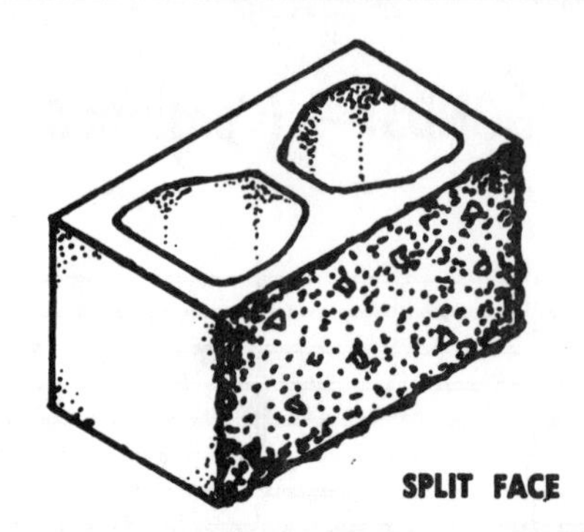

SPLIT FACE

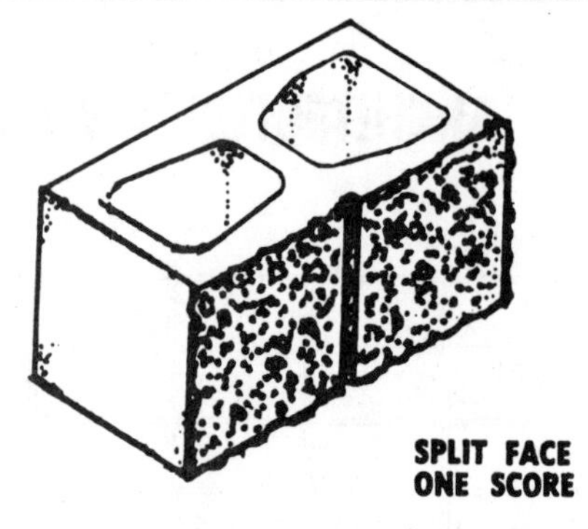

SPLIT FACE
ONE SCORE

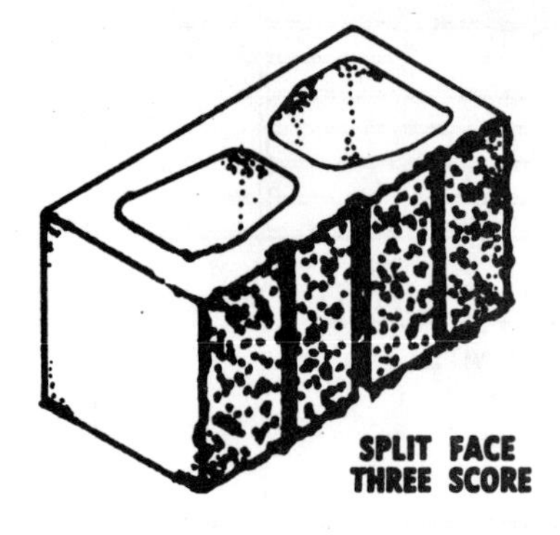

SPLIT FACE
THREE SCORE

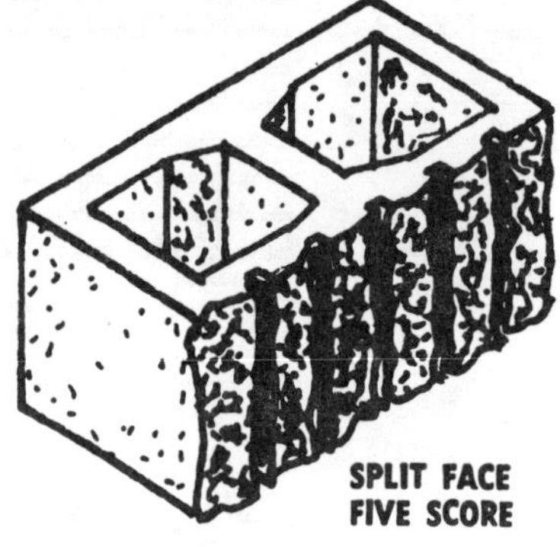

SPLIT FACE
FIVE SCORE

SPLIT FACE BLOCK

Split face block is manufactured as a unit that is normally made double and is literally split apart on a splitter; a machine which resembles a guillotine. The splitter has blades at the top and bottom (and sometimes at the sides) which exert pressure on the blocks, breaking them apart.

Many factors determine the look of the split face, both as to size variances and the amount of aggregate exposure. Split face block is intended to have a rougher texture than precision block. Various configurations of block such as fluted, scored, etc., will split in a different manner than a full split face. The vertical perpendicularity of scored and fluted split face block is subject to variation.

SPLIT FACE
THREE-WIDE SCORE

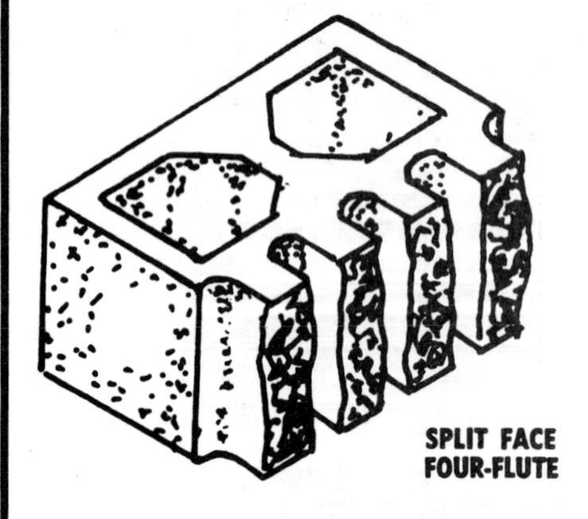

SPLIT FACE
FOUR-FLUTE

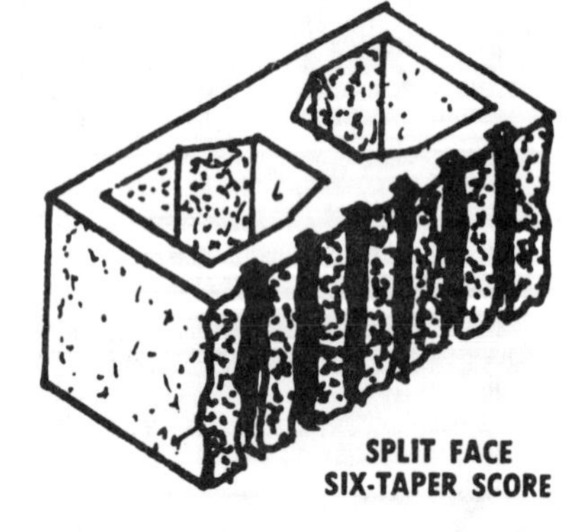

SPLIT FACE
SIX-TAPER SCORE

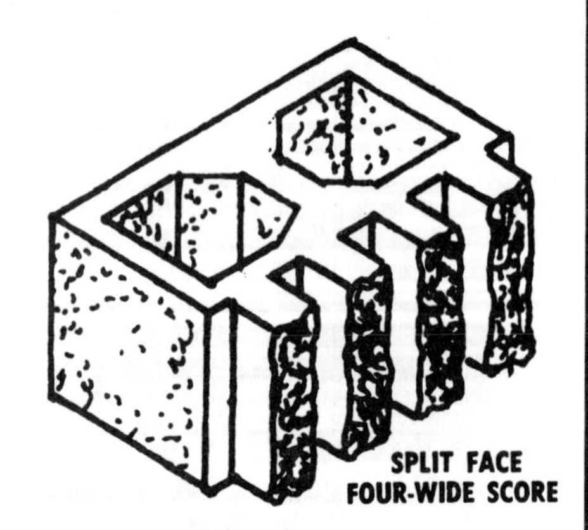

SPLIT FACE
FOUR-WIDE SCORE

NOTE: Split face units shown in this manual are a small sampling of the broad range of concrete masonry architectural units available from the industry on special order. Depths and widths of scores vary. Consult a local manufacturer for specific information.

TYPICAL DETAILS—LINTELS AND BOND BEAMS

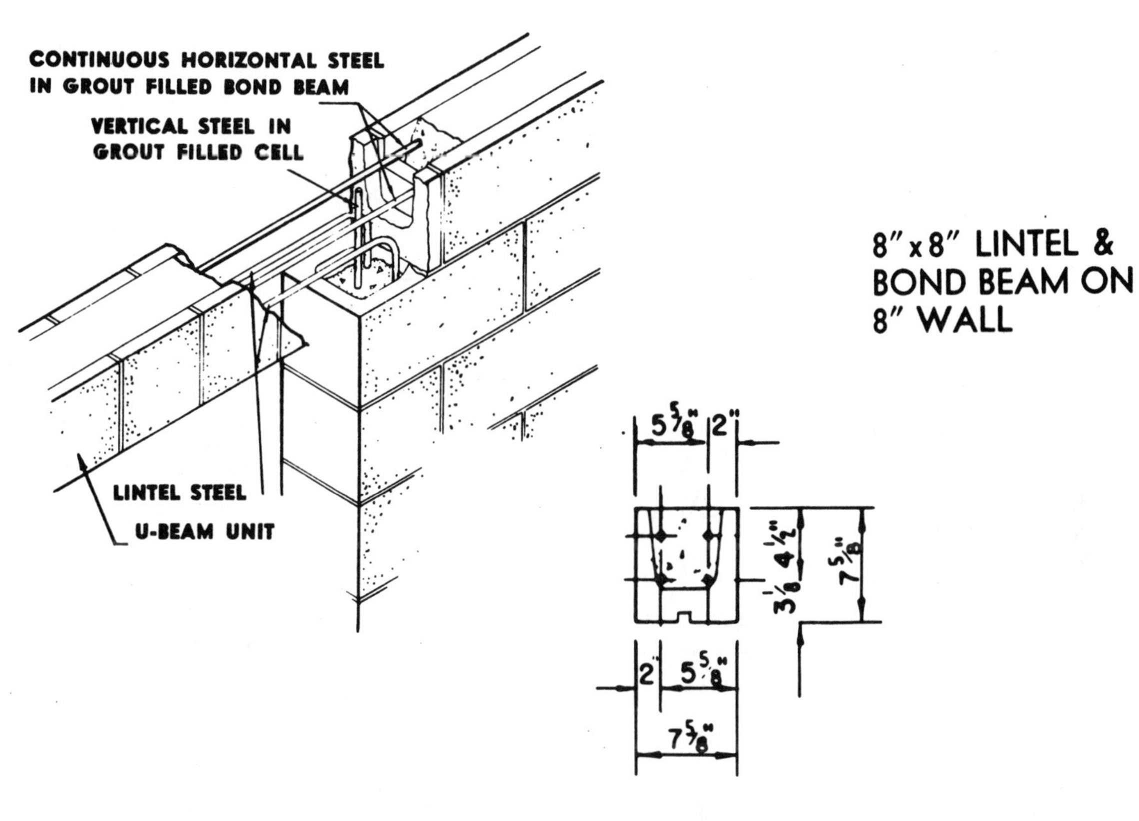

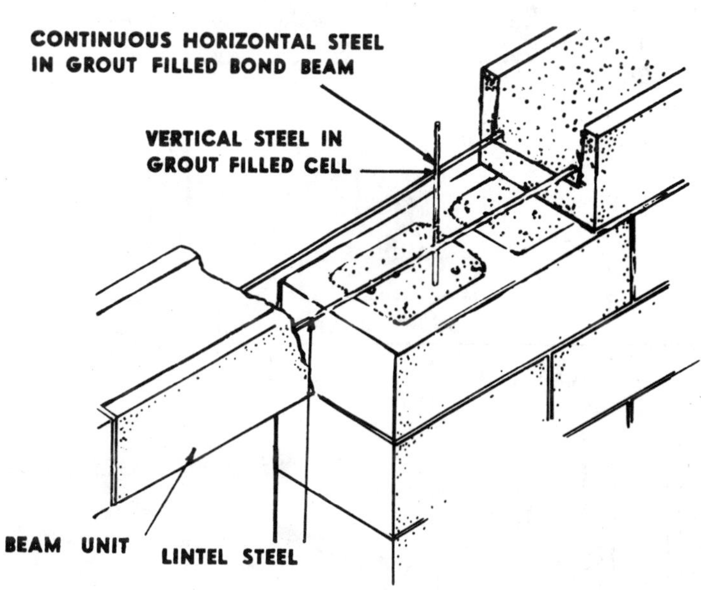

8″ x 16″
BOND BEAM ON
8″ WALL

METALS / WELDING | 05050

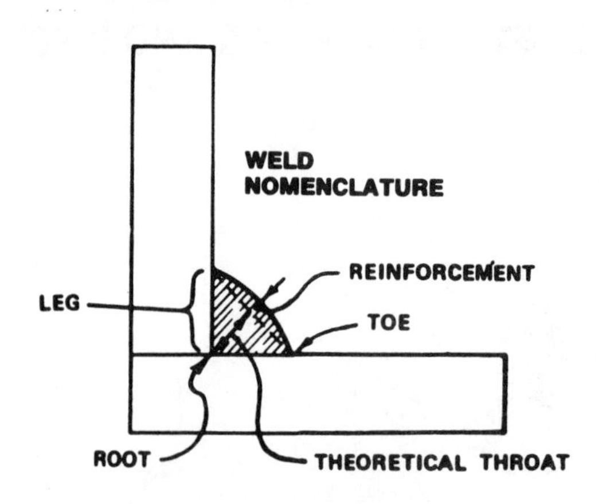

WELDED JOINTS

SQUARE BUTT

SINGLE VEE BUTT

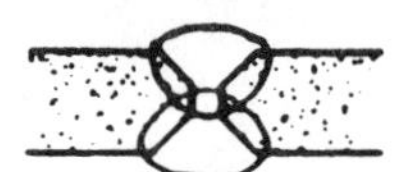
DOUBLE VEE BUTT

SINGLE U BUTT

DOUBLE U BUTT

SINGLE FILLET LAP

DOUBLE FILLET LAP

STRAP JOINT

SINGLE BEVEL TEE

DOUBLE BEVEL TEE

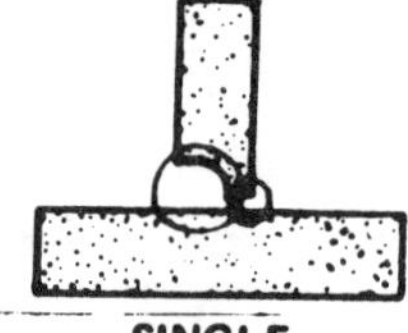
SINGLE J TEE

SQUARE TEE

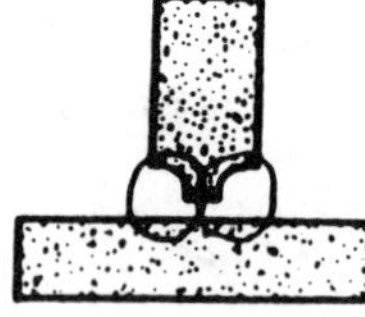
DOUBLE J TEE

CLOSED CORNER (FLUSH) JOINT

HALF OPEN CORNER JOINT

WELDING POSITIONS

FLAT (F)

HORIZONTAL (H)

VERTICAL (V)

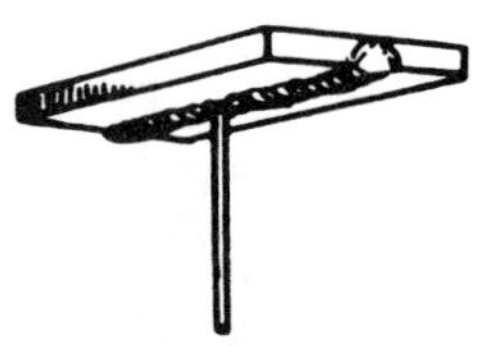
OVERHEAD (OH)

BOLTS IN COMMON USAGE

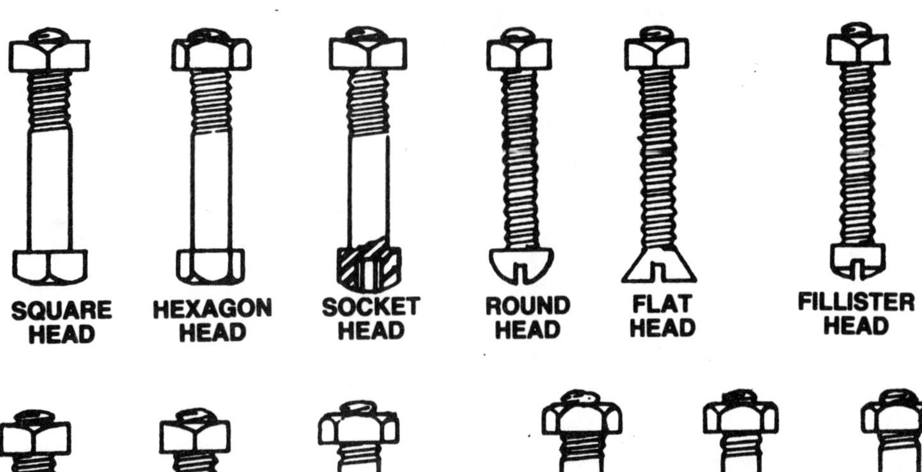

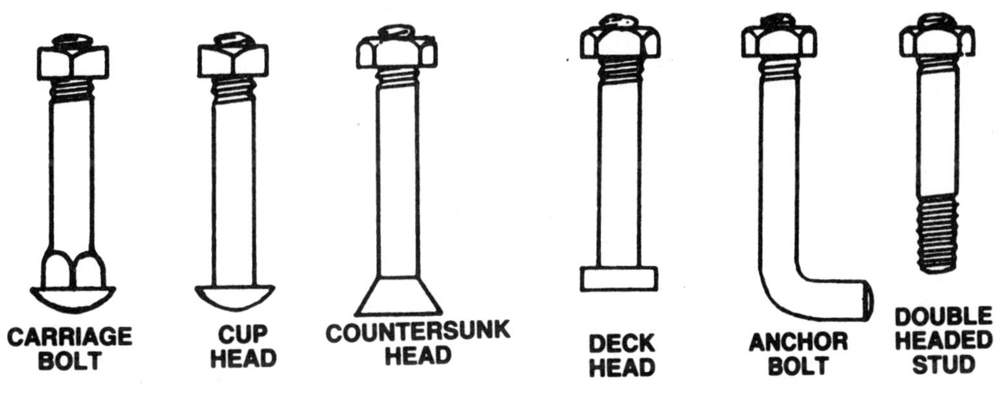

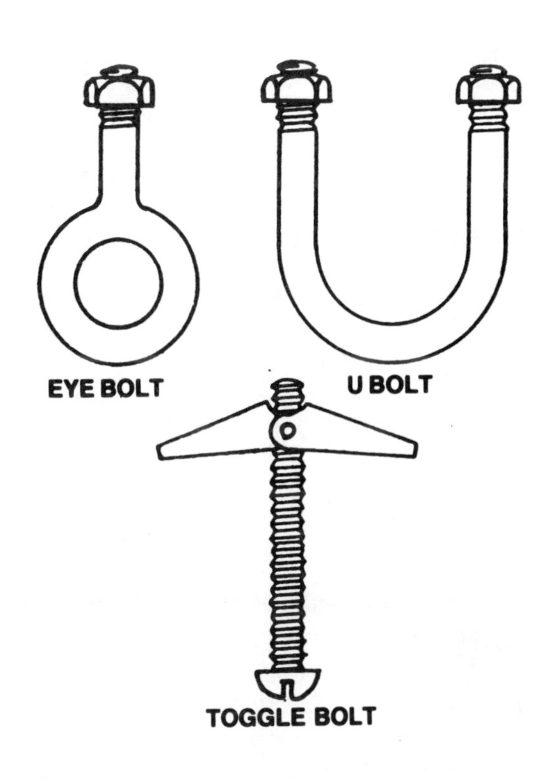

EXAMPLE OF SIMPLIFIED STRUCTURAL STEEL TAKEOFF METHOD

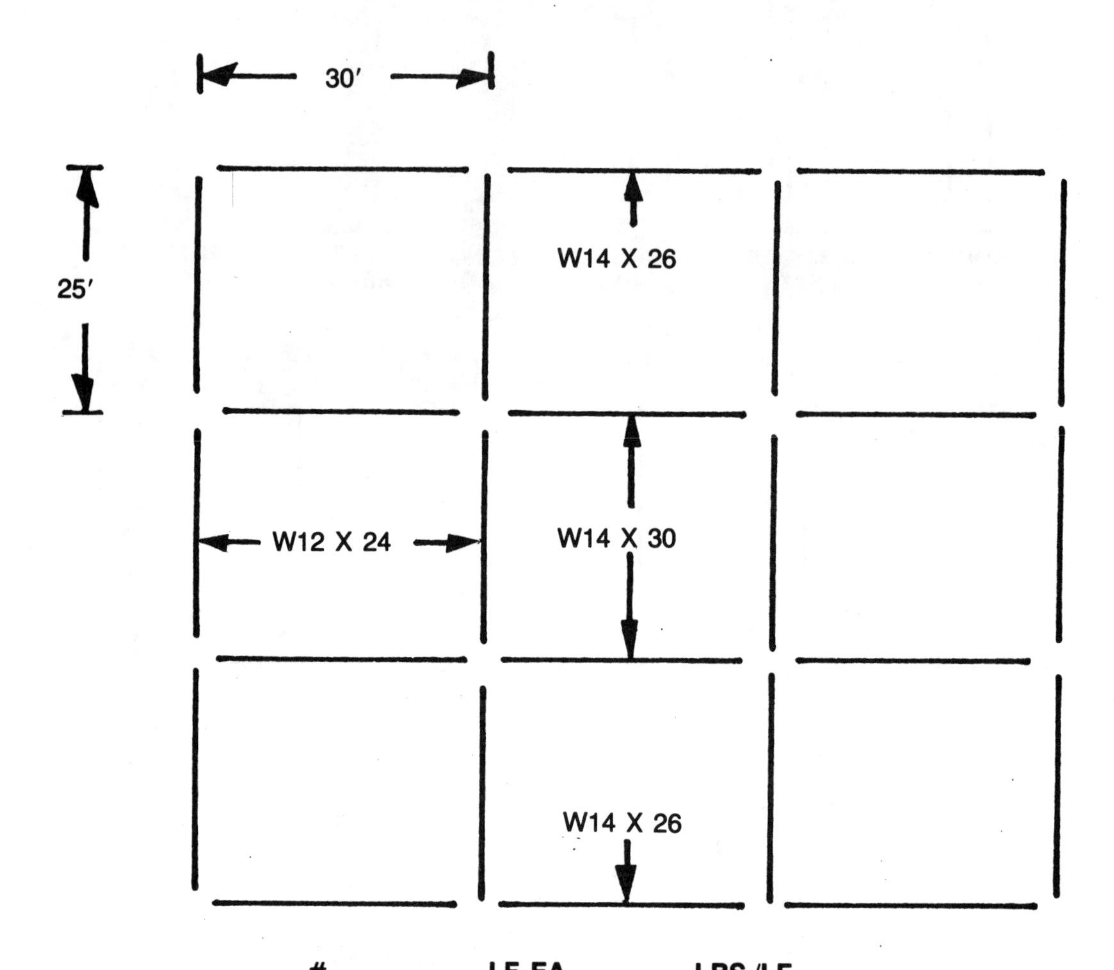

	#		LF. EA.		LBS./LF.			
W14 X 26	6	×	30	×	26	=	4,680	LBS.
W14 X 30	6	×	30	×	30	=	5,400	
W12 X 24	12	×	25	×	24	=	7,200	
							17,280	LBS.
							OR	
							9 ±	TONS

AFTER MAIN MEMBERS ARE ESTIMATED, ADD:

2 TO 3% FOR BASE PLATES
4 TO 5% FOR COLUMN SPLICES
4 TO 5% FOR MISCELLANEOUS COSTS

WOOD & PLASTICS / FASTENERS | 06050

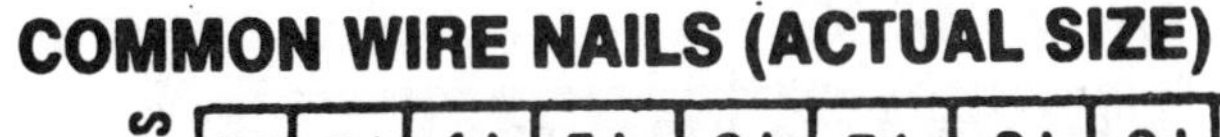

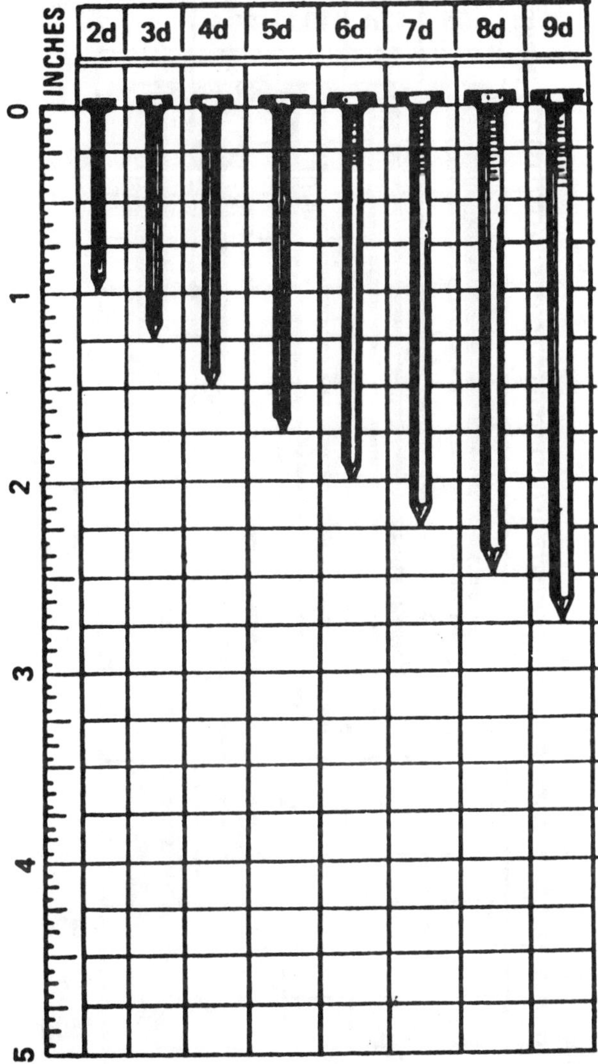

Cut Nails. Cut nails are angular-sided, wedge-shaped with a blunt point.

Wire Nails. Wire nails are round shafted, straight, pointed nails, and are used more generally than cut nails. They are stronger than cut nails and do not buckle as easily when driven into hard wood, but usually split wood more easily than cut nails. Wire nails are available in a variety of sizes varying from two penny to sixty penny.

Nail Finishes. Nails are available with special finishes. Some are galvanized or cadmium plated to resist rust. To increase the resistance to withdrawal, nails are coated with resins or asphalt cement (called cement coated). Nails which are small, sharp-pointed, and often placed in the craftsman's mouth (such as lath or plaster board nails) are generally blued and sterilized.

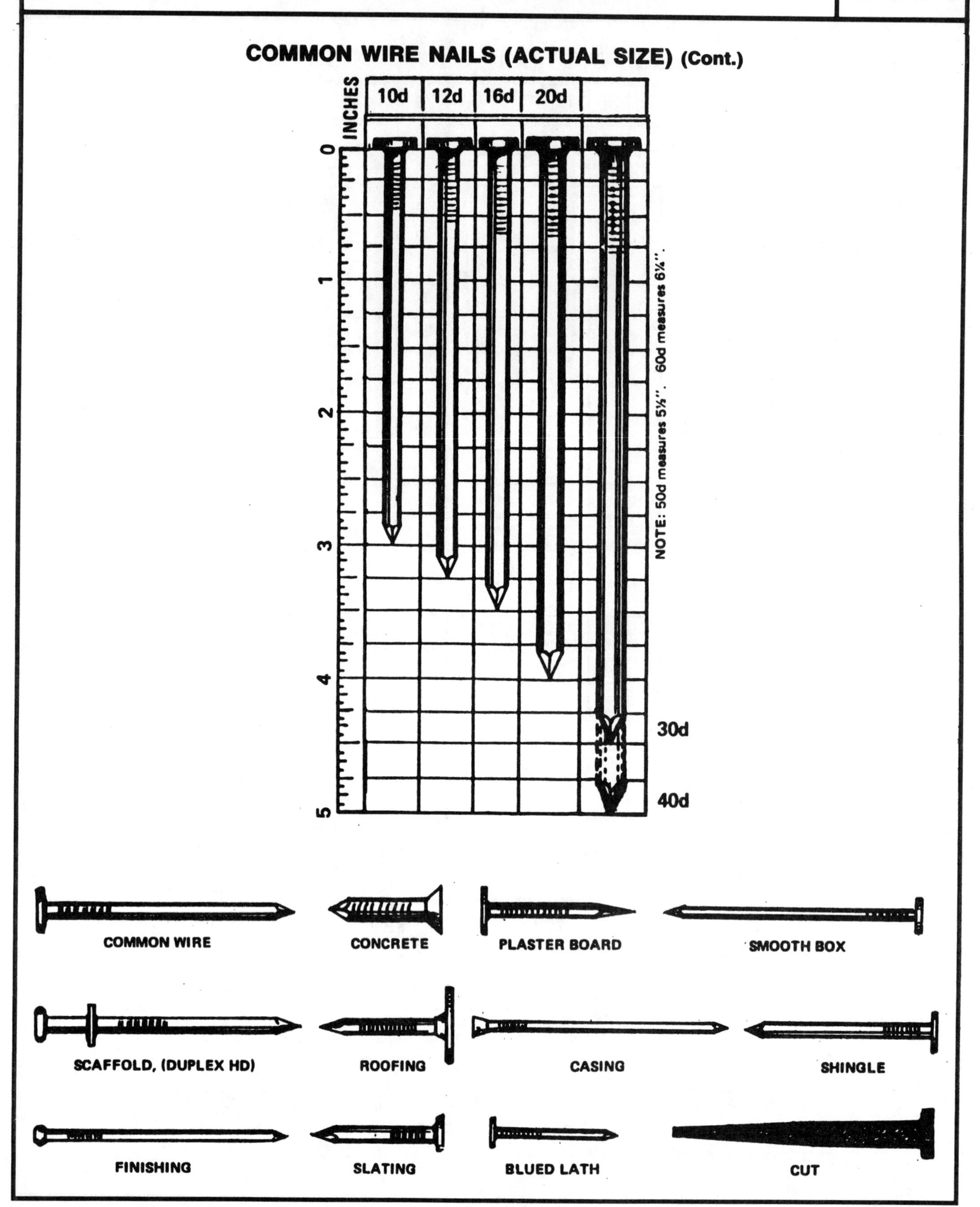
COMMON WIRE NAILS (ACTUAL SIZE) (Cont.)
INCHES
10d
12d
16d
20d
0
1
2
3
4
5
NOTE: 50d measures 5½". 60d measures 6¼".
30d
40d
COMMON WIRE
CONCRETE
PLASTER BOARD
SMOOTH BOX
SCAFFOLD, (DUPLEX HD)
ROOFING
CASING
SHINGLE
FINISHING
SLATING
BLUED LATH
CUT

PLYWOOD – BASIC GRADE MARKS

AMERICAN PLYWOOD ASSOCIATION (APA)

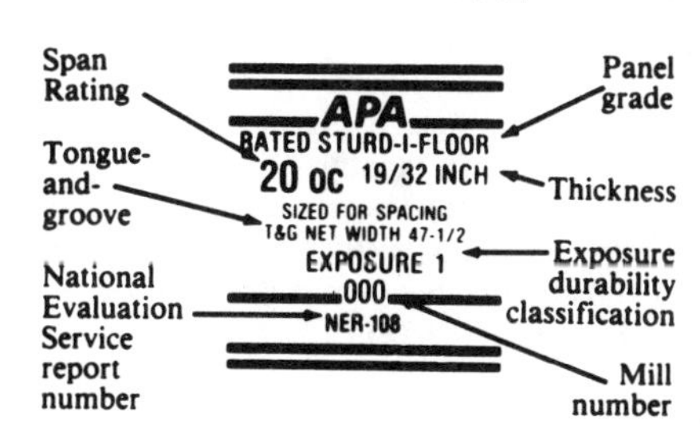

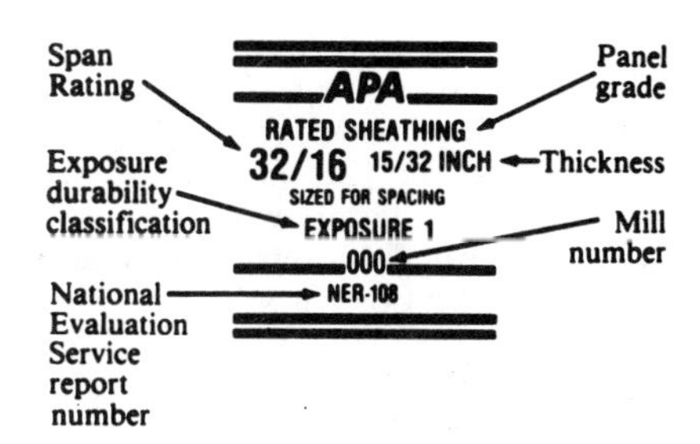

The American Plywood Association's trademarks appear only on products manufactured by APA member mills. The marks signify that the product is manufactured in conformance with APA performance standards and/or U.S. Product Standard PS 1-83 for Construction and Industrial Plywood.

APA
A-C GROUP 1
EXTERIOR
000
PS 1-83

APA A-C

For use where appearance of one side is important in exterior applications such as soffits, fences, structural uses, boxcar and truck linings, farm buildings, tanks, trays, commercial refrigerators, etc. ***Exposure Durability Classification:*** Exterior. ***Common Thicknesses:*** ¼, 11/32, ⅜, 15/32, ½, 19/32, ⅝, 23/32, ¾.

APA
A-D GROUP 1
EXPOSURE 1
000
PS 1-83

APA A-D

For use where appearance of only one side is important in interior applications, such as paneling, built-ins, shelving, partitions, flow racks, etc. ***Exposure Durability Classifications:*** Interior, Exposure 1. ***Common Thicknesses:*** ¼, 11/32, ⅜, 15/32, ½, 19/32, ⅝, 23/32, ¾.

APA
B-C GROUP 1
EXTERIOR
000
PS 1-83

APA B-C

Utility panel for farm service and work buildings, boxcar and truck linings, containers, tanks, agricultural equipment, as a base for exterior coatings and other exterior uses. ***Exposure Durability Classification:*** Exterior. ***Common Thicknesses:*** ¼, 11/32, ⅜, 15/32, ½, 19/32, ⅝, 23/32, ¾.

APA
B-D GROUP 2
INTERIOR
000
PS 1-83

APA B-D

Utility panel for backing, sides or builtins, industry shelving, slip sheets, separator boards, bins and other interior or protected applications. ***Exposure Durability Classifications:*** Interior, Exposure 1. ***Common Thicknesses:*** ¼, 11/32, ⅜, 15/32, ½, 19/32, ⅝, 23/32, ¾.

APA
PLYFORM
B-B CLASS I
EXTERIOR
000
PS 1-83

APA proprietary concrete form panels designed for high reuse. Sanded both sides and mill-oiled unless otherwise specified. Class I, the strongest, stiffest and more commonly available, is limited to Group 1 faces, Group 1 or 2 cross-bands, and Group 1, 2, 3 or 4 inner plies. Class II is limited to Group 1 or 2 faces (Group 3 under certain conditions) and Group 1, 2, 3 or 4 inner plies. Also available in HDO for very smooth concrete finish, in Structural I, and with special overlays. ***Exposure Durability Classification:*** Exterior. ***Common Thicknesses:*** 19/32, ⅝, 23/32, ¾.

Plywood panel manufactured with smooth, opaque, resin-treated fiber overlay providing ideal base for paint on one or both sides. Excellent material choice for shelving, factory work surfaces, paneling, built-ins, signs and numerous other construction and industrial applications. Also available as a 303 Siding with texture-embossed or smooth surface on one side only and Structural I. ***Exposure Durability Classification:*** Exterior. ***Common Thicknesses:*** 11/32, ⅜, 15/32, ½, 19/32, ⅝, 23/32, ¾.

SPECIALTY PANELS

HDO · A-A · G-1 · EXT-APA · 000 · PS1-83

Plywood panel manufactured with a hard, semi-opaque resin-fiber overlay on both sides. Extremely abrasion resistant and ideally suited to scores of punishing construction and industrial applications, such as concrete forms, industrial tanks, work surfaces, signs, agricultural bins, exhaust ducts, etc. Also available with skid-resistant screen-grid surface and in Structural I. ***Exposure Durability Classification:*** **Exterior.** ***Common Thicknesses:*** 3/8, 1/2, 5/8, 3/4.

MARINE · A-A · EXT-APA · 000 · PS1-83

Specialty designed plywood panel made only with Douglas fir or western larch, solid jointed cores, and highly restrictive limitations on core gaps and faces repairs. Ideal for both hulls and other marine applications. Also available with HDO or MDO faces. ***Exposure Durability Classification:*** Exterior. ***Common Thicknesses:*** 1/4, 3/8, 1/2, 5/8, 3/4.

Unsanded and touch-sanded panels, and panels with "B" or better veneer on one side only, usually carry the APA trademark on the panel back. Panels with both sides of "B" or better veneer, or with special overlaid surfaces (such as Medium Density Overlay), carry the APA trademark on the panel edge, like this:

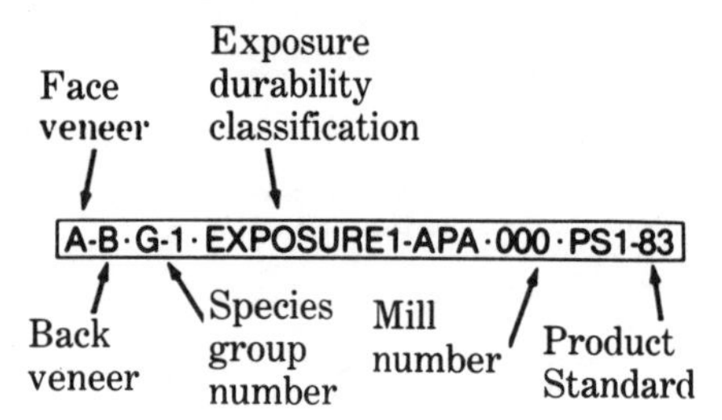

GLOSSARY OF LUMBER TERMS

Some of the words and terms used in the grading of lumber follow:

Bow. A deviation flatwise from a straight line drawn from end to end of the piece. It is measured at the point of greatest distance from the straight line.

Checks. A separation of the wood which normally occurs across the annual rings and usually as a result of seasoning.

Crook. A deviation edgewise from a straight line drawn from end to end of the piece. It is measured at the point of greatest distance from the straight line.

Cup. A deviation from a straight line drawn across the piece from edge to edge. It is measured at the point of greatest distance from the straight line.

Flat Grain. The annual growth rings pass through the piece at an angle of less than 45 degrees with the flat surface of the piece.

Mixed Grain. The piece may have vertical grain, flat grain or a combination of both vertical and flat grain.

Pitch. An accumulation of resin which occurs in separations in the wood or in the wood cells themselves.

Shake. A separation of the wood which usually occurs between the rings of annual growth.

Splits. A separation of the wood due to tearing apart of the wood cells.

Vertical Grain. The annual growth rings pass through the piece at an angle of 45 degrees or more with the flat surface of the piece.

Wane. Bark or lack of wood from any cause, except eased edges (rounded) on the edge or corner of a piece of lumber.

Warp. Any deviation from a true or plane surface, including crook, cup, bow or any combination thereof.

LUMBER GRADING

GRADING-MARK ABBREVIATIONS

GRADES

(Listed alphabetically — not by quality)

COM	Common
CONST	Construction
ECON	Economy
No. 1	Number One
SEL-MER	Select Merchantable
SEL-STR	Select Structural
STAN	Standard
UTIL	Utility

ALSC TRADEMARKS

CLIS	California Lumber Inspection Service
NELMA	Northeastern Lumber Manufacturers Association, Inc.
NH&PMA	Northern Hardwood & Pine Manufacturers Association, Inc.
PLIB	Pacific Lumber Inspection Bureau
RIS	Redwood Inspection Service
SPIB	Southern Pine Inspection Bureau
TP	Timber Products Inspection
WCLB	West Coast Lumber Inspection Bureau
WWP	Western Wood Products Association

SPECIES GROUPINGS

AF	Alpine Fir
DF	Douglas Fir
HF	Hem Fir
SP	Sugar Pine
PP	Ponderosa Pipe
LP	Lodgepole Pine
IWP	Idaho White Pine
ES	Engelmann Spruce
WRC	Western Red Cedar
INC CDR	Incense Cedar
L	Larch
LP	Lodgepole Pine
MH	Mountain Hemlock
WW	White Wood

MOISTURE CONTENT

S-GRN	Surfaced at a moisture content of more than 19%
S-DRY	Surfaced at a moisture content of 19% or less
MC-15	Surfaced at a moisture content of 15% or less.

Framing Estimating Rules of Thumb

For 16″ O.C. stud partitions figure 1 stud for every L.F. of wall; add for top and bottom plates.

For any type of framing, the quantity of basic framing members (in L.F.) can be determined based on spacing and surface area (S.F.):

12″ O.C.	1.2 L.F./S.F.
16″ O.C.	1.0 L.F./S.F.
24″ O.C.	0.8 L.F./S.F.

(Doubled-up members, bands, plates, framed openings, etc., must be added.)

Framing accessories, nails, joist hangers, connectors, etc., should be estimated as separate material costs. Installation should be included with framing. Rule of thumb allowance is 0.5 to 1.5% of lumber cost for rough hardware. Another is 30 to 40 pounds of nails per M.B.F.

BOARD FEET/LINEAR FEET FOR LUMBER

Nominal Size	Actual Size	Board Feet Per Linear Foot	Linear Feet Per 1000 Board Feet
1 × 2	¾ × 1½	.167	6000
1 × 3	¾ × 2½	.250	4000
1 × 4	¾ × 3½	.333	3000
1 × 6	¾ × 5½	.500	2000
1 × 8	¾ × 7¼	.666	1500
1 × 10	¾ × 9¼	.833	1200
1 × 12	¾ × 11¼	1.0	1000
2 × 2	1½ × 1½	.333	3000
2 × 3	1½ × 2½	.500	2000
2 × 4	1½ × 3½	.666	1500
2 × 6	1½ × 5½	1.0	1000
2 × 8	1½ × 7¼	1.333	750
2 × 10	1½ × 9¼	1.666	600
2 × 12	1½ × 11¼	2.0	500

Redwood. Redwood is a fairly strong and moderately lightweight material. The heartwood is red but the sapwood is white. One of the principal advantages of redwood is that the heartwood is highly resistant (but not entirely immune) to decay, fungus and insects. Standard Specifications require that all redwood used in permanent installations shall be "select heart." Grade marking shall be in accordance with the standards established in the California Redwood Association. Grade marking shall be done by, or under the supervision of the Redwood Inspection Service. (See Plate 21, Appendix.)

Redwood is graded for specific uses as indicated in the following table:

REDWOOD GRADING

Type of Lumber	Grade	Typical Use
Grades for Dimension Only Listed Here	Clear All Heart	Exceptionally fine, knot free, straight-grained timbers. This grade is used primarily for stain finish work of high quality.
	Clear	Same as Clear All Heart except that this grade may contain sound sapwood and medium stain.
	Select Heart	**This grade only is to be used in Agency work, unless otherwise specified in the plans or specifications.** It is sound, live heartwood free from splits or streaks with sound knots. It is generally used where the timber is in contact with the ground, as in posts, mudsills, etc.
	Select	
	Construction Heart	Slightly less quality than Select Heart. It may have some sapwood in the piece. Used for general construction purposes when redwood is needed.
	Construction Common	Same requirement as Construction Heart except that it will contain sapwood and medium stain. Its resistance to decay and insect attack is reduced.
	Merchantable	Used for fence posts, garden stakes, etc.
	Economy	Suitable for crating, bracing and temporary construction.

DOUGLAS FIR GRADING

Type of Lumber	Grade	Typical Use
Select Structural Joists and Planks	Select Structural	Used where strength is the primary consideration, with appearance desirable.
	No. 1	Used where strength is less critical and appearance not a major consideration.
	No. 2	Used for framing elements that will be covered by subsequent construction.
	No. 3	Used for structural framing where strength is required but appearance is not a factor.
Finish Lumber	Superior	For all types of uses as casings, cabinet, exposed members, etc., where a fine appearance is desired.
	Prime	
	E	
Boards (WCLIB)* *Grading is by West Coast Lumber Inspection Bureau rules, but sizes conform to Western Wood Products Assn. rules. These boards are still manufactured by some mills.	Select Merchantable	Intended for use in housing and light construction where a knotty type of lumber with finest appearance is required.
	Construction	Used for sub-flooring, roof and wall sheathing, concrete forms, etc. Has a high degree of serviceability.
	Standard	Used widely for general construction purposes, including subfloors, roof and wall sheathing, concrete forms, etc. Seldom used in exposed construction because appearance.
	Utility	Used in general construction where low cost is a factor and appearance is not important. (Storage shelving, crates, bracing, temporary scaffolding, etc.)

BOARD FEET CONVERSION TABLE

Nominal Size (In.)	ACTUAL LENGTH IN FEET								
	8	10	12	14	16	18	20	22	24
1 x 2		1⅔	2	2⅓	2⅔	3	3½	3⅔	4
1 x 3		2½	3	3½	4	4½	5	5½	6
1 x 4	2¾	3⅓	4	4⅔	5⅓	6	6⅔	7⅓	8
1 x 5		4⅙	5	5⅚	6⅔	7½	8⅓	9⅙	10
1 x 6	4	5	6	7	8	9	10	11	12
1 x 7		5⅝	7	8⅙	9⅓	10½	11⅔	12⅚	14
1 x 8	5⅓	6⅔	8	9⅓	10⅔	12	13⅓	14⅔	16
1 x 10	6⅔	8⅓	10	11⅔	13⅓	15	16⅔	18⅓	20
1 x 12	8	10	12	14	16	18	20	22	24
1¼ x 4		4⅙	5	5⅚	6⅔	7½	8⅓	9⅙	10
1¼ x 6		6¼	7½	8¾	10	11¼	12½	13¾	15
1¼ x 8		8⅓	10	11⅔	13⅓	15	16⅔	18⅓	20
1¼ x 10		10 5/12	12½	14 7/12	16⅔	18¾	20⅚	22 11/12	25
1¼ x 12		12½	15	17½	20	22½	25	27½	30
1½ x 4	4	5	6	7	8	9	10	11	12
1½ x 6	6	7½	9	10½	12	13½	15	16½	18
1½ x 8	8	10	12	14	16	18	20	22	24
1½ x 10	10	12½	15	17½	20	22½	25	27½	30
1½ x 12	12	15	18	21	24	27	30	33	36
2 x 4	5⅓	6⅔	8	9⅓	10⅓	12	13⅓	14⅔	16
2 x 6	8	10	12	14	16	18	20	22	24
2 x 8	10⅔	13⅓	16	18⅔	21⅓	24	26⅔	29⅓	32
2 x 10	13⅓	16⅔	20	23⅓	26⅔	30	33⅓	36⅔	40
2 x 12	16	20	24	28	32	36	40	44	48
3 x 6	12	15	18	21	24	27	30	33	36
3 x 8	16	20	24	28	32	36	40	44	48
3 x 10	20	25	30	35	40	45	50	55	60
3 x 12	24	30	36	42	48	54	60	66	72
4 x 4	10⅔	13⅓	16	18⅔	21⅓	24	26⅔	29⅓	32
4 x 6	16	20	24	28	32	36	40	44	48
4 x 8	21⅓	26⅔	32	37⅓	42⅔	48	53⅓	58⅔	64
4 x 10	26⅔	33⅓	40	46⅔	53⅓	60	66⅔	73⅓	80
4 x 12	32	40	48	56	64	72	80	88	96

MOISTURE PROTECTION / ROOFING 07000

SLOPE AREA CALCULATIONS

Rise and Run	Multiply Flat Area by	LF of Hips or Valleys per LF of Common Run
2 in 12	1.014	1.424
3 in 12	1.031	1.436
4 in 12	1.054	1.453
5 in 12	1.083	1.474
6 in 12	1.118	1.500
7 in 12	1.158	1.530
8 in 12	1.202	1.564
9 in 12	1.250	1.600
10 in 12	1.302	1.641
11 in 12	1.357	1.685
12 in 12	1.413	1.732

MOISTURE PROTECTION / DOWNSPOUTS 07600

DOWNSPOUT/VERTICAL LEADER CALCULATIONS

Roof Type	Slope	S.F. Roof/ Sq. In. Leader
Gravel	Less than ¼″ per foot	300
Gravel	Greater than ¼″ per foot	250
Metal or Shingle	Any	200

Alternate calculations:

$$\text{Diameter of downspout/leader} = 1.128\sqrt{\frac{\text{Area of drainage}}{\text{SF Roof/Sq. Inch}}}$$

TYPICAL MINIMUM SIZE OF VERTICAL CONDUCTORS AND LEADERS

Size of leader or conductor (Inches)	Maximum projected roof area (Square feet)
2	544
2½	987
3	1,610
4	3,460
5	6,280
6	10,200
8	22,000

TYPICAL MINIMUM SIZE OF ROOF GUTTERS

	Maximum projected roof area for gutters of various slopes			
Diameter gutter (Inches)	1/16 in. per Ft. slope (Sq. ft.)	⅛ in. per Ft. slope (Sq. ft.)	¼ in. per Ft. slope (Sq. ft.)	½ in. per Ft. slope (Sq. ft.)
3	170	240	340	480
4	360	510	720	1020
5	625	880	1250	1770
6	960	1360	1920	2770
7	1380	1950	2760	3900
8	1990	2800	3980	5600
10	3600	5100	7200	10000

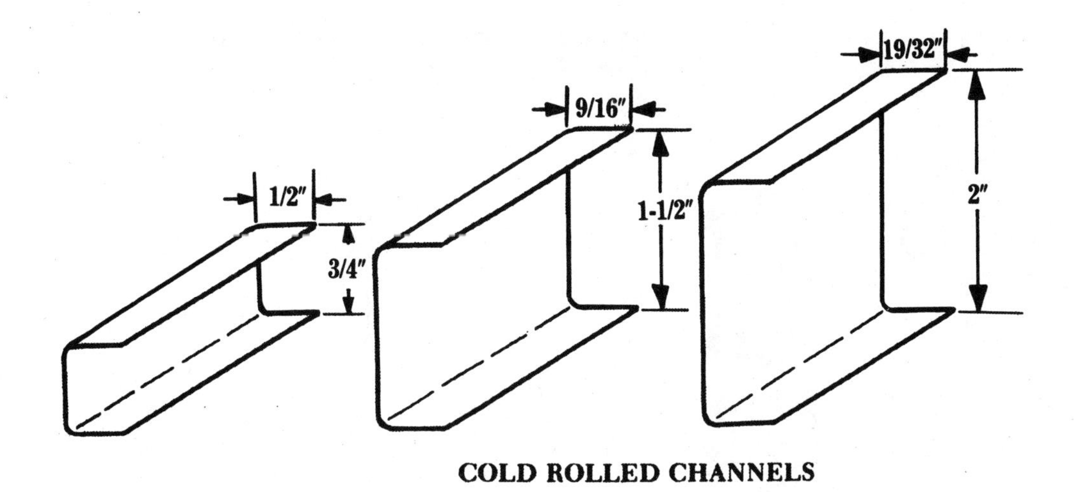

COLD ROLLED CHANNELS

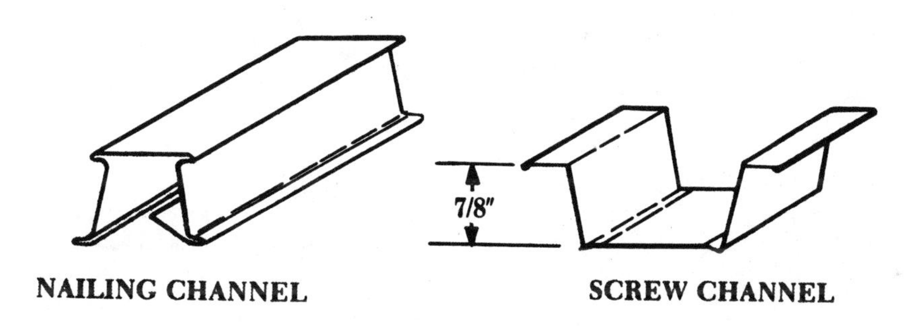

NAILING CHANNEL

SCREW CHANNEL

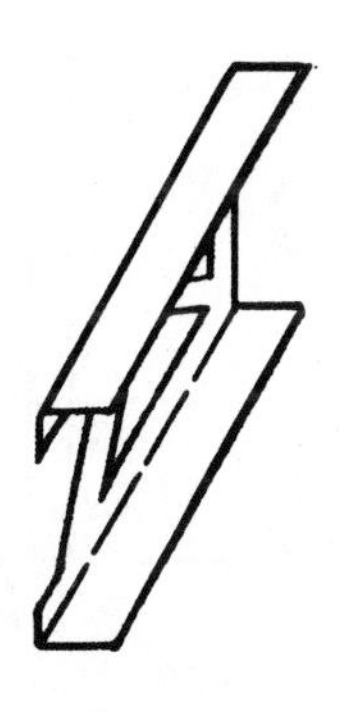
CHANNEL STUD

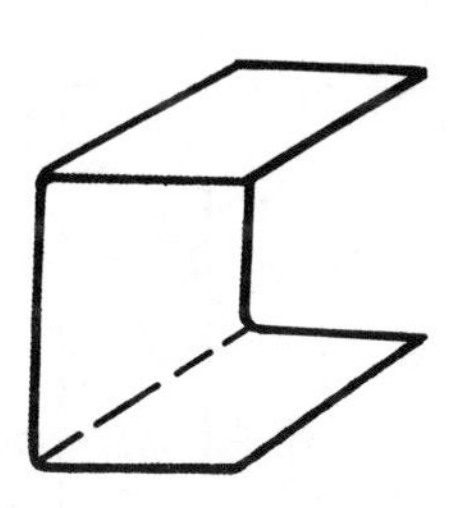
WIDE FLANGE CHANNEL

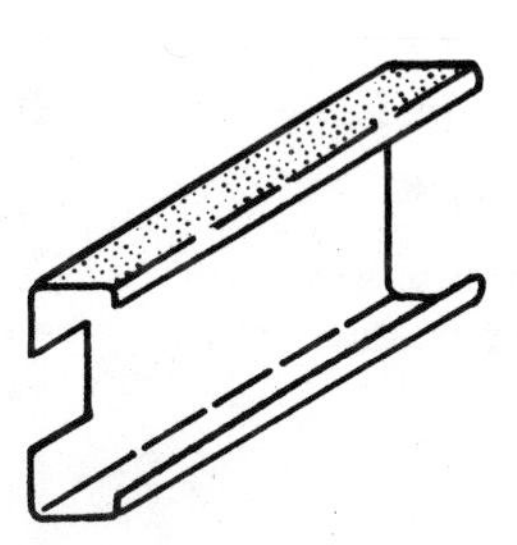
CEE STUD

STUDLESS SOLID PARTITION

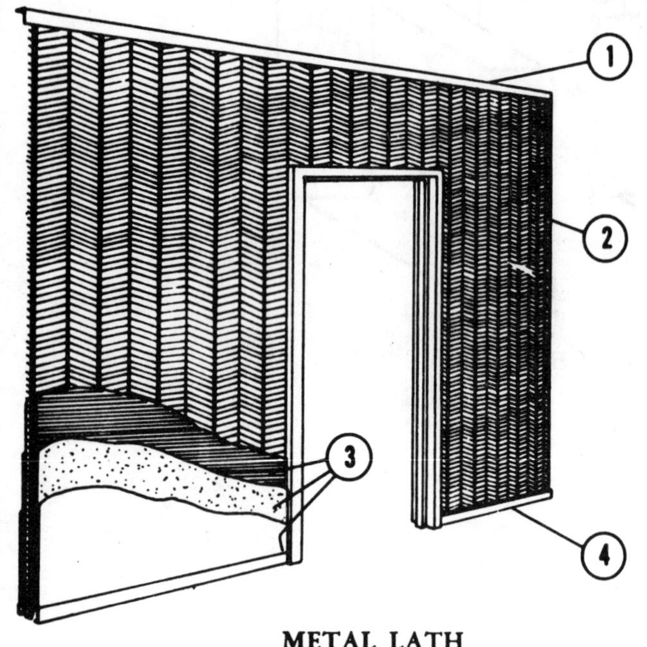

METAL LATH

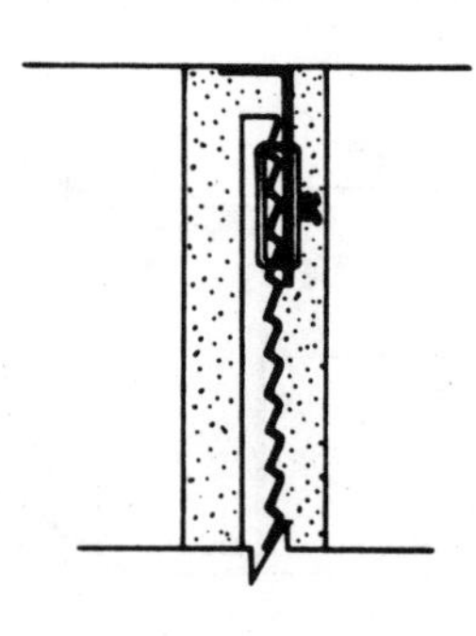

(1) Ceiling Runner
(2) Rib Metal Lath
(3) Plaster
(4) Combination Floor Runner and Screed

STUDLESS SOLID PARTITION

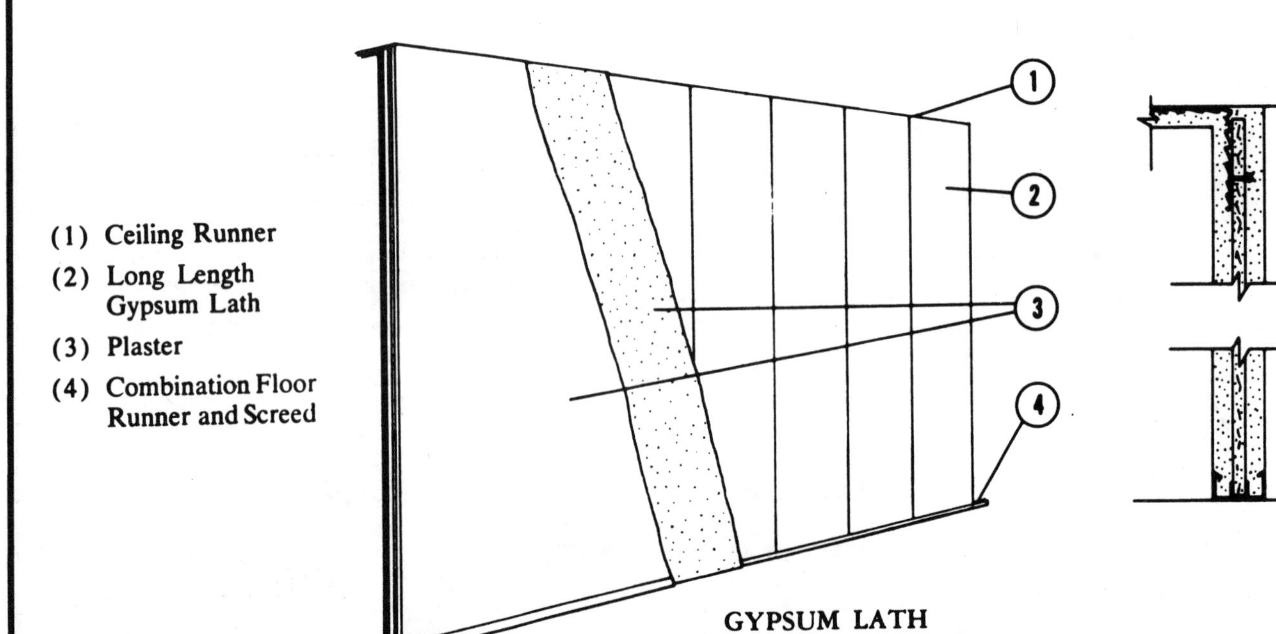

GYPSUM LATH

SUSPENDED CEILINGS

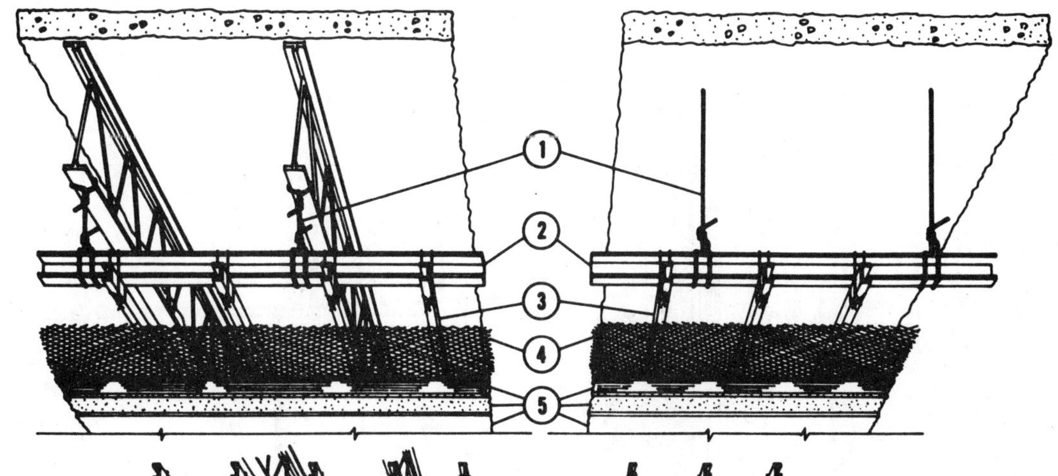

STEEL JOISTS

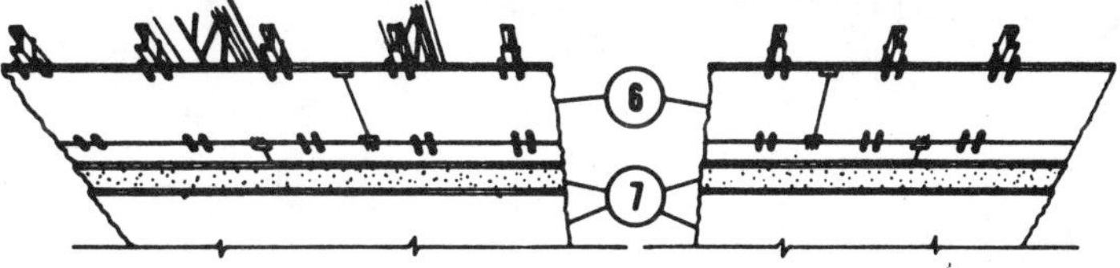

CONCRETE SLAB

(1) Hanger
(2) Main Runner Channel
(3) Furring Channel
(4) Metal or Wire Fabric Lath
(5) Plaster
(6) Gypsum Lath
(7) Plaster

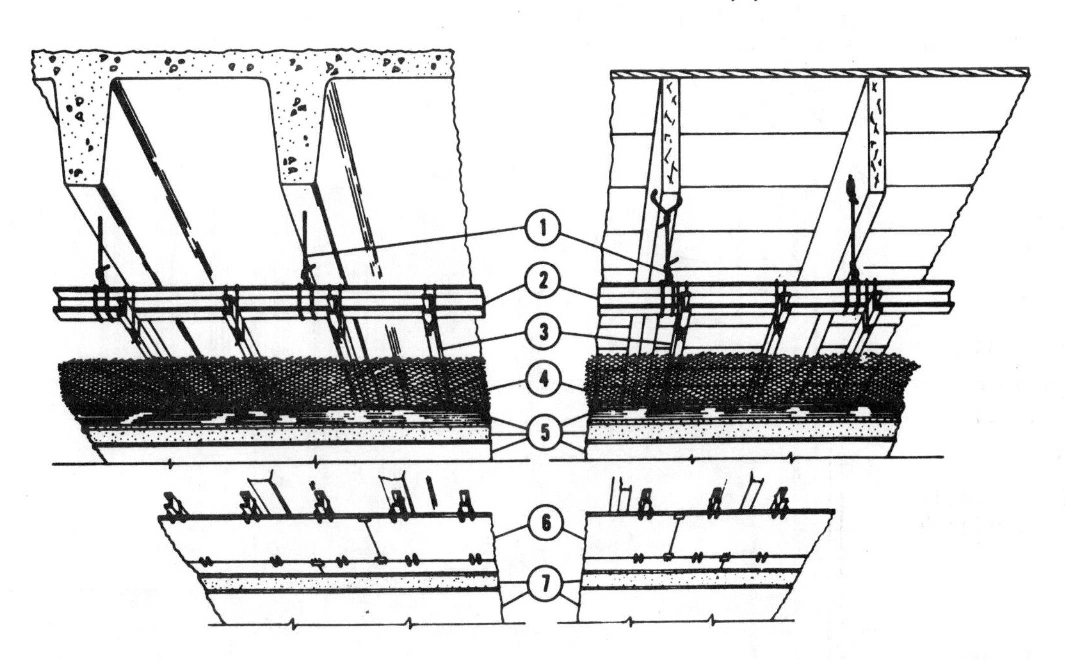

CONCRETE JOISTS

WOOD JOISTS

STEEL STUD
Hollow Partition

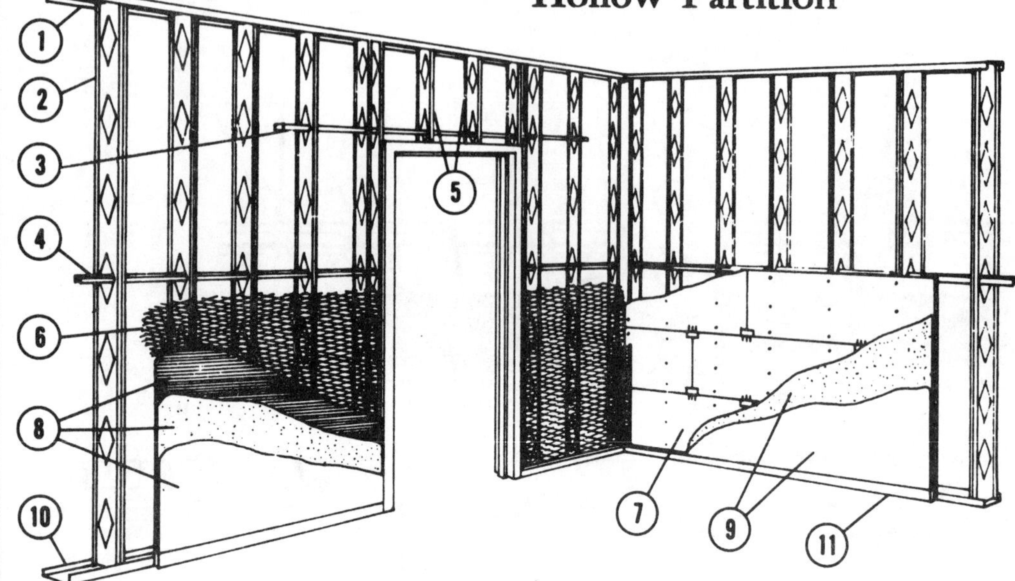

(1) Ceiling Runner Track
(2) Nailable Stud
(3) Door Opening Stiffener
(4) Partition Stiffener
(5) Jack Studs
(6) Metal or Wire Fabric Lath (screwed or wire tied)
(7) Gypsum Lath (nailed, clipped or screwed)
(8) Three Coats of Plaster (Scratch, Brown, Finish)
(9) Two Coats of Plaster (Brown, Finish)
(10) Floor Runner Track
(11) Flush Metal Base

SCREW STUD
Hollow Partition

(1) Ceiling Runner Track
(2) Screw Stud
(3) Door Opening Stiffener
(4) Partition Stiffener
(5) Jack Studs
(6) Metal or Wire Fabric Lath (screwed on)
(7) Gypsum Lath (screwed on)
(8) Three Coats of Plaster (Scratch, Brown, Finish)
(9) Two Coats of Plaster (Brown, Finish)
(10) Floor Runner Track
(11) Flush Metal Base

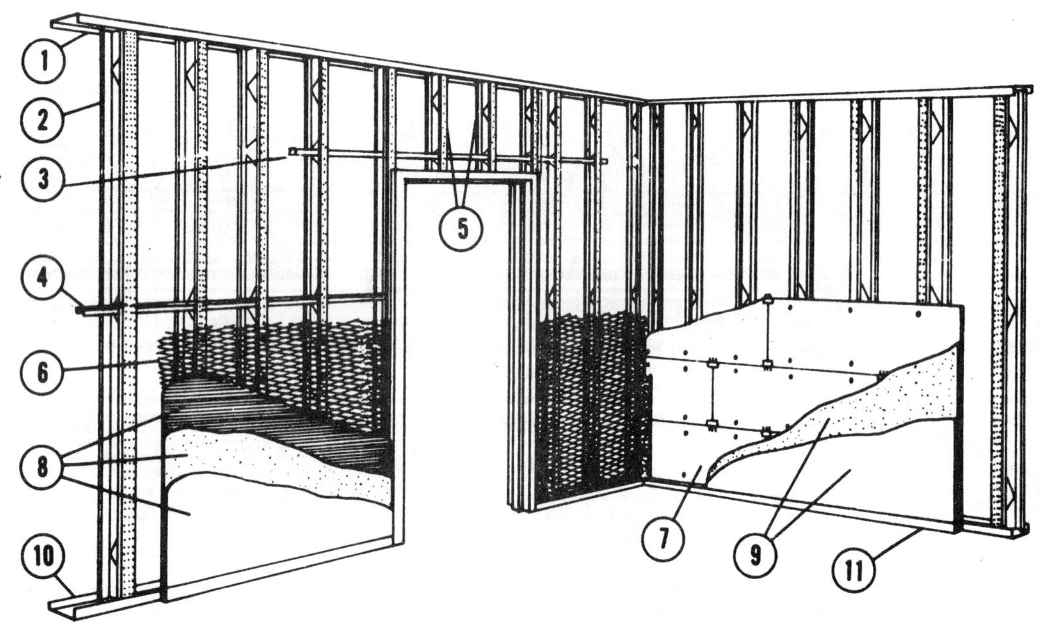

LOAD-BEARING HOLLOW PARTITION

Structural Stud

(1) Ceiling Runner Track
(2) Structural Stud (prefabricated)
(3) Structural Stud (nailable)
(4) Jack Studs
(5) Partition Stiffener (bridging)
(6) Metal or Wire Fabric Lath (wire-tied, nailed or stapled)
(7) Gypsum Lath (nailed or stapled)
(8) Three Coats of Plaster (Scratch, Brown, Finish)
(9) Two Coats of Plaster (Brown, Finish)

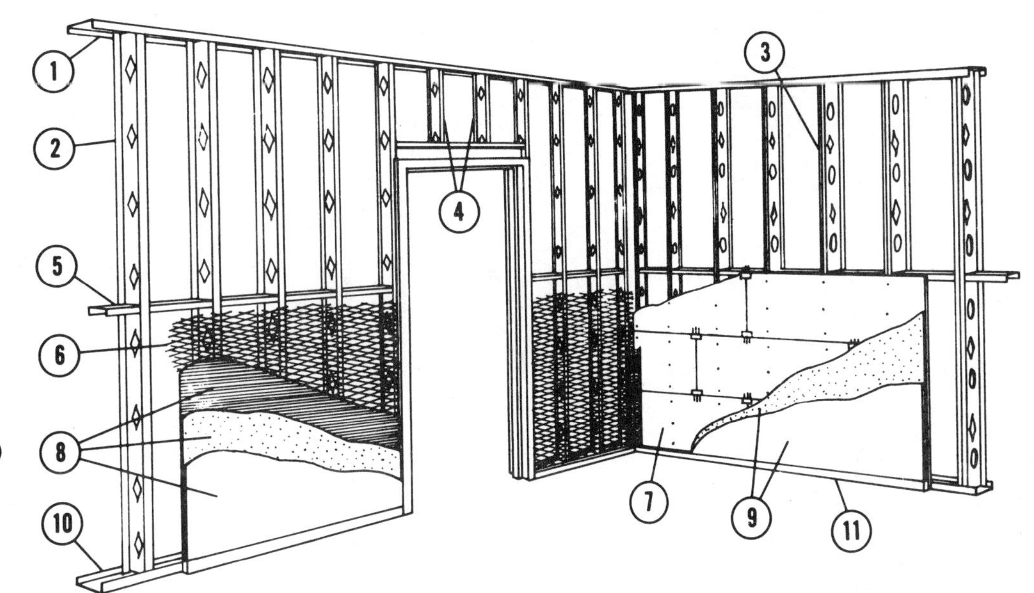

(10) Floor Runner Track
(11) Flush Metal Base

VERTICAL FURRING

With Studs

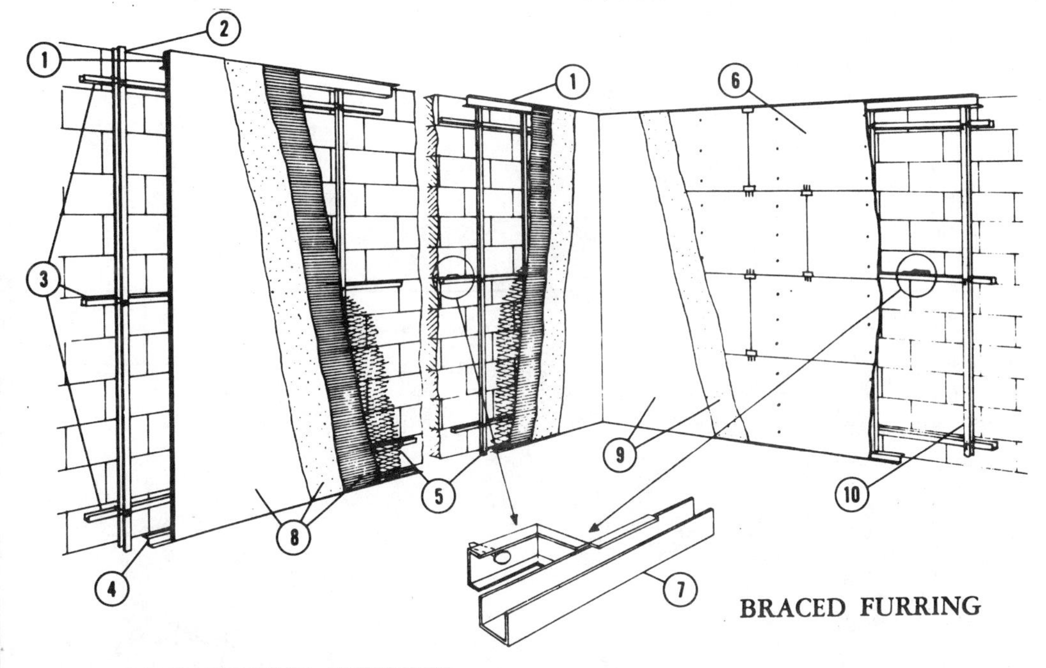

(1) Ceiling Runner
(2) Channel Stud
(3) Horizontal Stiffener
(4) Floor Runner
(5) Metal or Wire Fabric Lath
(6) Gypsum Lath
(7) Bracing
(8) Three Coats of Plaster (Scratch, Brown, Finish)
(9) Two Coats of Plaster (Brown, Finish)
(10) Screw Channel Studs

FREE STANDING FURRING

DOUBLE CHANNEL STUD
Hollow Partition

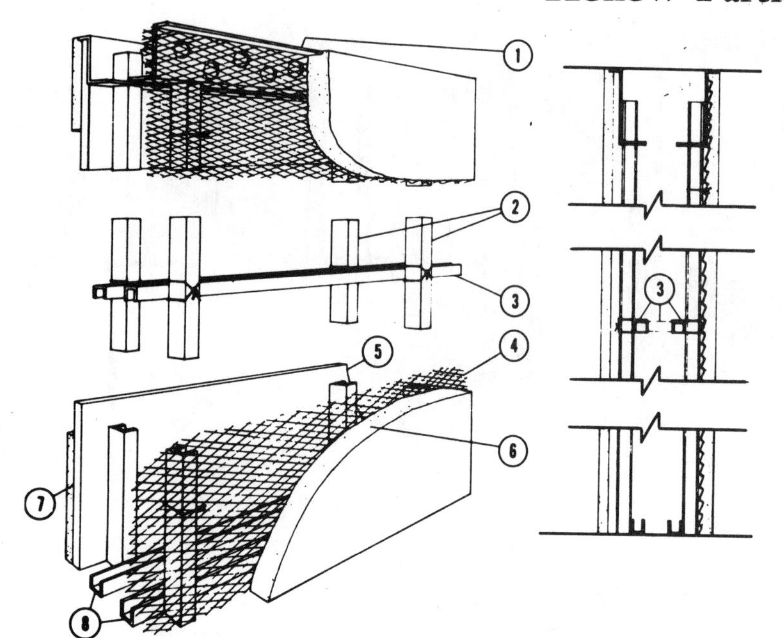

(1) Ceiling Runner
(2) Channel Studs
(3) Partition Stiffener and Channel Spacer
(4) Metal or Wire Fabric Lath (wire-tied)
(5) Gypsum Lath (clipped on)
(6) Three Coats of Plaster (Scratch, Brown, Finish)
(7) Two Coats of Plaster (Brown, Finish)
(8) Floor Runners (channel)

SINGLE CHANNEL STUD
Solid Partition

(1) Ceiling Runner
(2) Channel Stud
(3) Metal or Wire Fabric Lath (wire-tied)
(4) Plaster
(5) Combination Floor Runner and Screed

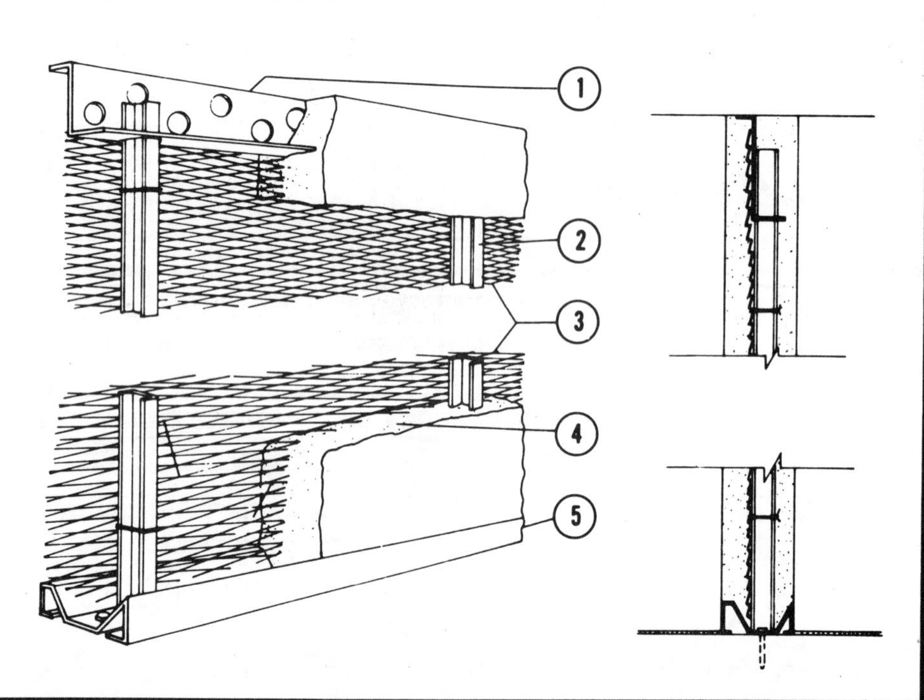

FURRED CEILINGS

STEEL JOISTS

CONCRETE JOISTS

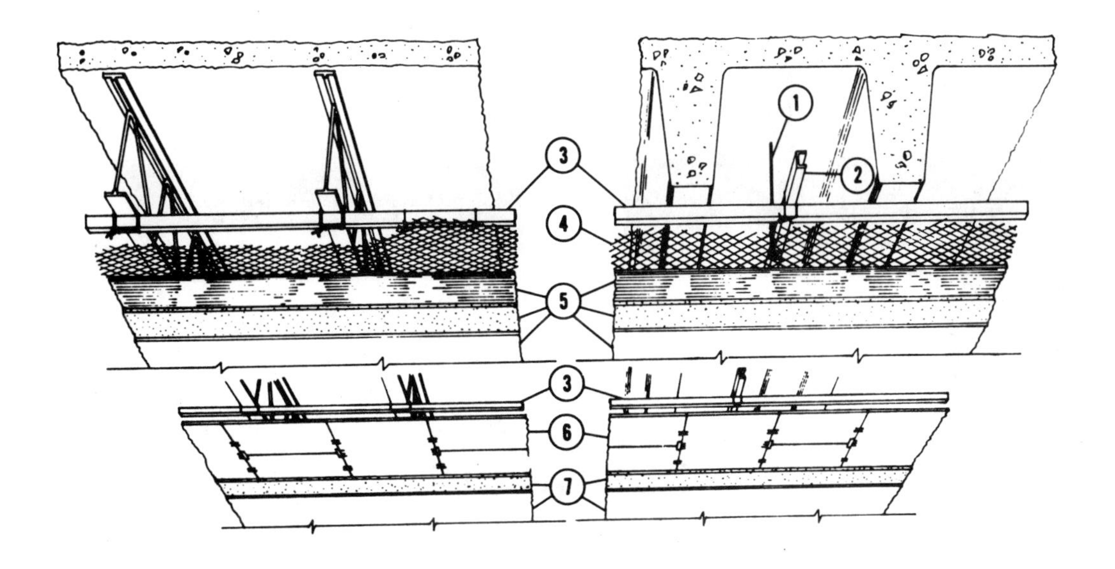

WOOD JOISTS

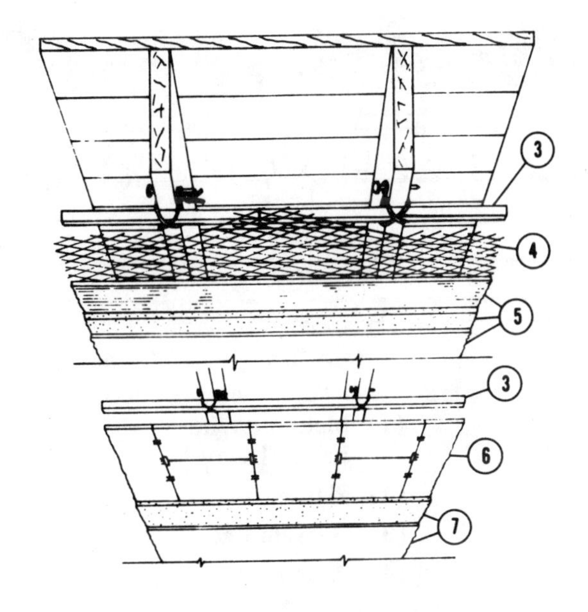

(1) Hanger
(2) ¾-inch Channel
(3) Cross Furring
(4) Metal or Wire Fabric Lath
(5) Plaster
(6) Gypsum Lath
(7) Plaster

CEILINGS

STEEL JOISTS   CONCRETE JOISTS

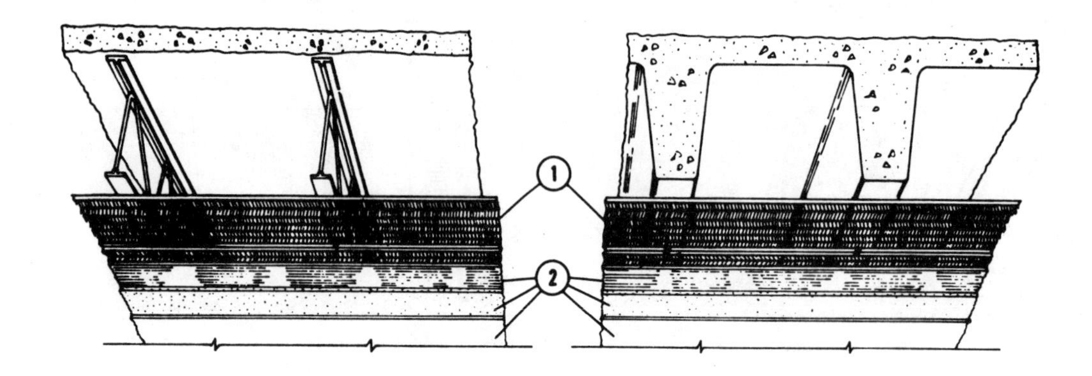

WOOD JOISTS

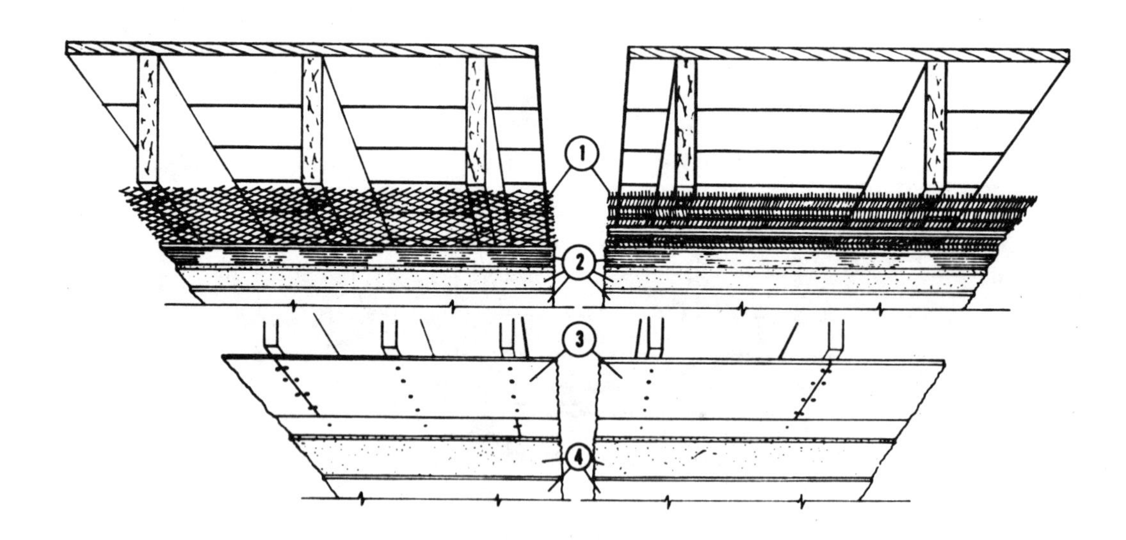

(1) Metal or Wire Fabric Lath
(2) Plaster
(3) Gypsum Lath
(4) Plaster

FINISHES / PLASTER ACCESSORIES | 09205

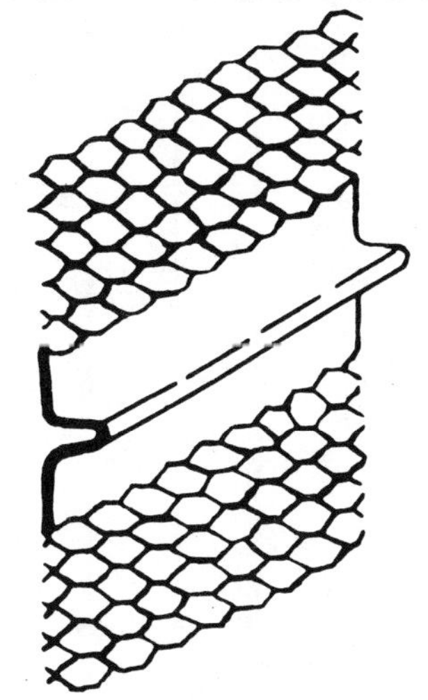

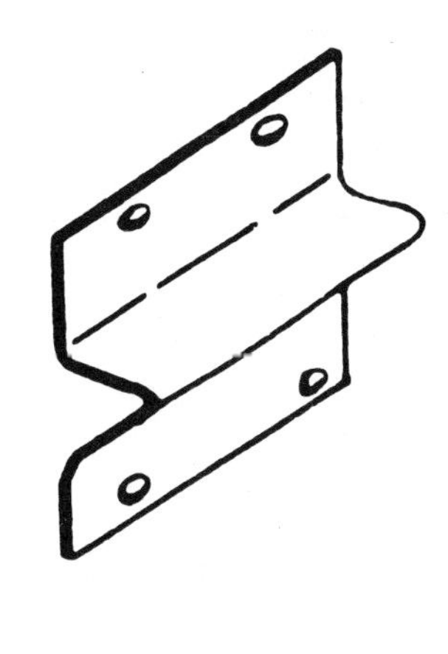

BASE OR PARTING SCREEDS

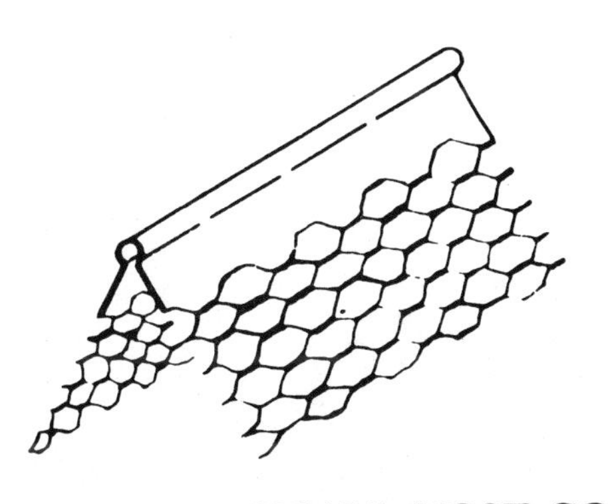

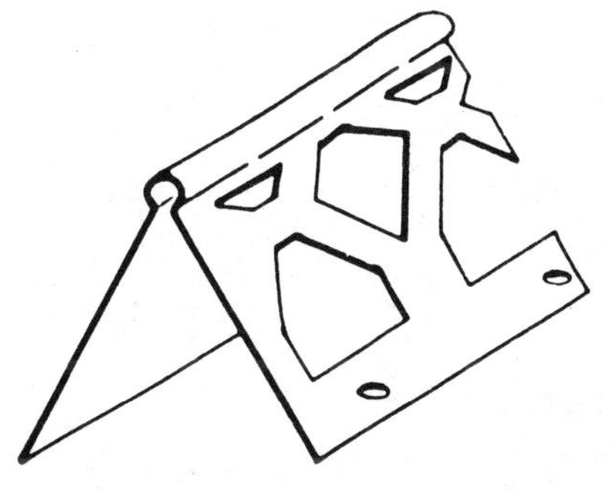

SMALL NOSE CORNER BEADS

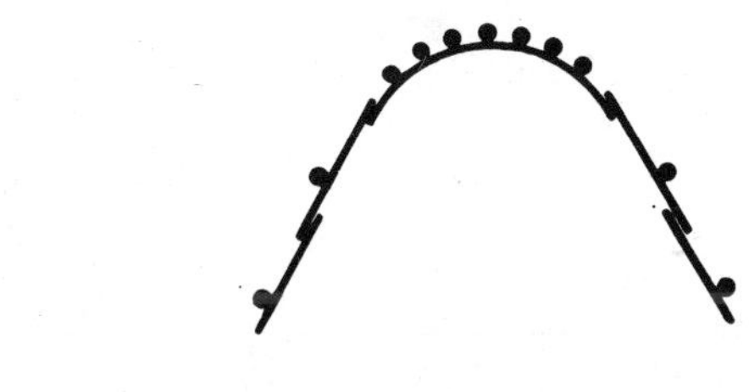

WIRE BULL NOSE
CORNER BEADS

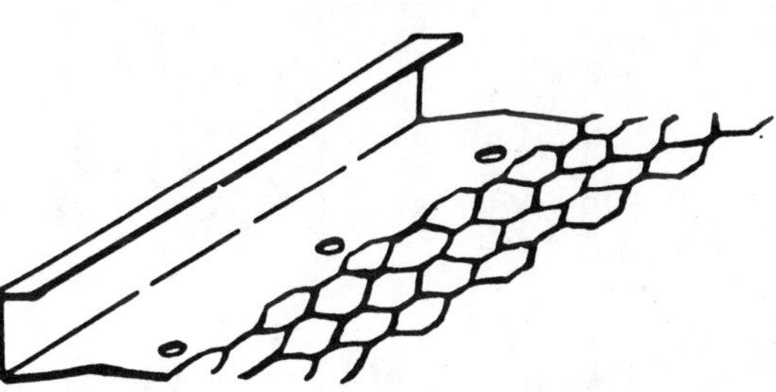

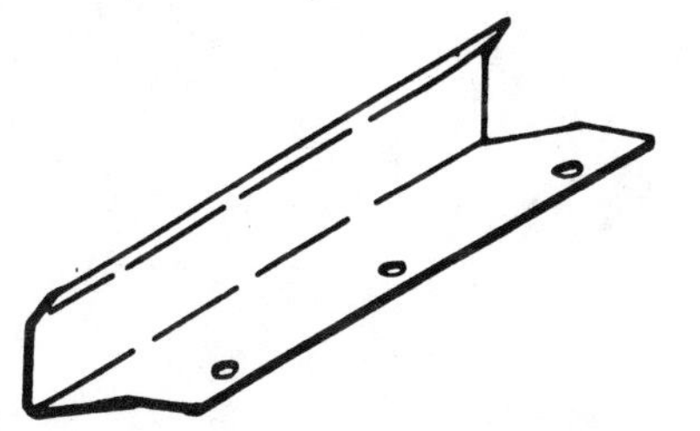

SQUARE CASING BEADS

FINISHES / METAL LATH

09205

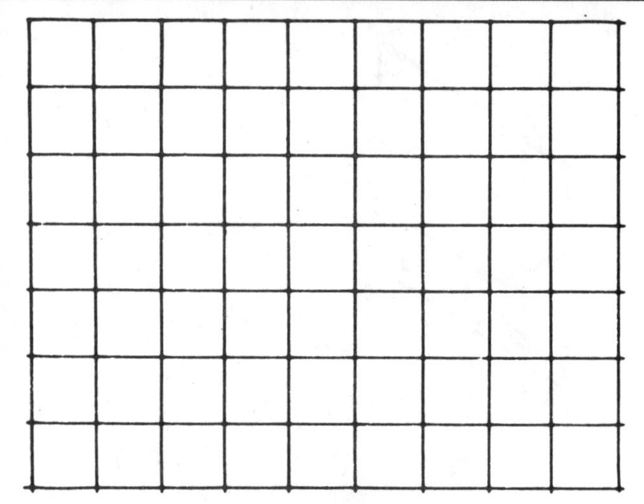

PLAIN WIRE FABRIC LATH

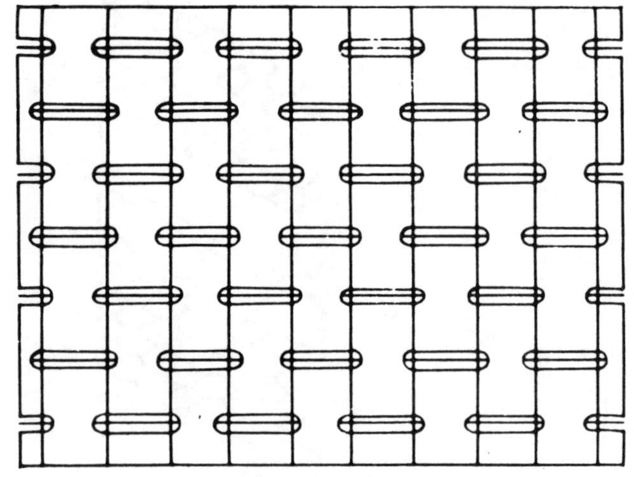

SELF-FURRING WIRE FABRIC LATH

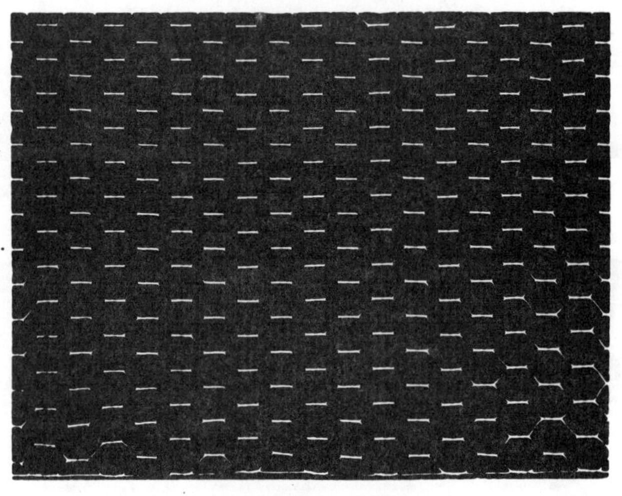

PAPER BACKED WOVEN WIRE FABRIC LATH

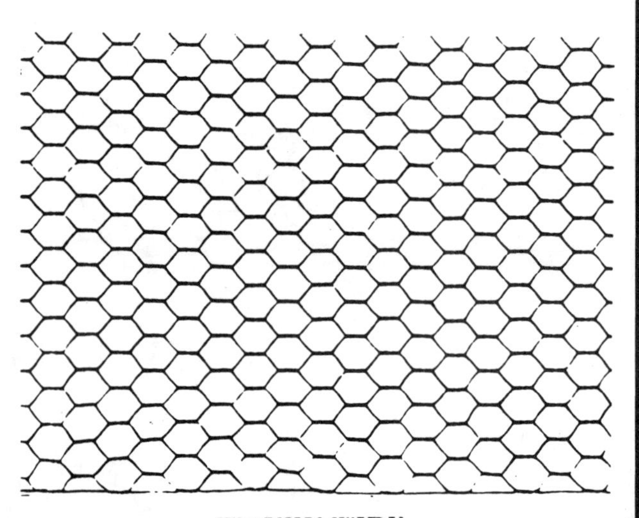

WOVEN WIRE FABRIC LATH (Also Available Self-Furred)

FLAT DIAMOND MESH METAL LATH

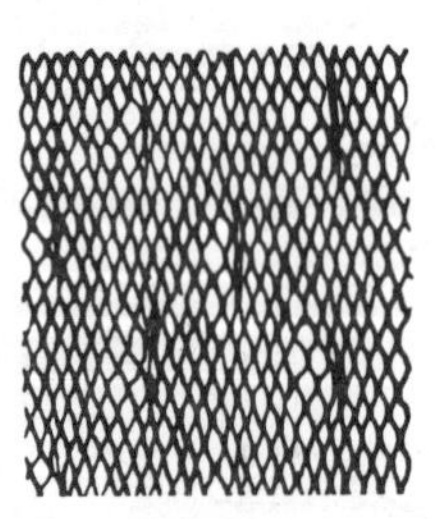

SELF-FURRING METAL LATH

FLAT RIB METAL LATH

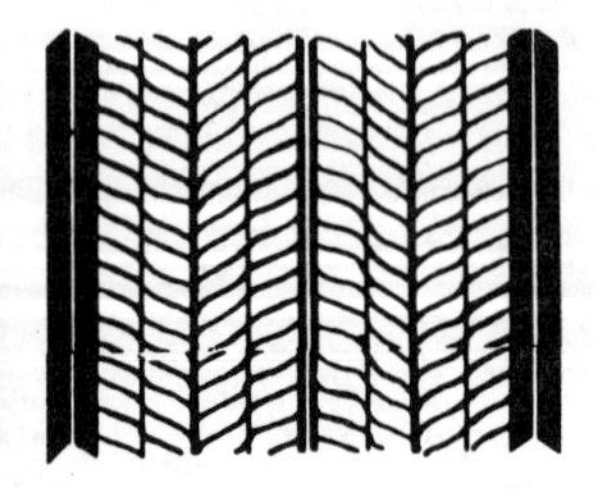

RIB METAL LATH

RIB METAL LATH

VERTICAL FURRING

Studless

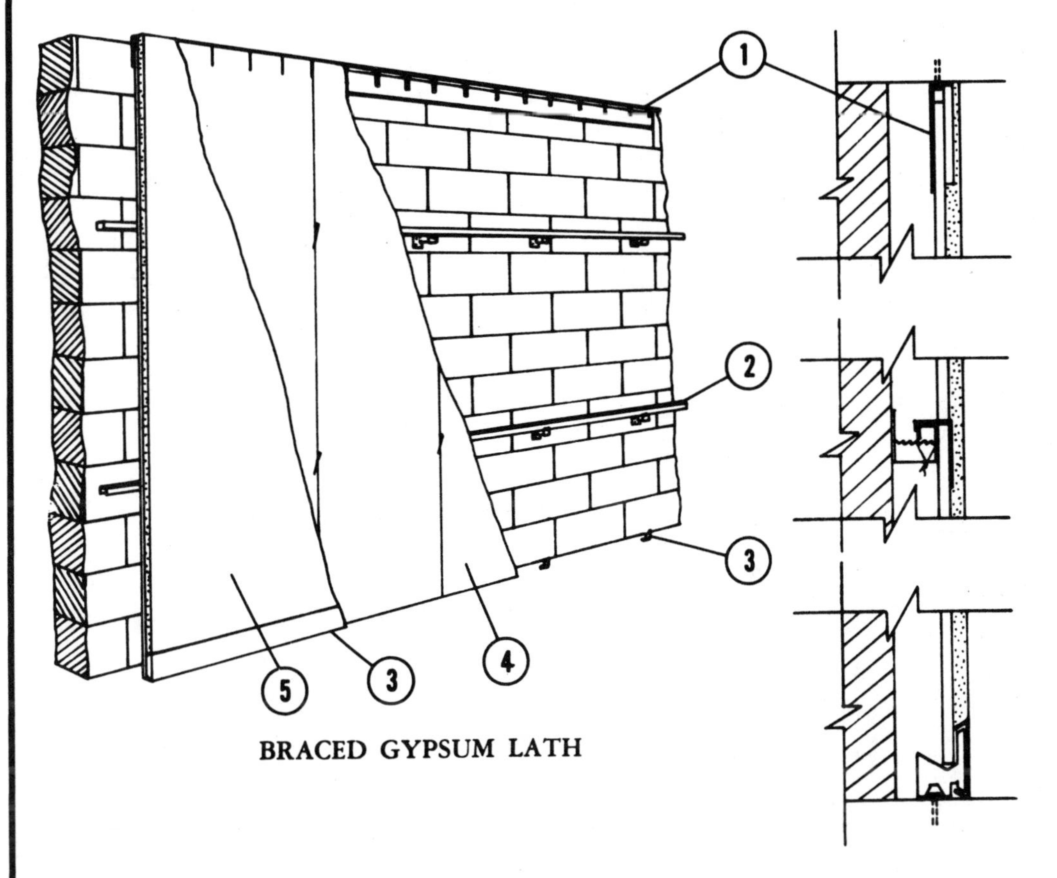

BRACED GYPSUM LATH

(1) Ceiling Runner

(2) Horizontal Stiffener (secured to bracing attachment)

(3) Metal Base and Clips

(4) Gypsum Lath

(5) Three Coats of Plaster (Scratch, Brown, Finish) (Minimum plaster thickness is ¾ inch.)

COLUMN FURRING

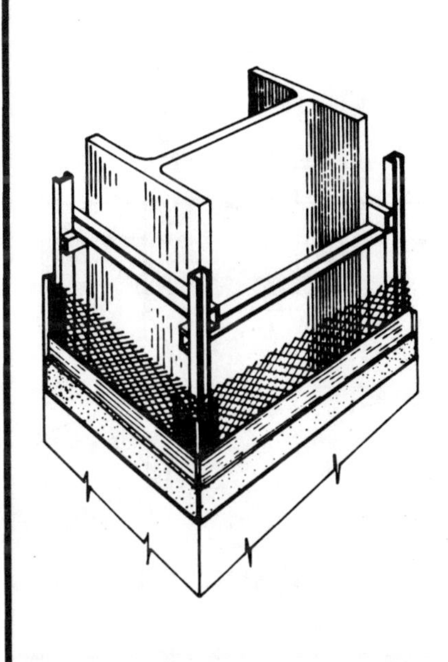

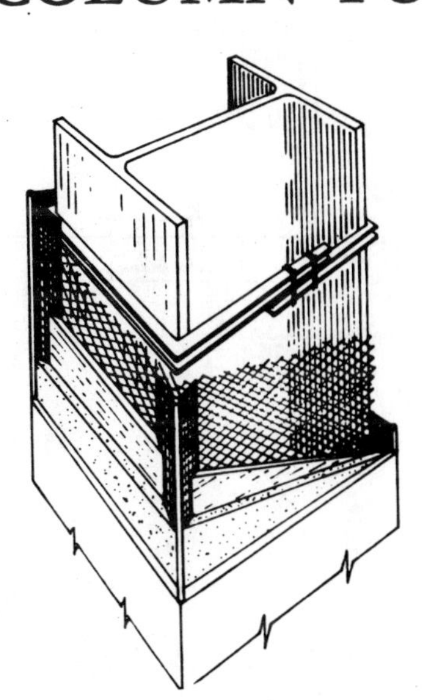

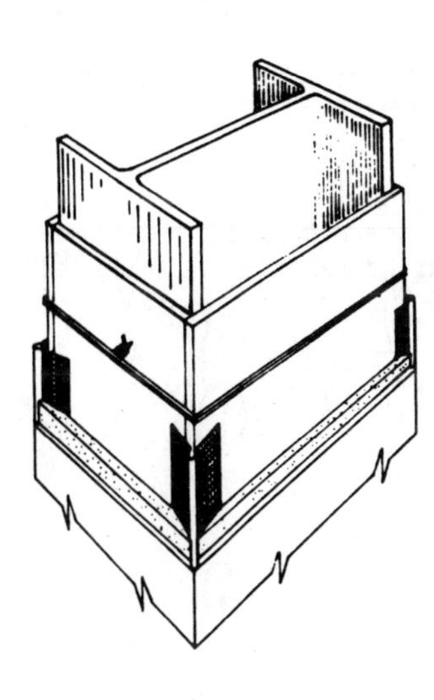

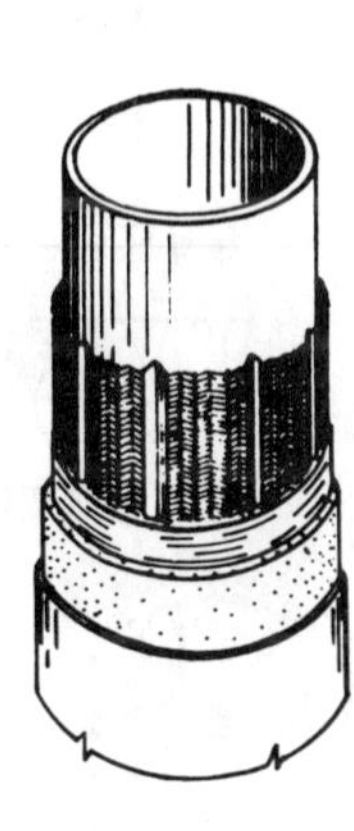

CONTACT FURRING

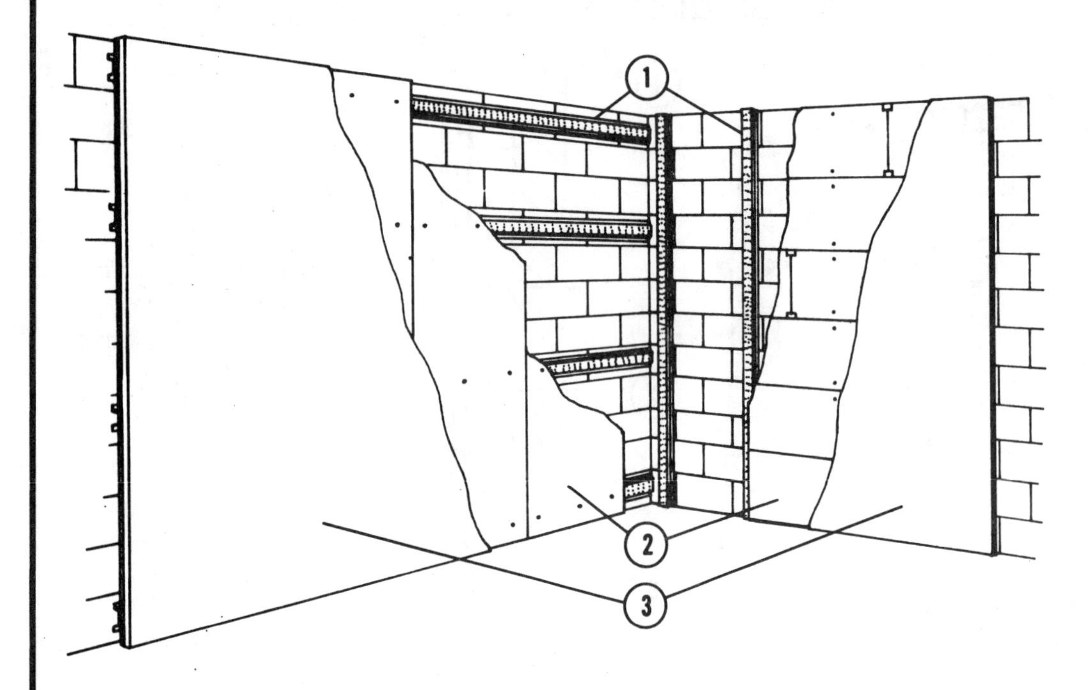

(1) Screw Channel
(2) Gypsum Lath (screwed on)
(3) Two Coats of Plaster (Brown, Finish)

FALSE BEAMS

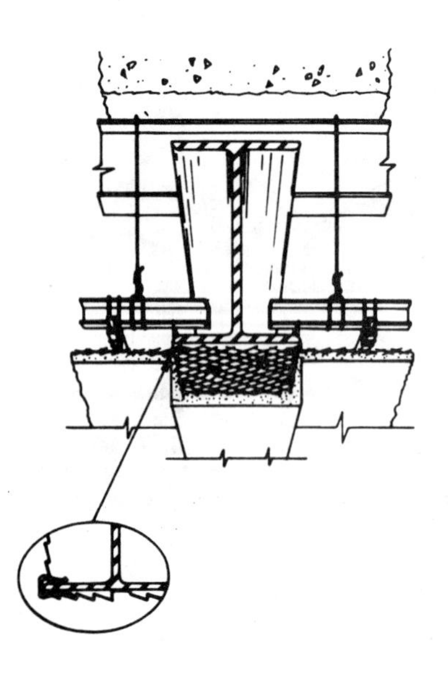

LATH DIRECT TO UNDERSIDE OF BEAM

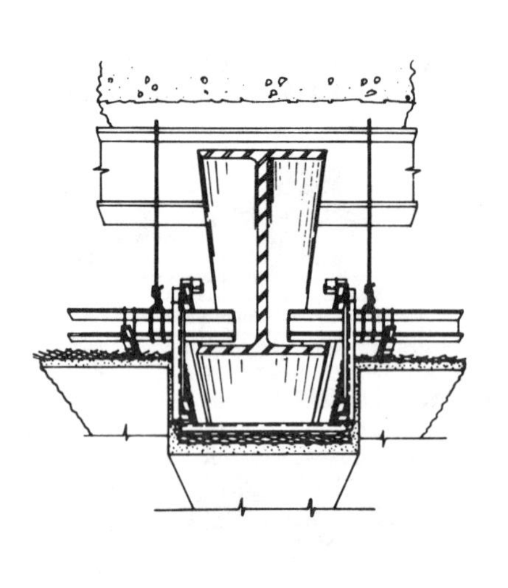

CHANNEL BRACKETS TO MAIN RUNNERS

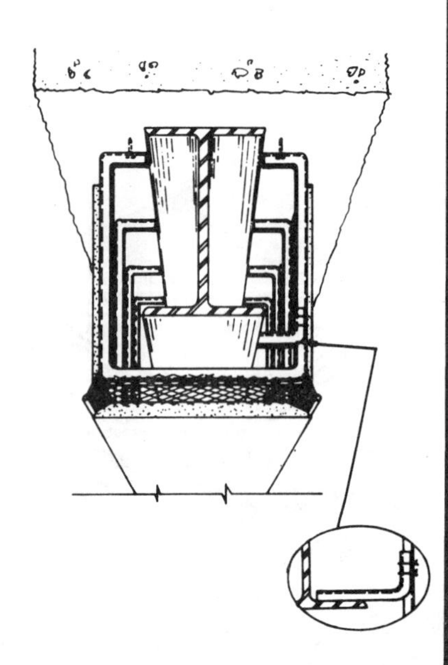

CHANNEL BRACKETS TO SLAB

MAXIMUM SPACING OF SUPPORTS FOR METAL LATH (Inches)

Type of Lath	Weight of Lath Lb. Per Sq. Yd.	WALLS AND PARTITIONS			CEILINGS	
		Wood Studs	Solid Partitions	Steel Studs, Wall Furring, etc.	Wood or Concrete	Metal
Diamond Mesh (flat expanded)	2.5	16	16	13½	12	12
	3.4	16	16	16	16	
Flat Rib	2.75	16	16	16	16	16
	3.4	19	24(3)	19	19	19
⅜″ Rib (1) (2)	3.4	24	(4)	24	24	24
	4.0	24	(4)	24	24	24
¾″ Rib	5.4	—	(4)	24(5)	36(6)	36(6)
Sheet Lath (2)	4.5	24	(4)	24	24	24

NOTE: Weights are exclusive of paper, fiber or other backing.

(1) 3.4 lb., ⅜″ Rib Lath is permissible under Concrete Joists at 27″ c.c.

(2) These spacings are based on a narrow bearing surface for the lath. When supports with a relatively wide bearing surface are used, these spacings may be increased accordingly, and still assure satisfactory work.

(3) This spacing permissible for Solid Partitions not exceeding 16′ in height. For greater heights, permanent horizontal stiffener channels or rods must be provided on channel side of partitions, every 6′ vertically, or else spacing shall be reduced 25%.

(4) For studless solid partitions, lath erected vertically.

(5) For interior wall furring or for application over solid surfaces for stucco.

(6) For contact or ceilings only.

TYPES OF LATH-ATTACHMENT TO WOOD AND METAL SUPPORTS

TYPE OF LATH	NAILS	MAXIMUM SPACING		SCREWS MAXIMUM SPACING		STAPLES Round or Flattened Wire				
									MAXIMUM SPACING	
	Type and Size	Vertical	Horizontal	Vertical	Horizontal	Wire Gauge No.	Crown	Leg	Vertical	Horizontal
		(In Inches)		(In Inches)			(In Inches)			
1. Diamond Mesh Expanded Metal Lath and Flat Rib Metal Lath	4d blued smooth box 1½ No. 14 gauge 7/32″ head (clinched) 1″ No. 11 gauge 7/16″ head, barbed 1½″ No. 11 gauge 7/16″ head, barbed	6 6 6	— — 6	6	6	16	¾	⅞	6	6
2. ⅜″ Rib Metal Lath and Sheet Lath	1½″ No. 11 ga. 7/16″ head, barbed	6	6	6	6	16	¾	1½	At Ribs	At Ribs
3. ¾″ Rib Metal Lath	4d common 1½″ No. 12½ gauge ¼″ head 2″ No. 11 gauge 7/16″ head, barbed	At Ribs	— At Ribs	At Ribs	At Ribs	16	¾	1⅝	At Ribs	At Ribs
4. Wire Fabric Lath	4d blued smooth box (clinched) 1″ No. 11 gauge 7/16″ head, barbed	6 6	— —	6	6	16	¾	⅞	6	6
	1½″ No. 11 gauge 7/16″ head, barbed 1¼″ No. 12 ga. ⅜″ head, furring 1″ no. 12 gauge ⅜″ head	6 6 6	6 6			16	7/16	⅞	6	6
5. ⅜″ Gypsum Lath	1⅛″ No. 13 gauge 12/61″ head, blued	8	8	8	8	16	¾	⅞	8	8
6. ½″ Gypsum Lath	1¼″ No. 13 gauge 12/61″ head, blued	8	8 6	8	8 6	16	¾	1⅛	8	8 6

STRESS RELIEF (CONTROL JOINTS)

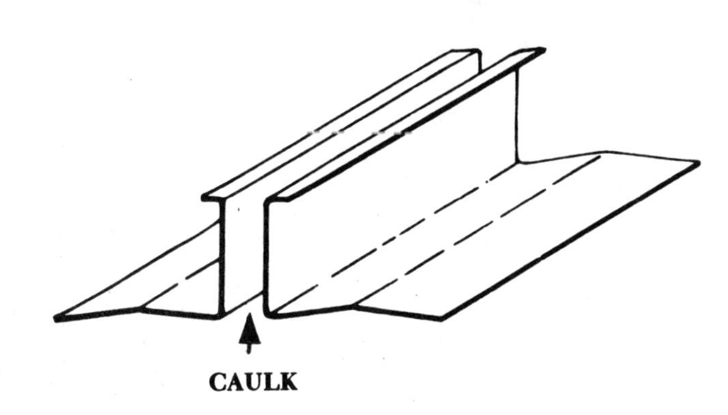

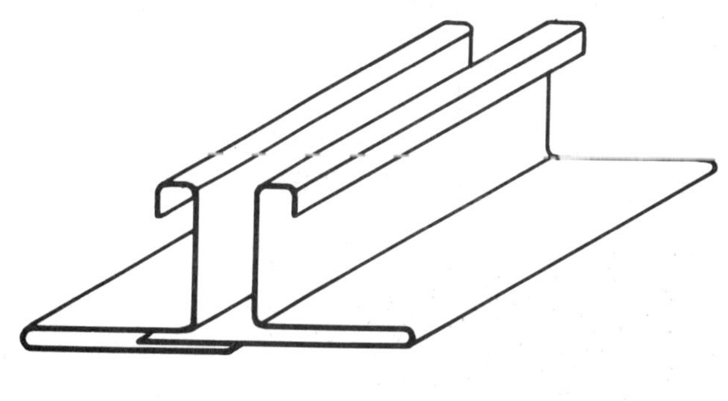

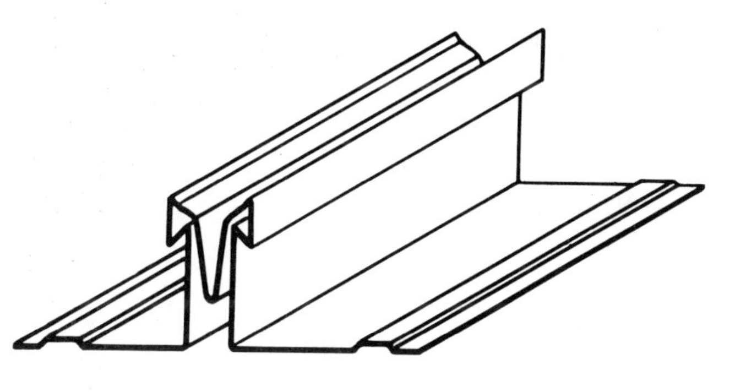

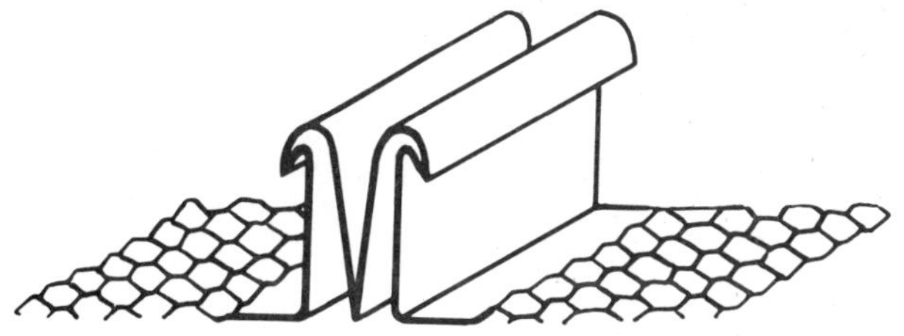

REVEALS

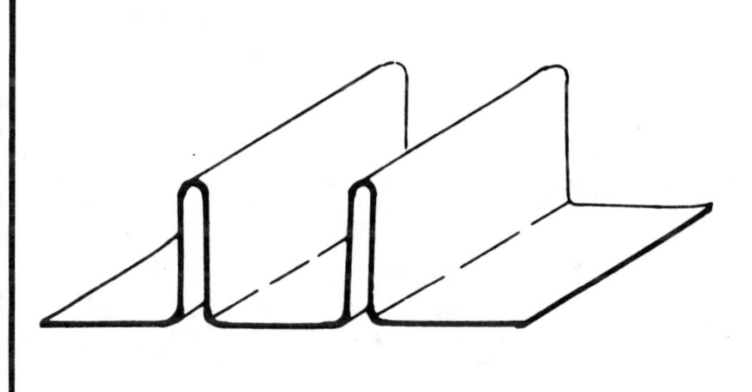

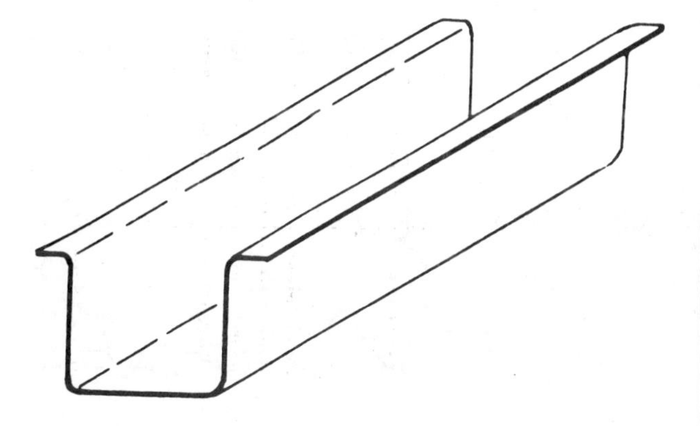

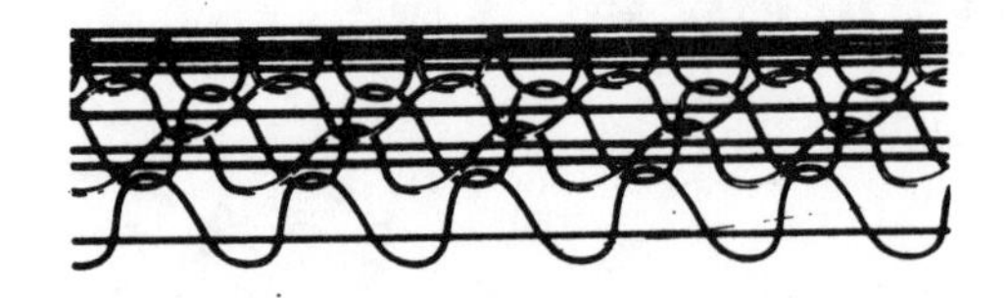

CORNER REINFORCEMENT (EXTERIOR) WIRE

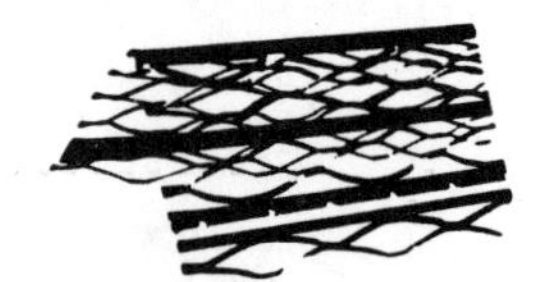

CORNER REINFORCEMENT (EXTERIOR) EXPANDED METAL

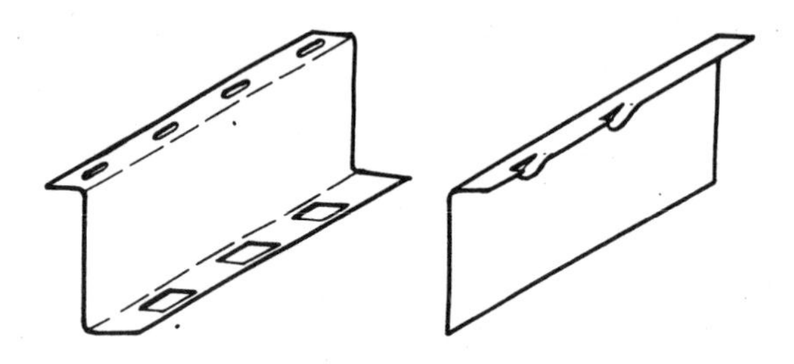

PARTITION RUNNERS (Z AND L SHAPE)

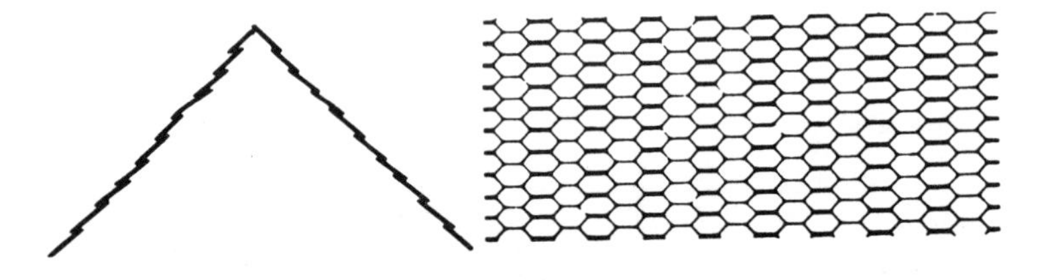

EXPANDED METAL CORNERITE

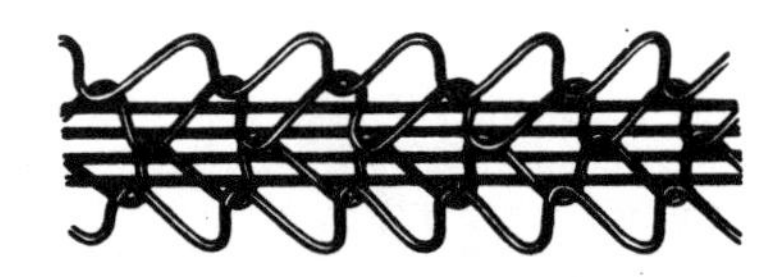

WIRE CORNERITE

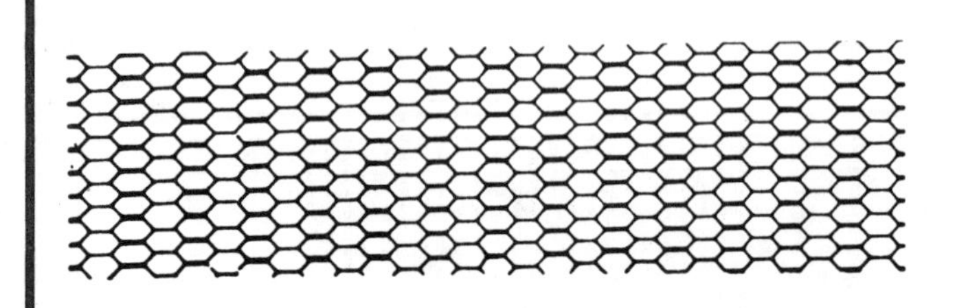

STRIP REINFORCEMENT (EXPANDED METAL)

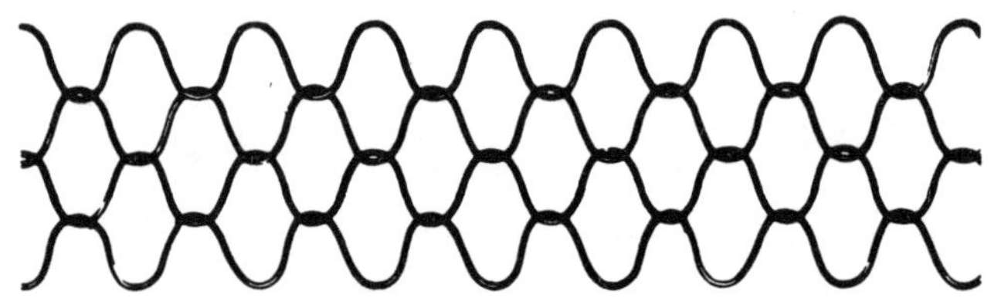

STRIP REINFORCEMENT (WIRE)

FINISHES / PLASTER ACCESSORIES 09205

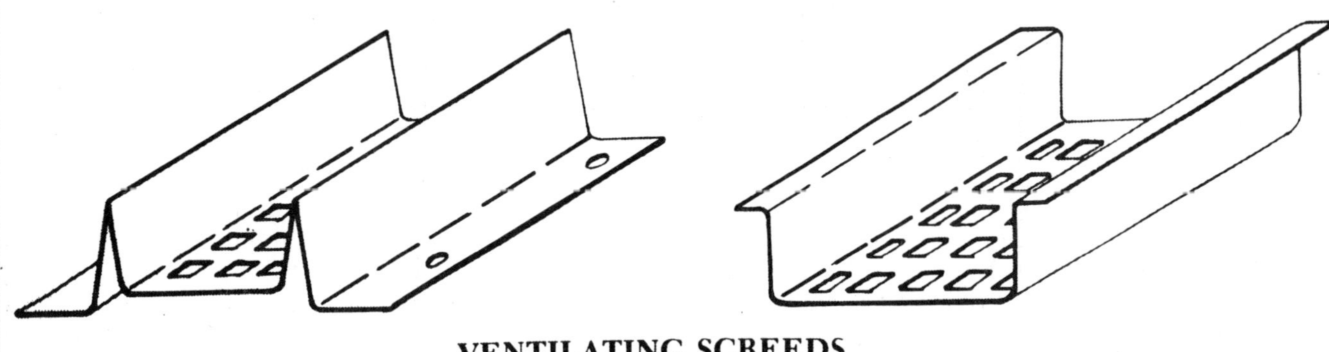

VENTILATING SCREEDS

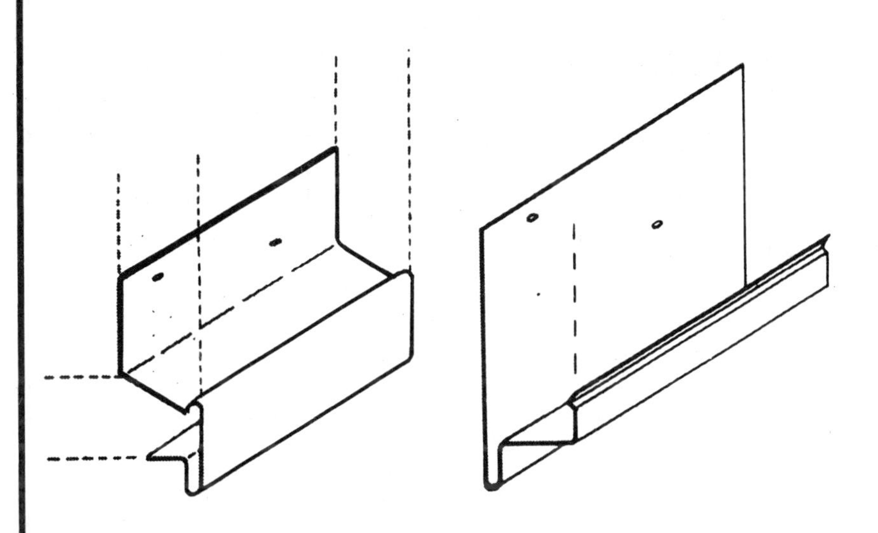

DRIP SCREEDS

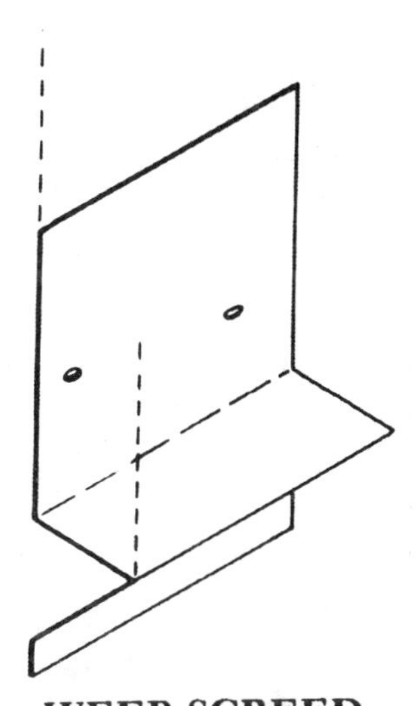

WEEP SCREED
(Also available with perforations)

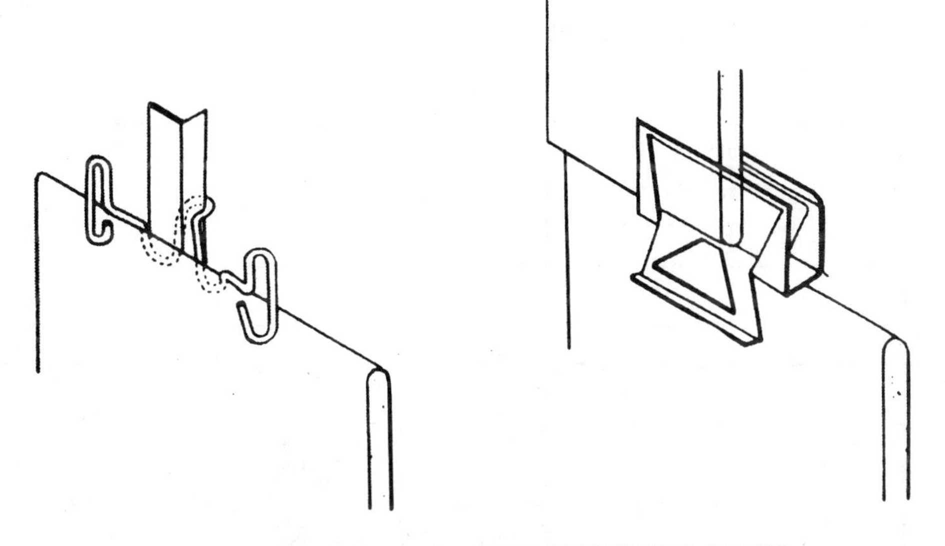

GYPSUM LATH ATTACHMENT CLIPS

VERTICAL FURRING

VERTICAL FURRING MEMBER	UNBRACED — STUD SPACING 24″	19″	16″	12″	BRACED — STUD SPACING 24″	19″	16″	12″
	Maximum Furring Heights				Maximum Vertical Distance Between Braces			
¾″ Channel	6′	7′	8′	9′	5′	5′	6′	7′
1½″ Channel	8′	9′	10′	12′	6′	7′	8′	9′
2″ Channel	9′	10′	11′	13′	7′	8′	9′	10′
2″ Prefab. Stud	8′	9′	10′	11′	6′	7′	8′	9′
2½″ Prefab. Stud	10′	11′	12′	14′	8′	9′	10′	11′
3¼″ Prefab. Stud	14′	16′	17′	20′	11′	13′	14′	16′

TYPES OF LATH—MAXIMUM SPACING OF SUPPORTS

TYPE OF LATH	Minimum Weight (psy), Gauge & Mesh Size	VERTICAL — WOOD	VERTICAL — METAL — Solid Plaster Partitions	VERTICAL — METAL — Other	HORIZONTAL — Wood or Concrete	HORIZONTAL — Metal
Expanded Metal Lath (Diamond Mesh)	2.5 3.4	16″ 16″	16″ 16″	12″ 16″	12″ 16″	12″ 16″
Flat rib Expanded Metal Lath	2.75 3.4	16″ 19″	16″ 24″	16″ 19″	16″ 19″	16″ 19″
Stucco Mesh Expanded Metal Lath	1.8 and 3.6	16″	—	—	—	—
⅜″ Rib Expanded Metal Lath	3.4 4.0	24″ 24″	—	24″ 24″	24″ 24″	24″ 24″
Sheet Lath	4.5	24″	—	24″	24″	24″
¾″ Rib Expanded Metal Lath (Not manufactured in West)	5.4	—	—	—	36″	36″
Wire Fabric Lath — Welded	1.95 lbs., 11ga., 2″ x 2″ 1.4 lbs., 16 ga., 2″ x 2″ 1.4 lbs., 18 ga., 1″ x 1″	24″ 16″ 16″	24″ 16″ —	24″ 16″ —	24″ 16″ —	24″ 16″ —
Wire Fabric Lath — Woven	1.4 lbs., 17 ga., 1½″ Hex. 1.4 lbs., 18 ga., 1″ Hex.	24″ 24″	16″ 16″	16″ 16″	24″ 24″	16″ 16″
⅜″ Gypsum Lath (plain)	—	16″	—	16″	16″	16″
(Large Size)	—	16″	—	16″	16″	16″
½″ Gypsum Lath (plain)	—	24″	—	24″	24″	24″
(Large Size)	—	24″	No supports; Erected vertically	24″	24″	16″
⅝″ Gypsum Lath (Large Size)	—	24″	No supports; Erected vertically	24″	24″	16″

PIPE WEIGHTS

CAST IRON PIPE

SERVICE WEIGHT			
Size, Inches	Weight of Pipe	Weight of Water	Total Weight-Lbs.
2″	3.8	1.45	5.3
3″	5.6	3.2	8.8
4″	7.5	5.5	13.0
5″	9.8	8.7	18.5
6″	12.4	12.5	24.9
8″	18.5	21.7	40.2

EXTRA HEAVY			
Size, Inches	Weight of Pipe	Weight of Water	Total Weight-Lbs.
2″	4.3	1.45	5.8
3″	8.3	3.2	11.5
4″	10.8	5.5	16.3
5″	13.3	8.7	22.0
6″	16.0	12.5	28.5
8″	26.5	21.7	48.2

STEEL PIPE

Pipe Size	W/40	H_2O/Lbs.	Total Lbs./L.F.
2″	3.65	1.45	5.1
2½″	5.79	2.07	7.86
3″	7.57	3.2	10.77
3½″	9.11	4.28	13.39
4″	10.8	5.51	16.31
5″	14.6	8.66	23.26
6″	18.0	12.5	30.5
8″	28.6	21.66	50.26
10″	40.5	34.15	74.65
12″			

SPRINKLER AREA CALCULATIONS

Typical maximum floor area allowed per system riser:

Light Hazard	Ordinary Hazard	Extra Hazard
52,000 S.F.	40,000-52,000 S.F.	25,000 S.F.

Typical maximum floor area coverage allowed per sprinkler head:

Light Hazard	Ordinary Hazard	Extra Hazard
130-200 S.F.	100-130 S.F.	90 S.F.

Typical maximum spacing between lines and sprinkler heads:

Light Hazard	Ordinary Hazard	Extra Hazard
12-15 feet	12-15 feet	12 feet

Note: This data is for estimating purposes only. Check all applicable codes and regulations for specific requirements.

SPRINKLER HEAD CALCULATIONS

Typical maximum quantity of sprinkler heads allowed by pipe size.

Light Hazard:

For sprinklers below ceiling:

Steel		Copper	
1 in. pipe	2 sprinklers	1 in. tube	2 sprinklers
1¼ in. pipe	3 sprinklers	1¼ in. tube	3 sprinklers
1½ in. pipe	5 sprinklers	1½ in. tube	5 sprinklers
2 in. pipe	10 sprinklers	2 in. tube	12 sprinklers
2½ in. pipe	30 sprinklers	2½ in. tube	40 sprinklers
3 in. pipe	60 sprinklers	3 in. tube	65 sprinklers
3½ in. pipe	100 sprinklers	3½ in. tube	115 sprinklers

For sprinklers above and below ceiling:

Steel		Copper	
1 in.	2 sprinklers	1 in.	2 sprinklers
1¼ in.	4 sprinklers	1¼ in.	4 sprinklers
1½ in.	7 sprinklers	1½ in.	7 sprinklers
2 in.	15 sprinklers	2 in.	18 sprinklers
2½ in.	50 sprinklers	2½ in.	65 sprinklers

Ordinary Hazard:

For sprinklers above ceiling:

Steel		Copper	
1 in. pipe	2 sprinklers	1 in. tube	2 sprinklers
1¼ in. pipe	3 sprinklers	1¼ in. tube	3 sprinklers
1½ in. pipe	5 sprinklers	1½ in. tube	5 sprinklers
2 in. pipe	10 sprinklers	2 in. tube	12 sprinklers
2½ in. pipe	20 sprinklers	2½ in. tube	25 sprinklers
3 in. pipe	40 sprinklers	3 in. tube	45 sprinklers
3½ in. pipe	65 sprinklers	3½ in. tube	75 sprinklers
4 in. pipe	100 sprinklers	4 in. tube	115 sprinklers
5 in. pipe	160 sprinklers	5 in. tube	180 sprinklers
6 in. pipe	275 sprinklers	6 in. tube	300 sprinklers

For sprinklers above and below ceiling:

Steel		Copper	
1 in.	2 sprinklers	1 in.	2 sprinklers
1¼ in.	4 sprinklers	1¼ in.	4 sprinklers
1½ in.	7 sprinklers	1½ in.	7 sprinklers
2 in.	15 sprinklers	2 in.	18 sprinklers
2½ in.	30 sprinklers	2½ in.	40 sprinklers
3 in.	60 sprinklers	3 in.	65 sprinklers

Extra Hazard:

For sprinklers below ceiling:

Steel		Copper	
1 in. pipe	1 sprinkler	1 in. tube	1 sprinkler
1¼ in. pipe	2 sprinklers	1¼ in. tube	2 sprinklers
1½ in. pipe	5 sprinklers	1½ in. tube	5 sprinklers
2 in. pipe	8 sprinklers	2 in. tube	8 sprinklers
2½ in. pipe	15 sprinklers	2½ in. tube	20 sprinklers
3 in. pipe	27 sprinklers	3 in. tube	30 sprinklers
3½ in. pipe	40 sprinklers	3½ in. tube	45 sprinklers
4 in. pipe	55 sprinklers	4 in. tube	65 sprinklers
5 in. pipe	90 sprinklers	5 in. tube	100 sprinklers
6 in. pipe	150 sprinklers	6 in. tube	170 sprinklers

Note: This data is for estimating purposes only. Check all applicable codes and regulations for specific requirements.

SPRINKLER HAZARD OCCUPANCIES

Typical Light Hazard Occupancies:

Churches
Clubs
Eaves and overhangs, if combustible construction with no combustible beneath
Educational
Hospitals
Institutional
Libraries, except large stack rooms
Museums
Nursing or Convalescent Homes
Office, including Data Processing
Residential
Restaurant seating areas
Theaters seating areas
Theaters and Auditoriums excluding stages and prosceniums
Unused attics

Typical Ordinary Hazard Occupancies (Group 1):

Automobile parking garages
Bakeries
Beverage manufacturing
Canneries
Dairy products manufacturing and processing
Electronic plants
Glass and glass products manufacturing
Laundries
Restaurant service areas

Typical Ordinary Hazard Occupancies (Group 2):

Cereal mills
Chemical plants — Ordinary
Cold Storage warehouses
Confectionery products
Distilleries
Leather goods mfg.
Libraries-large stack room areas
Mercantiles
Machine shops
Metal working
Printing and publishing
Textile mfg.
Tobacco Products mfg.
Wood product assembly

Typical Ordinary Hazard Occupancies (Group 3):

Feed mills
Paper and pulp mills
Paper process plants
Piers and wharves
Repair garages
Tire manufacturing
Warehouses (having moderate to higher combustibility of content, such as paper, household furniture, paint, general storage, whiskey, etc.)[1]
Wood machining

Typical Extra Hazard Occupancies (Group 1):

Combustible Hydraulic Fluid use areas
Die Casting
Metal Extruding
Plywood and particle board manufacting
Printing (using inks with below 100°F [37.8°C] flash points)
Rubber reclaiming, compounding, drying, milling, vulcanizing
Saw Mills
Textile picking, opening, blending, garnetting, carding, combining of cotton, synthetics, wool shoddy or burlap
Upholstering with plastic foams

Typical Extra Hazard Occupancies (Group 2):

Asphalt saturating
Flammable liquids spraying
Flow coating
Mobile Home or Modular Building assemblies (where finished enclosure is present and has combustible interiors)
Open Oil quenching
Solvent cleaning
Varnish and paint dipping

....

TYPICAL MINIMUM SIZE OF HORIZONTAL BUILDING STORM DRAINS AND BUILDING STORM SEWERS

Diameter of drain	Maximum projected area in square feet for various slopes		
	⅛ inch per feet slope	¼ inch per feet slope	½ inch per feet slope
3	822	1160	1644
4	1880	2650	3760
5	3340	4720	6680
6	5350	7550	10700
8	11500	16300	23000
10	20700	29200	41400
12	33300	47000	66600
15	59500	84000	119000

TYPICAL VENTILATION AIR REQUIREMENTS FOR SPECIAL USES

Occupancy Classification	Required ventilation air in cfm per human occupant
Special areas	
Lockers	2* (or 30 per locker)
Wardrobes	2*
Public bathrooms	40**
Private bathrooms	25**
Swimming pools	15 (per occupant)
Exitways and corridors	1½*

*Per square foot floor area.
**Per water closet or urinal.

TYPICAL VENTILATION AIR REQUIREMENTS FOR RETAIL USES

Occupancy Classification	Required ventilation air in cfm per human occupant
Mercantile	
Sales floors and showrooms (basement & grade floors)	7
Sales and showrooms (upper floors)	7
Storage areas	5
Dressing rooms	7
Malls	7
Shipping areas	15
Elevators	7
Supermarkets	
Meat processing rooms	5
Drugs stores	
Pharmacists' work rooms	20
Specialty shops	
Pet shops	1.0*
Florists	5
Greenhouses	5

*cfm per sq. ft. floor area.

TYPICAL VENTILATION AIR REQUIREMENTS FOR FACTORY AND INDUSTRIAL USES

Occupancy Classification	Required ventilation air in cfm per human occupant
Factory and industrial	
Metalworking & finishing	35
Automotive engine test	Require
Paint spray booths	Special
Picking, etching and plating lines	Exhaust
Degreasing booths	Systems
Sandblasting booths	
Chemicals and pharmaceuticals	
Dusty operations	30
Rooms containing potential gas emitters	20
Drying oven rooms	15
Fermentation rooms	15
Pillmaking booths	10
Packaging areas	10
Utility rooms	7
Computer rooms	7
Textiles-clothes manufacturer	15
Electronics and aerospace circuit board and soldering rooms	20
Wood products, papermaking	20
Brewing, distilling, wineries, bottling	20*
Food processing	20
Tobacco processing	20
Power plants	
Control rooms	10
Boiler rooms	35
Generator rooms	20
Sewage treatment plants	
Control rooms	10
Compressor/blower motor rooms	20
Glass and ceramic manufacturer	20
Agricultural	20

TYPICAL VENTILATION AIR REQUIREMENTS FOR BUSINESS USES

Occupancy Classification	Required ventilation air in cfm per human occupant
Business	
Banks	
(see offices)	
Vaults	5
Barber, beauty and health services	
Beauty shops (hair dressers)	25
Reducing salons	25
Sauna baths, steam rooms	5
barber shops	7
Photo studios	
Camera rooms, stages	5
Dark rooms	
Shoe repair shops	
Workrooms/trade areas	10
Offices	
General office space and showrooms	15
Conference rooms	25
Drafting/art rooms	7
Doctor's consultation rooms	10
Waiting rooms	10
Lithographing rooms	7
Diazo printing rooms	7
Computer rooms	5
Keypunch rooms	7
Communication	
TV/radio broadcasting booths, studios	30
Motion picture and TV stages	30
Pressrooms	15
Composing rooms	7
Engraving rooms	7
Telephone switchboard rooms (manual)	7
Telephone switchgear rooms (automatic)	7
Teletypewriter/facsimile rooms	5
Research institutes	
Laboratories:	
Light duty; non-chemical	15
Chemical	15
Heavy-duty	15
Radioisotope, chemical & biologically toxic	15
Machine shops	15
Dark rooms, spectroscopy rooms	10
Animal rooms	40
Veterinary hospitals	
Kennels, stalls	25
Operating rooms	25
Reception rooms	10

TYPICAL VENTILATION AIR REQUIREMENTS FOR INSTITUTIONAL USES

Occupancy Classification	Required ventilation air in cfm per human occupant
Institutional	
Prisons	
Cell blocks	7
Eating halls	15
Guard stations	7

APPLICATIONS FOR CONDUCTORS USED FOR GENERAL WIRING

	AMBIENT TEMPERATURE								
	60°C 140°F	75°C 167°F	85°C 185°F	90°C 194°F	110°C 230°F	200°C 392°F	Dry	Dry or Wet	FEATURES
R	X						X		Code Rubber
RH		X					X		Heat Resistant
RHH				X			X		More Heat Resistant
RW	X							X	Moisture Resistant
RH-RW	X							X	Moisture and Heat Resistant
		X					X		Moisture and Heat Resistant
RHW		X						X	Moisture and Heat Resistant
RU	X						X		Latex Rubber
RUH		X					X		Heat Resistant
RUW	X							X	Moisture Resistant
T	X						X		Thermoplastic
TW	X							X	Moisture Resistant
THHN				X			X		Heat Resistant
THW		X						X	Moisture and Heat Resistant
THWN		X						X	Moisture and Heat Resistant
MI			X					X	Mineral Insulated Metal Sheathed
V			X				X		Varnished Cambric
AVA					X		X		With Asbestos
AVB				X			X		With Asbestos
AVL					X			X	With Asbestos

This table does not include special condition conductors, thickness of conductor insulation, or reference to all outer protective coverings.

GENERAL CLASSIFICATION OF INSULATIONS:

A .... Asbestos
H .... Heat Resistant
MI .... Mineral Insulation
R .... Rubber
RU ....Latex Rubber
V ....Varnished Cambric
T ....Thermoplastic
W ....(Water) Moisture Resistant

WIRE AND SHEET METAL GAGES

(In Decimals of an Inch)

Name of Gage	American Wire Gage (A.W.G.) (Corresponds to Brown & Sharpe Gage)	Birmingham Iron Wire Gage (B.W.G.)	United States Standard Gage (U.S.S.G.)	
Principal Use	Electrical Wire & Non-Ferrous Sheet Metal	Iron or Steel Wire	Ferrous Sheet Metal	
Gage No.				Gage No.
00 00000				00 00000
0 00000	.5800			0 00000
00000	.5165	.500		00000
0000	.4600	.454		0000
000	.4096	.425		000
00	.3648	.380		00
0	.3249	.340		0
1	.2893	.300		1
2	.2576	.284		2
3	.2294	.259	23.91	3
4	.2043	.238	.2242	4
5	.1819	.220	.2092	5
6	.1620	.203	.1943	6
7	.1443	.180	.1793	7
8	.1285	.165	.1644	8
9	.1144	.148	.1495	9
10	.1019	.134	.1345	10
11	.0907	.120	.1196	11
12	.0808	.109	.1046	12
13	.0720	.095	.0897	13
14	.0641	.083	.0747	14
15	.0571	.072	.0673	15
16	.0508	.065	.0598	16
17	.0453	.058	.0538	17
18	.0403	.049	.0478	18
19	.0359	.042	.0418	19
20	.0320	.035	.0359	20
21	.0285	.032	.0329	21
22	.0253	.028	.0299	22
23	.0226	.025	.0269	23
24	.0201	.022	.0239	24
25	.0179	.020	.0209	25
26	.0159	.018	.0179	26
27	.0142	.016	.0164	27
28	.0126	.014	.0149	28
29	.0113	.013	.0135	29
30	.0100	.012	.0120	30
31	.0089	.010	.0105	31
32	.0080	.009	.0097	32
33	.0071	.008	.0090	33
34	.0063	.007	.0082	34
35	.0056	.005	.0075	35
36	.0050	.004	.0067	36
37	.0045		.0064	37
38	.0040		.0060	38
39	.0035			39
40	.0031			40

BNi Building News

Geographic Cost Modifiers

The costs as presented in this book attempt to represent national averages. Costs, however, vary among regions, states and even between adjacent localities.

In order to more closely approximate the probable costs for specific locations throughout the U.S., this table of Geographic Cost Modifiers is provided. These adjustment factors are used to modify costs obtained from this book to help account for regional variations of construction costs and to provide a more accurate estimate for specific areas. The factors are formulated by comparing costs in a specific area to the costs as presented in the Costbook pages. An example of how to use these factors is shown below. Whenever local current costs are known, whether material prices or labor rates, they should be used when more accuracy is required.

Cost Obtained from Costbook Pages X Location Cost Adjustment Factor = Adjusted Cost

For example, a project estimated to cost $125,000 using the Costbook pages can be adjusted to more closely approximate the cost in Los Angeles:

$125,000 X 1.07 = $133,750

Geographic Cost Modifiers

ALABAMA

City	Modifier
BIRMINGHAM	0.79
HUNTSVILLE	0.77
MOBILE	0.81
MONTGOMERY	0.75
TUSCALOOSA	0.75

ALASKA

City	Modifier
ANCHORAGE	1.30
JUNEAU	1.33
FAIRBANKS	1.37
NOME	1.43

ARIZONA

City	Modifier
FLAGSTAFF	0.91
PHOENIX	0.89
PRESCOTT	0.91
TUCSON	0.89
YUMA	0.87

ARKANSAS

City	Modifier
FAYETTEVILLE	0.73
FORTH SMITH	0.74
LITTLE ROCK	0.78
PINE BLUFF	0.75

CALIFORNIA

City	Modifier
ANAHEIM	1.02
BAKERSFIELD	0.97
LOS ANGELES	1.07
REDDING	0.92
RIVERSIDE	0.97
SACRAMENTO	1.00
SAN DIEGO	1.02
SAN JOSE	1.07
SAN FRANCISCO	1.12
SANTA BARBARA	1.07

COLORADO

City	Modifier
BOULDER	0.92
COLORADO SPRINGS	0.92
DENVER	0.95
GRAND JUNCTION	0.90
PUEBLO	0.90

CONNECTICUT

City	Modifier
BRIDGEPORT	1.03
HARTFORD	1.01
NEW LONDON	0.99
STAMFORD	1.06
WATERBURY	0.96

DELAWARE

City	Modifier
DOVER	0.89
WILMINGTON	0.91

FLORIDA

City	Modifier
JACKSONVILLE	0.80
MIAMI	0.86
ORLANDO	0.80
TAMPA	0.82
WEST PALM BEACH	0.84

GEORGIA

City	Modifier
ATLANTA	0.83
AUGUSTA	0.75
COLUMBUS	0.75
MACON	0.77
SAVANNAH	0.79

HAWAII

City	Modifier
HILO	1.31
HONOLULU	1.25
MAUI	1.28

IDAHO

City	Modifier
BOISE	0.92
LEWISTON	0.90
POCATELLO	0.88
TWIN FALLS	0.88

ILLINOIS

City	Modifier
CHICAGO	1.03
MOLINE	0.89
PEORIA	0.92
ROCKFORD	0.92
SPRINGFIELD	0.89

INDIANA

City	Modifier
FORT WAYNE	0.88
EVANSVILLE	0.88
GARY	0.99
INDIANAPOLIS	0.95
TERRE HAUTE	0.86

IOWA

City	Modifier
COUNCIL BLUFFS	0.85
DAVENPORT	0.87
DES MOINES	0.89
SIOUX CITY	0.85
WATERLOO	0.87

KANSAS

City	Modifier
DODGE CITY	0.77
SALINA	0.79
TOPEKA	0.83
WICHITA	0.81

KENTUCKY

City	Modifier
BOWLING GREEN	0.83
LEXINGTON	0.85
LOUISVILLE	0.89
PADUCAH	0.87

LOUISIANA

City	Modifier
BATON ROUGE	0.84
LAKE CHARLES	0.82
MONROE	0.78
NEW ORLEANS	0.86
SHREVEPORT	0.78

MAINE

City	Modifier
AUGUSTA	0.85
BANGOR	0.83
LEWISTON	0.87
PORTLAND	0.89

MARYLAND

City	Modifier
ANNAPOLIS	0.92
BALTIMORE	0.90
HAGERSTOWN	0.88
ROCKVILLE	0.95

MASSACHUSETTS

City	Modifier
BOSTON	1.01
LOWELL	0.94
FALL RIVER	0.91
SPRINGFIELD	0.94
WORCESTER	0.96

MICHIGAN

City	Modifier
DETROIT	0.95
GRAND RAPIDS	0.88
LANSING	0.88
MARQUETTE	0.86
SAGINAW	0.90

MINNESOTA

City	Modifier
DULUTH	0.93
MINNEAPOLIS	0.98
ROCHESTER	0.93
SAINT PAUL	0.98

MISSISSIPPI

City	Modifier
COLUMBUS	0.77
GULFPORT	0.79
JACKSON	0.81
VICKSBURG	0.79

MISSOURI

City	Modifier
JOPLIN	0.84
KANSAS CITY	0.86
SAINT LOUIS	0.90
SAINT JOSEPH	0.84
SPRINGFIELD	0.84

MONTANA

City	Modifier
BILLINGS	0.87
BUTTE	0.85
GREAT FALLS	0.87
HELENA	0.85
MISSOULA	0.85

Geographic Cost Modifiers

NEBRASKA	
GRAND ISLAND	0.81
LINCOLN	0.85
NORTH PLATTE	0.81
OMAHA	0.87

NEVADA	
CARSON CITY	0.95
LAS VEGAS	0.97
SPARKS	0.97
RENO	1.00

NEW HAMPSHIRE	
CONCORD	0.86
MANCHESTER	0.88
NASHUA	0.90

NEW JERSEY	
ATLANTIC CITY	0.97
CAMDEN	0.95
NEWARK	1.02
PATERSON	0.99
TRENTON	0.97

NEW MEXICO	
ALBUQUERQUE	0.86
CARLSBAD	0.84
LAS CRUCES	0.82
SANTA FE	0.92

NEW YORK	
ALBANY	0.87
BINGHAMTON	0.83
BUFFALO	0.90
LONG ISLAND	1.01
NEW YORK CITY	1.15
ROCHESTER	0.90
SYRACUSE	0.87
WATERTOWN	0.83
WHITE PLAINS	1.01

NORTH CAROLINA	
ASHEVILLE	0.75
CHARLOTTE	0.79
GREENSBORO	0.75
RALEIGH	0.77
WILMINGTON	0.74

NORTH DAKOTA	
BISMARK	0.86
FARGO	0.90
GRAND FORKS	0.88
MINOT	0.84

OHIO	
CINCINNATI	0.83
CLEVELAND	0.91
COLUMBUS	0.89
TOLEDO	0.89
YOUNGSTOWN	0.85

OKLAHOMA	
BARTLESVILLE	0.77
ENID	0.77
LAWTON	0.77
OKLAHOMA CITY	0.83
TULSA	0.81

OREGON	
EUGENE	0.90
MEDFORD	0.90
PORTLAND	0.95
SALEM	0.92

PENNSYLVANIA	
ALLENTOWN	0.91
HARRISBURG	0.85
PHILADELPHIA	0.98
PITTSBURGH	0.89
SCRANTON	0.87

RHODE ISLAND	
PAWTUCKET	0.95
PROVIDENCE	0.98
NEWPORT	0.98
WESTERLY	0.95
WOONSOCKET	0.95

SOUTH CAROLINA	
CHARLESTON	0.79
COLUMBIA	0.81
FLORENCE	0.79
GREENVILLE	0.81

SOUTH DAKOTA	
ABERDEEN	0.79
PIERRE	0.79
RAPID CITY	0.81
SIOUX FALLS	0.83
WATERTOWN	0.81

TENNESSEE	
CHATTANOOGA	0.77
JOHNSON CITY	0.75
KNOXVILLE	0.79
MEMPHIS	0.81
NASHVILLE	0.83

TEXAS	
AUSTIN	0.80
DALLAS	0.82
HOUSTON	0.86
LUBBOCK	0.78
SAN ANTONIO	0.80

UTAH	
LOGAN	0.83
OGDEN	0.85
PROVO	0.83
SALT LAKE	0.87

VERMONT	
BRATTLEBORO	0.85
BURLINGTON	0.94
RUTLAND	0.87

VIRGINIA	
ALEXANDRIA	0.94
LYNCHBURG	0.81
NORFOLK	0.83
RICHMOND	0.85
ROANOKE	0.81

WASHINGTON	
BELLINGHAM	0.99
SEATLE	1.06
SPOKANE	0.99
TACOMA	1.03
YAKIMA	0.96

WASHINGTON, D.C.	
DISTRICT	0.95

WEST VIRGINIA	
CHARLESTON	0.87
HUNTINGTON	0.89
MORGANTOWN	0.83
PARKERSBURG	0.83

WISCONSIN	
EAU CLAIRE	0.90
GREEN BAY	0.92
MADISON	0.90
MILWAUKEE	0.97
WAUSAU	0.90

WYOMING	
CASPER	0.89
CHEYENNE	0.91
ROCK SPRINGS	0.87
SHERIDAN	0.85

Square Foot Tables

The following Square Foot Tables list hundreds of actual projects for dozens of building types, each with associated building size, total square foot building cost and percentage of project costs for total mechanical and electrical components. This data provides an overview of construction costs by building type. These costs are for actual projects. The variations within similar building types may be due, among other factors, to size, location, quality and specified components, materials and processes. Depending upon all such factors, specific building costs can vary significantly and may not necessarily fall within the range of costs as presented. The data has been updated to reflect current construction costs.

BUILDING CATEGORY	***PAGE***

SQUARE FOOT TABLES

COMMERCIAL

AUTO DEALERSHIP

Project Size Gross S.F.	Project Cost $/S.F.	% Cost Mechanical	% Cost Electrical
7,700	90.20	16.7	9.4
16,100	51.10	10.2	15.9
20,000	53.70	12.9	23.4
26,300	54.70	12.5	22.0
43,600	43.40	19.4	13.2
53,600	72.30	12.5	11.5

BUSINESS CENTER

Project Size Gross S.F.	Project Cost $/S.F.	% Cost Mechanical	% Cost Electrical
3,900	54.00	12.0	9.2
9,900	52.20	9.1	7.6
54,400	34.90	3.4	12.2
135,000	38.00	8.2	1.5

CINEMA

Project Size Gross S.F.	Project Cost $/S.F.	% Cost Mechanical	% Cost Electrical
18,000	123.40	10.9	6.7
22,500 A	75.90	6.6	4.2

MALL/PLAZA

Project Size Gross S.F.	Project Cost $/S.F.	% Cost Mechanical	% Cost Electrical
9,700	38.20	15.0	13.3
10,500	61.10	8.0	13.5
16,300	55.60	9.4	10.6
26,900	55.40	15.0	7.0
36,000	46.90	10.0	11.0
36,300	53.20	12.4	8.2
44,720	55.60	18.3	12.4
59,100	54.50	9.8	9.5
60,000 R	55.90	10.0	9.5
64,100	56.20	22.4	18.4
66,000	64.30	13.5	11.5
67,400	46.80	21.0	15.0
73,500	67.60	14.9	6.9

MALL/PLAZA (Cont.)

Project Size Gross S.F.	Project Cost $/S.F.	% Cost Mechanical	% Cost Electrical
142,000	42.50	7.1	8.0
220,000	112.30	11.0	6.4
223,700	33.00	9.0	9.3
321,200	39.30	7.3	7.4
379,900	51.30	11.2	6.2
405,100	52.90	13.9	6.0
482,000	87.90	10.5	9.4
630,000	58.80	12.2	12.4

RESTAURANT

Project Size Gross S.F.	Project Cost $/S.F.	% Cost Mechanical	% Cost Electrical
4,300 R	97.10	6.9	8.7
4,400 R	142.90	14.6	8.5
5,800	116.10	28.0	10.6
6,800 A	135.80	7.0	11.1
7,360 R	146.10	16.0	6.5
9,600	145.20	24.7	13.1
10,000 R	151.00	21.0	10.0
10,100	130.30	28.6	18.4
10,600	239.90	20.4	6.4
22,900 R	155.90	15.8	16.9

RETAIL STORE

Project Size Gross S.F.	Project Cost $/S.F.	% Cost Mechanical	% Cost Electrical
1,000	146.10	12.8	6.7
3,000 R	130.90	14.3	10.5
12,300	179.40	14.0	10.0
30,000	90.60	15.6	26.2
61,300	46.90	13.3	13.0
115,000	64.90	14.6	11.3
154,700	91.90	11.2	12.4
314,700 R	77.50	13.8	9.4

A = Addition   R = Remodel

SQUARE FOOT TABLES

RESIDENTIAL

APARTMENTS

Project Size Gross S.F.	Project Cost $/S.F.	% Cost Mechanical	% Cost Electrical
3,700	54.30	14.9	4.4
13,900	74.40	7.9	7.2
19,200 R	118.10	45.3	7.5
19,700	69.20	7.4	10.4
23,700	76.00	10.6	4.3
26,500	72.00	25.2	12.8
35,100	58.90	16.4	5.6
54,000	101.90	23.3	13.1
62,700	74.00	17.0	9.0
67,300	67.60	13.8	8.4
70,600	38.10	18.1	7.8
75,300	84.00	13.2	8.1
75,600	81.80	14.5	8.9
72,200	70.90	18.4	10.9
77,600	91.70	26.9	14.3
88,100	85.10	15.3	9.3
89,500	79.80	10.7	11.1
94,100	48.10	8.5	6.8
96,000	60.00	17.0	13.3
102,000	104.10	17.8	12.5
103,200	47.30	12.1	8.9
103,600	76.20	19.1	9.0
105,200	97.10	14.9	8.9
106,200	69.90	12.8	9.3
110,900	67.20	15.6	8.4
111,800	95.40	17.3	7.4
115,900	71.50	12.5	8.6
117,200	44.70	12.9	8.0
119,000	61.40	15.6	8.4

APARTMENTS (Cont.)

Project Size Gross S.F.	Project Cost $/S.F.	% Cost Mechanical	% Cost Electrical
119,400	40.20	18.6	8.7
144,300	89.40	15.0	8.6
176,300	76.00	19.1	9.0
192,300	44.70	10.1	6.0
210,900	88.00	19.1	9.5
220,200	90.50	14.8	7.5
253,900	117.70	20.6	7.7
369,500	85.90	15.8	9.0

CONDOS/TOWNHOUSE

Project Size Gross S.F.	Project Cost $/S.F.	% Cost Mechanical	% Cost Electrical
8,600	77.10	8.5	4.4
16,700	62.20	13.2	6.5
18,000	130.90	14.6	9.7
18,400	67.10	12.9	5.8
74,800	60.90	13.8	5.2
111,700	76.90	9.2	7.1
150,300	85.80	15.9	7.9
278,800	134.10	14.1	7.9
1,109,900	61.70	9.8	4.7

SINGLE-FAMILY HOMES

Project Size Gross S.F.	Project Cost $/S.F.	% Cost Mechanical	% Cost Electrical
600 R	55.50	16.3	2.0
900	72.20	33.0	3.0
2,100	83.60	8.8	4.0
2,200	179.00	23.8	3.4
2,500	87.70	6.1	7.1
2,900	84.70	7.7	2.8
3,000	59.50	9.5	6.8
3,100	132.90	15.5	4.6
3,600	83.30	8.5	3.0
3,700	116.40	9.8	3.4

A = Addition R = Remodel

SQUARE FOOT TABLES

RESIDENTIAL (Cont.)

SINGLE-FAMILY HOMES (Cont.)

Project Size Gross S.F.	Project Cost $/S.F.	% Cost Mechanical	% Cost Electrical
4,200	101.10	15.5	5.7
4,600	160.70	9.0	4.7
5,200	208.80	8.3	6.0
5,700	122.30	7.4	3.7
5,700	83.20	8.7	12.6
21,300*	59.30	26.0	4.0
22,700*	55.10	27.0	4.6
45,000*	83.80	7.8	2.5
51,458*	60.70	11.0	5.0

*TOWNHOUSES

EDUCATIONAL

ADMINISTRATION (OFFICES)

Project Size Gross S.F.	Project Cost $/S.F.	% Cost Mechanical	% Cost Electrical
53,700	142.90	15.4	9.0

ATHLETIC FACILITY

Project Size Gross S.F.	Project Cost $/S.F.	% Cost Mechanical	% Cost Electrical
38,100	141.80	16.8	9.8
44,100	130.90	16.8	8.4
100,000	89.40	19.2	5.3
160,000	161.90	13.7	7.9
247,500	135.90	11.2	9.0
271,000	127.60	13.2	7.6
283,100	153.80	14.6	6.0

AUDITORIUM/PERFORMING ARTS

Project Size Gross S.F.	Project Cost $/S.F.	% Cost Mechanical	% Cost Electrical
9,900	229.00	17.5	29.2
17,800	209.40	11.4	16.1
29,200	207.20	16.2	10.5
62,700	150.70	13.0	12.9

CLASSROOM

Project Size Gross S.F.	Project Cost $/S.F.	% Cost Mechanical	% Cost Electrical
35,400	207.70	10.2	6.8
70,000	93.80	24.3	13.1
78,900	174.50	13.3	14.3
80,100	141.80	16.9	10.5
100,000	141.20	21.1	9.8
166,000	97.60	16.1	11.2
298,400	91.60	11.3	11.9

COMPLETE COLLEGE FACILITIES

Project Size Gross S.F.	Project Cost $/S.F.	% Cost Mechanical	% Cost Electrical
95,300	147.20	19.4	11.7
450,000	175.70	18.6	10.8

ELEMENTARY SCHOOL

Project Size Gross S.F.	Project Cost $/S.F.	% Cost Mechanical	% Cost Electrical
18,000	107.70	19.1	13.8
30,800	99.10	16.0	10.0
31,600	82.80	12.1	11.3
35,700	121.10	17.3	8.7
40,000	105.30	18.5	13.4
40,500	92.20	22.1	11.3
57,000	79.70	22.1	7.5
69,700	124.40	20.6	9.3
91,400	100.30	18.2	7.6

HIGH SCHOOL

Project Size Gross S.F.	Project Cost $/S.F.	% Cost Mechanical	% Cost Electrical
116,400	110.10	18.1	12.9
133,000	94.70	17.8	10.7
184,000	166.90	23.0	10.0
217,200 R	90.10	26.4	10.6
254,000 R	69.40	17.2	14.6
431,700	105.80	13.1	9.6

A = Addition R = Remodel

SQUARE FOOT TABLES

EDUCATIONAL (Cont.)

JUNIOR HIGH SCHOOL

Project Size Gross S.F.	Project Cost $/S.F.	% Cost Mechanical	% Cost Electrical
26,000	138.60	9.5	9.3
28,100	85.90	11.9	9.5
52,800	111.30	18.3	8.6
91,600	136.80	21.2	11.0
123,700	111.90	29.1	9.1

LABORATORY/RESEARCH

Project Size Gross S.F.	Project Cost $/S.F.	% Cost Mechanical	% Cost Electrical
9,200	239.90	25.4	4.5
80,300	188.70	20.2	15.5

LIBRARY

Project Size Gross S.F.	Project Cost $/S.F.	% Cost Mechanical	% Cost Electrical
6,900	125.80	17.9	10.3
8,200	115.50	18.8	10.6
12,000	149.70	19.4	15.9
15,000	130.90	15.7	18.7
16,300	110.80	11.3	7.4
28,600	99.20	15.7	9.0
30,100	134.70	13.0	11.0
37,700	105.80	14.6	8.0
43,500	89.40	16.6	6.7
47,900	156.70	29.7	9.0
51,400	160.10	13.1	12.4
63,400 A	120.40	13.5	8.3
64,000	112.70	10.9	11.4
74,000	127.60	17.2	8.0
75,600	169.50	12.5	16.7
176,000	93.90	11.2	9.8

SPECIAL NEEDS FUNCTION

Project Size Gross S.F.	Project Cost $/S.F.	% Cost Mechanical	% Cost Electrical
15,200	106.40	16.6	8.4
27,900	128.20	19.4	11.2

STUDENT CENTER/MULTIPURPOSE

Project Size Gross S.F.	Project Cost $/S.F.	% Cost Mechanical	% Cost Electrical
90,000	161.80	16.2	10.0
49,600	119.50	18.2	9.2
187,700	179.90	15.3	8.8
194,800	95.10	17.2	8.5

HOTEL/MOTEL

CONVENTION/CONFERENCE CENTER

Project Size Gross S.F.	Project Cost $/S.F.	% Cost Mechanical	% Cost Electrical
8,600 A	142.10	20.7	16.3
71,900	157.60	12.5	13.8
433,800	82.30	22.1	8.3

HOTEL

Project Size Gross S.F.	Project Cost $/S.F.	% Cost Mechanical	% Cost Electrical
19,900 A	81.20	16.8	5.8
25,875 R	72.80	11.8	10.5
48,400 A	145.70	23.6	8.2
64,300 R	196.30	20.3	10.1
104,200 A	91.20	15.0	8.8
108,040	77.90	13.0	7.0
110,100	100.80	18.4	10.5
132,000	188.40	16.4	5.4
135,900 A	116.50	15.2	7.7
144,100 A	130.90	19.3	11.5
231,000	141.60	15.2	8.4
449,800 A	81.80	13.1	7.0

HOTEL/INN

Project Size Gross S.F.	Project Cost $/S.F.	% Cost Mechanical	% Cost Electrical
57,400	94.50	13.0	10.0
73,000	61.90	24.4	18.1
75,900	82.30	16.5	7.6
162,000	99.10	17.5	8.0
197,000	94.80	15.7	7.7
277,900	69.60	18.8	9.4

A = Addition R = Remodel

SQUARE FOOT TABLES

INDUSTRIAL

MANUFACTURING

Project Size Gross S.F.	Project Cost $/S.F.	% Cost Mechanical	% Cost Electrical
14,300	80.50	13.0	7.0
18,500	121.40	20.5	13.5
26,600	43.40	6.1	12.1
31,400	95.90	18.9	17.7
33,400	55.40	23.1	15.8
37,300	41.30	3.0	24.0
43,400	55.10	14.7	11.7
45,400	89.40	15.9	15.0
79,800	92.90	16.0	14.5
81,100	99.20	8.2	8.4
137,400	62.30	41.6	13.6
179,600	71.90	26.3	13.6
186,000	67.20	19.5	11.7

RESEARCH AND DEVELOPMENT

Project Size Gross S.F.	Project Cost $/S.F.	% Cost Mechanical	% Cost Electrical
89,140	128.40	20.8	9.0
100,400	170.00	36.1	25.1
114,200	166.50	21.4	9.8
125,000	113.30	20.8	8.4
140,000	158.20	20.1	11.2

WAREHOUSE W/OFFICE

Project Size Gross S.F.	Project Cost $/S.F.	% Cost Mechanical	% Cost Electrical
14,000	37.00	6.5	9.2
19,000	33.10	2.3	1.8
19,700	43.50	9.5	7.0
31,200	37.40	7.0	7.0
40,500	47.80	6.6	10.5
62,000	60.60	10.8	10.0
96,200	35.10	2.2	6.3
105,000	34.10	5.3	11.1
149,800	35.60	14.5	8.6

WAREHOUSE W/OFFICE (Cont.)

Project Size Gross S.F.	Project Cost $/S.F.	% Cost Mechanical	% Cost Electrical
168,600	39.30	11.4	6.3
209,600	36.20	4.9	10.3
402,400	49.20	14.9	8.3

MEDICAL

EDUCATION CENTER

Project Size Gross S.F.	Project Cost $/S.F.	% Cost Mechanical	% Cost Electrical
35,400	220.30	10.2	6.8

HOSPITALS

Project Size Gross S.F.	Project Cost $/S.F.	% Cost Mechanical	% Cost Electrical
9,300 R	217.60	29.5	12.0
15,900 A	160.50	27.4	15.3
16,600 R	57.80	14.7	9.6
22,000	365.30	31.6	17.5
39,100	182.00	23.3	7.9
63,800	146.70	23.4	7.8
98,000 A	211.00	20.5	17.0
100,200 A	336.60	22.2	9.6
103,900 A	186.60	31.1	11.7
109,300	184.80	20.7	15.7
148,700	145.30	25.2	11.8
154,700	235.90	19.3	10.5
165,484	205.00	21.5	14.8
165,700 A	200.10	26.2	17.7
179,400 A	121.90	28.9	20.3
182,800	181.50	19.1	14.0
265,000	216.50	31.2	14.1
281,100	207.30	24.1	14.6
435,000	214.80	21.5	13.7
694,300	119.00	28.9	9.7
772,300	219.20	31.5	13.4

A = Addition R = Remodel

SQUARE FOOT TABLES

MEDICAL (Cont.)

MEDICAL OFFICES/CENTERS

Project Size Gross S.F.	Project Cost $/S.F.	% Cost Mechanical	% Cost Electrical
3,000	92.00	13.7	11.5
5,500	95.70	8.5	18.7
10,000	115.70	13.9	9.4
10,600	146.60	13.2	8.9
16,300	121.30	21.8	13.2
18,300	66.20	7.1	13.4
20,600	95.10	14.3	6.4
24,900	167.40	21.0	15.1
27,000	169.00	19.7	10.1
28,400	117.80	13.6	9.3
30,500	127.60	17.2	12.7
32.000	101.20	18.1	10.5
44,300	102.10	24.4	16.2
50,200	58.90	9.8	4.9
51,200	149.10	21.0	12.0
64,600	71.00	13.2	6.4
66,000	66.20	9.8	8.8
80,000	56.80	8.2	8.3
137,175	94.90	12.8	6.6

NURSING HOMES

Project Size Gross S.F.	Project Cost $/S.F.	% Cost Mechanical	% Cost Electrical
11,600 A	278.10	53.2	7.9
16,800	181.00	33.3	7.8
31,900 A	136.10	22.0	11.0
64,100	116.70	20.6	11.1
290,000	163.60	16.1	13.6

RESEARCH

Project Size Gross S.F.	Project Cost $/S.F.	% Cost Mechanical	% Cost Electrical
34,600	154.90	18.1	3.0

PUBLIC FACILITIES

ANIMAL CENTER

Project Size Gross S.F.	Project Cost $/S.F.	% Cost Mechanical	% Cost Electrical
20,000	196.30	20.7	4.6
39,100	146.70	22.9	6.8
44,300	139.70	8.1	5.8

AUTO DEALERSHIP

Project Size Gross S.F.	Project Cost $/S.F.	% Cost Mechanical	% Cost Electrical
7,700	91.50	16.7	9.4

BROADCASTING

Project Size Gross S.F.	Project Cost $/S.F.	% Cost Mechanical	% Cost Electrical
20,000	267.60	20.0	13.0
29,500	206.40	16.6	15.0
45,000 R	149.70	15.0	13.0

CIVIC CENTER

Project Size Gross S.F.	Project Cost $/S.F.	% Cost Mechanical	% Cost Electrical
6,000	147.40	9.2	2.8
23,900	185.20	12.3	10.9
34,400	84.40	3.5	17.8
69,800 A	221.20	17.7	8.8
206,500	119.60	15.0	11.0

CORRECTION FACILITIES

Project Size Gross S.F.	Project Cost $/S.F.	% Cost Mechanical	% Cost Electrical
44,600	178.80	20.9	13.1
66,000	109.10	15.0	23.0
257,800 A	188.80	20.9	10.7
360,000	127.20	32.7	13.2

FIRE STATION

Project Size Gross S.F.	Project Cost $/S.F.	% Cost Mechanical	% Cost Electrical
6,900	159.50	12.4	9.7
7,600	122.60	16.0	9.5
8,430	168.60	12.8	9.6
9,600	147.00	13.5	11.8

A = Addition R = Remodel

SQUARE FOOT TABLES

PUBLIC FACILITIES (Cont.)

GOVERMENT BUILDINGS

Project Size Gross S.F.	Project Cost $/S.F.	% Cost Mechanical	% Cost Electrical
12,300	114.40	13.5	10.7
23,500	187.50	11.1	15.6
27,300	148.00	19.5	9.3
31,600	166.60	27.1	11.3
46,600	156.80	17.4	13.8
72,100	170.80	24.4	10.7
78,200	141.70	19.7	15.2
332,900	137.50	16.2	14.2
364,100	180.00	14.6	12.9
771,000	198.90	17.6	11.3

MUSEUM

Project Size Gross S.F.	Project Cost $/S.F.	% Cost Mechanical	% Cost Electrical
27,600	137.40	17.8	14.1
30,100	157.80	18.6	7.9
43,264	147.00	11.1	8.3
63,000	159.10	8.8	18.1

PARKING GARAGE

Project Size Gross S.F.	Project Cost $/S.F.	% Cost Mechanical	% Cost Electrical
66,000	42.00	2.8	3.1
169,000	31.00	10.3	3.7
562,700	27.90	2.4	6.2

TRANSPORTATION

Project Size Gross S.F.	Project Cost $/S.F.	% Cost Mechanical	% Cost Electrical
7,300	252.50	15.1	3.2
14,300	202.80	13.3	16.6
23.000	140.10	9.6	13.7
35,500	116.70	1.0	19.0
49,100	174.00	35.5	11.6

TRANSPORTATION (Cont.)

Project Size Gross S.F.	Project Cost $/S.F.	% Cost Mechanical	% Cost Electrical
288,100	100.30	8.3	13.3
2,160,000	162.50	23.3	11.5

OFFICES

BANKS

Project Size Gross S.F.	Project Cost $/S.F.	% Cost Mechanical	% Cost Electrical
2,900	233.40	5.3	4.3
3,100	104.10	5.9	6.7
3,300	151.50	10.3	16.4
3,600	114.10	10.3	12.2
4,000	130.00	8.0	9.0
4,100	128.70	21.4	13.0
4,200	138.80	8.6	14.21
4,400	160.30	12.2	12.7
4,500	99.60	12.4	13.5
4,900	157.30	11.7	11.2
5,900	116.90	9.3	13.8
6,000	131.40	11.6	7.3
6,100	176.70	11.0	8.0
7,000	237.70	6.0	9.0
7,300	145.40	11.9	11.3
7,700	163.00	8.0	7.5
7,800	178.60	11.0	11.7
8,000	90.70	10.0	14.0
9,200	144.00	9.9	12.1
9,400	104.30	11.7	11.8
10,200	217.50	12.6	12.6
12,600	77.60	7.0	18.0

A = Addition R = Remodel

SQUARE FOOT TABLES

OFFICES (Cont.)

BANKS (Cont.)

Project Size Gross S.F.	Project Cost $/S.F.	% Cost Mechanical	% Cost Electrical
13,300	137.00	9.5	8.3
13,800	122.60	10.00	9.7
15,000	105.00	15.4	12.4
15,200	83.90	9.2	12.9
15,500	107.10	9.8	10.3
16,000	70.20	13.4	23.1
20,100	66.10	13.0	11.0
21,700	139.70	8.8	11.3
44,800	119.40	13.0	8.2
53,200	215.60	14.9	7.2
62,100	123.90	10.1	7.9
95,100	162.90	13.5	4.3

OFFICE BUILDINGS

Project Size Gross S.F.	Project Cost $/S.F.	% Cost Mechanical	% Cost Electrical
2,600	146.50	17.2	9.4
3,400	119.70	10.5	11.3
3,800	117.60	16.4	11.8
4,400	116.10	12.8	8.5
4,500	99.90	13.0	7.0
5,100	72.50	16.6	10.2
5,200	95.20	8.0	5.7
6,700	141.80	16.8	10.4
7,500	152.70	10.1	8.0
7,900	118.90	17.4	9.0
8,100	183.60	10.6	11.0
10,600	106.20	9.9	9.7
10,900	61.20	13.8	10.8

OFFICE BUILDINGS (Cont.)

Project Size Gross S.F.	Project Cost $/S.F.	% Cost Mechanical	% Cost Electrical
11,300	120.70	17.0	6.0
13,000	82.10	15.0	9.0
14,400	107.70	19.8	12.9
14,500	82.70	17.3	12.5
17,000	118.00	14.7	7.9
17,800 A	63.80	10.4	11.0
18,100	123.70	22.5	11.6
19,300	67.10	11.1	8.1
24,600	62.00	18.5	14.1
27,700	74.60	19.6	5.5
27,800	135.30	12.7	5.1
27,800	72.80	17.8	10.3
32,500 R	124.60	13.9	6.4
35,400	72.70	15.0	12.0
36.500	63.00	10.2	10.2
42,300	78.30	10.3	7.7
44,400	106.50	23.5	14.7
44,400	56.10	11.0	5.0
44,500	77.10	11.5	3.1
45,400	66.60	19.1	13.2
47,300	68.50	18.5	8.0
49,700	115.20	22.1	7.2
50,000	109.70	19.4	15.6
50,400	119.10	23.2	7.8
52,200	77.30	18.3	7.8
52,900	90.20	4.4	3.9
53,700	137.80	15.4	9.0

A = Addition R = Remodel

SQUARE FOOT TABLES

OFFICES (Cont.)

OFFICE BUILDINGS (Cont.)

Project Size Gross S.F.	Project Cost $/S.F.	% Cost Mechanical	% Cost Electrical
54,000	55.80	14.4	2.6
56,000	63.40	10.6	6.2
56,500	87.10	19.1	11.0
72,000 R	30.60	12.9	20.2
74,000	60.60	13.5	6.8
80,800	61.30	12.0	6.0
81,800	84.40	21.1	9.5
81,900	71.00	19.9	9.0
82,000	98.70	14.0	4.2
83,100	116.40	16.0	8.6
85,400	102.10	18.7	8.6
86,200	79.90	14.7	11.0
99,900	92.50	16.6	6.6
100,000	103.30	10.6	12.1
100,000	61.00	16.3	10.4
116,400	97.60	12.6	10.0
134,500	199.00	14.1	12.4
140,000	154.30	20.1	11.2
155,700	192.20	11.1	6.1
171,000	90.10	24.0	7.7
174,300	110.90	12.3	9.1
203,300	149.90	19.0	13.2
265,800	207.50	11.0	10.0
287,300	57.50	10.7	4.1
319,800	90.10	13.4	5.6
350,000	133.20	20.0	13.0
360,900	80.60	14.2	11.1
394,000	64.00	22.5	8.7

OFFICE BUILDINGS (Cont.)

Project Size Gross S.F.	Project Cost $/S.F.	% Cost Mechanical	% Cost Electrical
430,000	95.90	21.8	9.3
490,000	102.00	11.2	7.5
588,400	209.60	15.0	9.0
606,000	83.80	9.4	8.8
620,000	231.20	12.6	12.8
733,500	64.00	10.8	4.2

RECREATIONAL

ARENA

Project Size Gross S.F.	Project Cost $/S.F.	% Cost Mechanical	% Cost Electrical
315,200	239.90	10.4	8.00
385,800	137.40	13.2	8.80
727,000	124.20	14.9	7.40

HEALTH CLUB

Project Size Gross S.F.	Project Cost $/S.F.	% Cost Mechanical	% Cost Electrical
15,900	67.10	8.7	9.7
21,800	95.00	10.3	10.3
30,100	146.80	11.4	19.4
66,400	72.30	10.7	8.5

RECREATIONAL CENTER

Project Size Gross S.F.	Project Cost $/S.F.	% Cost Mechanical	% Cost Electrical
9,900	130.40	6.6	9.8
14,000	85.10	11.2	5.0
14,000	102.10	11.3	16.7
15,700	77.60	17.8	14.3
20,000	176.20	19.1	8.9
21,200	99.10	16.0	9.6

A = Addition   R = Remodel

SQUARE FOOT TABLES

RECREATIONAL (Cont.)

RECREATIONAL CENTER (Cont.)

Project Size Gross S.F.	Project Cost $/S.F.	% Cost Mechanical	% Cost Electrical
26,000	115.50	9.3	7.0
53,400 A	138.10	11.7	6.4
69,800	224.00	17.7	8.8

RELIGIOUS

CHURCH

Project Size Gross S.F.	Project Cost $/S.F.	% Cost Mechanical	% Cost Electrical
4,100	157.30	8.8	13.5
10,400 R	214.00	12.8	8.1
11,100	94.20	10.9	12.7
13,400	141.20	5.5	6.0
14,500	96.40	12.8	7.4
15,200	125.00	18.3	8.0
15,700	108.80	14.4	8.4

CHURCH (Cont.)

Project Size Gross S.F.	Project Cost $/S.F.	% Cost Mechanical	% Cost Electrical
16,000	177.00	14.1	9.6
20,900	110.40	16.0	14.0
21,500	121.20	7.3	8.0
22,900	109.40	12.1	9.5
30,600	82.80	15.5	8.0
42,700	93.50	18.6	7.7

MULTI-PURPOSE

Project Size Gross S.F.	Project Cost $/S.F.	% Cost Mechanical	% Cost Electrical
4,400	96.60	11.5	17.5
5,800 A	123.10	8.9	7.5
6,400	168.20	15.6	15.8
9,000	82.30	7.7	5.9
9,000	116.90	16.0	6.6
10,100	78.40	11.1	12.0
12,000	155.00	13.1	10.7
18,400	85.00	10.8	10.1
19,500	149.70	16.0	17.3

A = Addition R = Remodel

For more information subscribe to **Design Cost & Data**

BNi Building News

INDEX

INDEX

- C -

- D -

- E -

- F -

INDEX

- G -

- H -

INDEX

- I -

- J -

- K -

- L -

- M -

- N -

- O -

- P -

INDEX

- Q -

INDEX

INDEX

- T -

- U -

INDEX

Notes

Notes

Notes

Notes

Notes

Notes